Fundamentals of Thermodynamics and Applications

Ingo Müller · Wolfgang H. Müller

Fundamentals of Thermodynamics and Applications

With Historical Annotations and Many Citations from Avogadro to Zermelo

Springer

Prof. Ingo Müller
Institut für Prozess- und Verfahrenstechnik
Thermodynamik und Thermische Verfahrenstechnik
Technische Universität Berlin
Str. des 17. Juni 135
10623 Berlin
Germany
E-mail: ingo.mueller@alumni.tu-berlin.de

Prof. Dr. Wolfgang H. Müller
Fakultät V - Verkehrs- und Maschinensysteme
Technische Universität Berlin
Einsteinufer 5
D-10587 Berlin
Germany
E-mail: wolfgang.h.mueller@tu-berlin.de

ISBN 978-3-540-74645-4 e-ISBN 978-3-540-74648-5

DOI 10.1007/978-3-540-74648-5

Library of Congress Control Number: 2008942041

Typesetting: Scientific Publishing Services Pvt. Ltd., Chennai, India.
Cover Design: eStudio Calamar S.L.

Printed in acid-free paper

9 8 7 6 5 4 3 2 1

springer.com

Preface

Thermodynamics is the much abused slave of many masters • physicists who love the totally impractical Carnot process, • mechanical engineers who design power stations and refrigerators, • chemists who are successfully synthesizing ammonia and are puzzled by photosynthesis, • meteorologists who calculate cloud bases and predict föhn, boraccia and scirocco, • physico-chemists who vulcanize rubber and build fuel cells, • chemical engineers who rectify natural gas and distil fermented potato juice, • metallurgists who improve steels and harden surfaces, • nutrition counselors who recommend a proper intake of calories, • mechanics who adjust heat exchangers, • architects who construe – and often misconstrue – chimneys, • biologists who marvel at the height of trees, • air conditioning engineers who design saunas and the ventilation of air plane cabins, • rocket engineers who create supersonic flows, *et cetera*.

Not all of these professional groups need the full depth and breadth of thermodynamics. For some it is enough to consider a well-stirred tank, for others a stationary nozzle flow is essential, and yet others are well-served with the partial differential equation of heat conduction.

It is therefore natural that thermodynamics is prone to mutilation; different group-specific meta-thermodynamics' have emerged which serve the interest of the groups under most circumstances and leave out aspects that are not often needed in their fields. To stay with the metaphor of the abused slave we might say that in some fields his legs and an arm are cut off, because only one arm is needed; in other circumstances the brain of the slave has atrophied, because only his arms and legs are needed. Students love this reduction, because it enables them to avoid "nonessential" aspects of thermodynamics. But the practice is dangerous; it may backfire when a brain is needed.

In this book we attempt to exhibit the complete fundament of classical thermodynamics which consists of the equations of balance of mass, momentum and energy, and of constitutive equations which characterize the behavior of material bodies, mostly gases, vapors and liquids because, indeed, classical thermodynamics is often negligent of solids, – and so are we, although not entirely.

Many applications are treated in the book by specializing the basic equations; a brief look at the table of contents bears witness to that feature.

Modern thermodynamics is a lively field of research at extremely low and extremely high temperatures and for strongly rarefied gases and in nano-tubes, or nano-layers, where quantum effects occur. But such subjects are not treated in this book. Indeed, there is nothing here which is not at least 70 years old. We claim, however, that our presentation is systematic and we believe that classical thermodynamics should be taught as we present it. If it were, thermodynamics might shed the nimbus of a difficult subject which surrounds it among students.

Even classical thermodynamics is such a wide field that it cannot be fully described in all its ramifications in a relatively short book like this one. We had to resign ourselves to that fact. And we have decided to omit all discussion of • empirical state functions, • temperature dependent specific heats of liquids and ideal gases, and • irreversible secondary effects in engines. Such phenomena affect the neat analytical structure of thermodynamic problems, and we have excluded them, although we know full well that they are close to the hearts and minds of engineers who may even, in fact, consider incalculable irreversibilities of technical processes as the essence of thermodynamics. We do not share that opinion.

In the second half of the 19th century and early in the 20th century thermodynamics was at the forefront of physics, and eminent physicists and chemists like Planck, Einstein and Haber were steeped in thermodynamics; actually the formula $E = m\,c^2$, which identifies energy as it were, is basically a contribution to thermodynamics. We have made an attempt to enliven the text by a great many mini-biographies and historical annotations which are somewhat relevant to the development of thermodynamics or, in other cases, they illustrate early misconceptions which may serve to highlight the difficult emergence of the basic concepts of the field. A prologue has been placed in front of the main chapters in order to avoid going into subjects which are by now so commonplace that they are taught in high schools.

Colleagues, co-workers, and students have contributed to this work, some significantly, others little, but all of them something:

Manfred Achenbach, Giselle Alves, Jutta Ansorg, Teodor Atanackovic, Jörg Au, Elvira Barbera, Andreas Bensberg, Andreas Bormann, Michel Bornert, Tamara Borowski, Wolfgang Dreyer, Heinrich Ehrenstein, Fritz Falk, Semlin Fu, Uwe Glasauer, Anja Hofmann, Yongzhong Huo, Oliver Kastner, Wolfgang Kitsche, Gilberto M. Kremer, Thomas Lauke, I-shih Liu, Wilson Marquez jr., Angelo Morro, Olav Müller, André Musolff, Mario Pitteri, Daniel Reitebuch, Roland Rydzewski, Harsimar Sahota, Stefan Seelecke, Ute Stephan, Peter Strehlow, Henning Struchtrup, Manuel Torrilhon, Wolf Weiss, Krzysztof Wilmanski, Yihuan Xie, Huibin Xu, Giovanni Zanzotto.

Mark Warmbrunn has drawn the cartoons. Rudolf Hentschel and Marlies Hentschel have helped with the figures and part of the text.

Several teaching assistants have edited the text and converted it into Springer style: Matti Blume, Anja Klinnert, Volker Marhold, Christoph Menzel, Felix J. Müller.

Guido Harneit has given support with the computer.

Everyone of them deserves our sincere gratitude.

Berlin, in the summer of 2008

Ingo Müller & Wolfgang H. Müller

Contents

P Prologue on ideal gases and incompressible fluids 1
P.1 Thermal and caloric equations of state 1
• Ideal gases 1
• Incompressible fluid 1
P.2 "mol" 2
P.3 On the history of the equations of state 3
P.4 An elementary kinetic view of the equations of state for ideal gases; interpretation of pressure and absolute temperature 4

1 Objectives of thermodynamics and its equations of balance 7
1.1 Fields of mechanics and thermodynamics 7
1.1.1 Mass density, velocity, and temperature 7
1.1.2 History of temperature 7
1.2 Equations of balance 9
1.2.1 Conservation laws of thermodynamics 9
1.2.2 Generic equations of balance for closed and open systems 9
1.2.3 Generic local equation of balance in regular points 10
1.3 Balance of mass 11
1.3.1 Integral and local balance equations of mass 11
1.3.2 Mass balance and nozzle flow 11
1.4 Balance of momentum 12
1.4.1 Integral and local balance equations of momentum 12
1.4.2 Pressure 14
1.4.3 Pressure in an incompressible fluid at rest 14
1.4.4 History of pressure and pressure units 15
1.4.5 Applications of the momentum balance 16
• Buoyancy law of ARCHIMEDES 16
• Barometric steps in the atmosphere 17
• Thrust of a rocket 18
• Thrust of a jet engine 19
• Convective momentum flux 20
• Momentum balance and nozzle flow 21
• Bernoulli equation 23
• Kutta-Joukowski formula for the lift of an airfoil 24
1.5 Balance of energy 26
1.5.1 Kinetic energy, potential energy, and four types of internal energy 26
• Kinetic energy of thermal atomic motion. "Heat energy" 26
• Potential energy of molecular interaction. "Van der Waals energy" or "elastic energy" 27

• Chemical energy between molecules. "Heat of reaction" 27
• Nuclear energy 29
• Heat into work 29
1.5.2 Integral and local equations of balance of energy 29
1.5.3 Potential energy 31
1.5.4 Balance of internal energy 32
1.5.5 Short form of energy balance for closed systems 33
1.5.6 First Law for reversible processes. The basis of "pdV - thermodynamics" 34
1.5.7 Enthalpy and First Law for stationary flow processes 34
1.5.8 "Adiabatic equation of state" for an ideal gas – an integral of the energy balance 36
1.5.9 Applications of the energy balance 37
• Experiment of GAY-LUSSAC 37
• Piston drops into cylinder 38
• Throttling 40
• Heating of a room 40
• Nozzle flow 42
• Fan 46
• Turbine 47
• Chimney 48
• Thermal power station 50
1.6 History of the First Law 53
1.7 Summary of equations of balance 55

2. Constitutive equations 57
2.1 On measuring constitutive functions 57
2.1.1 The need for constitutive equations 57
2.1.2 Constitutive equations for viscous, heat-conducting fluids, vapors, and gas 57
2.2 Determination of viscosity and thermal conductivity 59
2.2.1 Shear flow between parallel plates. NEWTON's law of friction 59
2.2.2 Heat conduction through a window-pane 61
2.3 Measuring the state functions $p(\upsilon,T)$ and $u(\upsilon,T)$ 63
2.3.1 The need for measurements 63
2.3.2 Thermal equations of state 63
2.3.3 Caloric equation of state 64
2.3.4 Equations of state for air and superheated steam 66
2.3.5 Equations of state for liquid water 67
2.4 State diagrams for fluids and vapors with a phase transition 68
2.4.1 The phenomenon of a liquid-vapor phase transition 68
2.4.2 Melting and sublimation 70
2.4.3 Saturated vapor curve of water 70
2.4.4 On the anomaly of water 73

2.4.5 Wet region and (p, v)-diagram of water 75
2.4.6 3D phase diagram 75
2.4.7 Heat of evaporation and (h,T)–diagram of water 76
2.4.8 Applications of saturated steam 77
• The preservation jar 77
• Pressure cooker 78
2.4.9 Van der Waals equation 79
2.4.10 On the history of liquefying gases and solidifying liquids 81

3 Reversible processes and cycles "p dV thermodynamics" for the calculation of thermodynamic engines 83
3.1 Work and heat for reversible processes 83
3.2 Compressor and pneumatic machine. The hot air engine 84
3.2.1 Work needed for the operation of a compressor 84
3.2.2 Two-stage compressor 86
3.2.3 Pneumatic machine 86
3.2.4 Hot air engine 87
3.3 Work and heat for reversible processes in ideal gases. "Iso-processes" and adiabatic processes 88
3.4 Cycles 89
3.4.1 Efficiency in the conversion of heat to work 89
3.4.2 Efficiencies of special cycles 90
• Joule process 90
• Carnot cycle 91
• Modified Carnot cycle 92
• Ericson cycle 94
• Stirling cycle 96
3.5 Internal combustion cycles 96
3.5.1 Otto cycle 96
3.5.2 Diesel cycle 99
3.5.3 On the history of the internal combustion engine 101
3.6 Gas turbine 102
3.6.1 Brayton process 102
3.6.2 Jet propulsion process 103
3.6.3 Turbofan engine 104

4. Entropy 105
4.1 The Second Law of thermodynamics 105
4.1.1 Formulation and exploitation 105
• Formulation 105
• Universal efficiency of the Carnot process 105
• Absolute temperature as an integrating factor 107

• Growth of entropy ... 108
• (T,S)-diagram and maximal efficiency of the Carnot process ... 110
4.1.2 Summary ... 111
4.2 Exploitation of the Second Law ... 113
4.2.1 Integrability condition ... 113
4.2.2 Internal energy and entropy of a van der Waals gas and of an ideal gas ... 114
4.2.3 Alternatives of the Gibbs equation and its integrability conditions ... 115
4.2.4 Phase equilibrium. Clausius-Clapeyron equation ... 117
4.2.5 Phase equilibrium in a van der Waals gas ... 119
4.2.6 Temperature change during adiabatic throttling. Example: Van der Waals gas ... 120
4.2.7 Available free energies ... 123
4.2.8 Stability conditions ... 125
4.2.9 Specific heat c_p is singular at the critical point ... 126
4.3 A layer of liquid heated from below – onset of convection ... 127
4.4 On the history of the Second Law ... 131

5. Entropy as $S=k \ln W$ ... 135
5.1 Molecular interpretation of entropy ... 135
5.2 Entropy of a gas and of a polymer molecule ... 135
5.3 Entropy as a measure of disorder ... 139
5.4 Maxwell distribution ... 140
5.5 Entropy of a rubber rod ... 141
5.6 Examples for entropy and Second Law. Gas and rubber ... 143
5.6.1 Gibbs equation and integrability condition for liquids and solids ... 143
5.6.2 Examples for entropic elasticity ... 145
5.6.3 Real gases and crystallizing rubber ... 146
5.6.4 Free energy of gases and rubber. (p,V)- and (P,L)-curves. ... 148
5.6.5 Reversible and hysteretic phase transitions ... 150
5.7 History of the molecular interpretation of entropy ... 151

6. Steam engines and refrigerators ... 153
6.1 The history of the steam engine ... 153
6.2 Steam engines ... 155
6.2.1 The (T,S)-diagram ... 155
6.2.2 Clausius-Rankine process. The essential role of enthalpy ... 155
6.2.3 Clausius-Rankine process in a (T,S)-diagram ... 157
6.2.4 The (h,s)-diagram ... 159
6.2.5 Steam flow rate and efficiency of a power station ... 161
6.2.6 Carnotization ... 162

6.2.7 Mercury-water binary vapor cycle ... 163
6.2.8 Combined gas-vapor cycle ... 164
6.3 Refrigerator and heat pump ... 164
6.3.1 Compression refrigerator ... 164
6.3.2 Calculation for a cold storage room ... 165
6.3.3 Absorption refrigerator ... 166
6.3.4 Refrigerants ... 167
6.3.5 Heat pump ... 168

7. Heat Transfer ... 171
7.1 Non-Stationary Heat Conduction ... 171
7.1.1 The heat conduction equation ... 171
7.1.2 Separation of variables ... 171
7.1.3 Examples of heat conduction ... 172
- Heat conduction of an adiabatic rod of length L ... 172
- Heat conduction in an infinitely long rod ... 175
- Maximum of temperature of the heat-pole-solution ... 176
- Heat waves in the Earth ... 177

7.1.4 On the history of non-stationary heat conduction ... 179
7.2 Heat Exchangers ... 179
7.2.1 Heat transport coefficients and heat transfer coefficient ... 179
7.2.2 Temperature gradients in the flow direction ... 181
7.2.3 Temperatures along the heat exchanger ... 182
7.3 Radiation ... 184
7.3.1 Coefficients of spectral emission and absorption ... 184
7.3.2 Kirchhoff's law ... 186
7.3.3 Averaged emission coefficient and averaged absorption number ... 187
7.3.4 Examples of thermodynamics of radiation ... 190
- Temperature of the sun and its planets ... 190
- A comparison of radiation and conduction ... 192

7.3.5 On the history of heat radiation ... 193
7.4 Utilization of Solar Energy ... 194
7.4.1 Availability ... 194
7.4.2 Thermosiphon ... 195
7.4.3 Green house ... 196
7.4.4 Focusing collectors. The burning glass ... 198

8. Mixtures, solutions, and alloys ... 199
8.1 Chemical potentials ... 199
8.1.1 Characterization of mixtures ... 199
8.1.2 Chemical potentials. Definition and relation to Gibbs free energy ... 200
8.1.3 Chemical potentials; eight useful properties ... 201
8.1.4 Measuring chemical potentials ... 203

8.2 Quantities of mixing. Chemical potentials of ideal mixtures 204
8.2.1 Quantities of mixing 204
8.2.2 Quantities of mixing of ideal gases 206
8.2.3 Ideal mixtures 207
8.2.4 Chemical potentials of ideal mixtures 207
8.3 Osmosis 208
8.3.1 Osmotic pressure in dilute solutions. Van't Hoff's law 208
8.3.2 Applications of osmosis 210
- Pfeffer's tube 210
- A "perpetuum mobile" based on osmosis 212
- Physiological salt solution 213
- Osmosis as a competition of energy and entropy 213
- Desalination 215

8.4 Mixtures in different phases 216
8.4.1 Gibbs phase rule 216
8.4.2 Degrees of freedom 217
8.5 Liquid-vapor equilibrium (ideal) 218
8.5.1 Ideal Raoult law 218
8.5.2 Ideal phase diagrams for binary mixtures. 219
8.5.3 Evaporation in the (p,T)-diagram 221
8.5.4 Saturation pressure decrease and boiling temperature increase 222
8.6 Distillation, an application of Raoult's law 223
8.6.1 mol as a unit 223
8.6.2 Simple application of Raoult's law 224
8.6.3 Batch distillation 224
8.6.4 Continuous distillation and the separating cascade 227
8.6.5 Rectification column 229
8.7 Liquid-vapor equilibrium (real) 231
8.7.1 Activity and fugacity 231
8.7.2 Raoult's law for non-ideal mixtures 232
8.7.3 Determination of the activity coefficient 232
8.7.4 Determination of fugacity coefficients. 234
8.7.5 Activity coefficient and heat of mixing. Construction of a phase diagram 234
8.7.6 Henry coefficient 236
8.8 Gibbs free energy of a binary mixture in two phases 238
8.8.1 Graphical determination of equilibrium states 238
8.8.2 Graphical representation of chemical potentials 241
8.8.3 Phase diagram with unrestricted miscibility 241
8.8.4 Miscibility gap in the liquid phase 243
8.9 Alloys 243
8.9.1 (T, c_1)–diagrams 243
8.9.2 Solid solutions and the eutectic point 246

8.9.3 Gibbs phase rule for a binary mixture ... 247
8.10 Ternary Phase Diagrams ... 247
8.10.1 Representation ... 247
8.10.2 Miscibility gaps in ternary solutions ... 248

9. Chemically reacting mixtures ... 251
9.1 Stoichiometry and law of mass action ... 251
9.1.1 Stoichiometry ... 251
9.1.2 Application of stoichiometry. Respiratory quotient RQ ... 253
9.1.3 Law of mass action ... 253
9.1.4 Law of mass action for ideal mixtures and mixtures of ideal gases ... 254
9.1.5 On the history of the law of mass action ... 255
9.1.6 Examples for the law of mass action for ideal gases ... 256
- From hydrogen and iodine to hydrogen iodide and vice versa ... 256
- Decomposition of carbon dioxide into carbon monoxide and oxygen ... 257

9.1.7 Equilibrium in stoichiometric mixtures of ideal gases ... 258
9.2 Heats of reaction, entropies of reaction, and absolute values of entropies ... 260
9.2.1 The additive constants in u and s ... 260
9.2.2 Heats of reaction ... 262
9.2.3 Entropies of reaction ... 263
9.2.4 Le Chatelier's principle of least constraint ... 264
9.3 Nernst's heat theorem. The Third Law of thermodynamics ... 264
9.3.1 Third Law in Nernst's formulation ... 264
9.3.2 Application of the Third Law. The latent heat of the transformation gray→white in tin ... 265
9.3.3 Third Law in PLANCK's formulation ... 266
9.3.4 Absolute values of energy and entropy ... 267
9.4 Energetic and entropic contributions to equilibrium ... 267
9.4.1 Three contributions to the Gibbs free energy ... 267
9.4.2 Examples for minima of the Gibbs free energy ... 269
- $H_2 \rightarrow 2H$ ... 269
- $N_2 + 3H_2 \rightarrow 2NH_3$ Haber-Bosch synthesis of ammonia ... 270

9.4.3 On the history of the Haber-Bosch synthesis ... 271
9.5 The fuel cell ... 272
9.5.1 Chemical Reactions ... 272
9.5.2 Various types of fuel cells ... 273
9.5.3 Thermodynamics ... 274
9.5.4 Effects of temperature and pressure ... 276
9.5.5 Power of the fuel cell ... 276
9.5.6 Efficiency of the fuel cell ... 277

9.6 Thermodynamics of photosynthesis 278
9.6.1 The dilemma of glucose synthesis 278
9.6.2 Balance of particle numbers 279
9.6.3 Balance of energy. Why a plant needs lots of water 280
9.6.4 Balance of entropy. Why a plant needs air 282
9.6.5 Discussion 283

10. Moist air 285
10.1 Characterization of moist air 285
10.1.1 Moisture content 285
10.1.2 Enthalpy of moist air 285
10.1.3 Table for moist air 286
10.1.4 The (h_{1+x}, x)-diagram 288
10.2 Simple processes in moist air 289
10.2.1 Supply of water 289
10.2.2 Heating 290
10.2.3 Mixing 290
10.2.4 Mixing of moist air with fog 291
10.3 Evaporation limit and cooling limit 291
10.3.1 Mass balance and evaporation limit 291
10.3.2 Energy balance and cooling limit 292
10.4 Two Instructive Examples: Sauna and Cloud Base 294
10.4.1 A sauna is prepared 294
10.4.2 Cloud base 295
10.5 Rules of thumb 297
10.5.1 Alternative measures of moisture 297
10.5.2 Dry adiabatic temperature gradient 298
10.6 Pressure of saturated vapor in the presence of air 299

11. Selected problems in thermodynamics 301
11.1 Droplets and bubbles 301
11.1.1 Available free energy 301
11.1.2 Necessary and sufficient conditions for equilibrium 302
11.1.3 Available free energy as a function of radius 302
11.1.4 Nucleation barrier for droplets 304
11.1.5 Nucleation barrier for bubbles 305
11.1.6 Discussion 306
11.2 Fog and clouds. Droplets in moist air 306
11.2.1 Problem 306
11.2.2 Available free energy. Equilibrium conditions 307
11.2.3 Water vapor pressure in phase equilibrium 308
11.2.4 The form of the available free energy 308
11.2.5 Nucleation barrier and droplet radius 311
11.3 Rubber balloons 312
11.3.1 Pressure-radius relation 312

11.3.2 Stabililty of a balloon ... 315
11.3.3 A suggestive argument for the stability of a balloon ... 317
11.3.4 Equilibria between interconnected balloons ... 320
11.4 Sound ... 322
11.4.1 Wave equation ... 322
11.4.2 Solution of the wave equation, d'Alembert method ... 325
11.4.3 Plane harmonic waves ... 326
11.4.4 Plane harmonic sound waves ... 327
11.5 Landau theory of phase transitions ... 329
11.5.1 Free energy and load as functions of temperature and strain... Phase transitions of first and second order ... 329
11.5.2 Phase transitions of first order ... 329
11.5.3 Phase transitions of second order ... 332
11.5.4 Phase transitions under load ... 334
11.5.5 A remark on the classification of phase transitions ... 334
11.6 Swelling and shrinking of gels ... 335
11.6.1 Phenomenon ... 335
11.6.2 Gibbs free energy ... 337
11.6.3 Swelling and shrinking as function of temperature ... 340

12. Thermodynamics of irreversible processes ... 343
12.1 Single fluids ... 343
12.1.1 The laws of FOURIER and NAVIER-STOKES ... 343
12.1.2 Shear flow and heat conduction between parallel plates ... 345
12.1.3 Absorption and dispersion of sound ... 347
12.1.4 Eshelby tensor ... 349
12.2 Mixtures of Fluids ... 351
12.2.1 The laws of Fourier, Fick, and Navier-Stokes ... 351
12.2.2 Diffusion coefficient and diffusion equation ... 354
12.2.3 Stationary heat conduction coupled with diffusion and chemical reaction ... 356
12.3 Flames ... 358
12.3.1 Chapman-Jouguet equations ... 358
12.3.2 Detonations and flames ... 360
12.3.3 Equations of balance inside the flame ... 361
• Balance of fuel mass ... 362
• Energy conservation ... 362
12.3.4 Dimensionless equations ... 363
12.3.5 Solutions ... 364
12.3.6 On the precarious nature of a flame ... 366
12.4 A model for linear visco-elasticity ... 366
12.4.1 Internal variable ... 366
12.4.2 Rheological equation of state ... 368
12.4.3 Creep and stress relaxation ... 369

12.4.4 Stability conditions 371
12.4.5 Irreversibility of creep 371
12.4.6 Frequency-dependent elastic modulus and the complex elastic modulus 373
12.5 Shape memory alloys 374
12.5.1 Phenomena and applications 374
12.5.2 A model for shape memory alloys 378
12.5.3 Entropic stabilization 379
12.5.4 Pseudoelasticity 382
12.5.5 Latent heat 385
12.5.6 Kinetic theory of shape memory 387
12.5.7 Molecular dynamics 391

Index 395

P Prologue on ideal gases and incompressible fluids

P.1 Thermal and caloric equations of state

The systematic development of the thermodynamic theory and its applications begins in Chap. 1 of this book. Some of the applications concern ideal gases, notably air. Others concern nearly incompressible fluids, notably water. Therefore it is appropriate to have the equations for ideal gases and incompressible fluids available at the outset. There are two of them, the thermal equation of state and the caloric one.

The equations of state of ideal gases and of incompressible liquids are the results of the earliest researches in the field of thermodynamics. Their best-known pioneers are Robert BOYLE (1627-1691), Edmé MARIOTTE (1620-1684), Joseph Louis GAY-LUSSAC (1778-1850), James Prescott JOULE (1818-1889) and William THOMSON (Lord KELVIN, 1824-1907) and their work is nowadays a popular subject of the physics curricula in high schools.

Therefore in this somewhat advanced – or intermediate – book on thermodynamic we feel that we may assume those equations as known. We just list them in order to introduce notation and for future reference.

- *Ideal gases*

The *thermal equation of state* relates the pressure p of the gas to the mass density ρ and the absolute temperature T

$$p = \rho \frac{R}{M} T \text{, where } R = 8.314 \frac{\text{J}}{\text{mol·K}} \text{, and } M = \frac{m}{m_0} \frac{\text{g}}{\text{mol}} \tag{P.1}$$

denote the universal ideal gas constant and the molar mass, respectively; m is the atomic mass of the gas, and m_0 is a reference mass – nowadays $1/12$ of the mass of the most common carbon atom, the C^{12} isotope, so that $m_0 = 1.67 \cdot 10^{-27}\,\text{g}$ holds.

The *caloric equation of state* relates the specific internal energy u to the mass density and temperature

$$u = z \frac{R}{M}\left(T - T_R\right) + u_R \text{ where } z = \begin{cases} 3/2 & & \text{one} - \\ 5/2 & \text{for a} & \text{two} - \text{ atomic gas.} \\ 3 & & \text{more} - \end{cases} \tag{P.2}$$

For ideal gases u is independent of ρ, and it contains an arbitrary additive constant u_R, representing the internal energy at an arbitrary reference temperature T_R which is usually chosen as 25°C, or 298 K.

- *Incompressible fluid*

Incompressibility means

$$\rho = \text{const.} \tag{P.3}$$

The mass density is independent of either pressure or temperature. For water, which is incompressible to a good approximation, the value of ρ is $10^3\,\frac{\text{kg}}{\text{m}^3}$.

The caloric equation of state reads

$$u = c(T - T_R) + u_R, \tag{P.4}$$

where c is the specific heat. For water we have approximately $c = 4.18\frac{\text{kJ}}{\text{kg}\cdot\text{K}}$ but, indeed, c depends weakly on temperature, a fact which we often ignore.

P.2 "mol"

The "mol" – appearing in Section (P.1) as a unit – is a unit for the number of atoms or molecules. By $(\text{P.1})_3$ the mass of all atoms per mol is $M = \frac{m}{m_0}\frac{\text{g}}{\text{mol}}$ and therefore the number of atoms per mol is equal to

$$A = \frac{M}{m} = \frac{1}{m_0}\frac{\text{g}}{\text{mol}}, \text{ or } A = 6.023\cdot10^{23}\,\frac{1}{\text{mol}}. \tag{P.5}$$

This number is called AVOGADRO's number and most often denoted by A. One mol of all substances contains that same number of particles.

For a body of mass m, volume V and particle number N we may write

$$\rho = \frac{m}{V} = \frac{Nm}{V},$$

if the density is homogeneous throughout the body. In an ideal gas the thermal equation of state may then be expressed as

$pV = N\left(Rm_0\frac{\text{mol}}{\text{g}}\right)T$, and with the Boltzmann constant $k = Rm_0\frac{\text{mol}}{\text{g}} = 1.38\cdot10^{-23}\,\frac{\text{J}}{\text{K}}$

$$pV = NkT \tag{P.6}$$

This implies AVOGADRO's law: Equal volumes of different gases at the same pressure and temperature contain equally many particles.

The mol as a unit carries all the potential for confusion as any unit for a dimensionless quantity does. Physicists and most engineers would prefer to carry on without using the mol, but chemists and chemical engineers insist on it. And so the mol is a unit by international agreement along with meter, second, Newton, Kelvin, *etc.*

The approximate molar masses of a few common substances are as follows

$$\text{H}:1,\ \text{O}:16,\ \text{N}:14,\ C:12,\ \text{Ar}:40,\ \text{Cl}:35.5,\ \text{Na}:23;\ \text{all in } \frac{\text{g}}{\text{mol}} \tag{P.7}$$

Air is a mixture that behaves, under normal conditions, like an ideal gas. It consists of *

$$78.02\,\%\ N_2,\ 20.95\,\%\ O_2,\ 0.94\,\%\ \text{Ar},\ 0.03\,\%\ CO_2, \tag{P.8}$$

so that the mean mass of an "air molecule" is:

$$m_{\text{air}} = 0.7808\,m_{N_2} + 0.209\,m_{O_2} + 0.0094\,m_{\text{Ar}} + 0.0003\,m_{CO_2} = 28.96\,m_0. \tag{P.9}$$

---

* The percentages in (P.8) are "volume percent:" Before mixing – under the pressure p and temperature T – the constituents occupy 0.7808, 0.209, 0.0094, and 0.0003 parts of the volume of the air after mixing.

P.3 On the history of the equations of state

Robert Boyle, son of an Irish nobleman took a strong interest in the then new natural philosophy. He worked on chemistry and physics and repeated some of GALILEO's and GUERICKE's experiments. He was the first chemist to produce a gas other than air, namely hydrogen, by treating some metals with acids; but he did not name the gas. His contribution to thermodynamics is BOYLE's law: In a gas at constant temperature the volume is inversely proportional to pressure.

BOYLE was a zealous religious advocate – he had met God in a particularly intense thunderstorm while traveling in Switzerland – and in his lectures he defended Christianity against "notorious infidels, namely atheists, theists, pagans, jews and muslims."

Edmé Mariotte was as devout as BOYLE. He was in fact a priest and Prior of St. Martin sous Beaune. He rediscovered BOYLE's law and he extended it by noting that, at constant pressure, the volume of a gas grows with temperature t. BOYLE had not made that observation or, anyway, he did not mention it.

In that early age of natural science most scientists did not limit their attention to a single field of research and so MARIOTTE was also a keen meteorologist and a keen physiologist. He discovered the circulation of the earth's water supply and the "blind spot" of the eye.

Joseph Louis Gay-Lussac, about 100 years later than Mariotte, noticed that for a given pressure the volume of a gas is proportional to 273°C + t. This relation is also known as CHARLES's law, but GAY-LUSSAC found out a little more: The factor of proportionality is equal for all gases. GAY-LUSSAC's and Charles's findings foreshadowed the existence of a minimal temperature at t = - 273°C. Also GAY-LUSSAC found that gases combine chemically in simple volume proportions; thus he complemented DALTON's earlier observation according to which many chemical processes – between any constituents – combine by simple mass proportions.

In 1804 GAY-LUSSAC ascended to a height of 6 km (!) in a hot air balloon in an effort to determine changes of atmospheric pressure and temperature with height. He married a shop assistant when he caught her reading a chemistry book below the counter. GAY-LUSSAC is also known for his work on alcohol-water mixtures, which led to the "degrees GAY-LUSSAC" that were used in many countries to measure, classify, and tax alcoholic beverages.

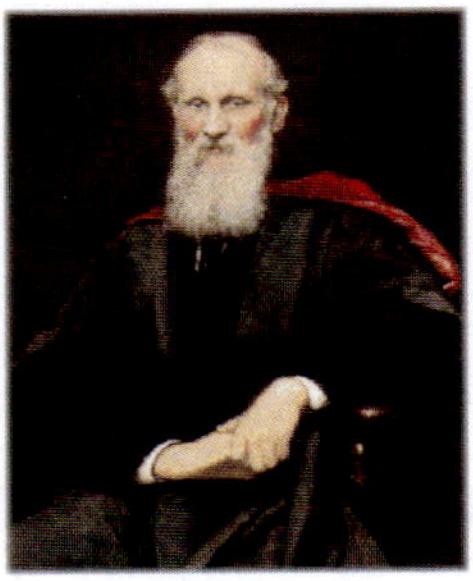

William Thomson (Lord Kelvin since 1892) completed the thermal equation of state of gases by suggesting the absolute temperature scale whose origin is put at t = - 273°C and whose degrees we call K in honor of LORD KELVIN. He accompanied the development of thermodynamics with deep interest and intelligent suggestions for more than half a century, and he promoted Joule, and made Joule's measurements of the *mechanical equivalent of heat* known to the scientific community.

KELVIN made a fortune through his connection with telegraph transmission through transatlantic cables.

James Prescott Joule was one of the pioneers of the First Law of thermodynamics. He constructed an apparatus in which a weight turned a paddle wheel immersed in water. The water became slightly warmer as a result of friction. Thus he determined the *mechanical equivalent of heat*: 772 foot-pounds of mechanical work raise the temperature of one pound of water by 1 degree Fahrenheit. In JOULE's honor the unit of energy is now called 1 Joule. After CLAUSIUS had conceived of the internal energy of ideal gases as independent of density or pressure, JOULE jointly with KELVIN found the Joule-Kelvin effect in air: Expanding air cools very slightly; the cooling is due to the work done to separate the molecules, and it occurs to the extent that the gas is not really ideal. GAY-LUSSAC with his cruder thermometers had missed the cooling effect.

Lorenzo Romano Amadeo Carlo Avogadro, Conte di Quarenga e di Cerreto (1776-1856) came from a distinguished family of lawyers and he himself studied ecclesiastical law and became a lawyer at the age of 16. He was, however, also much interested in physics and chemistry which he studied privately and successfully so that – in 1820 – he became a professor of natural philosophy at the University of Torino. He was able to see a common basis for DALTON's and GAY-LUSSAC's observations of simple proportions – by mass and volume, respectively – in chemical reactions and he concluded that equal volumes of different gases contain equally many molecules. This became known as AVOGADRO's law. Nowadays it is a simple corollary of the thermal equation of state.

P.4 An elementary kinetic view of the equations of state for ideal gases; interpretation of pressure and absolute temperature

We consider a monatomic ideal gas at rest in a rectangular box. The particle number density is $n = N/V$, and we assume n to be so small that the interaction forces between the molecules are negligible. The pressure on the walls results from the incessant bombardment of the walls by the atoms – at a rate of 10^9 collisions per second – and we present a strongly simplified model.

$-p\,\mathrm{d}A$ is the force of the wall element $\mathrm{d}A$ on the gas. This force must be equal to the change of momentum of the atoms which hit $\mathrm{d}A$ in the time interval $\mathrm{d}t$. The change of momentum is calculated as follows: For simplicity we assume that all atoms have the same speed and that $1/6$ of them fly in the 6 directions perpendicular to the walls. The momentum change of one atom on the wall element $\mathrm{d}A$ is then $-2mc$, *cf.* Fig. P1, and per time interval we will have $c\,\mathrm{d}A\,n/6$ such impacts on $\mathrm{d}A$. This is to say that all atoms in the cylindrical volume $c\,\mathrm{d}t\,\mathrm{d}A$ collide with the wall provided they fly forward toward $\mathrm{d}A$. Their momentum change is equal to

$$(-2mc)(c\,\mathrm{d}t\,\mathrm{d}A)\frac{n}{6} = -\rho\frac{1}{3}c^2\,\mathrm{d}A\,.$$

This expression must be equal to $-p\,\mathrm{d}A$ and we obtain

$$p = \rho\frac{1}{3}c^2\,.$$

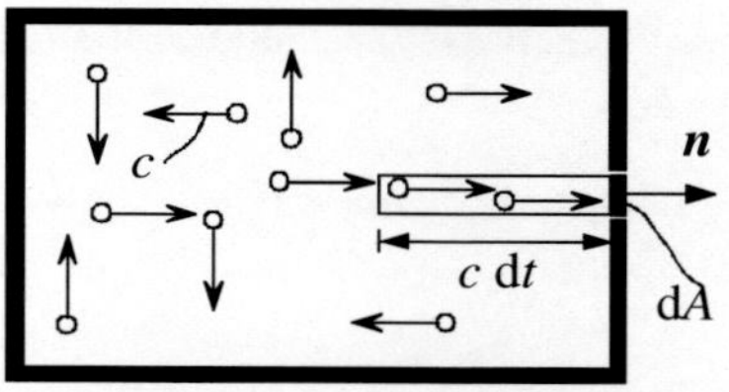

Fig. P1 Atomic interpretation of the pressure in an ideal gas

Comparison with the thermal equation of state provides

$$\frac{R}{M}T = \frac{1}{3}c^2 \quad \text{or} \quad \frac{3}{2}kT = \frac{m}{2}c^2\,, \tag{P.10}$$

so that the temperature is a measure for the kinetic energy of the atoms.

The energy density ρu of the model gas is obviously also related to the atomic kinetic energy. We have

$$\rho u = n\frac{m}{2}c^2 = \rho\frac{3}{2}\frac{R}{M}T\,. \tag{P.11}$$

This compares well with the caloric equation of state (P.2) for a monatomic gas. To be sure, T_R must be set to zero to make the analogy complete, and u_R must be zero as well, so that an atom at rest is assigned the energy zero.

The simple model for a gas described above was first introduced by **Daniel BERNOULLI** (1700-1782). He belonged to a family of eminent mathematicians. His father Johann and his uncle Jacob were both professors, and they knew that there was no money in mathematics. Therefore Daniel was destined to become a physician. But the genes were stronger and he drifted to mathematics during his life. His work on the kinetic theory of gases was largely ignored and the kinetic theory had to be reinvented – and improved – in the nineteenth century, a hundred years after Bernoulli. Daniel is also the author of the Bernoulli equation of fluid mechanics which we shall learn about in Paragraph 1.4.5.

In the eighteenth century scientific academies used to offer awards for the solution of certain unsolved mathematical or physical problems, and Daniel Bernoulli won several of them. In one of those he competed with his father. They both won and had to share the award money. It is said that this event made Johann Bernoulli break off relations with his son.

The equations (P.10) and (P.11) remain valid for more sophisticated gas models, except that in those models c must be replaced by the *mean* speed $\overline{c}$ of the atoms. The deviation of the caloric equation (P.11) from (P.2) for two- and more-atomic molecules is due to the fact that molecules can store energy in rotational motion.

A noteworthy consequence of (P.10) is the value of the (mean) speed of atoms. For helium with $M \approx 4\,\frac{\mathrm{g}}{\mathrm{mol}}$ and $T = 300\ \mathrm{K}$ the speed equals $\overline{c} = 1360\,\frac{\mathrm{m}}{\mathrm{s}}$. For heavier atoms it is smaller. The speed decrease is inversely proportional to $\sqrt{M}$, so that argon atoms with $M \approx 40\,\frac{\mathrm{g}}{\mathrm{mol}}$ move only about $1/3$ as fast as helium atoms.

1 Objectives of thermodynamics and its equations of balance

1.1 Fields of mechanics and thermodynamics

1.1.1 *Mass density, velocity, and temperature*

During a process the mass density, the velocity, and the temperature of a fluid are, in general, not homogeneous in space, nor are they constant in time. Therefore mass density, velocity, and temperature are called time-dependent *fields*.

Fluid mechanics proposes to calculate the fields of mass density $\rho(x_i,t)$, and velocity $w_j(x_i,t)$ in a fluid. *Thermodynamics* proposes to calculate the fields of mass density $\rho(x_i,t)$, velocity $w_j(x_i,t)$, *and* temperature $T(x_i,t)$ in the fluid. Therefore thermodynamics is more accurate than fluid mechanics: In addition to the motion of the fluid and its inertia, it takes into consideration how "hot" the fluid is.

On the interface between two bodies – the fluid and the wall of the container (say) – the temperature is continuous. This property defines temperature, it is the basis of all measurements of temperature by contact thermometers, and it is often referred to as the *Zeroth Law of Thermodynamics*.*

Most thermometers rely on the thermal expansion of the thermometric substance, often mercury. In this book we shall usually employ the Celsius scale – or centigrade scale – but often also the absolute or Kelvin scale. Both scales use the same degree of temperature such that melting ice and boiling water at normal pressure differ by 100 degrees. The values of temperature of these fix points are 0°C and 100°C, or 273.15K and 373.15K, respectively.

1.1.2 *History of temperature*

The word temperature is Latin in origin: *temperare* means to mix and, of course, mixing of two fluids of different temperatures is the easiest method to realize intermediate temperatures. Our skin is fairly sensitive to hot and cold in a certain range, but it is not perfect, because it cannot compensate for subjective circumstances. Therefore thermometers were developed in the 17th century. Gianfrancesco SAGREDO (1571-1620), a Venetian diplomat and pupil of GALILEO GALILEI (1564-1642) experimented with a thermometer in the early 17th century and he wondered at his observation that in winter time the air can be colder than ice or snow. Also in a letter to his former teacher GALILEI he expressed his surprise that

> ... *water from a well is actually colder in winter than in summer, although our senses judge otherwise.*

This is a neat example for the subjectivity of our senses. Indeed, if we bring water up from a well in wintertime and stick our hand into it, it feels warm. If we do the same in summertime, the water feels cool.

---

* By the time when the basic character of the definition of temperature was fully appreciated, the 1st and 2nd Laws of Thermodynamics had already been firmly labeled.

The genius of Galileo GALILEI as a physicist and mechanician is not really reflected in thermodynamics except that GALILEI categorically claimed – in his correspondence with SAGREDO – that he had invented the thermometer. There is some doubt concerning that claim although SAGREDO was enough of a diplomat to accept it. GALILEI immortalized SAGREDO in his famous narrative treatise *Il Dialogo dei Massimi Sistemi* which takes place in the palace of the SAGREDO family and in which SAGREDO plays the role of the intelligent dilettante. He can be seen as the central figure on the frontispiece of the book.

The introduction of scales with universally reproducible fix points ended the early phase of speculation and hapless observation. The most widely used scale today is the centigrade scale with one hundred equal degrees between the temperatures of boiling water and melting ice. This scale was introduced by Anders Cornelius CELSIUS (1701-1744) and Mårten STRÖMER (1707-1770). CELSIUS had counted downwards from boiling water at 0°C to melting ice at 100°C, because he wished to avoid negative temperatures in wintertime. STRÖMER reversed this direction.

Another traditional scale is the Fahrenheit scale, named after Gabriel Daniel FAHRENHEIT (1686-1736), which also uses melting ice (32°F) and boiling water (212°F) as fix points with 180°F in-between. This scale is still used in some countries, notably in the United States of America. The transformation formulae from Fahrenheit temperature (F) to Celsius temperature (C) and vice versa are given by $C = \frac{5}{9}(F - 32)$ and $F = \frac{9}{5}C + 32$.

Anders Cornelius CELSIUS

Mårten STRÖMER

Gabriel Daniel FAHRENHEIT

FAHRENHEIT also wanted to avoid negative temperatures and therefore in his scale the zero point occurs at -17.8°C. This is the freezing point of sea water and FAHRENHEIT may have thought that it surely would never get colder than that.

There are various quaint alternative propositions for fix points, such as • the melting point of butter, • the temperature of a deep cellar and • the temperature in the armpit of a healthy man.

A fairly complete account of the development of the thermometer, of temperature scales, and of fix points is given in the book by W.E. Knowles MIDDLETON* to which we refer the interested reader.

* W.E. Knowles Middleton, *A History of the Thermometer and its Use in Meteorology*. The Johns Hopkins Press, Baltimore, Md (1966).

The absolute scale was proposed by William THOMSON (Lord KELVIN). It relies on the existence of a lowest temperature which defines the zero point of the scale. The other fix point is the triple point of water which occurs at 0.01°C or 273.16 K.

1.2 Equations of balance

1.2.1 *Conservation laws of thermodynamics*

In order to determine the five fields $\rho(x_i,t)$, $w_j(x_i,t)$, and $T(x_i,t)$ we need five field equations, and these are based on the five conservation laws of mechanics and thermodynamics, namely the

- conservation law of mass, or *equation of continuity*,
- conservation law of momentum, or *NEWTON's Second Law*,
- conservation law of energy, or *First Law of Thermodynamics*.

Mathematically speaking all of these "laws" are balance equations. Therefore we may avoid repetition by formulating a generic equation of balance first and specializing it to specific cases later.

1.2.2 *Generic equations of balance for closed and open systems*

In general the elements of the surface ∂V of thermodynamic systems of volume V move with the velocity $u_j(x_i,t)$ which, as indicated by the variables x_i and t, are usually different for each element and may change over time. Consequently the shape, the size and the location of the system will change.

The equation of balance for a generic quantity   with the volume density $\rho\psi$ – *i.e.* the specific value ψ per mass – is an equation for its rate of change

$$\frac{\mathrm{d}}{\mathrm{d}t}\Psi = \frac{\mathrm{d}}{\mathrm{d}t}\int_V \rho\psi \,\mathrm{d}V . \tag{1.1}$$

The equation reads

$$\frac{\mathrm{d}}{\mathrm{d}t}\int_V \rho\psi \,\mathrm{d}V = -\int_{\partial V} \rho\psi(w_i - u_i)n_i \,\mathrm{d}A - \int_{\partial V} \phi_i \, n_i \,\mathrm{d}A + \int_V \rho(\pi + \zeta)\,\mathrm{d}V . \tag{1.2}$$

The first surface integral represents the *convective flux* through the surface, and the second one represents the *non-convective flux*; n_i is the outer unit normal on ∂V. The convective flux vector $\rho\psi(w_i - u_i)$ vanishes, if the surface moves with the velocity w_i of the fluid. ϕ_i is the non-convective flux density vector, the best-known example of which is the heat flux; the minus sign in front of the flux terms is chosen so that   increases, if the flux densities point *into* ∂V. The volume integral on the right hand side is taken over the *production density* $\rho\pi$, and the *supply density* $\rho\zeta$. Supplies are due to external forces, such as gravitation and inertial forces, and to absorption of radiation. In principle supplies may be "switched off" by taking suitable *external* measures. Production, however, is due to thermodynamic processes *inside* V, *e.g.* to internal friction or to heat conduction; in other words, it cannot be influenced *directly*. To be sure, in a conservation law the production *vanishes*, and it is this very property that defines a conserved quantity.

The system is called *closed*, if $w_i = u_i$ holds everywhere on ∂V. Then (1.2) reduces to

$$\frac{\mathrm{d}}{\mathrm{d}t}\int_V \rho\psi \,\mathrm{d}V = -\int_{\partial V} \phi_i \, n_i \,\mathrm{d}A + \int_V \rho(\pi + \zeta)\,\mathrm{d}V . \tag{1.3}$$

The system is called *open and at rest*, if $u_i = 0$ holds on ∂V. In this case V is independent of time and we may write for the left hand side in (1.2)

$$\frac{\mathrm{d}}{\mathrm{d}t}\int_V \rho\psi \,\mathrm{d}V = \int_V \frac{\partial \rho\psi}{\partial t}\,\mathrm{d}V \quad \text{to obtain} \tag{1.4}$$

$$\int_V \frac{\partial \rho\psi}{\partial t}\,\mathrm{d}V = -\int_{\partial V} (\rho\psi w_i + \phi_i)\, n_i \,\mathrm{d}A + \int_V \rho(\pi + \zeta)\,\mathrm{d}V . \tag{1.5}$$

By comparison of (1.3) through (1.5) we obtain

$$\frac{\mathrm{d}}{\mathrm{d}t}\int_V \rho\psi \,\mathrm{d}V = \int_V \frac{\partial \rho\psi}{\partial t}\,\mathrm{d}V + \int_{\partial V} \rho\psi w_i n_i \,\mathrm{d}A . \tag{1.6}$$

V on the left hand side is the time-dependent closed volume *V(t)*, while *V* on the right hand side is fixed; it is equal to $V(t_0)$ (say). The relation is known as the Reynolds transport theorem; it is the three-dimensional analogue of the Leibniz rule for the differentiation of a one-dimensional integral between variable limits.

1.2.3 *Generic local equation of balance in regular points*

We recall GAUSS's theorem for a smooth function $a(x_k)$. According to this theorem the surface integral of the function $a(x_k)\,n_i(x_k)$ over a closed surface is equal to the volume integral over the gradient field $\frac{\partial a}{\partial x_i}$

$$\int_{\partial V} a n_i \mathrm{d}A = \int_V \frac{\partial a}{\partial x_i}\mathrm{d}V . \tag{1.7}$$

If this rule is applied to the surface integral in (1.5) that equation may be written in the form

$$\int_V \left(\frac{\partial \rho\psi}{\partial t} + \frac{\partial(\rho\psi w_i + \phi_i)}{\partial x_i} - \rho(\pi + \zeta) \right) \mathrm{d}V = 0 , \tag{1.8}$$

provided that ρ, ψ, w_i, and ϕ_i are all smooth functions in *V*. Since (1.8) must hold for all V – even for infinitesimally small ones – the integrand itself must vanish. Thus in regular points, where the assumed smoothness holds, we obtain the generic equation of balance as a partial differential equation of the form

$$\frac{\partial \rho\psi}{\partial t} + \frac{\partial(\rho\psi w_i + \phi_i)}{\partial x_i} = \rho(\pi + \zeta). \tag{1.9}$$

1.3 Balance of mass

1.3.1 *Integral and local balance equations of mass*

In a closed system the mass is constant and therefore we have

$$\frac{\mathrm{d}}{\mathrm{d}t}\int_V \rho \,\mathrm{d}V = 0 . \tag{1.10}$$

Comparison with the generic equation (1.3) shows that in the case of the mass balance the generic quantities ψ, ϕ_i, π and ζ must be identified as shown in the following table

Ψ	ψ	ϕ_i	π	ζ
mass	1	0	0	0

Mass is a conserved quantity so that $\pi = 0$ holds. Also there is no way to supply mass to a system except through the surface, and this is why $\zeta = 0$ holds. And, obviously, a non-convective mass flux cannot exist on a surface which moves with the mass.

With the assignments of the table the mass balance of an open system at rest reads according to (1.5):

$$\int_V \frac{\partial \rho}{\partial t}\,\mathrm{d}V + \int_{\partial V} \rho w_i \, n_i \,\mathrm{d}A = 0 , \tag{1.11}$$

and the local equation (1.9) is reduced to

$$\frac{\partial \rho}{\partial t} + \frac{\partial \rho w_i}{\partial x_i} = 0 . \tag{1.12}$$

1.3.2 *Mass balance and nozzle flow*

We consider a flow through a nozzle as shown in Fig. 1.1. The flow is assumed to be stationary which is to say that at each fixed point x_i the density and the velocity are constant. Formally this means that $\frac{\partial}{\partial t} = 0$ holds.

The dashed line in the figure represents an open volume at rest. It consists of two cross-sections, A^{I} and A^{II}, and the *mantle surface* A^{M}. The mass balance (1.11) reads

$$\int_{\partial V} \rho w_i \, n_i \,\mathrm{d}A = 0 . \tag{1.13}$$

w_i is perpendicular to n_i on the mantle so that on that surface $w_i n_i = 0$ holds, which means that the mantle does not contribute in (1.13). Thus the equation reduces to

$$\int_{A^{\mathrm{I}}} \rho w_i \, n_i \,\mathrm{d}A + \int_{A^{\mathrm{II}}} \rho w_i \, n_i \,\mathrm{d}A = 0 . \tag{1.14}$$

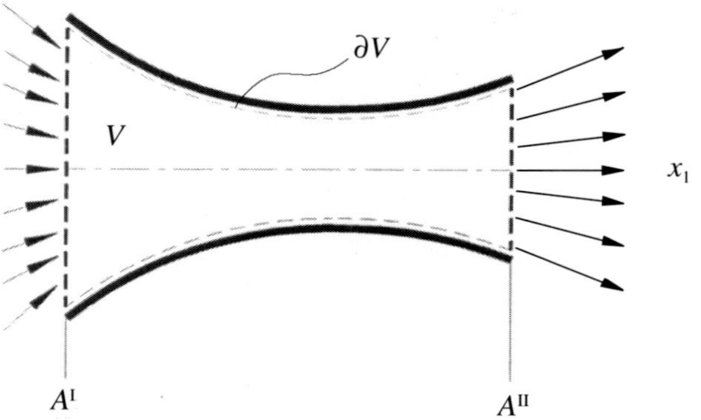

Fig. 1.1 Nozzle flow

In particular, if we assume one-dimensionality of the flow, meaning that ρ and w_1 are homogeneous on a cross-section, we obtain from (1.14)

$$(\rho\, w_1 A)\big|^{\mathrm{I}} = (\rho\, w_1 A)\big|^{\mathrm{II}}, \tag{1.15}$$

since on A^{I} the normal vector equals $(-1\,,0\,,0)$, while on A^{II} it reads $(1\,,0\,,0)$. Thus the *mass transfer rate*

$$\dot{m} = \rho w_1 A \tag{1.16}$$

is equal in magnitude for all cross-sections: In general ρ, w_1, and A are all different for the cross-sections. The product $\rho\, w_1 A$, however, is a constant. It seems that this property has given rise to the name continuity equation for the mass balance, an expression which is often used in fluid mechanics.

1.4 Balance of momentum

1.4.1 Integral and local balance equations of momentum

NEWTON's law of motion states that the momentum $\int\limits_V \rho w_j \mathrm{d}V$ of a closed system is changed by the forces acting on the system. These forces are of two types

i.) stress forces or tractions on the surface $\int\limits_{\partial V} t_j \mathrm{d}A = \int\limits_{\partial V} t_{ji} n_i \mathrm{d}A$

ii.) body forces $\int\limits_V \rho f_j \mathrm{d}V$.

$t_j \mathrm{d}A$ is the j-component of the stress force on the surface element $\mathrm{d}A$. The element is equilibrated by the forces $t_{j1}\, \mathrm{d}A_1$ and $t_{j2}\, \mathrm{d}A_2$, *cf.* Fig. 1.2. Thus, by $\mathrm{d}A_1 = \mathrm{d}A \cos\alpha$ and $\mathrm{d}A_2 = \mathrm{d}A \sin\alpha$ we have

$$t_j = t_{j1} \cos\alpha + t_{j2} \sin\alpha = t_{ji} n_i \,.$$

t_{ji} is called the stress tensor. The figure illustrates the significance of the components t_{ji} albeit – for simplicity – only in a two-dimensional case.

In general the stress tensor t_{ji} has nine different components, while the stress force has only three. It is thus clear that the stress tensor t_{ji} at some place cannot be determined by the stress force t_j at that place; we need the stress force on the whole surface. In most fluids – and always in this book – the stress tensor is symmetric so that $t_{ji} = t_{ij}$ holds, and then it has six independent components. Exceptions are fluids with an intrinsic spin, such as liquid crystals.

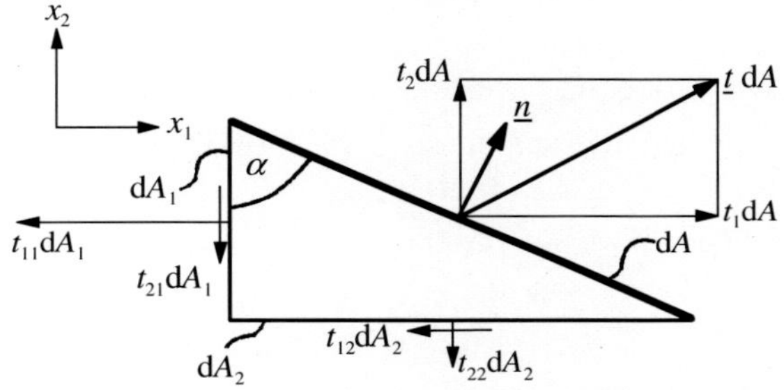

Fig. 1.2 On the components t_{11}, t_{12}, t_{21}, and t_{22} of the stress tensor in two dimensions

f_j is the specific body force. In all cases relevant to this book the body force is the gravitational force. And, if this force is accounted for at all*, the coordinate system is chosen such that

$f_j = (0, 0, -g)$, where $g = 9.8067 \frac{\mathrm{m}}{\mathrm{s}^2}$ is the gravitational acceleration.

Thus NEWTON's law of motion for a closed system may be written as

$$\frac{\mathrm{d}}{\mathrm{d}t} \int_V \rho \, w_j \, \mathrm{d}V = \int_{\partial V} t_{ji} \, n_i \, \mathrm{d}A + \int_V \rho \, f_j \, \mathrm{d}V \, . \tag{1.17}$$

It has the form of an equation of balance for a closed system and the comparison with the generic equation (1.3) leads to the following assignments.

Ψ	ψ	ϕ_i	π	ζ
momentum	ρw_j	$-t_{ji}$	0	f_j

The production density of momentum vanishes, because momentum is a conserved quantity, *i.e.* in the absence of forces – on the surface and within the body – the momentum is constant.

With the assignments of the table we obtain for an open system at rest

* The effect of gravitation on a gas or on a vapor may often be neglected because their densities are small.

$$\int_V \frac{\partial \rho\, w_j}{\partial t}\,\mathrm{d}V + \int_{\partial V} \left(\rho\, w_j w_i - t_{ji}\right) n_i \,\mathrm{d}A = \int_V \rho\, f_j \,\mathrm{d}V\,, \tag{1.18}$$

and locally in regular points

$$\frac{\partial \rho w_j}{\partial t} + \frac{\partial \left(\rho\, w_j w_i - t_{ji}\right)}{\partial x_i} = \rho\, f_j\,. \tag{1.19}$$

1.4.2 *Pressure*

In a fluid at rest, or with a homogeneous velocity field, the stress force $t_{ji}n_i\mathrm{d}A$ on an area element $\mathrm{d}A$ is perpendicular to the element for all orientations of the element. This means that

$$t_{ji}n_i = -p\, n_j \tag{1.20}$$

holds for all n_k. p is called the pressure. By setting n_k=(1,0,0), n_k=(0,1,0), and n_k=(0,0,1) we conclude that the stress tensor for the fluid has the form

$$t_{ji} = -p\delta_{ji}\,,\ \text{where}\ \delta_{ji} = \begin{pmatrix} 1 & 0 & 0 \\ 0 & 1 & 0 \\ 0 & 0 & 1 \end{pmatrix}, \tag{1.21}$$

and we say that the stress tensor is *isotropic*. The minus sign in (1.20) and (1.21) is due to a convention by which the force of a container wall on the fluid is considered positive. Thus obviously $p\,\mathrm{d}A$ is the force exerted by the fluid on the element $\mathrm{d}A$ of the wall.

If the velocity field of the fluid is not homogeneous so that velocity gradients exist, the stress tensor does have off-diagonal elements and its diagonal components may be different. We say that $t_{ji} + p\delta_{ji}$ – if unequal to zero – represents the viscous forces in a fluid.

1.4.3 *Pressure in an incompressible fluid at rest*

With $w_j = 0$, $t_{ji} = -p\delta_{ji}$, and $f_j = (0, 0, -g)$ the local balance of momentum (1.19) reduces to the three equations

$$\frac{\partial p}{\partial x_1} = 0\,,\ \frac{\partial p}{\partial x_2} = 0\,,\ \frac{\partial p}{\partial x_3} = -\rho g\,. \tag{1.22}$$

Hence p depends only on x_3 and – since $\rho = \text{const}$ holds in an incompressible fluid – we have

$$p(x_3) = p_0 - \rho\, g\, x_3\,, \tag{1.23}$$

where p_0 is the pressure at $x_3 = 0$. Therefore the pressure increases linearly with increasing depth. In water with $\rho = 10^3\,\frac{\text{kg}}{\text{m}^3}$ the pressure increases by 1 bar roughly for every 10 meters of depth.

1.4.4 *History of pressure and pressure units*

Evangelista TORRICELLI (1608-1647) was GALILEI's last pupil and he was given the problem to investigate the function of a suction pump. Fig. 1.3 shows a sketch of such a pump. When the piston is lifted, the water below the piston rises along with the piston. This phenomenon was explained at that time by the so-called *horror vacui*, the perceived horror of nature of an empty space. Indeed, if the water did not follow, a vacuum would develop beneath the piston. And yet, water could not be raised by more than about 10 m in this manner and, if that was attempted, a vacuum *did* appear below the piston.

TORRICELLI solved the problem. He argued that air has weight and this weight pushes on the free surface of the water everywhere adjacent to the pump. When the piston is raised, the air pressure pushes the water into the cylinder of the pump through the valve. This works until the pressure of the lifted water equals the pressure of the air.

TORRICELLI tested his theory by filling a pipe of a height of 1 m with mercury and stuck the open end vertically into a water basin. Mercury has a larger density than water – $\rho_{\mathrm{Hg}} = 13.595 \frac{\mathrm{g}}{\mathrm{cm}^3}$ – and, indeed, some mercury flowed out of the pipe but a column of roughly the height $H_{\mathrm{Hg}} = 760\,\mathrm{mm}$ remained. Above the mercury column an almost empty space had formed – now called a TORRICELLI vacuum – filled only with a minute amount of mercury vapor.

From this observation the maximum height of the water column in a suction pump could be calculated. If mercury and water exert the same pressure, (1.23) implies

$$\rho_{\mathrm{Hg}} H_{\mathrm{Hg}} = \rho_{\mathrm{H_2O}} H_{\mathrm{H_2O}}, \text{ hence } H_{\mathrm{H_2O}} = 10.33\ \mathrm{m}\,.$$

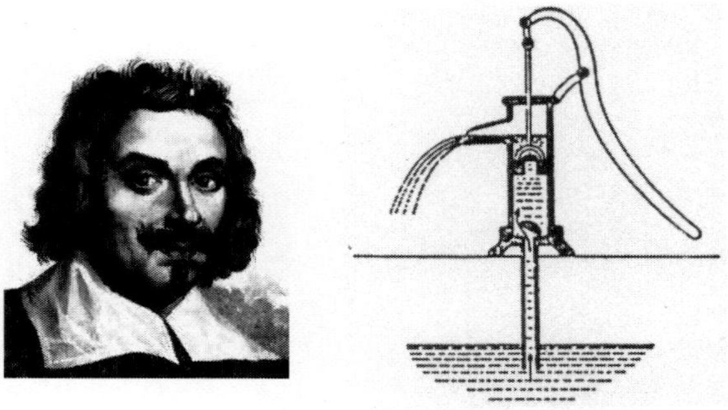

Fig. 1.3 Evangelista TORRICELLI. A suction pump

The weight of the mercury column of $760\,\mathrm{mm}$ height – or of water with $10.33\,\mathrm{m}$ height – must be equal to the weight of air on the same cross-section.

For a rough estimate of the thickness of the atmospheric shell of the earth one may assume that the density of air is independent of height and equal to the value $\rho_{\mathrm{air}} = 1.293 \frac{\mathrm{kg}}{\mathrm{m}^3}$ at the surface. In that way H_{air} would follow from

$$\rho_{\mathrm{Hg}} H_{\mathrm{Hg}} = \rho_{\mathrm{air}} H_{\mathrm{air}} \quad \text{as} \quad H_{\mathrm{air}} = 8000\ \mathrm{m}\,.$$

Although the assumption of a homogeneous air density is bad, the estimate shows that the atmosphere is quite thin, at least compared to the radius of the earth.

The pressure of 760 mm mercury at 0 °C is called a physical atmosphere, abbreviated as 1 atm. We have

$$1\,\text{atm} = \rho_{\text{Hg}} g\, H_{\text{Hg}} = 1.01325 \cdot 10^5 \frac{\text{N}}{\text{m}^2}.$$

Along with the physical atmosphere, the technical atmosphere was frequently used in earlier times. It was defined as

$$1\,\text{at} = 1\frac{\text{kp}}{\text{cm}^2},$$

where a kilopond is the weight of 1 kg under the influence of the gravitational acceleration $g = 9.8067\frac{\text{m}}{\text{s}^2}$ of the earth. Consequently

$$1\,\text{at} = 0.9807 \cdot 10^5 \frac{\text{N}}{\text{m}^2}.$$

Nowadays, by international agreement, we use the unit Pascal or bar,

$$1\,\text{Pa} = 1\frac{\text{N}}{\text{m}^2} \quad \text{and} \quad 1\,\text{bar} = 10^5 \frac{\text{N}}{\text{m}^2}.$$

1.4.5 *Applications of the momentum balance*

- *Buoyancy law of ARCHIMEDES*

The force on a body exerted by a liquid in which the body is submerged is

$$F_j = \int_{\partial V} t_{ji}\, n_i\, \text{d}A.$$

If the liquid is at rest so that $t_{ji} = -p\delta_{ji}$ holds, we have

$$F_j = -\int_{\partial V} p\, n_j\, \text{d}A,$$

and, by GAUSS's theorem (1.6)*

$$F_j = -\int_V \frac{\partial p}{\partial x_j}\, \text{d}V \quad \text{and with} \quad \frac{\partial p}{\partial x_j} = (0, 0, -\rho g),\ \textit{cf.}\ (1.22)$$

$$F_j = mg\,(0, 0, 1),$$

where $m = \int_V \rho\, \text{d}V$ is the mass of the liquid displaced by water.

Therefore the buoyancy force exerted by a fluid on a body submerged in it is equal to the weight of the displaced fluid. This is ARCHIMEDES' law. Note that this law holds for compressible as well as incompressible fluids. ARCHIMEDES in his often quoted experiment was concerned with water.

* The proper argument is more subtle than this, because, after all, there is no fluid inside ∂V.

ARCHIMEDES (ca. 287-ca.212 B.C.) found the law of buoyancy when his king – Hieron of Siracusa, Sicily – asked him whether his new crown was pure gold, as commissioned, or contained an admixture of cheap copper or silver. ARCHIMEDES measured the volume of the crown by dipping it into water and watching the rise of the water level. He compared that to the rise of the level when a piece of gold of the same weight was immersed. They say that the successful solution of the task made him cry out *EUREKA!*, Greek for I've got it!

ARCHIMEDES is considered the greatest scientist of ancient times. He made some progress with the rectification of the circle and the determination of the number . When his town was conquered by Roman soldiers he was killed on the beach while hunched over some circles drawn in the sand, or so they say.

- *Barometric steps in the atmosphere*

Meteorologists postulate a standard atmosphere for their calculations. In their model the temperature drop with increasing height is characterized by the gradient

$$\gamma = \frac{0.65\ \mathrm{K}}{100\ \mathrm{m}}. \tag{1.24}$$

The value of temperature on the ground is $T_G = 15\ °\mathrm{C}$, or $T_G = 288\ \mathrm{K}$, so that at a height of 10 km we have a temperature of -50°C. After that – at greater heights – the temperature remains constant, at least within the range of heights interesting to meteorologists.

The pressure drop must satisfy the differential equation $(1.22)_3$ just like in an incompressible fluid. But now, in air, we have, by virtue of (P.2)

$$\rho = \frac{p}{\frac{R}{M}T} \quad \text{and, by (1.24),} \quad T = T_G\left(1 - \frac{\gamma}{T_G}x_3\right). \tag{1.25}$$

Thus we obtain in the lower atmosphere – up to 10 km –

$$\frac{1}{p}\frac{\mathrm{d}p}{\mathrm{d}x_3} = -\frac{g}{\frac{R}{M}T_G}\frac{1}{1 - \frac{\gamma}{T_G}x_3} \quad \text{or, by integration from the ground upwards,}$$

$$p(x_3) = p_G\left(1 - \frac{\gamma}{T_G}x_3\right)^{\frac{1}{R/M}\frac{g}{\gamma}}. \tag{1.26}$$

Up to a height of $x_3 = 5\ \mathrm{km}$ the second term in the parentheses is smaller than 0.1 and, therefore, we may approximate the pressure formula by expansion into a Taylor series with only one relevant term:

$$p(x_3) \approx p_G\left(1 - \frac{g}{\frac{R}{M}T_G}x_3\right) = p_G - \delta\, x_3\ , \quad \text{where } \delta = \frac{100\ \mathrm{N/m^2}}{8.42\ \mathrm{m}}. \tag{1.27}$$

For $p_G = 1\,\text{bar}$ and $T_G = 288\,\text{K}$ this means that the air pressure decreases by 1 hPa for every 8.4 meters in gain of altitude. This difference in height is called the *barometric step*. It is the basis for the traditional altitude measurement in airplanes.

Sometimes we are interested in the pressure drop of an atmosphere which has a homogeneous temperature. This case may either be treated directly using $(1.22)_3$ and $(1.25)_1$, or the result may be obtained from (1.26) in the limit $\gamma \to 0$. With the approximation

$$1 - \frac{\gamma}{T_G} x_3 \approx \exp\left(-\frac{\gamma}{T_G} x_3\right)$$

we obtain

$$p(x_3) = p_G \exp\left(-\frac{Mg}{RT} x_3\right), \tag{1.28}$$

a formula known as the *barometric altitude equation*.

- *Thrust of a rocket*

For simplicity we assume that a rocket moves through vacuum and we neglect gravitation. Its mass $m(t)$ changes with time, because fuel is burned and expelled at the constant rate $\mu = -\frac{\mathrm{d}m}{\mathrm{d}t}$. The expelled gases move at the constant speed a with respect to the rocket. We wish to determine the acceleration $\frac{\mathrm{d}w}{\mathrm{d}t}$ of the rocket.

For this purpose we use the momentum balance (1.17) and choose the control volume V that contains the rocket and the expelled gases, *cf.* Fig. 1.4. In that case the rate of change of momentum is equal to zero and we have

$$\frac{\mathrm{d}}{\mathrm{d}t} \int_V \rho \, w \, \mathrm{d}V = 0 .$$

The momentum in V consists of the momentum $m(t)\,w(t)$ of the rocket, – including the momentum of its remaining fuel at time t –, *and* of the momentum of the expelled gas. The gas that was expelled between times τ and $\tau + \mathrm{d}\tau$ has the speed $w(\tau) - a$ and the momentum $\mu\,\mathrm{d}\tau\,(w(\tau) - a)$, so that the burned fuel as a whole has the momentum $\mu \int_{t_0}^{t} (w(\tau) - a)\,\mathrm{d}\tau$ where t_0 is the time of ignition.

Therefore the momentum balance reads

$$\frac{\mathrm{d}[m(t)\,w(t)]}{\mathrm{d}t} + \frac{\mathrm{d}}{\mathrm{d}t}\left[\mu \int_{t_0}^{t} (w(\tau) - a)\,\mathrm{d}\tau\right] = 0 .$$

And, by LEIBNIZ's rule for differentiation of integrals over a variable domain, we have

$$\frac{\mathrm{d}[m(t)\,w(t)]}{\mathrm{d}t}+\mu\left(w(t)-a\right)=0 \quad \text{or, by} \quad \mu=-\frac{\mathrm{d}m}{\mathrm{d}t},$$

$$m(t)\frac{\mathrm{d}\,w(t)}{\mathrm{d}t}=\mu\,a\,. \tag{1.29}$$

Fig. 1.4 On the thrust of a rocket
Saturn 1B-carrier rocket with Apollo capsule. First launch on February 26, 1966. Weight without fuel 84 t, fuel weight 506 t, payload 17 t. Rate of fuel consumption of first stage $2600\,\frac{\mathrm{kg}}{\mathrm{s}}$, initial thrust $7.1\cdot 10^{6}\ \mathrm{N}$.

This is the equation of motion of the rocket. The right hand side is called the thrust of the rocket. From the data given in the caption of Fig. 1.4 we conclude that the Saturn 1B rocket had an expulsion speed $a=2730\frac{\mathrm{m}}{\mathrm{s}}$, far larger than the speed of sound.

With $m(t)=m(t_0)-\mu(t-t_0)$ and $w(t_0)=0$ we obtain by integration

$$w(t)=a\ln\frac{m(t_0)}{m(t_0)-\mu(t-t_0)},$$

so that the acceleration increases in time until the fuel is used up.

- *Thrust of a jet engine*

In a jet engine air enters the inlet of a nozzle at atmospheric pressure; fuel is injected into that air and ignited, whereupon the combustion gases – still mostly air – leave the nozzle with a large velocity and, again, with atmospheric pressure. We consider stationary conditions and place ourselves in the frame where the engine is at rest, so that the incoming air moves with the speed of the jet, *cf.* Fig. 1.5. The force R is the air resistance – or drag -- and, in stationary conditions, it is equilibrated by the thrust F, so that $R+F=0$ holds. We calculate F in terms of the mass transfer rate and the relevant speeds. For the purpose we use the momentum balance (1.18) in which we ignore viscous stresses and gravitation, so that the 1-component of the equation reads

$$\int_{\partial V} \left(\rho\, w_1 w_j n_j - t_{1j}\, n_j\right) \mathrm{d}A = 0\,.$$

The integral over the stress is composed of two parts; one part is the shear force R on the support of the engine – the resistance – and the other one results from the pressure of the air

$$\int_{\partial V} - t_{1j}\, n_j\, \mathrm{d}A = -R + \int_{\partial V} p n_1 \mathrm{d}A\,.$$

The integral over the velocities vanishes except on the cross sectional areas A^{II} and A^{I}, where $n_j = (\pm 1{,}0{,}0)$. If we assume that the pressure is homogeneous on ∂V – and equal to the atmospheric pressure –, the integral over pn_1 vanishes, and we obtain

$$-R - \int_{A^I} \rho\, w_1^2 \mathrm{d}A + \int_{A^{II}} \rho\, w_1^2 \mathrm{d}A = 0\,.$$

If ρ and w_1 are homogeneous on A^{I} and A^{II} we thus obtain for the thrust of the jet engine

$$F = \dot{m}(w_1|^{\mathrm{I}} - w_1|^{\mathrm{II}})\,; \qquad (1.30)$$

obviously it points opposite to x_1, since $w_1|^{\mathrm{II}}$ is greater than $w_1|^{\mathrm{I}}$.

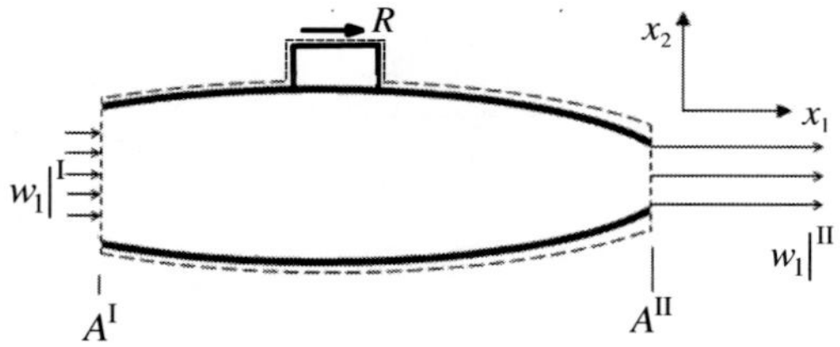

Fig. 1.5 On the thrust of a jet engine

The result (1.30) for the thrust is typical for conclusions drawn from the equations of balance or, here, from the momentum balance. It relates the thrust to the difference of velocities before and behind the engine and there is no reference – in the result – to the way in which the velocity difference comes about. Thus, it may be the result of the acceleration of the air in a combustion chamber, – as implied in the above argument –, or it may be due to the acceleration by a propeller; which is the case for a propeller driven aircraft.

- *Convective momentum flux*

We hold a rigid plate of area A out of the window of a car, perpendicular to the road. The plate must be supported from behind by a force F in order to offset the momentum flux of the horizontally onrushing air, *cf.* Fig. 1.6. The flow is stationary. For the purpose of a rough estimate we set $t_{ij} = -p\delta_{ij}$ so that viscous forces are neglected. In this case the 1-component of the momentum balance (1.18) reads

$$\int_{\partial V} (\rho\, w_1 w_i n_i + p\, n_1)\, \mathrm{d}A = 0\,.$$

n_1 is zero on the mantle surface A^{M} of ∂V and equal to ± 1 on the right and on the left surfaces A^{R} and A^{L}, respectively. Also on A^{L} we have $w_1 = 0$ and on A^{R} the pressure p equals p_0 and $w_i n_i$ is equal to w_1. Hence it follows – if A^{R} has the area A – that the force which the onrushing air exerts on the plate is given by

$$\int_{A^{\mathrm{L}}} p\, \mathrm{d}A = \left(p_0 + \rho\, w_1^2\right) A + \int_{A^{\mathrm{R}}} \rho\, w_1 w_i n_i\, \mathrm{d}A\,.$$

We neglect the product $w_1 w_i n_i$ on the mantle surface, because at the points where w_1 is large, $w_i n_i$ is small, and *vice versa*. Thus

$$\int_{A^{\mathrm{L}}} p\, \mathrm{d}A - p_0 A \approx \rho\, w_1^2\, A\,.$$

Since p_0 is the pressure behind the plate – as well as far in front of it – the difference $\int_{A^{\mathrm{L}}} p\, \mathrm{d}A - p_0 A$ determines the force F which thus turns out to be equal to $\rho\, w_1^2\, A$. For an air density $\rho = 1.29 \frac{\mathrm{kg}}{\mathrm{m}^3}$, and the speed $w_1 = 100 \frac{\mathrm{km}}{\mathrm{h}}$ we obtain $\rho\, w_1^2\, A \approx 10^3\,\mathrm{Pa}$. The force on a palm-sized plate of area $A = 100\,\mathrm{cm}^2$ is therefore $10\,\mathrm{N}$, equivalent to a weight of $1\,\mathrm{kg}$.

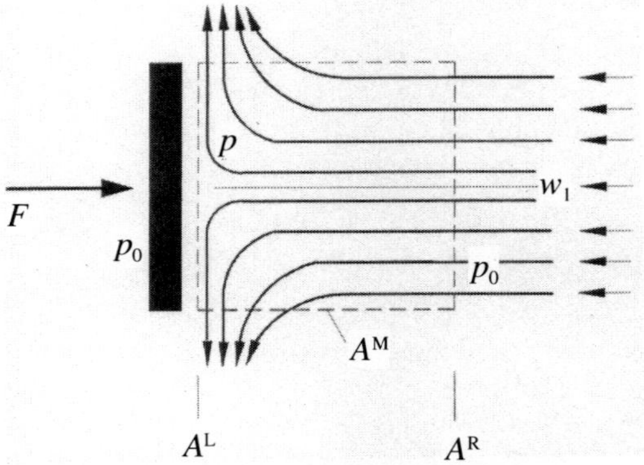

Fig. 1.6 On convective momentum flux

- *Momentum balance and nozzle flow*

We consider the stationary flow of Fig. 1.1 and apply the momentum balance (1.18) to the volume with the dashed surface ∂V. The 1-component of the momentum balance is of particular interest and, once again, we ignore friction forces so that $t_{ij} = -p\delta_{ij}$ holds. Consequently the balance reduces to the equation

$$\int_{\partial V} (\rho\, w_1 w_i n_i + p\, n_1)\, \mathrm{d}A = 0\,.$$

The mantle gives no contribution to the first term in the integral, because on the mantle we have $w_i n_i = 0$. On the cross-sections A^{I} and A^{II} we have $n_i = (\mp 1, 0, 0)$, respectively. We assume that ρ, w_1, and p are homogeneous on a cross-section and obtain

$$-\left(\rho^{\mathrm{I}}\, {w_1^{\mathrm{I}}}^2 + p^{\mathrm{I}}\right) A^{\mathrm{I}} + \left(\rho^{\mathrm{II}}\, {w_1^{\mathrm{II}}}^2 + p^{\mathrm{II}}\right) A^{\mathrm{II}} + \int_{A^M} p\, n_1\, \mathrm{d}A = 0\,, \tag{1.31}$$

where A^M denotes the surface of the mantle.

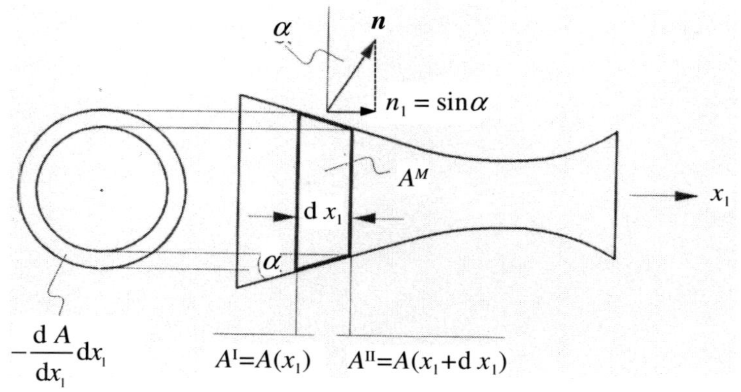

Fig. 1.7 On the momentum balance for nozzle flow

This equation does not permit an immediate comparison of the states on the cross-sections I and II, since the pressure on the mantle surface is unknown. Despite this we may derive an interpretable equation from (1.31) by considering two immediately adjacent cross-sections, *cf.* Fig. 1.7. In this case we may use an expansion into a Taylor series and write

$$\left(\rho^{\mathrm{II}}\, {w_1^{\mathrm{II}}}^2 + p^{\mathrm{II}}\right) A^{\mathrm{II}} - \left(\rho^{\mathrm{I}}\, {w_1^{\mathrm{I}}}^2 + p^{\mathrm{I}}\right) A^{\mathrm{I}} = \frac{\mathrm{d}\left(\rho\, w_1^2 + p\right)A}{\mathrm{d}x_1}\mathrm{d}x_1\,. \tag{1.32}$$

The integral over the mantle may then be written as

$$\int_{A^M} p\, n_1\, \mathrm{d}A = p\, n_1 A^M = p \sin\alpha\, A^M = p\left(-\frac{\mathrm{d}A}{\mathrm{d}x_1}\mathrm{d}x_1\right), \tag{1.33}$$

because $A^M \sin\alpha$ is equal to the decrease of the cross-sectional area. In Fig. 1.7 this circumstance is illustrated.

Combination of (1.32), (1.33) with (1.31) provides

$$\frac{\mathrm{d}\rho\, w_1^2 A}{\mathrm{d}x_1}+\frac{\mathrm{d}\, pA}{\mathrm{d}x_1}-p\frac{\mathrm{d}\,A}{\mathrm{d}x_1}=0 \quad \text{and with} \quad \rho\, w_1 A=\text{const.}$$

$$\frac{\mathrm{d}\, w_1^2/2}{\mathrm{d}x_1}=-\frac{1}{\rho}\frac{\mathrm{d}\, p}{\mathrm{d}x_1}. \tag{1.34}$$

For a long narrow nozzle, in which w_2 and w_3 are both much smaller than w_1, we have $w_1^2 \approx w^2$, and therefore we may interpret (1.34) by saying that the *specific* kinetic energy of the flow increases along the nozzle to the extent in which the pressure drops.

- *Bernoulli equation*

We specialize the local momentum balance (1.19) for a stationary incompressible and non-viscous fluid in a gravitational field:

$$\rho=\text{const}\,,\; t_{ij}=-p\delta_{ij}\,,\; \frac{\partial w_i}{\partial x_i}=0\,,\; f_j=(0,0,-g)=-\frac{\partial\, g x_3}{\partial x_j} \tag{1.35}$$

where $(1.35)_3$ follows for $\rho=\text{const}$ from the mass balance (1.12). Thus we obtain

$$\rho\, w_i\frac{\partial w_j}{\partial x_i}+\frac{\partial p}{\partial x_j}+\frac{\partial\, \rho g x_3}{\partial x_j}=0\,,$$

or, after scalar multiplication by w_j

$$w_i\frac{\partial}{\partial x_i}\left(\frac{\rho}{2}w^2+p+\rho\, g\, x_3\right)=0\,. \tag{1.36}$$

This is the Bernoulli equation. It states that the combination

$$\frac{\rho}{2}w^2+p+\rho\, g\, x_3$$

is constant along a streamline of the flow field, *i.e.* a line in the direction of w_j. In particular: The pressure is small where the speed is large.

The Bernoulli equation may be used to measure the speed of the flow of a liquid in a pipe (say) by measuring pressures, *cf.* Fig. 1.8. Indeed, for known values of p_0 and w_0 – at the inlet (say) – we have

$$w=\sqrt{w_0^2+\frac{2}{\rho}(p_0-p)}\,,$$

and $p_0-p=\rho\, g\,(H_0-H)$ is measured by the height of the liquid in the monitoring tubes.

Another well-known application of the Bernoulli equation concerns the conversion of kinetic into potential energy. If we use a hose to spout water vertically upwards, $\frac{\rho}{2}w^2+\rho\, g\, x_3$ is constant, since the pressure is nearly the same everywhere in the water jet. If w_E is the exit speed from the hose we may calculate the height h that is reached by the water from

$$\frac{1}{2} w_E^2 = gh \, .$$

If the end of the hose is squeezed flat, the speed increases in inverse proportion to the exit area and the height h increases.

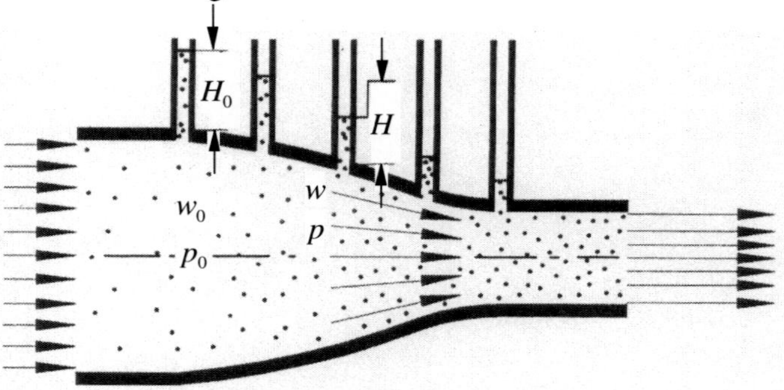

Fig. 1.8 Measuring pressure in a pipe

- *Kutta-Joukowski formula for the lift of an airfoil*

A plane plate tilted by an angle α to an incoming flow of air with speed V, *cf.* Fig. 1.9, experiences a force L in the vertical direction which is called the lift force. This force keeps an airplane aloft, and we proceed to calculate it as follows. The force F of the flow on the plate results from the difference of pressure on the upper and on the lower side, and we write by using the Bernoulli equation

$$F = -\oint p\,(x, y=0)\, b \, \mathrm{d}x = \oint \frac{\rho}{2} w^2 (x, y=0)\, b \, \mathrm{d}x \, , \tag{1.37}$$

where b is the span of the airfoil. The integrals are contour integrals – in clockwise direction – along the streamline that runs along the sides of the foil.

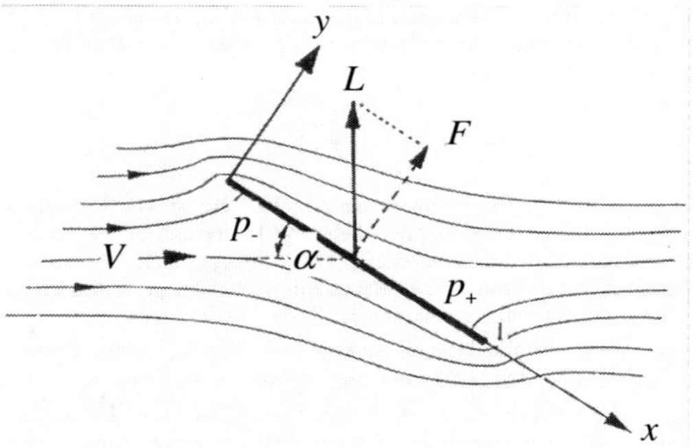

Fig. 1.9 On the lift force of a plane airfoil

We write

$$w^2(x, y=0) = [V\cos\alpha + \delta\, w(x, y=0)]^2$$
$$\approx V^2 \cos^2\alpha + 2V\cos\alpha\, \delta\, w(x, y=0)$$

$$= -V^2 \cos^2 \alpha + 2V \cos\alpha\, w(x, y=0),$$

provided δw is small compared to V. Insertion into (1.37) gives

$$F = \rho\, b\, V \cos\alpha\, \Gamma \text{ with the circulation } \Gamma = \oint w\,(x, y=0)\,\mathrm{d}x\,.$$

This force is decomposed into a drag force in the direction of the plate and a lift force L perpendicular to the incoming air. Thus we obtain the equation of Kutta-Joukovsky

$$L = \rho\, b\, V\, \Gamma\,,$$

which relates the lift force of the airfoil to the circulation Γ of the flow field.

Martin Wilhelm KUTTA (1867-1944) was an engineer with strong interests in mathematics. He taught at the Universities of Munich, Jena, Aachen, and Stuttgart. And he is best known nowadays for his work on the approximate solution of ordinary differential equations, which led to the now famous Runge-KUTTA method. That work was done in his doctoral thesis in 1900.

When airplanes were still in their infancy, and flight heavier than air was largely still a mystery, KUTTA became interested in the question why the planes stay aloft and – in his habilitation thesis – he discovered the formula for the lift acting on a moving airfoil.

Nikolai Egorovich ZHUKOWSKII (or ZHUKOVSKY, or JOUKOVSKY) (1847-1921) purchased one of the only eight glider planes which Otto VON LILIENTHAL, the pioneer of flight heavier than air, sold in his small firm in Berlin. As a physicist and mathematician JOUKOVSKY studied the working of the plane and came up with the correct expression for the lift force independently of KUTTA. JOUKOVSKI was interested in all aspects of flying, he taught the world's first course on aviation theory at Moscow Technical University, and – during World War I – he formulated a theory of bombing from airplanes.

An important problem of aerodynamics is the calculation of Γ for a given profile of an airfoil. We will not get into this. It may suffice to state without proof that the tilted plate of Fig. 1.9 has the circulation $\Gamma = \pi\, l\, V \sin\alpha$. Therefore its lift force is given by

$$L = 2\pi\, \alpha \frac{\rho}{2} V^2 b\, l\,,$$

if the angle of inclination is small. L is proportional to the angle – as long as α is small – and to the area $b\,l$ of the plate, and to the kinetic energy $\frac{\rho}{2} V^2$ of the incoming flow.

1.5 Balance of energy

1.5.1 *Kinetic energy, potential energy, and four types of internal energy*

- *Kinetic energy of thermal atomic motion. "Heat energy"*

We know from mechanics of mass points and of rigid bodies that the kinetic energy of a moving body may be converted into potential energy and *vice versa*. A suggestive example for this energy conversion is a car of mass $m = 2\,\mathrm{t}$ which moves with a speed of 36 km/h toward a hill. Initially it has a kinetic energy $mv^2/2 = 10^5\,\mathrm{J}$ and – running uphill – it may reach a height H of approximately $5\,\mathrm{m}$, ignoring friction and with $g \approx 10\,\mathrm{m/s}^2$. We may say that the kinetic energy $mv^2/2$ of the car has been converted into the potential energy mgH. The kinetic energy can be recovered by letting the car roll back downhill. In this experiment the sum of the kinetic and of the potential energy is conserved.

Gottfried Wilhelm Leibniz (1646-1716) is said to have been the last man to take all knowledge for his province. He is sometimes called the Aristotle of the 17th century. Leibniz was an able mathematician, a successful diplomat, and an illustrious philosopher. In the latter capacity he advocated the idea that this world is the best of all possible worlds, – a doctrine that few people were able to appreciate.

Amongst his feats in mathematical physics is the recognition that mechanical energy – potential energy and kinetic energy – is conserved.

It was only in the mid-nineteenth century when it was recognized that the energy is still conserved when the car comes to a stop by running against a tree. In this case the original kinetic energy of the car is distributed among the atoms of the tree and of the car, and a large part of it can be found as increased kinetic energy of the disordered atomic motion, the *thermal motion*. The rest has been converted into inter-atomic potential energy which is increased by the deformation of the car and the tree. After the collision the energy cannot be *seen* anymore with our coarse senses, because we cannot make out the atoms. We say that the original kinetic energy has been converted into *internal energy*. The part that has gone into the kinetic energy of the atoms may be *felt* as "heat." Indeed, the temperature is a measure for the (mean) kinetic energy of the atoms in a body that is at rest as a whole, *cf.* Section P.3.

In particular, for water we have: If the kinetic energy of the molecules of 1 g is raised by 1 J the temperature rises by 1/4.18 K; for air this is approximately 1 K, and for iron 2.2 K.

If $10^5\,\mathrm{J}$, the energy of the moving car, is fully converted into kinetic energy of the atoms of 1 liter of water, the water temperature is raised by about 25°C. A hammer of mass of 1 kg hitting a piece of iron of 10 g at a speed of 40 m/s will raise its temperature by 180°C; three blows with the hammer bring the iron to red heat.

- *Potential energy of molecular interaction. "Van der Waals energy" or "elastic energy"*

Molecules attract each other at long distances with so-called van der Waals forces – electric in nature. If we want to separate two molecules, work is required in order to increase the potential energy of the pair. That potential energy is part of the internal energy of a body.

The best known and most common phenomenon, which is accompanied by an increase of van der Waals energy, is evaporation. Upon evaporation of 1 g of water – at 1 atm – the water absorbs 2078 J while the temperature is unchanged. The energy 10^5 J of the above-mentioned car would thus be able to effect the evaporation of about 50 g of water.

The van der Waals forces do not only resist separation by their attraction. They also repel two atoms when they are very close. Therefore, if a sheet of metal is bent, the van der Waals energy grows everywhere: In the parts of the sheets that experience a tension *and* in the parts that are compressed. It is common to speak of the elastic energy in this context. In an accident between two cars part of the energy is converted into the elastic energy of the deformed bodywork; car builders attempt to maximize the deformability of the car body so that little or no energy is left to be absorbed by the bones and skins of the passengers, so to speak.

- *Chemical energy between molecules. "Heat of reaction"*

A third contribution to the internal energy is provided by the chemical bonds. Thus the splitting of water molecules into hydrogen and oxygen atoms according to the reaction $H_2O \rightarrow H_2 + \frac{1}{2}O_2$ requires $1.6 \cdot 10^4$ J per gram of water – eight times as much as evaporation. The 10^5 J of the moving car would be able to split no more than 6g of water. Inversely, 1.6 10^4 J are set free when 6 g of water are created by chemical reaction of an oxygen-hydrogen gas mixture. Under normal conditions this reaction occurs explosively and the new energy appears as kinetic energy of the molecules. We speak of the *heat of reaction*.

Food contains substances that release energy during the process of digestion. This energy maintains the circulation of blood and the body temperature. Humans need about 8 10^6 J per day and this is 80 times (!) the energy of the car. Another way of appreciating the human energy consumption is to compare it with a 100 W light bulb; both need the same energy.

The energy needs of animals are covered by a reaction of the oxygen in air with the carbon and the hydrogen of the food. What emerges is carbon dioxide CO_2 and water. There are three types of nutrients:

- Carbohydrates, for which glucose $C_6H_{12}O_6$ is a good representative, *cf.* Fig. 1.10.*

---

* The name carbohydrate was introduced by the chemist GAY-LUSSAC. He believed glucose to consist of six carbon atoms in a row, each one linked to a water molecule in the manner of hydrates. This is totally incorrect but GAY-LUSSAC cannot be blamed. In his time the knowledge of chemical valence bonds was not available yet.

- Proteins, which are polymers of different amino acids connected by peptide links. There is a large number of these proteins – differing by the number and the sequence of amino acids and by spatial configuration – and all are too complex to be represented by a simple chemical formula.
- Fats or lipids. A typical example is trioleine $C_{54}H_{104}O_6$, *cf.* Fig. 1.10. Basically fats are formed by esterification of glycerol $C_3H_8O_3$ and fatty acids.

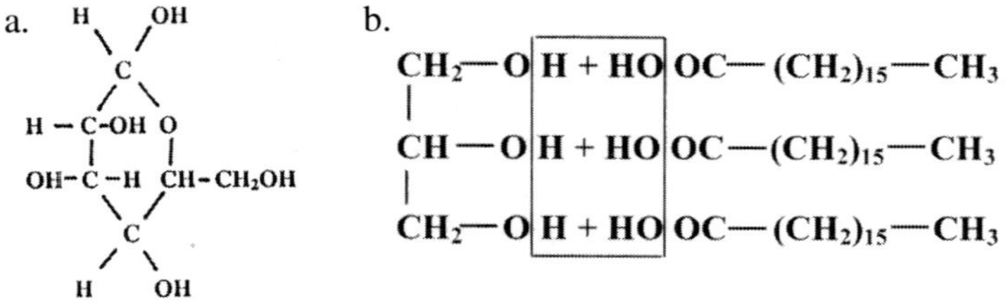

Fig. 1.10 a. Glucose molecule
b. Esterification of glycerol with three oleic acids. Trioleine

In comparison with carbohydrates the molecules of lipids contain less oxygen atoms for the extant number of carbon and hydrogen. Therefore they can deliver more energy by combining with oxygen. And, indeed, if we write the specific energetic gain into the chemical equation for the combustion of glucose and trioleine we obtain

$$C_6H_{12}O_6 + 6\,O_2 \rightarrow 6\,CO_2 + 6\,H_2O - 15.5\tfrac{\text{kJ}}{\text{g}},$$

$$C_{54}H_{104}O_6 + 77\,O_2 \rightarrow 54\,CO_2 + 52\,H_2O - 39.5\tfrac{\text{kJ}}{\text{g}}. \qquad (1.38)$$

Thus the combustion of fat delivers considerably more than twice as much energy as the combustion of carbohydrate. The heats of reaction of proteins lie somewhere between these values. Those heats are measured by burning the dried substances and registering the heat produced. The burning is usually accompanied by a flame.

In the body of animals the combustion occurs – obviously without a flame – by mediation of enzymes and very slowly. This, however, does not alter the heats of reaction. The fact that open combustion in a flame and biological combustion in a living organism are energetically equivalent was proved toward the end of the nineteenth century by careful measurements. Thus the old theory that organic reactions require a special spark of life, a *vis viva*, was put to rest.

The mass of glucose to be burned by humans to cover their energy needs of $8\cdot 10^6$ J per day equals 516 g by (1.38). For fat this is only 203 g. Therefore fat is the more effective energy storage for bad times. This is why the body accumulates fat in good times.

Chemical bonds are different in strength but for most of the well-known stoichiometric reactions they lie in the range between a few dozen and a few hundred kJ per mol. There is one bond, however, which stands out as particularly weak – about 1 kJ per mol – and this is called a hydrogen bond, because it is formed by a hydrogen atom. This bond is responsible for the *shape* of organic molecules, such as enzymes and other proteins, which is important for their proper functioning. The hydrogen bond is so weak that it can be broken by the thermal motion when the body temperature is raised only slightly. This fact is responsible for the sensitiveness of biological tissue to a rise in temperature, and its easy corruption. Also the hydrogen bond in cold water and ice makes for the *anomaly of water* by which ice is lighter than liquid water.

- *Nuclear energy*

Nuclear energy is the energy provided by the bonds of the particles in the nuclei of the atoms. This energy can be converted into the energy of thermal motion – *i.e.* heat – by the fission of large nuclei and the fusion of small ones. Thus the fusion of two heavy hydrogen nuclei – proton plus neutron – to a helium nucleus provides a heat of reaction of $4.53 \cdot 10^{-12}$ J. If the human organism were able to achieve that fusion – somehow enzymatically – it could cover its energy needs of $8 \cdot 10^6$ J per day by ingesting about $4 \cdot 10^{18}$ hydrogen atoms with a total mass of about $13 \cdot 10^{-6}$ g – a very light diet indeed.

Of course, this does not work – not in this world and not outside a science fiction story. On earth the fusion of hydrogen to helium has only worked in the hydrogen bomb so far. However, the sun covers its energy needs that way. Per day it radiates off $3 \cdot 10^{31}$ J and, if this energy is to be covered by the fusion of hydrogen, the sun must consume hydrogen at the rate of $5 \cdot 10^{13}$ t per day, an amount that is small enough, when compared to the total mass of the sun, to guarantee a stationary energy output for billions of years.

- *Heat into work*

The fact that the kinetic or the potential energy of a moving body can be converted into "heat energy" without changing value, was recognized in the mid 19th century. This discovery was considered so important that it became known as the *First Law of Thermodynamics*. The history of the discovery is briefly reviewed in Sect. 1.6.

Now, if the kinetic energy of the rolling car can be converted into heat – by running the car against a tree (say) – the question arose as to whether the opposite might not also be possible, *i.e.* whether it is possible to convert heat to work and kinetic energy. And indeed, one of the major subjects of technical thermodynamics is the construction and study of *heat engines* which produce work from heat. In later chapters we shall investigate such possibilities.

1.5.2 *Integral and local equations of balance of energy*

The internal energy U, and K, the kinetic energy, may be written as integrals over the corresponding densities

$$U + K = \int_V \rho \left(u + \tfrac{1}{2} w^2\right) \mathrm{d}V \,. \tag{1.39}$$

u and $\frac{1}{2}w^2$ are the specific values of the internal and of the kinetic energy, respectively, *i.e.* energies per unit mass. The energy balance for a closed system determines the rate of change of $U+K$.

The rate of change of the energy may result from the working $\dot{W}$, or power, of stress and of body force, and we have, by Paragraph 1.4.1

$$\dot{W} = \int_{\partial V} t_{ij} n_j w_i \, \mathrm{d}A + \int_V \rho f_j w_j \, \mathrm{d}V \,. \tag{1.40}$$

If $\dot{W} > 0$, work is done to the system. Otherwise, *i.e.* for $\dot{W} < 0$, the system performs work on its surroundings.

Another contribution to the rate of change of energy is the heating $\dot{Q}$. It consists of the flux of internal energy – or heat flux—through the surface ∂V and of absorption or emission of radiant energy either on the surface or in the interior of V. We write

$$\dot{Q} = -\int_{\partial V} q_i n_i \, \mathrm{d}A + \int_V \rho z \, \mathrm{d}V \,. \tag{1.41}$$

q_i is the flux density vector of internal energy and – in a single body -- we call it the *heat flux*. The minus sign has been chosen so that $U+K$ grows, if q_i points into V. z is the specific value of the absorbed heat radiation. $\dot{Q}$ is called the *heating*. If $\dot{Q} = 0$, the system is called *adiabatic* (Greek: *a* ="not" and *diabainein* = "to transfer"). For $\dot{Q} > 0$ heat is provided to the system. Otherwise, *i.e.* for $\dot{Q} < 0$, the system loses heat.

Radiation will largely be ignored in this book, except for occasional remarks and except for Sect. 7.3 which deals exclusively with some simple aspects of radiation. A complete account of the interaction of matter with the electro-magnetic field of radiation is an extensive subject in its own right and it is not treated in this book.

The equations (1.39) through (1.41) may be combined to form the balance of energy of a closed system

$$\frac{\mathrm{d}}{\mathrm{d}t} \int_V \rho \left(u + \tfrac{1}{2} w^2\right) \mathrm{d}V = \int_{\partial V} \left(t_{ji} w_j - q_i\right) n_i \, \mathrm{d}A + \int_V \rho \left(f_i w_i + z\right) \mathrm{d}V \,. \tag{1.42}$$

Comparison with the generic balance (1.3) reveals the following correspondence.

Ψ	ψ	ϕ_i	π	ζ
energy	$u + \frac{1}{2} w^2$	$-t_{ji} w_j + q_i$	0	$f_i w_i + z$

With this the alternative generic equations of balance are easily written. For open systems at rest we have

$$\int_V \frac{\partial \rho\left(u+\frac{1}{2}w^2\right)}{\partial t}\mathrm{d}V + \int_{\partial V}\left[\left(u+\frac{1}{2}w^2\right)w_i - t_{ji}w_j + q_i\right]n_i\,\mathrm{d}A = \int_V \rho\left(f_i w_i + z\right)\mathrm{d}V \qquad (1.43)$$

and in regular points we have

$$\frac{\partial \rho\left(u+\frac{1}{2}w^2\right)}{\partial t} + \frac{\partial}{\partial x_i}\left[\left(u+\frac{1}{2}w^2\right)w_i - t_{ji}w_j + q_i\right] = \rho\left(f_i w_i + z\right). \qquad (1.44)$$

Without body force and absorption of radiation and without the fluxes on the surface, the energy of a closed system is constant. This is to say that the energy is a conserved quantity, the production of which is zero.

The three equations (1.42), (1.43), and (1.44) represent different forms of the conservation law of energy or the *First Law of Thermodynamics*.

1.5.3 *Potential energy*

If the body force f_i can be written as the gradient $-\frac{\partial \varphi}{\partial x_i}$ of a time-independent potential φ – *e.g.* in the gravitational field, where φ is equal to gx_3 – we say that the force field is conservative. In this case the working of the body force may be rewritten as

$$\int_V \rho f_i w_i\,\mathrm{d}V = -\int_V \rho \frac{\partial \varphi}{\partial x_i} w_i\,\mathrm{d}V = -\int_V \frac{\partial \rho \varphi w_i}{\partial x_i}\mathrm{d}V + \int_V \varphi \frac{\partial \rho w_i}{\partial x_i}\mathrm{d}V .$$

Employing GAUSS's theorem and the mass balance (1.12) we obtain

$$\int_V \rho f_i w_i\,\mathrm{d}V = -\int_V \frac{\partial \rho \varphi}{\partial t}\mathrm{d}V - \int_{\partial V} \rho \varphi w_i n_i\,\mathrm{d}A \quad \text{or, by (1.6)}$$

$$\int_V \rho f_i w_i\,\mathrm{d}V = -\frac{\mathrm{d}}{\mathrm{d}t}\int_V \rho \varphi \mathrm{d}V .$$

Insertion into (1.42) through (1.44) provides

$$\frac{\mathrm{d}}{\mathrm{d}t}\int_V \rho\left(u+\varphi+\frac{1}{2}w^2\right)\mathrm{d}V = \int_{\partial V}\left(t_{ji}w_j - q_i\right)n_i\,\mathrm{d}A + \int_V \rho z\,\mathrm{d}V ,$$

$$\int_V \frac{\partial \rho\left(u+\varphi+\frac{1}{2}w^2\right)}{\partial t}\mathrm{d}V + \int_{\partial V}\left[\rho\left(u+\varphi+\frac{1}{2}w^2\right)w_i - t_{ji}w_j + q_i\right]n_i\,\mathrm{d}A = \int_V \rho z\,\mathrm{d}V \qquad (1.45)$$

$$\frac{\partial \rho\left(u+\varphi+\frac{1}{2}w^2\right)}{\partial t} + \frac{\partial \rho\left(u+\varphi+\frac{1}{2}w^2\right)w_i - t_{ji}w_j + q_i}{\partial x_i} = \rho z .$$

If $z = 0$ holds, *i.e.* if there is no absorption or emission of radiation, this equation may be considered as a conservation law for the sum of internal energy, *potential energy*

$$E_{\text{pot}} = \int_V \rho \varphi \mathrm{d}V \qquad (1.46)$$

and kinetic energy.

1.5.4 *Balance of internal energy*

The energy balance (1.44) contains a redundancy, since implicitly it relates the rate of change of internal energy u to the rates of change of mass density ρ and velocity $\boldsymbol{w}$. Indeed, the rates of change of the latter two quantities are already determined by the equations of balance of mass and momentum, *cf.* (1.12) and (1.19). In order to get rid of the redundancy we start with (1.44) in the form

$$\frac{\partial \rho u}{\partial t}+\frac{\partial}{\partial x_i}(\rho u w_i+q_i)-t_{ji}\frac{\partial w_j}{\partial w_i}-\rho z$$

$$+\frac{\partial \frac{\rho}{2}w^2}{\partial t}+\frac{\partial}{\partial x_i}\left(\frac{\rho}{2}w^2 w_i\right)-\frac{\partial t_{ji}}{\partial x_i}w_j-\rho f_i w_i=0 \qquad (1.47)$$

and rewrite the first two terms of the second line as follows

$$\frac{\partial \frac{\rho}{2}w^2}{\partial t}+\frac{\partial}{\partial x_i}\left(\frac{\rho}{2}w^2 w_i\right)=\frac{1}{2}w^2\underbrace{\left(\frac{\partial \rho}{\partial t}+\frac{\partial \rho w_i}{\partial x_i}\right)}_{\substack{=0 \text{ by mass} \\ \text{balance (1.12)}}}+\rho w_j\left(\frac{\partial w_j}{\partial t}+w_i\frac{\partial w_j}{\partial x_i}\right)$$

$$=w_j\left(\frac{\partial \rho w_j}{\partial t}+\frac{\partial \rho w_j w_i}{\partial x_i}\right)-w^2\overbrace{\left(\frac{\partial \rho}{\partial t}+\frac{\partial \rho w_j}{\partial x_j}\right)}^{\substack{=0 \text{ by mass} \\ \text{balance (1.12)}}}.$$

Hence follows by insertion into (1.47)

$$\frac{\partial \rho u}{\partial t}+\frac{\partial}{\partial x_i}(\rho u w_i+q_i)-t_{ji}\frac{\partial w_j}{\partial w_i}-\rho z$$

$$+w_j\underbrace{\left[\frac{\partial \rho w_j}{\partial t}+\frac{\partial}{\partial x_i}(\rho w_j w_i-t_{ji})-\rho f_j\right]}_{=0 \text{ by momentum balance (1.19)}}.$$

There remains

$$\frac{\partial \rho u}{\partial t}+\frac{\partial}{\partial x_i}(\rho u w_i+q_i)=t_{ji}\frac{\partial w_j}{\partial x_i}+\rho z \qquad (1.48)$$

and this is the balance of internal energy in regular points. Comparison with the generic equation of balance (1.9) provides correspondences as shown in the table.

Ψ	ψ	ϕ_i	π	ζ
internal energy	u	q_i	$t_{ji}\frac{\partial w_j}{\partial x_i}$	z

Note that here we have a production term for the first time. Indeed, the internal energy is not conserved, because it can be converted into kinetic energy, or it can emerge from the destruction of kinetic energy, *cf.* Paragraph 1.5.1.

The integral equations of balance of the internal energy in closed and open systems now result easily from the equations (1.3) and (1.5)

$$\frac{\mathrm{d}}{\mathrm{d}\,t}\int_V \rho\, u\,\mathrm{d}V + \int_{\partial V} q_i n_i\,\mathrm{d}A = \int_V t_{ji}\frac{\partial w_j}{\partial x_i}\,\mathrm{d}V + \int_V \rho\, z\,\mathrm{d}V\,, \tag{1.49}$$

$$\int_V \frac{\partial \rho\, u}{\partial t}\,\mathrm{d}V + \int_{\partial V}(\rho\, u\, w_i + q_i)n_i\,\mathrm{d}A = \int_V t_{ji}\frac{\partial w_j}{\partial x_i}\,\mathrm{d}V + \int_V \rho\, z\,\mathrm{d}V\,. \tag{1.50}$$

1.5.5 *Short form of energy balance for closed systems*

We recall (1.39) through (1.41), and (1.46), where symbols were introduced for energy, working, and heating, *viz.*

$$\begin{aligned}
U+K &= \int_V \rho\left(u+\frac{1}{2}w^2\right)\mathrm{d}V\,,\\
\dot{W} &= \underbrace{\int_{\partial V} t_{ij}n_j w_i\,\mathrm{d}A}_{\dot{W}_{\text{stress}}} + \underbrace{\int_V \rho\, f_j w_j\,\mathrm{d}V}_{-\frac{\mathrm{d}E_{\text{pot}}}{\mathrm{d}t}}\,,\\
\dot{Q} &= -\int_{\partial V} q_i n_i\,\mathrm{d}A + \int_V \rho\, z\,\mathrm{d}V\,,\\
E_{\text{pot}} &= \int_V \rho\,\varphi\,\mathrm{d}V\,.
\end{aligned} \tag{1.51}$$

We also define the *internal working*, the production density of internal energy

$$\dot{W}_{\text{int}} = \int_{\partial V} t_{ji}\frac{\partial w_j}{\partial x_i}\mathrm{d}V\,. \tag{1.52}$$

With these definitions the equations of balance (1.42), $(1.45)_1$, and (1.49) for total energy and internal energy may be written in the compact forms

$$\begin{aligned}
\frac{\mathrm{d}(U+K)}{\mathrm{d}t} &= \dot{W}+\dot{Q}\,,\\
\frac{\mathrm{d}(U+E_{\text{pot}}+K)}{\mathrm{d}t} &= \dot{W}_{\text{stress}}+\dot{Q}\,,\\
\frac{\mathrm{d}U}{\mathrm{d}t} &= \dot{W}_{\text{int}}+\dot{Q}\,,
\end{aligned} \tag{1.53}$$

which are easy to memorize. All of these equations are equivalent; they represent different forms of the *First Law of Thermodynamics* – appropriate for closed systems.

The internal working $\dot{W}_{\text{int}}$ can generally not be prescribed, because it depends on the stress tensor and on the velocity gradient *inside* V. Neither of these can be controlled from the boundary.

1.5.6 *First Law for reversible processes. The basis of "pdV - thermodynamics"*

In this paragraph we ignore gravity. Then, if viscous forces can be neglected, so that $t_{ji} = -p\delta_{ji}$ holds, and if p is homogeneous in V and on ∂V, the working $\dot{W}_{\text{stress}}$ and the internal working $\dot{W}_{\text{int}}$ are equal, because we have

$$\dot{W}_{\text{stress}} = \int_{\partial V} t_{ji} w_j n_i \,\mathrm{d}A = -\int_{\partial V} p\, w_j n_j \,\mathrm{d}A = -p \int_{\partial V} w_j n_j \,\mathrm{d}A = -p\frac{\mathrm{d}V}{\mathrm{d}t}, \qquad (1.54)$$

$$\dot{W}_{\text{int}} = \int_V t_{ji} \frac{\partial w_j}{\partial x_i} \mathrm{d}V = -\int_V p \frac{\partial w_i}{\partial x_i} \mathrm{d}V = -p \int_{\partial V} \frac{\partial w_i}{\partial x_i} \mathrm{d}V = -p\frac{\mathrm{d}V}{\mathrm{d}t}. \qquad (1.55)$$

Note that the last integrals of both chains of equations are equal because of GAUSS's theorem. Also note that $\int_{\partial V} w_j n_j \,\mathrm{d}A$ is equal to the rate of change of V.

We conclude from the equality of $\dot{W}$ and $\dot{W}_{\text{int}}$, and $(1.53)_{1,3}$ that in a process with $t_{ji} = -p\delta_{ji}$, and homogeneous p, the rate of change of the kinetic energy is negligible. The work done in the time $\mathrm{d}t$ then reads

$$\dot{W}_{\text{stress}}\,\mathrm{d}t = \dot{W}_{\text{int}}\,\mathrm{d}t = -p\,\mathrm{d}V. \qquad (1.56)$$

This expression is the hallmark of *reversible* processes. The two equations $(1.53)_{1,3}$ read in this case

$$\dot{Q} = \frac{\mathrm{d}U}{\mathrm{d}t} + p\frac{\mathrm{d}V}{\mathrm{d}t}. \qquad (1.57)$$

Thus, when $\frac{\mathrm{d}U}{\mathrm{d}t}$ and $\frac{\mathrm{d}V}{\mathrm{d}t}$ are reversed, *i.e.* replaced by $-\frac{\mathrm{d}U}{\mathrm{d}t}$ and $-\frac{\mathrm{d}V}{\mathrm{d}t}$, the heating is also reversed.

The First Law for reversible processes (1.57) is the basis for many approximate calculations in technical thermodynamics. Such calculations ignore viscous effects and shear stresses as they might appear in turbulent motion, pipe flow, and boundary layers, – and they assume homogeneous pressures.

1.5.7 *Enthalpy and First Law for stationary flow processes*

We recall the First Law (1.57) of $p\mathrm{d}V$-thermodynamics and apply it to an isobaric heating process, *i.e.* heating under a constant pressure. In this case the First Law may be written as

$$\dot{Q} = \frac{\mathrm{d}(U + pV)}{\mathrm{d}t},$$

so that isobaric heating leads to an increase of the *enthalpy* (Greek: en = "inside" and thalpos = "heat") or "heat content." The enthalpy is usually denoted by the letter H

$$H = U + pV \quad \text{or, specifically} \quad h = u + \frac{p}{\rho}; \qquad (1.58)$$

the enthalpy comes closest among all energetic quantities to what the layman calls "heat".

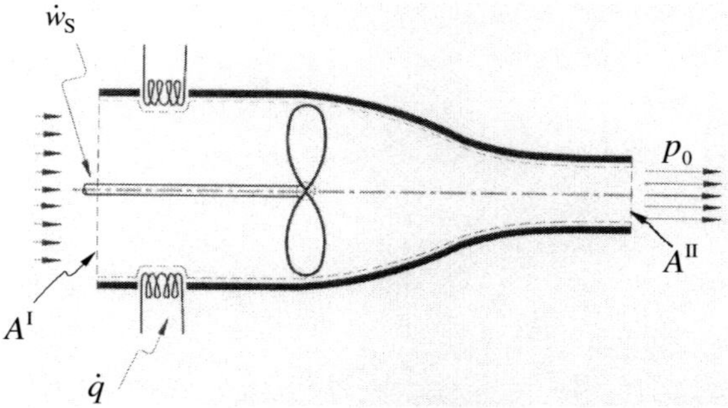

Fig. 1.11 A fan (schematic)

The specific enthalpy plays an important role in stationary flows of fluids that are driven by a turbine, or a fan, and heated, *cf.* Fig. 1.11, which shows a schematic picture of a fan. We consider the First Law in the form $(1.45)_2$, ignore radiation and apply the equation to the dashed surface ∂V shown in the figure

$$\int_{\partial V} \left[\rho\left(u+\varphi+\tfrac{1}{2}w^2\right)w_i - t_{ji}w_j + q_i\right] n_i \,\mathrm{d}A = 0 \,. \qquad (1.59)$$

Everywhere on ∂V, except where the turbine shaft is cut, we neglect shear stresses and put $t_{ji} = -p\,\delta_{ji}$. But inside the shaft elastic shear stresses are transmitting work to the turbine from a motor, or transmitting work from the turbine to a generator. We call this working of the shaft $\dot{W}_S$ and, as always, we put $\dot{Q} = -\int_{\partial V} q_i n_i \,\mathrm{d}A$.* We ignore the heating through the cross-sections A^{I} and A^{II}. The terms with $w_i n_i$ do not contribute over the mantle surface, since that scalar product is equal to zero there. Thus, assuming that ρ, w_1, u, p, and φ are homogeneous throughout the cross-sections, we obtain

$$-\rho\left(u+\varphi+\frac{p}{\rho}+\tfrac{1}{2}w^2\right)w_1 A\bigg|_{A^{I}} + \rho\left(u+\varphi+\frac{p}{\rho}+\tfrac{1}{2}w^2\right)w_1 A\bigg|_{A^{II}} = \dot{W}_S + \dot{Q} \,. \qquad (1.60)$$

Note that $w_i n_i = -w_1$ holds on the left cross-section A^{I}, while $w_i n_i = +w_1$ holds on the right cross-section A^{II}. By (1.16) the mass transfer $\dot{m} = \rho\, w_1 A$ has the same value on both cross-sections so – with $\dot{q} = \frac{\dot{Q}}{\dot{m}}$, and $\dot{w}_S = \frac{\dot{W}_S}{\dot{m}}$, and $h = u + \frac{p}{\rho}$ – we obtain

* Radiation is usually ignored, because it plays a very minor role in many technical processes. In Sect. 7.3 we do discuss radiation as a phenomenon in its own right, but otherwise we mostly ignore it.

$$-\left(h+\varphi+\tfrac{1}{2}w^2\right)\Big|_{A^{\mathrm{I}}}+\rho\left(h+\varphi+\tfrac{1}{2}w^2\right)w_1 A\Big|_{A^{\mathrm{II}}}=\dot{w}_{\mathrm{S}}+\dot{q}\,. \tag{1.61}$$

This form of the First Law is well-suited for the study of stationary flow problems through pipes, nozzles, and turbines which have only one cross-section of inflow and one of outflow.

1.5.8 *"Adiabatic equation of state" for an ideal gas – an integral of the energy balance*

In slow – reversible – processes the kinetic energy may be neglected and the formula (1.57) of $p\mathrm{d}V$-thermodynamics holds true. In particular in a reversible *adiabatic* process we have

$$\frac{\mathrm{d}U}{\mathrm{d}t}=-p\frac{\mathrm{d}V}{\mathrm{d}t}\,,$$

or, by $\mathrm{d}U=m\,z\frac{R}{M}\mathrm{d}T$ and $pV=m\frac{R}{M}T$ for ideal gases

$$\frac{\mathrm{d}\ln T^{z}}{\mathrm{d}t}=-\frac{\mathrm{d}\ln V}{\mathrm{d}t}\,.$$

By integration we obtain the equations

$$V\,T^{z}=\text{const}\quad\text{or, by}\quad pV=m\frac{R}{M}T:\quad\frac{p}{T^{z+1}}=\text{const}\,,\quad pV^{\frac{z+1}{z}}=\text{const}\,.$$

These are sometimes called adiabatic equations of state for an ideal gas. Mostly one replaces z by $\kappa=(z+1)/z$ (*cf.* Sect. 2.3) and writes

$$V^{\kappa-1}T=\text{const}\,,\quad\frac{p}{T^{\frac{\kappa}{\kappa-1}}}=\text{const}\,,\quad pV^{\kappa}=\text{const}\,. \tag{1.62}$$

These equations define *adiabates* in a (T,V)-, or a (p,T)-, or a (p,V)-diagram. Since $\kappa>1$ holds, the pressure increases more rapidly during an adiabatic compression than during an isothermal one. We have

$$p\sim\frac{1}{V^{\kappa}}\quad\text{instead of}\quad p\sim\frac{1}{V}\,.$$

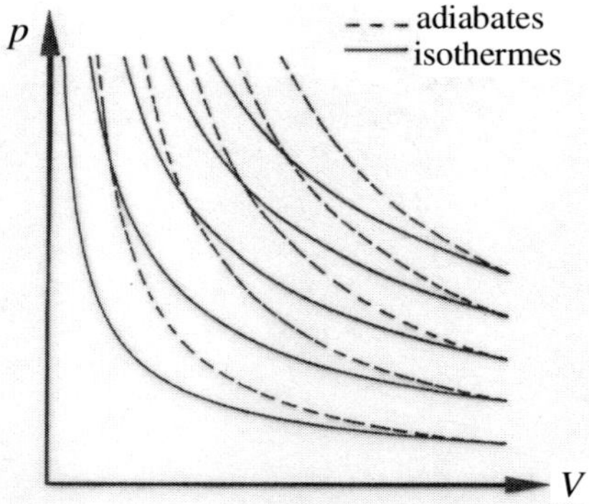

Fig. 1.12 Adiabates are steeper than isotherms

Therefore the adiabates in a (p,V)- diagram are steeper than isotherms, *cf.* Fig. 1.12. It follows that an adiabatic compression heats a gas while an adiabatic expansion cools it.

The values of κ for one-, two-, and more-atomic ideal gases are

$$\kappa = \begin{cases} 5/3 = 1.66 & \text{one -} \\ 7/5 = 1.4 \quad \text{for} & \text{two - atomic gases}. \\ 4/3 = 1.33 & \text{more -} \end{cases}$$

1.5.9 *Applications of the energy balance*

- *Experiment of Gay-Lussac*

Gay-Lussac has performed a simple experiment which may be used to show that the specific internal energy of ideal gases is independent of the density. The experimental set-up and the execution of the experiment are illustrated in Fig. 1.13. A slide valve is moved so as to allow a compressed gas to expand and flow into an initially evacuated chamber. The whole system is adiabatically isolated and its wall is kept fixed. Observation shows that the temperatures before and after the process, T_i and T_f, are equal.

For exploitation of the experiment we use the First Law in the form

$$\frac{\mathrm{d}(U+K)}{\mathrm{d}t} = \dot{Q} + \dot{W}$$

applied to the whole container. Both $\dot{Q}$ and $\dot{W}$ vanish because of adiabaticity and since the surface is at rest; gravity plays no role in this case.

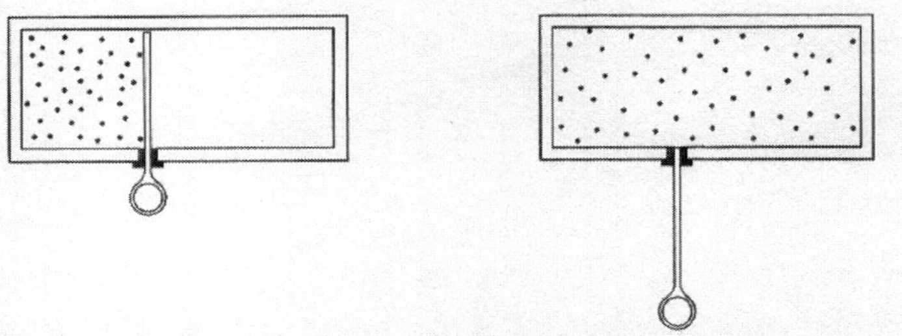

Fig. 1.13 Experiment of Gay-Lussac. Initial and final states

Initially and at the end the kinetic energy K is zero and hence it follows

$$U_\mathrm{f} = U_\mathrm{i} \quad \text{or, since } T_\mathrm{f} = T_\mathrm{i} \text{ holds } u(\rho_\mathrm{f},T) = u(\rho_\mathrm{i},T).$$

ρ_i is arbitrary, it depends on the position of the valve. Therefore the equation requires that u is independent of ρ

$$u = u(T). \tag{1.63}$$

The flow of the gas in the Gay-Lussac experiment is a complicated process and it is instructive to consider it in detail rather than regard only the beginning and the end: When the valve is opened the gas expands in a strongly non-homogeneous manner and its temperature decreases non-homogeneously, since the expansions are essentially reversibly adiabatic. Shear stresses and heat conduction play no role in the short term. However, there is a strongly turbulent motion, *i.e.* the internal energy has partly been converted into kinetic energy. As time goes on, the turbulent motion decreases in intensity by internal friction and viscosity, and eventually it comes to rest. The kinetic energy goes back into internal energy and eventually the temperature recovers a homogeneous value by heat conduction. This value is identical with the original one.

Actually GAY-LUSSAC's instruments were somewhat coarse. When JOULE and THOMSON (Lord KELVIN) repeated the experiment forty years later with the best thermometers of the time – and for an initial pressure of 20 bar—they found a slight cooling of the gas, indicating that air is not quite an ideal gas.

- *Piston drops into cylinder*

We consider air of mass m_A in a cylinder of cross-section A which is closed off at the height H_i by a movable piston of mass m_P. The initial temperature in the cylinder is T_i, the initial pressure p_i, and the constant external pressure is denoted by p_0. We assume that $(p_i - p_0) \neq m_P g$ holds so that the piston must be kept in position by a clamp in order to realize the initial state. After release of the piston there will be a period of adjustment of the enclosed air – generally with turbulent motion and with non-homogeneous pressure and temperature – before the final state is reached. We calculate T_f and H_f, the final values of temperature and of height, respectively. For simplicity we assume that the cylinder wall and the piston are adiabatically insulated and we ignore the working of gravitation on the air.*

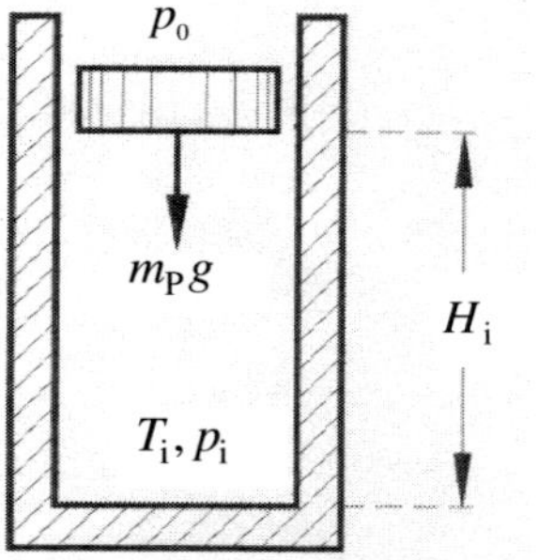

Fig. 1.14 Piston drops into cylinder

We employ the First Law for the gas and the piston in the form

$$\frac{d(U+K)}{dt} = \dot{Q} + \dot{W},$$

* It is entirely possible to do the analysis without any of these simplifications, but it is more cumbersome. And the results are qualitatively the same ones.

where $\dot{Q}$ vanishes because of adiabaticity. $\dot{W}$ has two terms, because it contains the working of the external pressure p_0 and the working of gravity on the piston

$$\dot{W} = \int_{\partial V} t_{ji} w_j n_i \,\mathrm{d}A + \int_V \rho f_i w_i \,\mathrm{d}V \quad \text{with} \quad t_{ji} = -p_0 \delta_{ji} \text{ and } f_i = (0, 0, -g),$$

$$\dot{W} = -(p_0 A + m_\mathrm{P} g)\frac{\mathrm{d}H}{\mathrm{d}t}.$$

Once again – as in the GAY-LUSSAC experiment – the kinetic energy K vanishes at the beginning and at the end of the process. Therefore integration over the whole duration of the process between times t_i and t_f provides

$$m_\mathrm{A} z \frac{R}{M_\mathrm{A}} T_\mathrm{f} + (p_0 A + m_\mathrm{P} g) H_\mathrm{f} = m_\mathrm{A} z \frac{R}{M_\mathrm{A}} T_\mathrm{i} + (p_0 A + m_\mathrm{P} g) H_\mathrm{i}. \tag{1.64}$$

The momentum balance applied to the piston in the final equilibrium gives

$$p_\mathrm{f} A = p_0 A + m_\mathrm{P} g \tag{1.65}$$

and, moreover, we have to observe the thermal equation of state at the end of the process

$$p_\mathrm{f} A H_\mathrm{f} = m_\mathrm{A} \frac{R}{M_\mathrm{A}} T_\mathrm{f}. \tag{1.66}$$

The equations (1.64) through (1.66) represent three equations for the three unknowns T_f, H_f, and p_f. A simple calculation leads to

$$T_\mathrm{f} = \frac{z}{z+1} T_\mathrm{i} + \frac{z}{z+1} \frac{p_0 A + m_\mathrm{P} g}{m_\mathrm{A} \frac{R}{M_\mathrm{A}}} H_\mathrm{i},$$

$$H_\mathrm{f} = \frac{1}{z+1} H_\mathrm{i} + \frac{z}{z+1} \frac{m_\mathrm{A} \frac{R}{M_\mathrm{A}}}{p_0 A + m_\mathrm{P} g} T_\mathrm{i}. \tag{1.67}$$

As an example we choose $V_\mathrm{i} = 1\,\mathrm{l}$, $H_\mathrm{i} = 10\,\mathrm{cm}$, $T_\mathrm{i} = 300\,\mathrm{K}$, $p_\mathrm{i} = p_0 = 1\,\mathrm{bar}$, and $z = 5/2$ for air. If the piston has the mass $m_\mathrm{P} = 10^3\,\mathrm{kg}$ we obtain $T_\mathrm{f} = 1140\,\mathrm{K}$ and $H_\mathrm{f} = 3.5\,\mathrm{cm}$. Thus the load of one ton of the piston has compressed the air down to 35% of its original volume and has thus converted its potential energy into internal energy such that the temperature has increased nearly four times.

An interesting limit case is the infinitely heavy piston with $m_\mathrm{P} \to \infty$. In this case we have

$$T_\mathrm{f} \to \infty \quad \text{but} \quad H_\mathrm{f} \to \tfrac{2}{7} H_\mathrm{i}.$$

Thus, even an infinite pressure cannot reduce the volume of the gas to zero, because the temperature reaches the value infinity before that. And the infinitely hot gas resists a further decrease in volume.

It is instructive to compare this irreversible compression with the reversible adiabatic compression considered in Paragraph 1.5.8. In that case the piston was guided slowly – *i.e.* reversibly – downwards from outside. Thus a part of its decrease in potential energy was gained as work rather than dissipated into heat. Note that in the reversible adiabatic process V tends to zero as p tends to infinity, *cf.* $(1.62)_3$. This is in marked contrast to the present irreversible process. In both cases, though, T tends to infinity.

- *Throttling*

Throttling occurs when a gas, a vapor, or a liquid are pushed or sucked through a small hole in the cross-section of a pipe, *cf.* Fig. 1.15. It is a method for lowering the pressure in a flowing substance. Behind the obstruction the flow is turbulent at first but then it settles down to a laminar flow again with $t_{ji} = -p\delta_{ji}$. We assume the process to be adiabatic and stationary and apply the First Law in the form (1.59) in order to compare conditions in the cross-sections A^{I} and A^{II}. The change of potential energy can be ignored if the flow is horizontal, and we assume the kinetic energy to be negligible compared to the internal energy. The mantle part of ∂V gives no contribution, if we assume that the velocity vanishes on the wall. Thus (1.59) reduces to

$$u^{\mathrm{I}} + \frac{p^{\mathrm{I}}}{\rho^{\mathrm{I}}} = u^{\mathrm{II}} + \frac{p^{\mathrm{II}}}{\rho^{\mathrm{II}}}\,, \quad \text{or} \quad h^{\mathrm{I}} = h^{\mathrm{II}}. \tag{1.68}$$

This is to say that adiabatic throttling is an isenthalpic process.

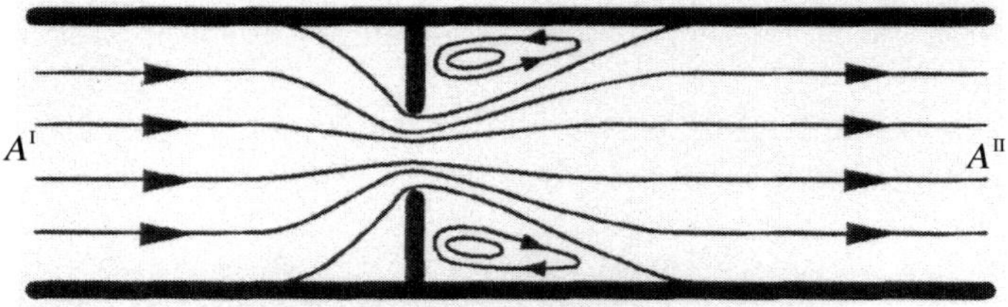

Fig. 1.15 Throttle valve (schematic)

Note that in throttling we have an irreversible adiabatic process. The enthalpy of the flow is first converted into turbulent motion of the fluid behind the valve, and, – as that motion dies out by internal friction and heat conduction –, enthalpy is reestablished. In an ideal gas the temperatures before and (far) behind the valve are equal. This is not so, however, in vapors and liquids, *cf.* Paragraph 4.2.6.

The throttling process in detail has much in common with the experiment of GAY-LUSSAC described above, except that now the beginning and final states are stationary flows rather than states of rest.

- *Heating of a room*

The heating of a room by a stove provides a good example for an open system at rest, because the pressure remains constant so that the density must decrease with increasing temperature. This means that air escapes the room through cracks and slits in the doors and windows. We consider the heating

$$\dot{Q} = -\int_{\partial V} q_i n_i \,\mathrm{d}A$$

as constant and ask for the change of temperature in time.

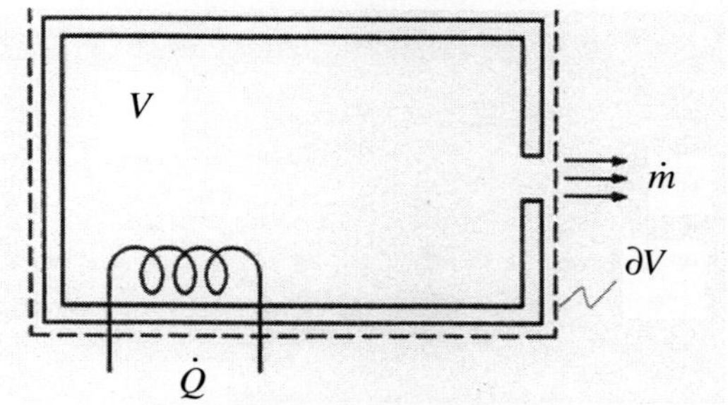

Fig. 1.16 Heating of a room

Because of the slowness of the heating process it is reasonable to assume that pressure, density, and temperature are homogeneous in the room. We ignore gravity, radiation, and viscosity so that the stress tensor is $t_{ji} = -p\delta_{ji}$. Accordingly the equations of balance of mass (1.11) and energy (1.43) read (note that the time derivative may be inside or outside the integrals, since V is time-independent; we put it outside)

$$\frac{\mathrm{d}}{\mathrm{d}t}\int_V \rho\,\mathrm{d}V + \int_{\partial V} \rho w_i n_i\,\mathrm{d}A = 0 \quad\Rightarrow\quad \frac{\mathrm{d}m}{\mathrm{d}t} = -\rho w_i n_i A\,,$$

$$\frac{\mathrm{d}}{\mathrm{d}t}\int_V \rho\left(u + \frac{1}{2}w^2\right)\mathrm{d}V + \int_{\partial V}\left[\rho\left(u+\frac{1}{2}w^2\right)w_i - t_{ji}w_j + q_i\right]n_i\,\mathrm{d}A = 0$$

$$\Rightarrow\quad \frac{\mathrm{d}\,mu}{\mathrm{d}t} + \rho\left(u + \frac{p}{\rho} + \frac{1}{2}w^2\right)w_i n_i A = \dot{Q}\,,$$

where A is the small area through which air escapes. We neglect the kinetic energy of the escaping air and obtain by elimination of $\rho w_i n_i A$ between the two equations

$$\frac{\mathrm{d}\,mu}{\mathrm{d}t} - \frac{\mathrm{d}m}{\mathrm{d}t}\left(u + \frac{p}{\rho}\right) = \dot{Q} \quad\Rightarrow\quad m\frac{\mathrm{d}u}{\mathrm{d}t} - \frac{\mathrm{d}m}{\mathrm{d}t}\frac{p}{\rho} = \dot{Q}\,.$$

With the equations of state for air, *cf.* P1 and P.2, we have $\frac{\mathrm{d}u}{\mathrm{d}t} = \frac{5}{2}\frac{R}{M_\mathrm{A}}\frac{\mathrm{d}T}{\mathrm{d}t}$ and $\frac{p}{\rho} = \frac{R}{M_\mathrm{A}}T$ and thus we obtain

$$\frac{5}{2}m\frac{R}{M_\mathrm{A}}\frac{\mathrm{d}T}{\mathrm{d}t} - \frac{\mathrm{d}m}{\mathrm{d}t}\frac{R}{M_\mathrm{A}}T = \dot{Q}\,.$$

Hence by $pV = m\frac{R}{M_\mathrm{A}}T = \mathrm{const}$

$$\frac{7}{2}m\frac{R}{M_\mathrm{A}}\frac{\mathrm{d}T}{\mathrm{d}t} = \dot{Q} \quad \text{and by} \quad m\frac{R}{M_\mathrm{A}} = \frac{pV}{T}$$

$$\frac{7}{2}pV\frac{\mathrm{d}\ln T}{\mathrm{d}t} = \dot{Q}\,.$$

Since $\dot{Q}$ is independent of time by assumption, integration provides

$$T(t) = T(0)\exp\left(\frac{\dot{Q}t}{7/2\, pV}\right).$$

The temperature increases exponentially, since the constant heating is used to increase the temperature of an ever smaller mass of air. However, from the point of view of heating costs this is not desirable, because the escaping air has already been heated.

Obviously the heating does not lead to an exponentially growing temperature in reality. Indeed, when the temperature rises, the efflux of heat through walls and windows provides a constant temperature in the end. This effect is ignored in the present calculation.

- *Nozzle flow*

Once again – as in Paragraph 1.3.2 and 1.4.5 – we consider the stationary flow through a nozzle, *cf.* Fig. 1.1. At this time we are interested in the energy balance of that flow. We assume for simplicity that $q_i = 0$ holds on ∂V, and that gravitation, radiation, and shear stresses may be neglected. Thus we have $t_{ji} = -p\,\delta_{ji}$ and (1.43) implies

$$\int_{\partial V} \rho\left(u + \frac{p}{\rho} + \frac{1}{2}w^2\right) w_i n_i \,\mathrm{d}A = 0\,.$$

On the mantle surface we have $w_i n_i = 0$, and therefore it is only the cross-sections A^{I} and A^{II} that contribute to the integral. This is a special case of the First Law for stationary flow processes, *cf.* (1.59), (1.60), and we obtain for $w^2 \approx w_1^2$

$$h + \frac{1}{2}w_1^2 = \text{const.} \tag{1.69}$$

In words: The increase of kinetic energy of the nozzle is fed by a decrease of enthalpy; or the "heat content" is converted into kinetic energy.

We recall the equations of balance for mass (1.15), (1.16), and momentum (1.34) for the nozzle flow and rewrite them along with the balance of energy (1.69)

$$\boxed{\rho\, w_1 A = \dot{m} = \text{const.}}\,, \quad \frac{\mathrm{d}\frac{1}{2}w_1^2}{\mathrm{d}x_1} = -\frac{1}{\rho}\frac{\mathrm{d}p}{\mathrm{d}x_1}\,, \quad \boxed{h + \frac{1}{2}w_1^2 = \text{const.}} \tag{1.70}$$

For ideal gases the differential equation $(1.70)_2$ can be integrated. For this purpose we eliminate w_1^2 between $(1.70)_{2,3}$ and use the thermal and the caloric equation of state. Hence follows

$$\frac{\mathrm{d}h}{\mathrm{d}x_1} = \frac{1}{\rho}\frac{\mathrm{d}p}{\mathrm{d}x_1} \quad \text{and with } h = u + \frac{p}{\rho} = \frac{\kappa}{\kappa+1}\frac{R}{M}(T - T_{\mathrm{R}}) + \left(u_{\mathrm{R}} + \frac{R}{M}T_{\mathrm{R}}\right)$$

and $\rho = \dfrac{p}{R/M\,T}$

$$\frac{\mathrm{d}\ln T^{\frac{\kappa}{\kappa-1}}}{\mathrm{d}x_1}=\frac{\mathrm{d}\ln p}{\mathrm{d}x_1} \quad\Rightarrow\quad \boxed{\frac{T}{p^{\frac{\kappa-1}{\kappa}}}=\text{const.}\ ,\quad \frac{p}{\rho^{\kappa}}=\text{const.}} \tag{1.71}$$

These are the "adiabatic equations of state," *cf.* Paragraph 1.5.8. They hold for the nozzle flow, since by assumption the flow is reversible and adiabatic.

The equations shown in frames are the basic equations for the nozzle flow. We may use them to calculate the efflux from a pressure vessel, or from the combustion chamber of a rocket, where pressure and temperature have the values p_{C} and T_{C}, while w_1^{C} is zero. We assume that the cross-section $A(x_1)$ is a known function of x_1 and ask for w_1 and p as functions of x_1.

First of all we determine w_1 as a function of p. From $(1.70)_3$ we have

$$w_1=\sqrt{2(h_{\mathrm{C}}-h)}=\sqrt{2\frac{\kappa}{\kappa-1}\frac{R}{M}T_{\mathrm{C}}\left(1-\frac{T}{T_{\mathrm{C}}}\right)} \quad \text{or, by } (1.71)_2$$

$$w_1=\sqrt{2\frac{R}{M}T_{\mathrm{C}}}\sqrt{\frac{\kappa}{\kappa-1}\left(1-\left(\frac{p}{p_{\mathrm{C}}}\right)^{\frac{\kappa-1}{\kappa}}\right)}\,. \tag{1.72}$$

Equation $(1.71)_3$ gives ρ as a function of p

$$\rho=\rho_{\mathrm{C}}\left(\frac{p}{p_{\mathrm{C}}}\right)^{\frac{1}{\kappa}}, \tag{1.73}$$

and so we have ρw_1 as a function of p

$$\rho\, w_1=\frac{\sqrt{2}p_{\mathrm{C}}}{\sqrt{\frac{R}{M}T_{\mathrm{C}}}}\underbrace{\left(\frac{p}{p_{\mathrm{C}}}\right)^{\frac{1}{\kappa}}\sqrt{\frac{\kappa}{\kappa-1}\left(1-\left(\frac{p}{p_{\mathrm{C}}}\right)^{\frac{\kappa-1}{\kappa}}\right)}}_{\psi=\psi\left(\frac{p}{p_{\mathrm{C}}}\right)}\,. \tag{1.74}$$

The brace identifies the *flux function* ψ, as indicated, which is a function of $\frac{p}{p_{\mathrm{C}}}$, whose form for $\kappa=1.66$ – a monatomic gas – is graphically represented in Fig. 1.17. It has a maximum at the so-called critical values

$$\frac{p^*}{p_{\mathrm{C}}}=\left(\frac{2}{\kappa+1}\right)^{\frac{\kappa}{\kappa-1}} \quad\text{and}\quad \psi^*=\left(\frac{2}{\kappa+1}\right)^{\frac{1}{\kappa-1}}\sqrt{\frac{\kappa}{\kappa-1}}\,, \tag{1.75}$$

whose significance will be discussed below.

By (1.74) the mass balance $(1.70)_1$ may be written as

$$\dot{m} = \frac{\sqrt{2}p_C}{\sqrt{\frac{R}{M}T_C}}\psi\left(\frac{p(x_1)}{p_C}\right)A(x_1) = \frac{\sqrt{2}p_C}{\sqrt{\frac{R}{M}T_C}}\psi\left(\frac{p(x_1^E)}{p_C}\right)A(x_1^E), \tag{1.76}$$

where x_1^E is the position of the end of the nozzle. Equation $(1.76)_2$ is the required equation for $p(x_1)$. It is implicit and, of course, $A(x_1)$ must be known before a solution can be given. Fig. 1.18 shows the pressure function in a nozzle which is a rotational hyperboloid

$$A(x_1) = A(0)\left(1 + \frac{x_1^2}{l_1^2}\left[\frac{A_{min}}{A(0)} - 1\right]\right).$$

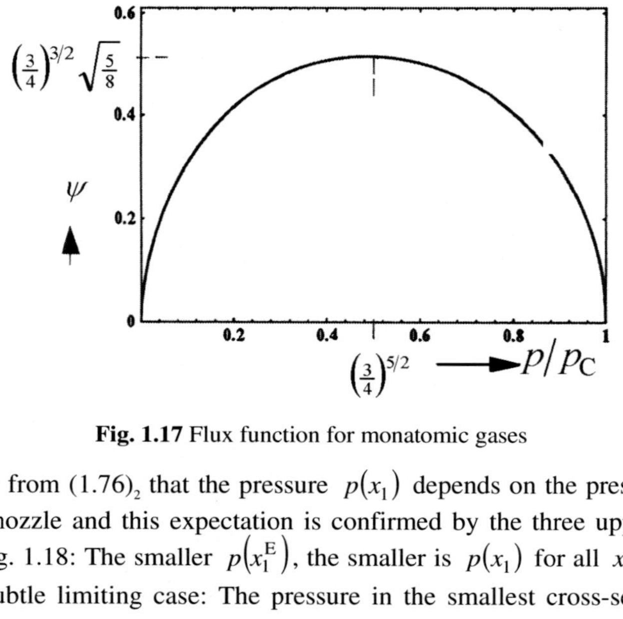

Fig. 1.17 Flux function for monatomic gases

It is clear from $(1.76)_2$ that the pressure $p(x_1)$ depends on the pressure $p(x_1^E)$ behind the nozzle and this expectation is confirmed by the three upper pressure curves in Fig. 1.18: The smaller $p(x_1^E)$, the smaller is $p(x_1)$ for all x_1. However, there is a subtle limiting case: The pressure in the smallest cross-section A_{min} cannot drop below the critical pressure p^*. Indeed, if $p(l_1)$ were smaller than p^*, the flux function ψ would have decreased in the narrowing part of the nozzle as Fig. 1.18 shows. This, however, is impossible, since ψA is constant by $(1.76)_1$ so that with decreasing ψ the cross-sectional area A must *increase*.

For $p(l_1) = p^*$ a bifurcation occurs. Either the pressure increases in the widening part of the nozzle or it continues to decrease. The dashed lines in Fig. 1.18 represent that alternative. In the former case the state (ψ, p) of the gas moves downward from the maximum ψ^* along the right-hand branch. In the latter case the state (ψ, p) moves downward along the left branch of $\psi(x_1)$. Which of these possibilities is realized will depend on the pressure $p\left(x_1^{\mathrm{E}}\right)$ and on whether that pressure is p_1^{E} or p_2^{E}, *cf.* Fig. 1.18.

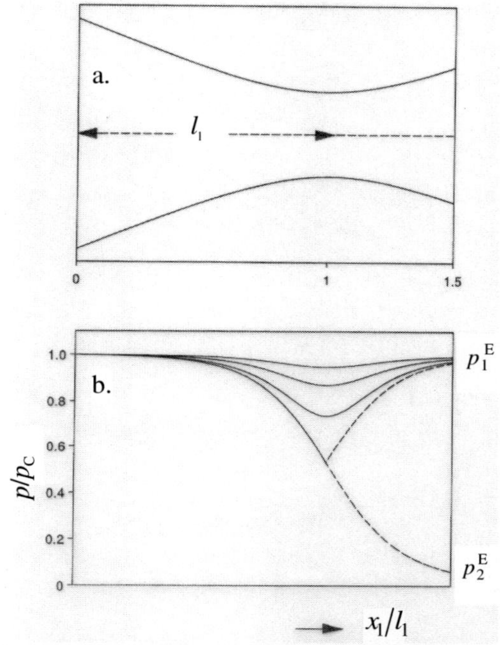

Fig. 1.18 a. Hyperbolic cross-section of the nozzle; **b.** Pressure $p(x_1)$

Now, since the pressure in the narrowest cross-section cannot fall below p^*, the mass transfer $\dot{m}$ has a maximum at

$$\dot{m}_{\max} = \frac{\sqrt{2}\,p_{\mathrm{C}}}{\sqrt{\frac{R}{M}T_{\mathrm{C}}}}\psi^* A_{\min} . \tag{1.77}$$

More mass cannot pass through the nozzle irrespective of the pressure $p\left(x_1^{\mathrm{E}}\right)$. For an engineer it is important to know this limit when he designs a safety valve through which a given amount of mass must exit in order to avoid catastrophic pressure increases in a vessel.

The physical reason for the limiting mass transfer may be recognized by a calculation of the velocity w_1^* at p^* by (1.72) and $(1.75)_1$

$$w_1^* = \sqrt{2\frac{\kappa}{\kappa+1}\frac{R}{M}T_{\mathrm{C}}} \quad \text{or, by } (1.71)_2 \text{ and } (1.75)_1$$

$$w_1^* = \sqrt{\kappa\frac{R}{M}T^*} . \tag{1.78}$$

This expression identifies the speed of sound in the narrowest cross-section, *cf.* Paragraph 11.4. Thus the limited efflux of mass becomes clear: The pressure vessel remains unaffected by end pressures $p_E < p_1^E$. It is true that information about lowering p_E below p_1^E propagates into the nozzle at the speed of sound, but the flow moves in the other direction with the speed of sound in the narrowest cross-section – or even supersonically in the widening part of the nozzle. Therefore the information about $p_E < p_1^E$ never arrives in the pressure vessel.

There remains the question what happens if p_E lies in the interval $p_2^E < p_E < p_1^E$. Observation shows that in this case a shock appears in the widening part of the nozzle, *i.e.* a sharp increase of pressure and a sharp decrease of velocity from supersonic to subsonic flow. This shock is an irreversible phenomenon and our equations (1.60), (1.70) do not cover it. We leave that subject to books on gas dynamics.

A nozzle with narrowing *and* widening parts is important for rocket engines, where the thrust shall be maximized which, by (1.29), equals $\dot{m}\, w_1\left(x_1^E\right)$. It is true that $\dot{m}$ is limited by (1.77), but $w_1\left(x_1^E\right)$ may grow far beyond the speed of sound in the widening part. We recall in this context the Saturn 1B rocket of Fig. 1.4 which has an exhaust speed of $2730\,\frac{\text{m}}{\text{s}}$.

The widening part of the nozzle is needed for such high speeds, because while the pressure decreases, the mass density decreases even faster and the cross-section must grow to accommodate the passage of an ever more voluminous gas. A nozzle like this is called *Laval nozzle.*

Carl Gustaf Patrik DE LAVAL (1845-1913) is sometimes called the Swedish EDISON, because toward the end of his life he owned more than a thousand patents and, like EDISON, he was head of a firm in which over a hundred engineers developed his ideas. His career as an inventor began in the dairy industry, where he contributed a centrifugal cream separator and a vacuum milking machine. Other inventions concerned lighting, – the great challenge of the time – and electrometallurgy. His greatest achievement, however, is the steam turbine. He realized that the partially expanded steam needs space because of its small density and accordingly is best sent through a *diffusor*, a nozzle with a widening cross section.

- *Fan*

In a fan the flow of air is accelerated by a propeller and heated, *cf.* Fig. 1.11. The pressure p_0 before and behind the fan is equal to the atmospheric pressure. For the mass transfer $\dot{m}$ we have

$$\dot{m} = \rho_I w_1^I A^I = \rho_{II} w_1^{II} A^{II} = \frac{p_0}{\frac{R}{M} T_{II}} w_1^{II} A^{II} . \tag{1.79}$$

For given ρ_I, w_1^I, and for given areas A^I, A^{II} this is one equation for the determination of the unknowns w_1^{II} and T_{II}.

A second equation follows from the First Law (1.60) for stationary flow processes. Since the fan is held horizontally, φ does not change between A^{I} and A^{II}. And since the flow velocity is assumed to be far less than the speed of sound $\sqrt{\kappa \frac{R}{M} T}$, the specific kinetic energy $\frac{1}{2} w_1^2$ may be neglected in comparison with

$$h = u + \frac{p}{\rho} = \frac{\kappa}{\kappa-1} \frac{R}{M} (T - T_{\mathrm{R}}) + \left(u_{\mathrm{R}} + \frac{R}{M} T_{\mathrm{R}}\right).$$

Thus (1.60) reads

$$\frac{\kappa}{\kappa-1} \frac{R}{M} (T_{\mathrm{II}} - T_{\mathrm{I}}) = \frac{\dot{W}_{\mathrm{S}} + \dot{Q}}{\dot{m}}.$$

For given values of $\dot{W}_{\mathrm{S}}$ and $\dot{Q}$ this equation provides a second equation for w_1^{II} and T_{II}, in addition to (1.79). Solving both equations we obtain

$$\frac{R}{M} T_{\mathrm{II}} = \frac{R}{M} T_{\mathrm{I}} + \frac{\dot{W}_{\mathrm{S}} + \dot{Q}}{\frac{\kappa}{\kappa-1} \dot{m}} \quad \text{and} \quad w_1^{\mathrm{II}} = \frac{\dot{m}}{p_0 A_{\mathrm{II}}} \left(\frac{R}{M} T_{\mathrm{I}} + \frac{\dot{W}_{\mathrm{S}} + \dot{Q}}{\frac{\kappa}{\kappa-1} \dot{m}} \right),$$

where $\dot{m} = \rho_{\mathrm{I}} w_1^{\mathrm{I}} A_{\mathrm{I}}$. We put

$$p_0 = 1\,\text{bar}, \rho_{\mathrm{I}} = 1.29 \tfrac{\text{kg}}{\text{m}^3}, T_{\mathrm{I}} = 293\,\text{K}, A^{\mathrm{I}} = 2A^{\mathrm{II}} = 10\,\text{cm}^3, w_1^{\mathrm{I}} = 2 \tfrac{\text{m}}{\text{s}},$$

$$\dot{Q} = 100\,\text{W}, \dot{W}_{\mathrm{S}} = 10\,\text{W}$$

and obtain $\dot{m} = 2.58 \cdot 10^{-3} \frac{\text{kg}}{\text{s}}$, and with $\kappa = 1.4$, $M = 29 \frac{\text{g}}{\text{mol}}$, appropriate for air

$$T_{\mathrm{II}} = 335\,\text{K} \quad \text{and} \quad w_1^{\mathrm{II}} = 4.96 \tfrac{\text{m}}{\text{s}}.$$

Thus we conclude that the application of heating and working has accelerated the internal and the kinetic energy of the air flow.

For this case it is possible to make a rough check of the assumption that the kinetic energy is much smaller than the internal one. Indeed our values show that

$$\tfrac{1}{2}\left(w_1^{\mathrm{II}}\right)^2 = 12.3 \tfrac{\text{m}^2}{\text{s}^2}, \quad \text{while} \quad \frac{5}{2} \frac{R}{M} T_{\mathrm{II}} = 20.8 \cdot 10^4 \tfrac{\text{m}^2}{\text{s}^2}.$$

- *Turbine*

Steam expands during the passage from the steam vessel to the condenser. The expansion takes place in a nozzle, *cf.* Fig. 1.19 , in whose widening part a turbine wheel is installed. The turbine wheel is set into rotation by the flow, and the working $\dot{W}_{\mathrm{S}}$ may be taken off from the rotating shaft. Before and behind the turbine – in the cross-sections A^{I} and A^{II} – the kinetic energy of the flow is negligible compared to the internal energy or enthalpy. The potential φ plays no essential role and $\dot{q} = 0$, if the turbine is adiabatically isolated.

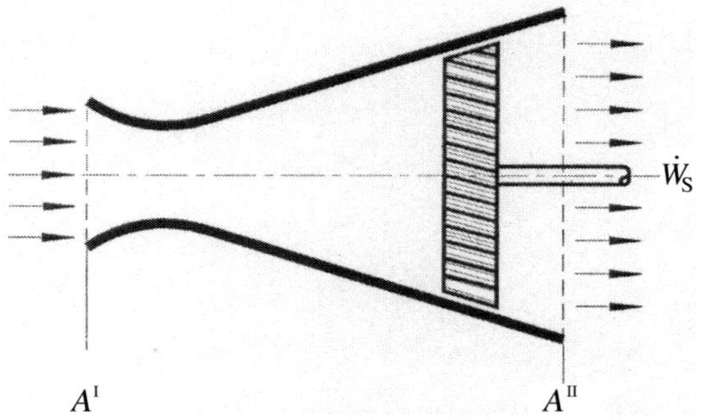

Fig. 1.19 A turbine (schematic)

Under these circumstances the First Law (1.60) reads

$$\frac{\dot{W}_S}{\dot{m}} = h_{II} - h_I \tag{1.80}$$

so that the working of the turbine is equal to the loss of the enthalpy of the steam.

- *Chimney*

A chimney is more than a hole in the roof through which the smoke can escape. Indeed, the chimney provides the air, which is heated in the fireplace, with the possibility for an adiabatic ascent. The inflowing air that replaces the rising air maintains the fire by supplying oxygen. The heating from the fireplace occurs mostly by radiation.

We consider the stationary flow through a chimney with cross-section A and height H, *cf.* Fig. 1.20. Ignoring viscous stresses and heat conduction we obtain the energy balance (1.60) with $\varphi = g\,x_3$ in the form

$$\left(h + \tfrac{1}{2}w^2 + g\,x_3\right)_t - \left(h + \tfrac{1}{2}w^2 + g\,x_3\right)_b = 0, \tag{1.81}$$

where the indices t and b characterize the conditions on top of the chimney and at the bottom. We wish to determine the mass flux $\dot{m} = \rho_b w_b A_b = \rho_t w_t A$ and start by calculating w_t from (1.81). A_b is the area of the vertical mantle of a cylinder surrounding the fireplace. It is much bigger than the cross-section A of the chimney and we may therefore neglect w_b^2 compared to w_t^2. Thus with

$$h = \frac{\kappa}{\kappa - 1}\frac{R}{M}(T - T_R) + u_R + \frac{p_R}{\rho_R}$$

and $\kappa = \frac{7}{5}$, appropriate for air, (1.81) assumes the form

$$\frac{7}{2}\frac{R}{M}T_b\left(1 - \frac{T_t}{T_b}\right) + \frac{1}{2}w_t^2 + gH = 0. \tag{1.82}$$

T_b and T_t are temperatures at the lower and at the upper ends of the chimney. They may be quite different from the temperature T of the surrounding air, but the pressures p_b and p_t are equal to the barometric pressures, *cf.* (1.28) and

Fig. 1.20. The pressure p_b is equal to the atmospheric ground pressure p_G. Because of adiabaticity we have

$$\frac{T_t}{T_b}=\left(\frac{p_t}{p_G}\right)^{\frac{\kappa-1}{\kappa}}=\exp\left(-\frac{2gH}{7\frac{R}{M}T}\right). \tag{1.83}$$

Thus by elimination of T_t/T_b between (1.82), $(1.83)_2$ we obtain

$$w_t=\sqrt{7\frac{R}{M}T_b\left[1-\exp\left(-\frac{2gH}{7\frac{R}{M}T}\right)\right]-2gH}\,. \tag{1.84}$$

It remains to calculate ρ_t. We have

$$\rho_t=\rho_b\left(\frac{p_t}{p_G}\right)^{\frac{1}{\kappa}}\underset{\text{with }\rho_b T_b=\frac{p_G}{\frac{R}{M}}}{=}\frac{p_G}{\frac{R}{M}T_b}\exp\left(-\frac{5gH}{7\frac{R}{M}T}\right).$$

Hence it follows from $\dot{m}=\rho_t w_t A$

$$\dot{m}=\frac{p_G A}{\frac{R}{M}T_b}\exp\left(-\frac{5gH}{7\frac{R}{M}T}\right)\sqrt{7\frac{R}{M}T_b\left[1-\exp\left(-\frac{2gH}{7\frac{R}{M}T}\right)\right]-2gH}\,.$$

For heights H less than 1 km we have $\frac{gH}{\frac{R}{M}T}<<1$ and for this approximation we obtain

$$\dot{m}=\frac{p_G A}{\sqrt{\frac{R}{M}T}}\exp\left(-\frac{5gH}{7\frac{R}{M}T}\right)\sqrt{2\frac{R}{M}\frac{T}{T_b}}\sqrt{\frac{T_b}{T}-1}\,. \tag{1.85}$$

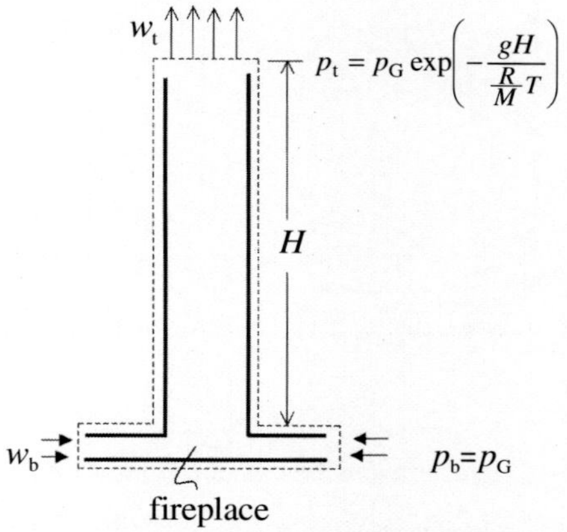

Fig. 1.20 A chimney (schematic). *T* is the outside temperature, assumed homogeneous

If we put $p_G = 1\ \text{bar}$, $T = 300\ \text{K}$, $A = 400\ \text{cm}^2$, $H = 10\ \text{m}$ we obtain

$$\dot{m} = 0.66\,\frac{\text{m}^3}{\text{s}}\,\frac{T}{T_b}\sqrt{\frac{T_b}{T} - 1}\,,$$

and we conclude that $\dot{m}$ increases for $1 < \frac{T_b}{T} \leq 2$; for larger values of T_b it decreases. Hence it follows that a chimney "draws" best for $T_b = 600\text{K}$. The initial increase of $\dot{m}$ results from a better buoyancy of the heated air, and the eventual decrease for $T_b > 2T$ is due to the very small density of the heated air. Its volume is too large to pass the narrow chimney comfortably.

- *Thermal power station*

Fig. 1.21 shows the schematic picture of a thermal power station. Its central part is the chimney, and therefore the thermodynamic treatment of the power station is much like that of the chimney. There are, however, two essential differences:

- The rising air drives a turbine and thus produces the working $\dot{W}_S$.
- There is no fireplace. The heating of the air is achieved by solar radiation which heats the ground below a collector area A_r of radius r. The heated ground in turn heats the air above it, which flows radially inward. Thus its temperature rises from the outside temperature T to the temperature T_b of the air before it enters the turbine.

We retain much of the notation from the treatment of the chimney. However, w_b^2 cannot be neglected.

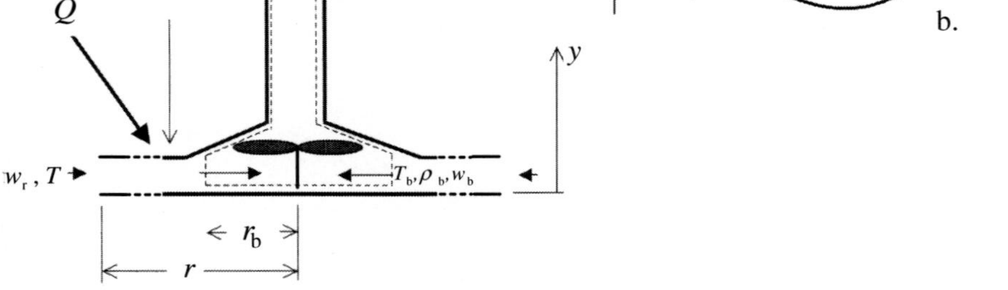

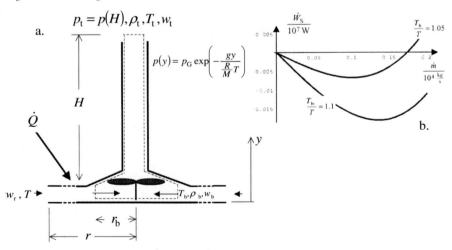

Fig. 1.21 a. Thermal power station (schematic).
b. Dependence of $\dot{W}_S$ on $\dot{m}$ for the parameters of the Manzanares plant (*cf.* end-of-section)

The heating $\dot{Q}$ of the inflowing air can be related to the temperature T_b and to $\dot{m}$ as follows

$$\dot{Q} = \dot{m}\left[h_b - h_r + \tfrac{1}{2}\left(w_b^2 - w_r^2\right)\right] \text{ and with } h = \tfrac{7}{2}\tfrac{R}{M}(T - T_R) + h_R \text{ and } w_r^2 << w_b^2$$

$$\dot{Q} = \dot{m}\left[\frac{7}{2}\frac{R}{M}T\left(\frac{T_b}{T} - 1\right)\frac{1}{2}w_b^2\right] \text{ and with } w_b^2 = \frac{\dot{m}}{\rho_b A_b} = \frac{\frac{R}{M}T_b}{p_G}\frac{\dot{m}}{A_b}$$

$$\dot{Q} = \dot{m}\left[\frac{7}{2}\frac{R}{M}T\left(\frac{T_b}{T} - 1\right)\frac{1}{2}\left(\frac{\frac{R}{M}T_b}{p_G}\right)^2\left(\frac{\dot{m}}{A_b}\right)^2\right]. \tag{1.86}$$

The working $\dot{W}_S$ transmitted by the turbine shaft will now be calculated from the First Law (1.60) for stationary flows applied to the dashed surface shown in Fig. 1.21. We wish to calculate $\dot{W}_S$ as a function of $\dot{m}$ and T

$$\dot{m}\left[\frac{7}{2}\frac{R}{M}T_b\left(\frac{T}{T_b} - 1\right) + \tfrac{1}{2}\left(w_t^2 - w_b^2\right) + gH\right] = \dot{W}_S .$$

We assume an isothermal outside atmosphere with temperature T and therefore we have

$$T_t = T_b\left(\frac{p_t}{p_G}\right)^{\frac{\kappa-1}{\kappa}} = T_b \exp\left(-\frac{2gH}{7\frac{R}{M}T}\right),$$

$$\rho_t = \rho_b\left(\frac{p_t}{p_G}\right)^{\frac{1}{\kappa}} = \frac{p_G}{\frac{R}{M}T_b}\exp\left(-\frac{5gH}{7\frac{R}{M}T}\right).$$

Thus, if we replace w_t by $\dfrac{\dot{m}}{\rho_t A}$ and $w_b = \dfrac{\dot{m}}{\rho_b A_b}$, we obtain

$$\dot{W}_S = \dot{m}\left[\frac{7}{2}\frac{R}{M}T_b(\exp(-\frac{2gH}{7\frac{R}{M}T}) - 1) + \frac{1}{2}\dot{m}^2\left(\frac{\frac{R}{M}T_b}{p_G A}\right)^2\left(\exp\left(\frac{10}{3}\frac{gH}{\frac{R}{M}T}\right) - \frac{A^2}{A_b^2}\right) + gH\right]$$

This is already the desired relation between $\dot{W}_S$, and $\dot{m}$, T_b. We simplify it by the assumption $gH << \frac{R}{M}T$ and obtain

$$\dot{W}_S = \dot{m}\left[-\left(\frac{T_b}{T} - 1\right)gH + \frac{1}{2}\left(\frac{\frac{R}{M}T}{p_G A}\frac{T_b}{T}\right)^2\left(1 - \frac{A^2}{A_b^2}\right)\dot{m}^2\right]. \tag{1.87}$$

For two values of T_b this function is plotted in Fig. 1.21. It has a minimum for

$$\dot{m}_{\min} = \sqrt{\tfrac{2}{3}}\, p_G A \frac{\sqrt{gH}}{\frac{R}{M}T} \frac{1}{\sqrt{1-\frac{A^2}{A_b^2}}} \frac{T}{T_b} \sqrt{\frac{T_b}{T}-1}\,.$$

The minimal working – *i.e.* maximal output of the station – is therefore given by

$$\dot{W}_{\min} = -\sqrt{\tfrac{2}{3}}\, p_G A \frac{\sqrt{gH}^3}{\frac{R}{M}T} \frac{1}{\sqrt{1-\frac{A^2}{A_b^2}}} \frac{T}{T_b} \sqrt{\frac{T_b}{T}-1}^{\,3}\,. \tag{1.88}$$

For this case of maximal output the temperature T_b may be calculated for a given amount $\dot{Q}$ of solar heating by insertion of $\dot{m}_{\min}$ into (1.86)

$$\dot{Q} = \tfrac{7}{2}\sqrt{\tfrac{2}{3}}\, p_G A \sqrt{gH}\, \frac{T}{T_b} \sqrt{\frac{T_b}{T}-1}^{\,3} \left[1 - \frac{2}{21} \frac{gH}{\frac{R}{M}T} \frac{\frac{A}{A_b}}{1-\frac{A^2}{A_b^2}}\right]. \tag{1.89}$$

The bracket on the right hand side may be replaced by 1 – again for $gH \ll \frac{R}{M}T$ – and therefore the efficiency of the station in the point of maximal output is given by

$$e = \frac{\dot{W}_{\min}}{\dot{Q}} = \frac{4}{21} \frac{gH}{\frac{R}{M}T} \frac{1}{\sqrt{1-\frac{A^2}{A_b^2}}}\,. \tag{1.90}$$

It increases linearly with H .

In Manzanares, Spain, they have built a thermal power station with the following data • chimney height $H = 200\,\text{m}$, • chimney cross-section $A = \pi(5\,\text{m})^2$, • height of collector roof $h = 2\,\text{m}$, • collector diameter 250 m, • turbine diameter 20 m, hence $A_b = \pi(20\,\text{m})h$.

We put $\dot{Q} = 500\frac{\text{W}}{\text{m}^2}\pi r^2$. This is a realistic value as long as the sun is shining; hence follows from (1.88) through (1.90)

$$\frac{T_b}{T} = 1.09\;,\quad \dot{W}_{\min} = -136\,\text{kW}\;,\quad e = 0.56\,.$$

At this time they are planning a station with a collector diameter of $3.6\,\text{km}$, a chimney height of $950\,\text{m}$ and a working of $100\,\text{MW}$.

1.6 History of the First Law

Benjamin THOMSON (1753-1814), later Graf RUMFORD – by the grace of the Elector KARL THEODOR OF BAVARIA – was the first to express doubts about the then popular caloric theory of heat.

By that theory – introduced by the great chemist Antoine Laurent LAVOISIER (1743-1792) – heat was a weightless fluid, the "caloric". LAVOISIER counted it among the elements along with light and the real elements of our present time. The heat emerging by taking chips off a metal in a lathe (say) was interpreted as a release of caloric.

Antoine Laurent LAVOISIER introduced exact measurement by weighing and quantitative analysis into chemical experimentation. In this way he was able to determine that diamond is pure carbon, because when it is burned in a closed box an appropriate amount of CO_2 appears. This and many other discoveries are reported in his book "Elementary Treatise on Chemistry" published in 1789.

LAVOISIER was guillotined in 1792 because of his involvement in tax collection under the *ancien régime*. Before the execution many of his friends and colleagues argued with the authorities that he should be spared because of his great achievements in chemistry and because of further important results to be expected from his work. However, the detaining officer decided that "the revolution does not need chemistry".

Benjamin THOMSON, later Graf RUMFORD, was an American who took the side of the British in the war of independence. He fought for them so that after the defeat of the British he had to leave the new United States of America. He went to Europe and left his mark in England, Bavaria, and France. In Bavaria he bored cannon barrels for the elector and noticed that a dull drill produced more heat than a sharp-edged one, although no chips appeared. In a report about this observation he stated that the heating of the barrel *equals that of nine big wax candles.* He became more quantitative later when he published his results: The total weight of ice water that could be heated to 180°F in 2 hours and 30 minutes by the barrel amounted to 26 pounds.

Although RUMFORD's value comes out about 30% too high, his numbers provide the first reasonable determination of the *mechanical equivalent of heat.* It was ignored for another 40 years until the 1840's. In England RUMFORD founded the Royal Institution, a kind of postgraduate school and in France he married LAVOISIER's widow.

Robert Julius MAYER (1814-1878) was the first person to express in any generality the idea that heat is a form of energy and that energy is conserved. He included the energy source of the sun in his considerations as well as the tides and life functions. In the manner of the 19th century he expressed his discovery in a Latin slogan

Ex nihilo nil fit, nil fit ad nihilum.

In his first published work MAYER says: … thus it turns out that the heating of a certain weight of water from 0°C to 1°C corresponds to the lowering of an equal weight by 367 m:

$$1°\text{heat} = 1\text{ g at a height of } \begin{cases} 367\,\text{m} \\ 1130\,\text{Parisian feet} \end{cases}$$

MAYER's reasoning is correct, but his value is about 15% too low, because his information about the specific heat of air was inaccurate. Nowadays we might say

$$1\ \text{cal} = 4.18\ \text{J},$$

because 1 Joule = 1 Nm is the unit for all types of energy including heat energy. The calorie is now also obsolete as a unit. It used to be the heat needed to raise the temperature of 1 g of water of 15°C by 1°C.

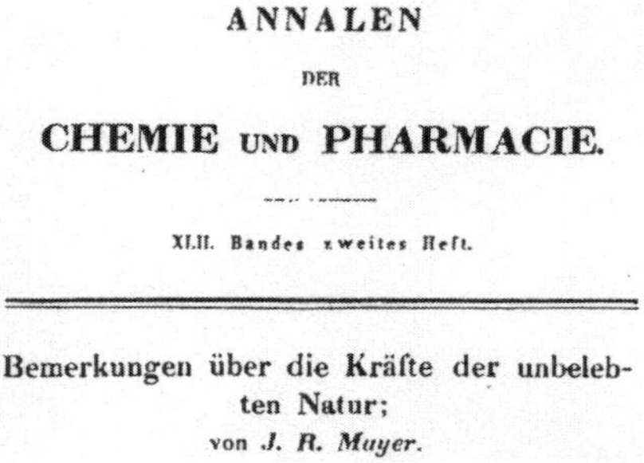

ANNALEN

DER

CHEMIE UND PHARMACIE.

XLII. Bandes zweites Heft.

Bemerkungen über die Kräfte der unbelebten Natur;

von *J. R. Mayer.*

Robert Prescott JOULE made extensive measurement and eventually he came up with what MAYER expressed as

$$1°\text{heat} = 1\ \text{g at a height of} \begin{cases} 425\ \text{m} \\ 1308\ \text{Parisian feet}\ . \end{cases}$$

JOULE suggested to measure temperatures above and below a waterfall and he calculated that the water at the bottom of the Niagara falls should be 0.2°F warmer than at the top.

Hermann Ludwig Ferdinand HELMHOLTZ (1818-1889) is the third person often credited with the discovery of the First Law of thermodynamics. He had a clear notion that the kinetic energy of a body falling on the ground is conserved but converted into the kinetic energies of the molecules of the body, *i.e.* heat energy. He used this notion to explain the energy source of the sun by a contraction of the star, and came up with the result that 25 million years ago the sun had a radius equal to the radius of the Earth's orbit. It followed that the Earth could not be older than 25 million years, a prediction that threw out all reasonable expectations of geologists and biologists. As it turned out, however, HELMHOLTZ was wrong: The heat of the sun is not gravitational in origin but nuclear. He could not have known that, of course.

It can easily be argued that the First Law, the conservation of energy, was the most important discovery of the 19th century. And yet, all three discoverers had great difficulty to make their ideas known. MAYER had to grovel to VON LIEBIG to have his paper published and JOULE's work was rejected by the Royal Society. Eventually he presented it to an audience of laymen and got it published in the Manchester Courier by reluctant editors, and only because his brother was a music critic of the staff of that daily newspaper. HELMHOLTZ's paper was rejected as containing "nothing but philosophy" and he published it privately as a pamphlet.

1.7 Summary of equations of balance

We summarize the generic equations of balance and their special forms for mass, momentum, energy, and internal energy in the following form

closed system $$\frac{\mathrm{d}}{\mathrm{d}t}\int_V \rho\psi\,\mathrm{d}V + \int_{\partial V} \phi_i\, n_i\,\mathrm{d}A = \int_V \rho(\pi+\zeta)\mathrm{d}V$$

open system at rest $$\int_V \frac{\partial\rho\psi}{\partial t}\mathrm{d}V + \int_{\partial V} (\rho\psi\, w_i + \phi_i) n_i\,\mathrm{d}A = \int_V \rho(\pi+\zeta)\mathrm{d}V$$

local in regular points $$\frac{\partial\rho\psi}{\partial t} + \frac{\partial(\rho\psi\, w_i + \phi_i)}{\partial x_i} = \rho(\pi+\zeta)\,.$$

Ψ	ψ	ϕ_i	π	ζ
mass	1	0	0	0
momentum	w_j	$-t_{ji}$	0	f_j
energy	$u+\frac{1}{2}w^2$	$-t_{ji}w_j + q_i$	0	$f_i w_i + z$
internal energy	u	q_i	$t_{ji}\dfrac{\partial w_j}{\partial x_i}$	z

2 Constitutive equations

2.1 On measuring constitutive functions

2.1.1 *The need for constitutive equations*

We recall the objective of thermodynamics which is the determination of the five fields

$$\text{mass density } \rho(x_i,t),\ \text{velocity } w_j(x_i,t),\ \text{temperature } T(x_i,t). \tag{2.1}$$

For this purpose we need five equations and we choose the five equations of balance of mass, momentum, and (internal) energy

$$\begin{aligned}
&\frac{\partial \rho}{\partial t}+\frac{\partial \rho w_i}{\partial x_i}=0\,,\\
&\frac{\partial \rho w_j}{\partial t}+\frac{\partial\left(\rho\, w_j w_i - t_{ji}\right)}{\partial x_i}=\rho\, f_j\,,\\
&\frac{\partial \rho\, u}{\partial t}+\frac{\partial}{\partial x_i}\left(\rho\, u\, w_i + q_i\right)=t_{ji}\frac{\partial w_j}{\partial x_i}+\rho\, z\,.
\end{aligned} \tag{2.2}$$

We shall assume that f_j and z are given functions of x_i and t. In our applications f_j is the gravitational force $f_j=(0,0,-g)$, and z is usually set equal to zero. Even then, the equations (2.2) are not sufficient for the determination of ρ, w_j, and T, because they contain additional fields, namely the stress tensor t_{ji}, the specific internal energy u, and the heat flux q_i; and the temperature does not occur at all.

Therefore we need additional equations which relate t_{ji}, u, and q_i to the basic fields ρ, w_j, and T and these are called the *constitutive equations of thermodynamics.* Indeed, experience with the thermodynamic substances – usually fluids, vapors, and gases – has taught us that stress, internal energy, and heat flux depend on the fields ρ, w_j, and T in a *materially dependent* manner.

2.1.2 *Constitutive equations for viscous, heat-conducting fluids, vapors, and gas*

There is a large class of fluids, vapors, and gases, which obey constitutive equations as follows

$$\begin{aligned}
&u=u(\rho,T)\,,\\
&t_{ji}=-p(\rho,T)\delta_{ji}+\lambda(\rho,T)\frac{\partial w_n}{\partial x_n}\delta_{ji}
\end{aligned}$$

$$+\eta(\rho,T)\left(\frac{\partial w_i}{\partial x_j}+\frac{\partial w_j}{\partial x_i}-\frac{2}{3}\frac{\partial w_n}{\partial x_n}\delta_{ji}\right),$$

$$q_i=-\kappa(\rho,T)\frac{\partial T}{\partial x_i}. \qquad (2.3)$$

Accordingly the specific internal energy u and the pressure p are functions of ρ and T, and these relations are called *caloric and thermal equations of state*. These functions must be determined experimentally by measuring pressures, densities, temperatures, and specific heats. We shall soon come back to this.

The leading term in the stress t_{ji} is the pressure p. In addition to the pressure the stress contains a *viscous* term that depends linearly on velocity gradients. The factor η in (2.3) is called *(shear) viscosity* and the factor λ is called *bulk viscosity*. Both depend generally on T and – weakly – on ρ. In the next section we shall see how η can be measured. The example given there will also provide a suggestive interpretation of the constitutive relation for the stress. The bulk viscosity λ plays a role in expansive and compressive flows. Its value is often negligible. In gases λ vanishes, and we shall not encounter this coefficient in this book until much later.

The heat flux q_i is proportional to the temperature gradient $\frac{\partial T}{\partial x_i}$ and points in the opposite direction. The factor of proportionality κ is called the *thermal conductivity*. Generally it depends on T and – again weakly – on ρ. In the next section we shall see how κ can be measured.

A simple form of the constitutive equation for the stress t_{ji} goes back to NEWTON, who had already introduced the notion of viscosity. Therefore, fluids which satisfy $(2.3)_2$ are called Newtonian fluids. NEWTON was interested to know whether and how the moon's orbit was affected by friction with the ether in which he believed. The full equation $(2.3)_2$ is called the NAVIER-STOKES law after two scientists in the 19th century who have formulated it and who have used it for the calculation of flows. Fluids which do not obey the NAVIER-STOKES equations are called non-Newtonian. They include polymer melts, asphalt, and most bodily fluids, *e.g.* saliva.

The constitutive equation $(2.3)_3$ for the heat flux is called FOURIER's law after the 19th century scientist who discovered it. We shall come back to FOURIER's treatment of heat conduction in a later chapter, *cf.* Sect. 7.1.

If the viscosities η and λ, and the thermal conductivity κ, and the equations of state $u(\rho,T)$ and $p(\rho,T)$ are known, one obtains five explicit field equations by elimination of t_{ji}, u, and q_i between (2.2) and (2.3). These five equations may, in principle, be used for the determination of the five fields ρ, w_j, and T appropriate to given boundary and initial data. A solution is called a *thermodynamic process*.

However, the field equations are non-linear coupled partial differential equations and it is practically impossible to solve them for an arbitrary initial and boundary value problem. Therefore the thermodynamicist simplifies the problem by various assumptions which are valid under certain conditions. The simplifications may use one or more of the following strategems:

- Linearization of the field equations;
- Assumption of the approximate homogeneity of the solution, *i.e.*, independence of x_n ;
- Consideration of stationary processes, *i.e.* independence of t ;
- Assumption of homogeneity of the solution in cross-sections of a flow, *i.e.* a flow through a nozzle or a pipe;
- Consideration of especially simple boundary conditions;
- Exploitation of suspected symmetries of the solution.

All processes calculated and presented in this book rely on the application of one or more of these simplifications. Typical examples are the following calculations of shear flow and of heat conduction.

2.2 Determination of viscosity and thermal conductivity

2.2.1 *Shear flow between parallel plates. Newton's law of friction*

We consider a stationary shear flow between two infinite plates of which the lower one is at rest, while the upper one moves with the velocity V , *cf.* Fig. 2.1.

The fluid is supposed to stick to the plates, a condition that is universally known as "no slip."

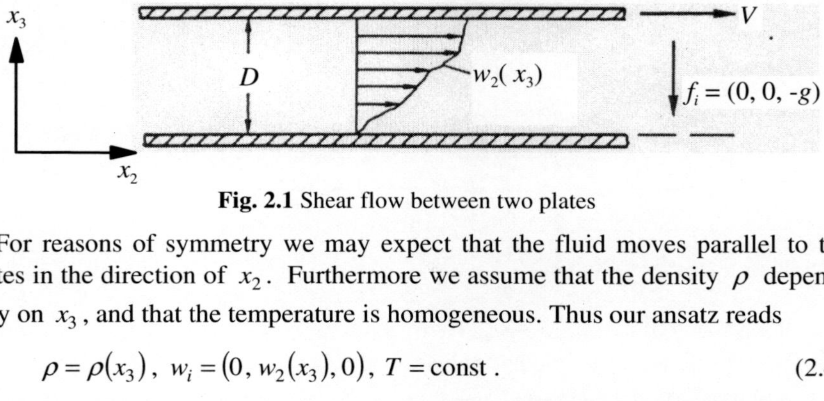

Fig. 2.1 Shear flow between two plates

For reasons of symmetry we may expect that the fluid moves parallel to the plates in the direction of x_2. Furthermore we assume that the density ρ depends only on x_3, and that the temperature is homogeneous. Thus our ansatz reads

$$\rho = \rho(x_3), \; w_i = (0, w_2(x_3), 0), \; T = \text{const} . \qquad (2.4)$$

The form of the functions $\rho(x_3)$ and $w_2(x_3)$ must be calculated from the field equations.*

The mass balance $(2.2)_1$ is identically satisfied by (2.4). The momentum balance $(2.2)_2$ must be supplemented by the stress $(2.3)_2$ which in the present case has the form

---

* The method of assuming part of the solution – as in (2.4) – is known as the *semi-inverse method*.

$$t_{ji} = -p(\rho,T)\delta_{ji} + \eta \begin{bmatrix} 0 & 0 & 0 \\ 0 & 0 & \frac{\partial w_2}{\partial x_3} \\ 0 & \frac{\partial w_2}{\partial x_3} & 0 \end{bmatrix} . \tag{2.5}$$

Therefore the 1-component of the momentum balance is identically satisfied. The viscosity η is assumed constant, and the 2- and 3-components of the momentum balance read

$$\text{2-component: } \frac{\partial^2 w_2}{\partial x_3^2} = 0 ,$$

$$\text{3-component: } \frac{\partial p}{\partial \rho}\frac{\mathrm{d}\rho}{\mathrm{d}x_3} = -\rho g . \tag{2.6}$$

By $(2.6)_1$ the velocity w_2 is a linear function of x_3 and with the no-slip conditions at $x_3 = 0$ and $x_3 = D$ we obtain

$$w_2(x_3) = \frac{V}{D}x_3 \text{, hence } t_{23} = \eta\frac{V}{D} . \tag{2.7}$$

The velocity is independent of the fluid properties, but the shear stress t_{23} depends on the material coefficient η, the viscosity. Therefore the viscosity may be measured by measuring the shear force necessary to maintain the relative velocity V between the plates.* Some values of η are listed in Table 2.1. The expression $(2.7)_2$ for the shear stress is called NEWTON's law of friction, because NEWTON was the first to find this relation experimentally.

Table 2.1 Viscosity η in 10^{-5} Ns/m^2. Note that the viscosity depends strongly on temperature. For gases it grows with increasing temperature while it drops for liquids. For some liquids the drop is drastic, *e.g.* for glycerin.

T [°C]	air	water	glycerin
0	1.7	180	$12\cdot 10^6$
70	1.8	100	$15\cdot 10^4$
100	2.2	30	1480

The solution of the 3-component $(2.6)_2$ of the momentum balance may be used to calculate $\rho(x_3)$. For that calculation we need to know the thermal equation of state $p = p(\rho,T)$. For ideal gases we obtain the barometric stratification explained in Paragraph 1.4.5, *cf.* (1.28).

The conservation law of energy $(2.2)_3$ is a precarious condition in the present case. The left hand side vanishes for the fields (2.4) but the shear flow produces internal energy, because we have

* Actual viscosity measurements are not performed between plates but between rotating cylinders.

$$t_{ji}\frac{\partial w_i}{\partial x_j}=\eta\left(\frac{V}{D}\right)^2 .$$

Therefore the energy balance reads

$$0=\eta\left(\frac{V}{D}\right)^2+\rho z$$

so that the assumed solution (2.4) requires that the internal energy production is radiated off. This, however, is a totally unrealistic requirement. In reality the radiation will be minimal and negligible. Instead the fluid will be heated, *i.e.* the temperature grows in the flow direction, and the semi-inverse ansatz (2.4) contradicts the field equation $(2.2)_3$. In fluid mechanics this phenomenon is usually not noticed, since fluid mechanics ignores the energy balance, and the effect is small enough that it does not affect viscosity measurements appreciably. Yet, this discussion illustrates the pitfalls of semi-inverse assumptions. We shall come back to this problem later, *cf.* Paragraph 11.1.2.

2.2.2 *Heat conduction through a window-pane*

We consider a stationary temperature field in air which is enclosed between two glass plates, *cf.* Fig. 2.2. We assume that the temperature depends on x_3 only and that the thermal conductivity κ is independent of T. In this case elimination of q_3 between the energy balance $(2.2)_3$ and FOURIER's law $(2.3)_3$ provides

$$\frac{\mathrm{d}^2T}{\mathrm{d}x_3^2}=0 ,$$

if radiation is ignored. It follows that the temperature is a *linear* function of x_3 in each one of the three regions

$$T_J=\alpha_J x_3+\beta_J ,\ (J=\mathrm{I},\mathrm{II},\mathrm{III}) .$$

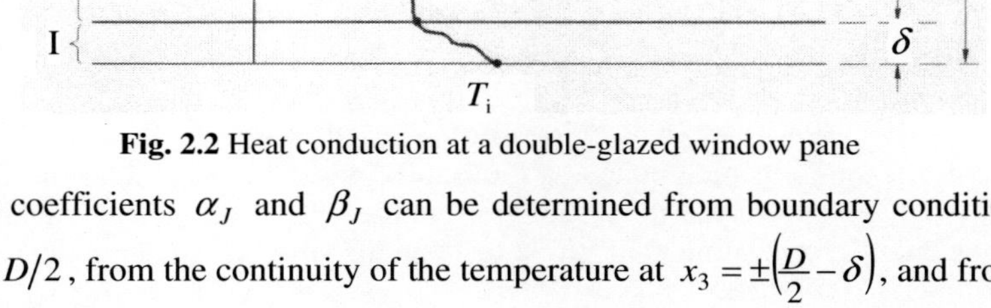

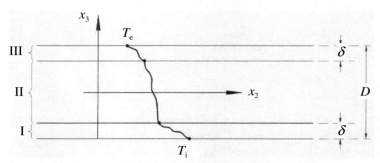

Fig. 2.2 Heat conduction at a double-glazed window pane

The coefficients α_J and β_J can be determined from boundary conditions at $x_3=\pm D/2$, from the continuity of the temperature at $x_3=\pm\left(\frac{D}{2}-\delta\right)$, and from the requirement that the heat fluxes

$$q_J=-\kappa_J\frac{\mathrm{d}T}{\mathrm{d}x_3}=-\kappa_J\alpha_J$$

are continuous at $x_3 = \pm\left(\frac{D}{2} - \delta\right)$. Thus a simple calculation yields

$$T_{\mathrm{I}} = T_{\mathrm{i}} - \frac{\kappa_{\mathrm{II}}}{\kappa_{\mathrm{I}}} \frac{T_{\mathrm{i}} - T_{\mathrm{e}}}{D - \left(1 - \frac{\kappa_{\mathrm{II}}}{\kappa_{\mathrm{I}}}\right) 2\delta} \left(x_3 + \frac{D}{2} \right),$$

$$T_{\mathrm{II}} = \frac{T_{\mathrm{i}} + T_{\mathrm{e}}}{2} - \frac{T_{\mathrm{i}} - T_{\mathrm{e}}}{D - \left(1 - \frac{\kappa_{\mathrm{II}}}{\kappa_{\mathrm{I}}}\right) 2\delta} x_3, \tag{2.8}$$

$$T_{\mathrm{III}} = T_{\mathrm{e}} - \frac{\kappa_{\mathrm{II}}}{\kappa_{\mathrm{I}}} \frac{T_{\mathrm{i}} - T_{\mathrm{e}}}{D - \left(1 - \frac{\kappa_{\mathrm{II}}}{\kappa_{\mathrm{I}}}\right) 2\delta} \left(x_3 - \frac{D}{2} \right),$$

The common heat flux of all three regions reads

$$q = \frac{\kappa_{\mathrm{II}}}{D - \left(1 - \frac{\kappa_{\mathrm{II}}}{\kappa_{\mathrm{I}}}\right) 2\delta} \left(T_{\mathrm{i}} - T_{\mathrm{e}} \right). \tag{2.9}$$

If $\kappa_{\mathrm{I}} = \kappa_{\mathrm{II}} = \kappa$ holds, all three temperature functions are reduced to a *single* linear function

$$T = \frac{T_{\mathrm{i}} + T_{\mathrm{e}}}{2} - \frac{T_{\mathrm{i}} - T_{\mathrm{e}}}{D} x_3, \text{ hence } q = \frac{\kappa}{D} \left(T_{\mathrm{i}} - T_{\mathrm{e}} \right). \tag{2.10}$$

Therefore one may measure the thermal conductivity by determining the heat flux necessary to maintain the temperatures T_{i} and T_{e} on the sides of a plate of thickness D. Some measured values are given in Table 2.2 for different (mean) temperatures.

Table 2.2. Thermal conductivity κ in 10^{-2} N/(sK)

T [°C]	air	water	glass	steel
0	2.41	55.4		2000
30	2.60	61.1	1100	through
100	3.14	68.1		6000

By (2.9) the heat flux through a double-glazed window-pane with $D = 1\,\mathrm{cm}$, $\delta = 0.2\,\mathrm{cm}$, $\kappa_{\mathrm{I}} = 11\,\mathrm{N/(sK)}$, $\kappa_{\mathrm{II}} = 2.5 \cdot 10^{-2}\,\mathrm{N/(sK)}$, and for temperatures $T_{\mathrm{i}} = 20°\mathrm{C}$, $T_{\mathrm{e}} = 10°\mathrm{C}$ amounts to $q = 42\,\mathrm{W/m^2}$. For a single pane of thickness 0.2 cm, by $(2.10)_2$, that heat flux would have the value $q = 55 \cdot 10^3\,\mathrm{W/m^2}$. In other words: The single pane conducts a thousand times more heat than the double-glazed pane. Therefore the now common double-glazed panes – or isolating panes – save energy for heating. The effect rests on the small thermal conductivity of air in comparison to glass.

We use the same effect when we conserve body heat by clothing: The air trapped in the woolen mesh of a sweater, and between the sweater and the body, isolates our body from the environment.

2.3 Measuring the state functions $p(v,T)$ and $u(v,T)$

2.3.1 *The need for measurements*

Ideal gases and incompressible fluids are the only substances for which we know the thermal and caloric equations of state

$$p(v,T) \text{ and } u(v,T)$$

as explicit analytic functions, *cf.* (P.1), (P.2), (P.3), and (P.4). For all other substances, *e.g.* vapors and compressible fluids, the state functions must be determined experimentally by measurements.

Having said this we must qualify: Truly incompressible fluids do not exist, nor do truly ideal gases. Thus for instance air exhibits a slight dependence of the internal energy u on the density ρ, as put in evidence by the JOULE-THOMSON effect, see above, Paragraph 1.5.9. Therefore we must consider the state equations reported in the Prologue as idealizations. They are good enough, however, for the calculation of most technical processes, at least with regard to the leading phenomena, and they provide the right orders of magnitude for the results.

2.3.2 *Thermal equations of state*

There is little to say about the experimental determination of the thermal equation of state. That equation connects three measurable quantities, namely p, v, and T: One adjusts two of those and measures the third one. In principle this is easy, but in practice it is still a laborious process since the adjusted pair – v and T (say) – must vary over the whole range of values that is of interest. Still, when this work is done, one obtains tables or graphs of the thermal equation of state

$$p = p(v,T), \text{ or } v = v(p,T), \text{ or } T = T(v,p). \tag{2.11}$$

In actual practice these functions are obtained by

- fixing T and registering the relative change of v created by a small change of p, or
- fixing p and measuring the relative change of v created by a small change of , or
- fixing v and measuring the relative change of p associated by a small change of T.

In this manner one determines the *coefficients of state*: compressibility κ_T, thermal expansion α, and tension coefficient β

$$\kappa_T = -\frac{1}{v}\left(\frac{\partial v}{\partial p}\right)_T, \quad \alpha = \frac{1}{v}\left(\frac{\partial v}{\partial T}\right)_p, \quad \beta = \frac{1}{p}\left(\frac{\partial p}{\partial T}\right)_v. \tag{2.12}$$

All of these coefficients generally depend on (v,T), or (p,T), or (v,p). But not all of them are independent, because for a generic function z of two variables x and y the following mathematical identity must hold

$$\left(\frac{\partial z}{\partial x}\right)_y \left(\frac{\partial x}{\partial y}\right)_z \left(\frac{\partial y}{\partial z}\right)_x = -1. \tag{2.13}$$

For the thermal equation of state (2.11) this implies

$$\left(\frac{\partial p}{\partial \upsilon}\right)_T \left(\frac{\partial \upsilon}{\partial T}\right)_p \left(\frac{\partial T}{\partial p}\right)_\upsilon = -1 \text{ or } \alpha = p\kappa_T\beta \,. \tag{2.14}$$

Thus, if any two of the coefficients of state have been measured, the third one may be calculated.

Once the coefficients of state have been measured the state functions (2.11) may be obtained be integration.

For ideal gases the coefficients of state read by (P.1)

$$\kappa_T = \frac{1}{p}, \ \alpha = \frac{1}{T}, \ \beta = \frac{1}{T} \tag{2.15}$$

and, of course, $(2.14)_2$ is satisfied.

2.3.3 *Caloric equation of state*

The experimental determination of the caloric equation of state is far from easy, even in principle. The principal difficulty lies in the fact that the internal energy u is not a measurable quantity. However, the two derivatives

$$\left(\frac{\partial u}{\partial T}\right)_\upsilon \text{ and } \left(\frac{\partial u}{\partial \upsilon}\right)_T \tag{2.16}$$

can be measured, or rather they can be expressed through measurable quantities. After that has been done, u follows by integration to within an additive constant.

The measurable quantities to which the derivatives (2.16) may be related are the *specific heat capacities* c_υ and c_p.* The heating $\dot{q}$ is applied to a unit of mass in the time interval $\mathrm{d}t$. This leads to a temperature change $\mathrm{d}T$ according to

$$\dot{q}\,\mathrm{d}t = c\,\mathrm{d}T \,, \tag{2.17}$$

where the factor of proportionality c is the specific heat. Equation (2.17) defines that quantity. c is different, if the heating occurs at constant υ, or at constant p. Accordingly we have c_υ and c_p. Given $\dot{q}\,\mathrm{d}t$ both can be measured by registering $\mathrm{d}T$.

In order to determine the derivatives (2.16) of u from such measurements we use the First Law in the form (1.57) applied to a unit of mass

$$\dot{q}\,\mathrm{d}t = \mathrm{d}u + p\,\mathrm{d}\upsilon \text{ and with } u = u(\upsilon,T) \tag{2.18}$$

$$\dot{q}\,\mathrm{d}t = \underbrace{\left(\frac{\partial u}{\partial T}\right)_\upsilon}_{c_\upsilon} \mathrm{d}T + \left[\left(\frac{\partial u}{\partial \upsilon}\right)_T + p\right] \mathrm{d}\upsilon \,. \tag{2.19}$$

If υ is fixed during the heating, *i.e.* if $\mathrm{d}\upsilon = 0$ holds, comparison of (2.17) and (2.19) provides

* Most often they are simply called *specific heats* at constant υ and p, respectively.

$$\boxed{\left(\frac{\partial u}{\partial T}\right)_{\upsilon} = c_{\upsilon}} \tag{2.20}$$

as indicated in (2.19).

On the other hand, at constant pressure, the form (2.19) is not appropriate for the determination of the specific heat. In order to make it appropriate one must replace $\mathrm{d}\upsilon$ by means of the thermal equation of state as follows

$$\mathrm{d}\upsilon = \left(\frac{\partial \upsilon}{\partial T}\right)_p \mathrm{d}T + \left(\frac{\partial \upsilon}{\partial p}\right)_T \mathrm{d}p\,.$$

Thus one obtains

$$\dot{q}\,\mathrm{d}t = \underbrace{\left\{c_{\upsilon} + \left[\left(\frac{\partial u}{\partial \upsilon}\right)_T + p\right]\left(\frac{\partial \upsilon}{\partial T}\right)_p\right\}}_{c_p}\mathrm{d}T + \left[\left(\frac{\partial u}{\partial \upsilon}\right)_T + p\right]\left(\frac{\partial \upsilon}{\partial p}\right)_T \mathrm{d}p\,. \tag{2.21}$$

and therefore

$$\boxed{\left(\frac{\partial u}{\partial \upsilon}\right)_T = \frac{c_p - c_{\upsilon}}{\left(\frac{\partial \upsilon}{\partial T}\right)_p} - p}\,. \tag{2.22}$$

On the right hand side of the equations in the frames we have only measurable quantities. Therefore the derivatives $\left(\frac{\partial u}{\partial T}\right)_{\upsilon}$ and $\left(\frac{\partial u}{\partial \upsilon}\right)_T$ can be calculated and then – by integration – the caloric equation of state $u = u(\upsilon,T)$ may be determined to within an additive constant. Of course c_{υ} and c_p must be measured for all pairs (υ,T) and this requires many laborious and difficult measurements. All caloric measurements are difficult, because the experimentalist must be certain that the applied heat really gets into the fluid and is not lost for the heating of the container or the surrounding air.

It is possible to express c_p more easily than by the expression indicated in (2.21). Indeed, we recall the definition of enthalpy $h = u - Ts$ and write (2.18) in the form

$$\dot{q}\,\mathrm{d}t = \mathrm{d}h - \upsilon\,\mathrm{d}p \quad \text{or with } h = h(T,p) \tag{2.23}$$

$$\dot{q}\,\mathrm{d}t = \underbrace{\left(\frac{\partial h}{\partial T}\right)_p}_{c_p}\mathrm{d}T + \left[\left(\frac{\partial h}{\partial p}\right)_T - \upsilon\right]\mathrm{d}p\,. \tag{2.24}$$

Hence follows

$$\boxed{\left(\frac{\partial h}{\partial T}\right)_p = c_p}\,. \tag{2.25}$$

We may summarize the equations (2.20) and (2.25) by saying that c_υ is the temperature derivative of the internal energy u, and that c_p is the temperature derivative of the enthalpy h. Note, however, that in the first case υ is kept constant and in the second case p.

In particular for ideal gases with, *cf.* (P.2)

$$u = z\frac{R}{M}(T - T_R) + u_R \text{ and}$$

$$h = (z+1)\frac{R}{M}(T - T_R) + u_R + \frac{p_R}{\rho(p_R, T_R)}$$

we obtain for the specific heats and their quotient and difference

$$c_\upsilon = z\frac{R}{M},\ c_p = (z+1)\frac{R}{M},$$

$$\text{hence } \kappa \equiv \frac{c_p}{c_\upsilon} = \frac{z+1}{z},\ c_p - c_\upsilon = \frac{R}{M}. \tag{2.26}$$

Both specific heats are constant in this case, *i.e.* independent of either ρ and T, and their quotient κ is greater than 1. Later we shall see that $c_p > c_\upsilon$ holds for *all* substances – not only for ideal gases. The reason is easily understood – at least for gases – because for $\mathrm{d}\upsilon = 0$ it is only the internal energy that is increased by heating, while for $\mathrm{d}p = 0$ the enthalpy is increased. We then have

$$\mathrm{d}h = \mathrm{d}u + \mathrm{d}(p\upsilon) \underset{p=\text{const.}}{=} \mathrm{d}u + p\,\mathrm{d}\upsilon$$

so that, in addition to increasing u, the heating must also provide the work $p\,\mathrm{d}\upsilon$ of the pressure when the volume is increased. Therefore the rise of the temperature is smaller for constant p, than for constant υ.* Later we shall see that $c_p > c_\upsilon$ holds for *all* substances – not only for ideal gases.

2.3.4 *Equations of state for air and superheated steam*

Air is not a bad example for an ideal gas – at least not for normal pressures and temperatures – but it is not perfect. Indeed, when careful thermal and caloric measurements are made with air it turns out that $p\upsilon$ is not really proportional to T and independent of υ. Nor are u and h truly linear functions of T and independent of υ. We recall that JOULE and KELVIN registered a tiny cooling in GAY-LUSSAC's expansion experiment which indicates that u depends weakly on ρ, or υ, or p.

* The argument about $c_p > c_\upsilon$ is more subtle for liquids, particularly for those liquids, where the volume *drops* during isobaric heating. This, however, is a rare phenomenon. Water between 0°C and 4°C is a liquid showing such behavior.

Still we always use the ideal gas laws for air and consider the mistake as negligible.

Unlike air, superheated steam of water is not well described by the ideal gas laws, at least not at high pressures. No analytic functions are available for the representation of the equations of state over a large range of the variables. But there are tables of values in the thermodynamic handbooks, like those of Landolt-Börnstein.* Table 2.4 presents a small excerpt of one such table.

Table 2.3 Thermal state function $\upsilon(p,T)$ for superheated steam (υ in l/kg).

T [°C] / p [bar]	400	500	600	700	800
370	2.0102	6.357	8.899	10.832	12.516
380	1.9717	6.096	8.614	10.515	12.166

In order to appreciate the deviation from the ideal gas law, it is easiest to calculate the expression $\frac{p\upsilon}{\frac{R}{M}T}$ with $M = 18\,\frac{\text{g}}{\text{mol}}$ – appropriate for water – for the triples of values presented in the table. Thus one obtains values between 0.24 and 0.93, instead of 1, which would be the proper value, if the steam were an ideal gas. For this reason we need to refer to tables, or to diagrams when calculating the efficiencies of steam engines (say).

2.3.5 *Equations of state for liquid water*

As indicated in the Prologue liquid water is nearly incompressible with a density $\rho = 10^3\,\frac{\text{kg}}{\text{m}^3}$. Indeed, the unit "kg" is defined such that the water density can be assigned that round number.

In earlier times the unit of heat energy was 1 cal, and this was also determined in terms of water properties. One wrote the specific heat of water of 15°C as $c = 1\text{cal}/(\text{gK})$. This has now changed: There is only one unit for all forms of energy, namely the Joule, and we write $c = 4.18\,\text{J}/(\text{gK})$ for water.

In reality, neither the density of liquid water nor the specific heats are constants. Rather they are slowly varying functions of p and T. At least they vary slowly in the range of pressures and temperatures encountered in hydraulic machines. For the calculation of the functioning of these machines it is usually sufficient to use the constant values of ρ and c given above. To the extent that the compressibility of water may be neglected, the difference between c_p and c_υ can also be neglected, *cf.* Paragraph 4.2.3.

If the engineer notices that the pressure in his machines is so high that the density of water is affected, he must consult tables of values published in handbooks like those of LANDOLT-BÖRNSTEIN which were already cited above. Reliable

---

* Landolt-Börnstein: Group IV Physical Chemistry Numerical Data and Functional Relationships in Science and Technology, Vol. 4, High-Pressure Properties of Matter. Springer 1980.

analytic functions of $\rho(p,T)$ and $c(p,T)$ representing the values in these tables are not available. However, it is often sufficient to approximate such functions linearly.

2.4 State diagrams for fluids and vapors with a phase transition

2.4.1 *The phenomenon of a liquid-vapor phase transition*

Figs. 2.3 and 2.4 show in schematic form the phenomena and the states that occur in an isobaric evaporation and in an isothermal condensation. The former case occurs during sustained heating and the latter case requires cooling. During evaporation and condensation, pressure and temperature remain constant. The liquid and the vapor phase coexist and the contents of the cylinder is called *wet steam*, short for *saturated liquid-vapor mixture*. In wet steam the phases are often not as clearly separated as they are shown in the figures. Rather liquid droplets may float in the vapor and wet steam may thus look like a fog.

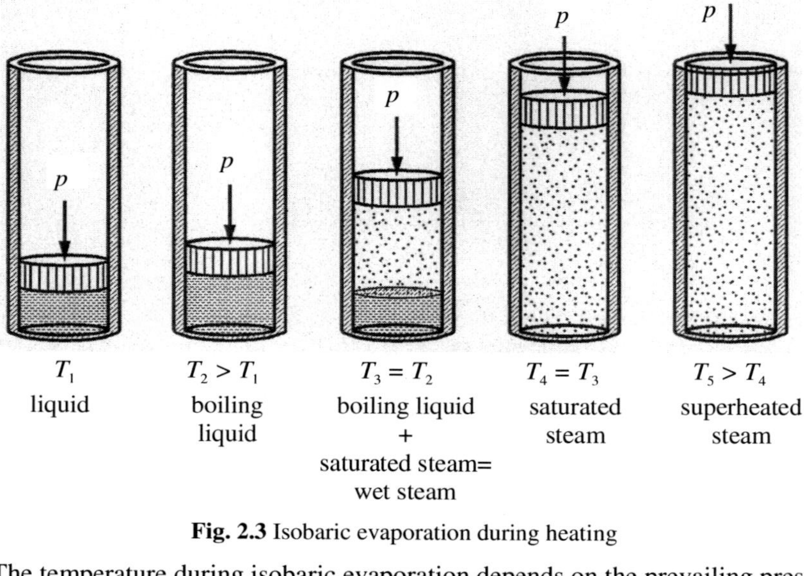

Fig. 2.3 Isobaric evaporation during heating

The temperature during isobaric evaporation depends on the prevailing pressure and the pressure during isothermal condensation depends on the prevailing temperature. Thus for wet steam we have

$$p = p(T). \tag{2.27}$$

Therefore the densities, or specific volumes v' and v'', respectively, of boiling liquid and saturated steam also depend on one variable only, either T or p, and we may write*

$$v' = v'(T) \text{ and } v'' = v''(T). \tag{2.28}$$

---

* It is common practice to characterize quantities related to the boiling liquid and to saturated vapor by ´ and ´´, respectively.

The equations (2.27), (2.28) may be called thermal equations of state for wet steam, although this expression is not much in use. Note that v' and v'' are unchanged during the course of the phase transition; *e.g.* they are independent of the amount of liquid that has already been evaporated.

The heat supplied to the wet steam from the beginning to the end of the evaporation again depends only on temperature, or pressure and we may write the specific *heat of evaporation* as

$$r = r(T). \tag{2.29}$$

As long as the evaporation proceeds at constant pressure – as indicated in Figs. 2.3, 2.4 – the heat of evaporation increases the enthalpy; *cf.* Paragraph 1.5.7. Therefore we have

$$r(T) = h'' - h', \tag{2.30}$$

where h' and h'' are the specific enthalpies of the boiling liquid and the saturated steam, respectively. Both are functions of T, or p.

In the course of an isobaric – and isothermal – evaporation or condensation, when the *steam content* $x = \frac{m''}{m}$ is changed by Δx, the corresponding fraction $\Delta x\, r(T)$ of the heat of evaporation needs to be applied.

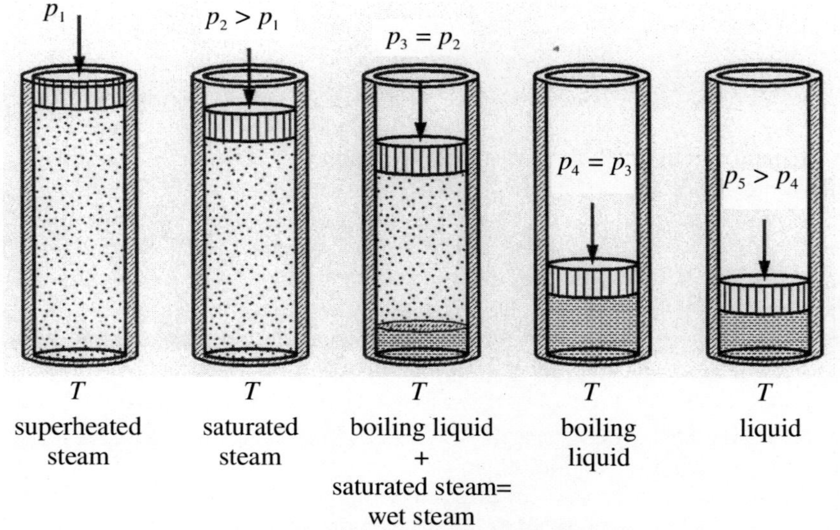

Fig. 2.4 Isothermal condensation during cooling

In the wet steam, particularly if it looks like a fog, we may define a specific volume $v = V/m$ and write

$$V = V' + V'' \quad \Big|\cdot \frac{1}{m}$$

$$\frac{V}{m} = \frac{m'}{m}\frac{V'}{m'} + \frac{m''}{m}\frac{V''}{m''}, \text{ or, with } v = \frac{V}{m}, \quad v' = \frac{V'}{m'}, \quad v'' = \frac{V''}{m''}$$

$$v = (1-x)v' + xv'' \tag{2.31}$$

so that v is the weighted sum of v' and v'' with the liquid content $1-x$ and the steam content x as weight factors.

Analogously one obtains

$$u = (1-x)u' + xu'' \text{ and } h = (1-x)h' + xh'', \tag{2.32}$$

where u', h' and u'', h'' are the specific energies and enthalpies of the boiling liquid and of the saturated steam, respectively. All of these are, of course, functions of T, or p, *i.e.* of only one variable, since $p = p(T)$ holds.

2.4.2 *Melting and sublimation*

The transitions solid-liquid, *i.e.* melting or freezing, and solid-vapor, *i.e.* sublimation or de-sublimation, are qualitatively quite similar to the transition liquid-vapor: The melting pressure and the sublimation pressure are functions of T only and so are the specific volumes of the melting or sublimating solid and of the freezing liquid and de-sublimating vapor. Also, the specific heats of melting and sublimation depend only on T, or p, just like the heat of evaporation.

2.4.3 *Saturated vapor curve of water*

The function $p(T)$ in (2.27) cannot be expressed analytically. It is measured for a substance, and its graphical representation in a (p,T)-diagram is called the *saturated vapor curve* or the *vapor pressure curve*. All such curves – for different substances – are different in value, but qualitatively similar: They are all convex and increase monotonically; and all start in the triple point Tr and end in the critical point C, *cf.* Fig. 2.5 a.

Fig. 2.5 b shows melting and sublimation curves of water in addition to the saturated vapor curve in a schematic form. The three curves have one common point – the triple point – where they intersect. In that point the boiling liquid, the melting solid, and the saturated vapor can coexist in any proportion.

The critical point marks the end of the saturated vapor curve. For higher pressures there is no clearly identifiable phase transition at a definite pair of values (p,T). Therefore the vapor pressure curve may be bypassed in going from point 1 to point 2, *cf.* Fig. 2.5 a. Thus one arrives at superheated vapor from a liquid without ever experiencing a separation of phases.

The saturated vapor pressures for different temperatures may be read off from Table 2.4. The positive slope of the saturated vapor curve implies that water remains liquid to higher temperatures for higher pressures. Conversely for low pressures – at a great height (say) – water boils at lower temperatures. Thus it takes a minute longer to boil an egg in Mexico City. The town lies at a height of 2200 m and therefore the air pressure is 260 barometric steps lower than at sea level, *cf.* Paragraph 1.4.5. By Table 2.4 the temperature of boiling water is then only about 92°C.

For water the melting curve has a negative slope. This is one aspect of the so-called *anomaly of water*. The negative slope means that the melting point of ice decreases for increasing pressure. Thus the high pressure under the skids of ice-skates melts the ice – provided that it is not too cold – and the melted water serves as a lubricating layer so that there is very little friction. They say, however, that on sub-arctic roads, where the ice is very cold, a car can drive safely along an icy road, because no melting occurs.

The sublimation of ice – *i.e.* the direct transition from the solid phase to water vapor – may be observed on dry and cold winter days. What matters is that the vapor pressure – not the air pressure, of course – is smaller than 0.0061 bar, and that the sun shines in order to provide the heat of sublimation. Under these conditions the snow on the ground sublimates and a frozen towel on the clothes line dries without ever becoming wet during that process.

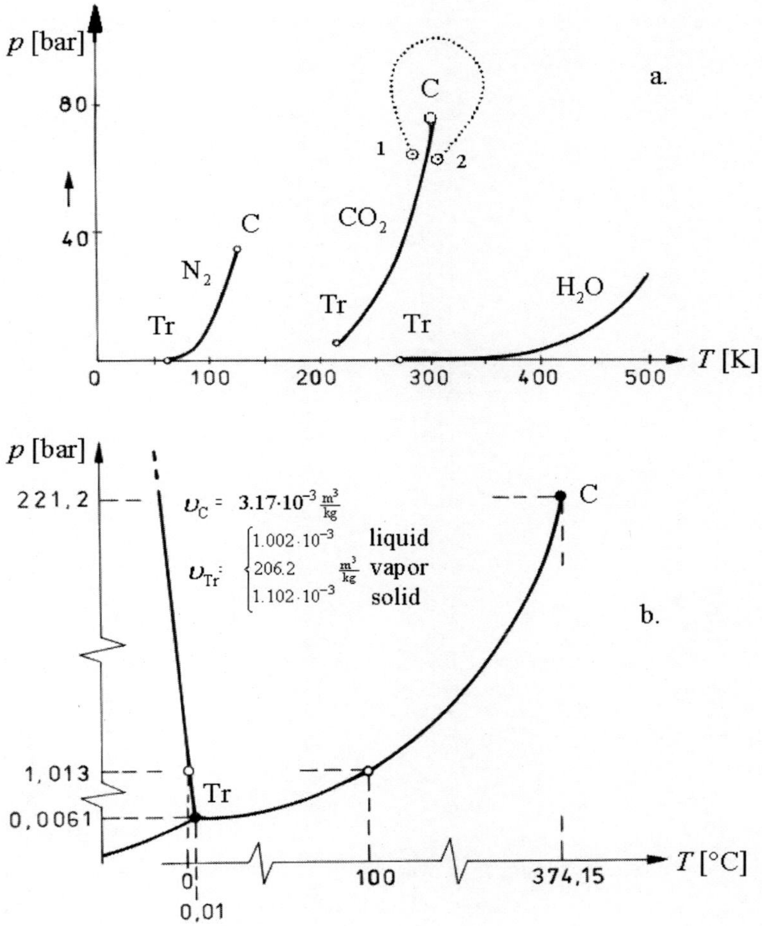

Fig. 2.5 a. Three saturated vapor curves
b. Phase separation lines of water (schematic)

By Fig. 2.5 a carbon dioxide has a triple point pressure of about 5 bar – actually 5.17bar. Therefore solid CO_2 does not melt, it sublimates for rising temperatures under normal circumstances. Solid CO_2 is used as "dry ice" for cooling purposes, and thus one avoids that the cooled material is soaked by the melt; after all, there is no melt.

Table 2.4 Important values for boiling liquid water and saturated water vapor

t °C	p bar	v' dm³/kg	v'' m³/kg	h' kJ/kg	h'' kJ/kg	r kJ/kg
0,01	0,006112	1,0002	206,2	0,00	2501,6	2501,6
5	0,008718	1,0000	147,2	21,01	2510,7	2489,7
10	0,01227	1,0003	106,4	41,99	2519,9	2477,9
15	0,01704	1,0008	77,98	62,94	2529,1	2466,1
20	0,02337	1,0017	57,84	83,86	2538,2	2454,3
25	0,03166	1,0029	43,40	104,77	2547,3	2442,5
30	0,04241	1,0043	32,93	125,66	2556,4	2430,7
35	0,05622	1,0060	25,24	146,56	2565,4	2418,8
40	0,07375	1,0078	19,55	167,45	2574,4	2406,9
45	0,09582	1,0099	15,28	188,35	2583,3	2394,9
50	0,12335	1,0121	12,05	209,26	2592,2	2382,9
55	0,1574	1,0145	9,579	230,17	2601,0	2370,8
60	0,1992	1,0171	7,679	251,09	2609,7	2358,6
65	0,2501	1,0199	6,202	272,02	2618,4	2346,3
70	0,3116	1,0228	5,046	292,97	2626,9	2334,0
75	0,3855	1,0259	4,134	313,94	2635,4	2321,5
80	0,4736	1,0292	3,409	334,92	2643,8	2308,8
85	0,5780	1,0326	2,829	355,92	2652,0	2296,5
90	0,7011	1,0361	2,361	376,94	2660,1	2283,2
95	0,8453	1,0399	1,982	397,99	2668,1	2270,2
100	1,0133	1,0437	1,673	419,1	2676,0	2256,9
110	1,4327	1,0519	1,210	461,3	2691,3	2230,0
120	1,9854	1,0606	0,8915	503,7	2706,0	2202,3
130	2,701	1,0700	0,6681	546,3	2719,9	2173,6
140	3,614	1,0801	0,5085	589,1	2733,1	2144,0
t	p	v'	v''	h'	h''	r
150	4,760	1,0908	0,3924	632,2	2745,4	2113,2
160	6,181	1,1022	0,3068	675,5	2756,7	2081,2
170	7,920	1,1145	0,2426	719,1	2767,1	2048,0
180	10,027	1,1275	0,1938	763,1	2776,3	2013,2
190	12,551	1,1415	0,1563	807,5	2784,3	1976,8
200	15,549	1,1565	0,1272	852,4	2790,9	1938,5
210	19,077	1,173	0,1042	897,5	2796,2	1898,7
220	23,198	1,190	0,08604	943,7	2799,9	1856,2
230	27,976	1,209	0,07145	990,3	2802,0	1811,7
240	33,478	1,229	0,05965	1037,6	2802,2	1764,6
250	39,776	1,251	0,05004	1085,8	2800,4	1714,6
260	46,943	1,276	0,04213	1134,9	2796,4	1661,5
270	55,058	1,303	0,03559	1185,2	2789,9	1604,6
280	64,202	1,332	0,03013	1236,8	2780,4	1543,6
290	74,461	1,366	0,02554	1290,0	2767,6	1477,6
300	85,927	1,404	0,02165	1345,0	2751,0	1406,0
310	98,700	1,448	0,01833	1402,4	2730,0	1327,6
320	112,89	1,500	0,01548	1462,6	2703,7	1241,1
330	128,63	1,562	0,01299	1526,5	2670,2	1143,6
340	146,05	1,639	0,01078	1595,5	2626,2	1030,7
350	165,35	1,741	0,00880	1671,9	2567,7	895,7
360	186,75	1,896	0,00694	1764,2	2485,4	721,3
370	210,54	2,214	0,00497	1890,2	2342,8	452,6
374,15	221,20	3,17	0,00317	2107,4	2107,4	0,0

Fig. 2.6 shows a (p,T)-diagram of water with the saturated vapor curve and some isochors. In the vapor and liquid regions the isochors have positive slopes, *i.e.* water and water vapor "attempt" to expand under increasing temperatures, as most other substances also do. This is not true, however, below 3.98°C, where

water attempts to contract with increasing temperature. This is another aspect of the *anomaly of water*, but the scale of Fig. 2.6 is too coarse to capture this.

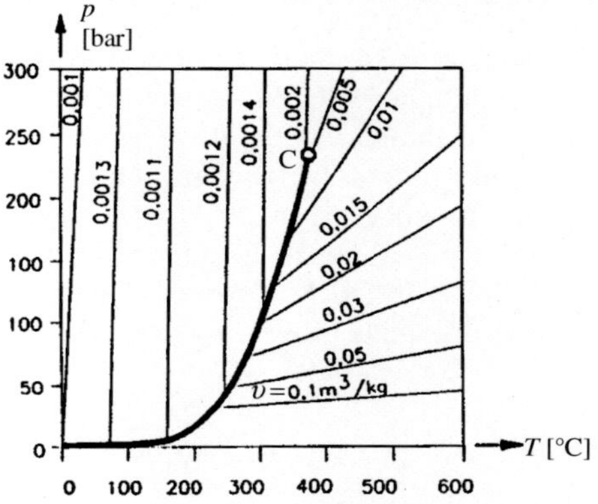

Fig. 2.6 Isochors in a (p,T)–diagram of water

2.4.4 *On the anomaly of water*

In cold weather – under cold air – a lake cools on the surface and, if the lake were to contain a "normal" liquid, such as methane,* the cold layer sinks down from the surface and drives the warmer liquid from the bottom up to the surface to be cooled there. In this manner – by *convection* – a cooling methane lake attains a homogeneous temperature eventually, at least when the cooling process is slow enough, and lasts long enough.

For water this is the same as long as the temperature is above 3.98°C, the temperature at which water has the highest density and the smallest specific volume.

When the cooling water is below that temperature, the cold water at the top does not sink, rather it stays at the surface and the water at the bottom remains warmer for a long time until the slow process of *conduction* homogenizes the temperature.

When the liquid *freezes* due to the cooling at the surface of the methane lake, the methane ice sinks down to the bottom, because in "normal" substances the solid phase is denser than the liquid phase. Thus in such a "normal" substance the ice accumulates at the bottom. A methane lake solidifies from the bottom upward and the last remaining liquid lies at the surface.

For water this is different. Water ice has only 92% of the density of liquid water. Therefore the ice floats on top and the water below will freeze only slowly – upon sustained cooling of the surface. Thus a water lake freezes from the top downward, and the freezing takes a long time, because the cold temperature spreads downward by the slow process of heat conduction. In this way lakes of a sizable depth do not freeze all through from top to bottom, and so marine life can survive through the winter.

---

* Science fiction and NASA have it that such lakes occur on Saturn's moon Titan.

The climatic effects of the anomaly of water are considerable. If the ice were sinking to the ground upon freezing during wintertime, it could not be reached by the warming rays of the sun in spring, and most of the water in lakes and in the polar oceans would be permanently frozen. Our Earth would thus be a cold planet indeed, and life as we know it could not have evolved.

The reason for the anomaly of liquid water and water ice lies in the molecular structure of water. That structure makes a hydrogen bond possible between two molecules. The bond keeps the hydrogen atoms in one water molecule at a fairly large distance from the oxygen molecule of its neighbors. Since the bond is weak, it "melts" at higher temperatures so that the "normal" behavior of a liquid, the positive thermal expansion, can dominate in water above 3.98°C.

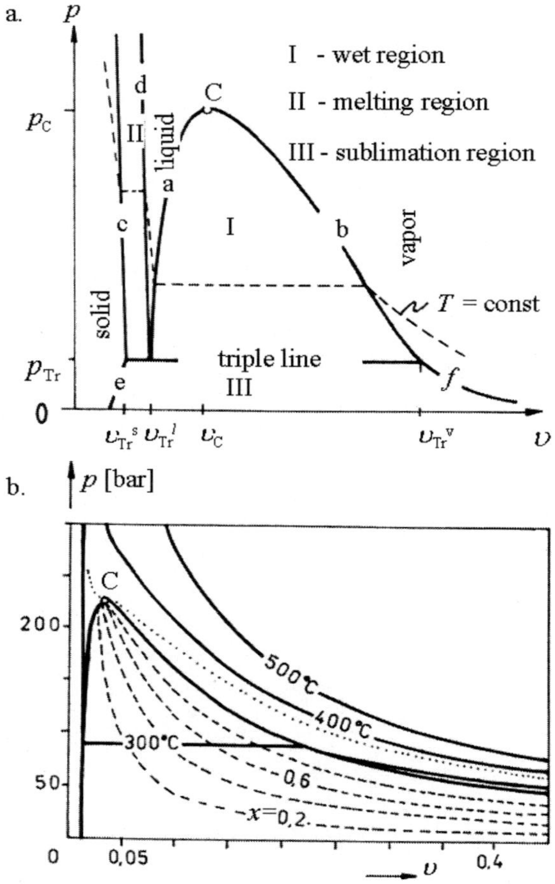

Fig. 2.7 a. Phase separation lines in the (p, v)-diagram (schematic)
a, b – evaporation and dew line
c, d – melting and freezing line
e, f – sublimation and de-sublimation line
b. (p, v)-diagram of water (without melting and sublimation regions)

2.4.5 Wet region and (p, v)-diagram of water

By (2.28) the specific volumes v' and v'' of a boiling liquid and of saturated steam depend only on temperature just as the pressure does. Elimination of the temperature between these relations provides

$$v' = v'(p) \quad \text{and} \quad v'' = v''(p).$$

These two relations are shown in Fig. 2.7 as boiling line and dew line, respectively, in a schematic (p,v)-diagram. Just like the (p,T) -diagram, the (p,v)-diagram exhibits universal traits in the shapes of the boiling and dew lines. For the critical pressure p_C, where phase separation does not occur, we have $v'(p_C) = v''(p_C)$. For smaller pressures the range between $v'(p)$ and $v''(p)$ defines the *wet region*. Analogously there exists a *melting region* and a *sublimation region*, both shown schematically in Fig. 2.7 a.

As anticipated in Paragraph 2.4.4 the anomaly of water implies that ice is lighter than liquid water. Therefore, for water the freezing line lies to the right of the melting line. This feature is not represented in the schematic diagram of Fig. 2.7 a which therefore represents the properties of a "normal" substance.

Inside the wet steam region v is given by the weighted sum of v' and v'' with $(1-x)$ and x as the weight factors in the manner of (2.31). The isolines $x = \text{const}$ are shown as dashed lines in 2.7 b. Since in wet steam the pressure is determined by temperature, the isotherms run horizontally in the wet steam region.

For high temperatures the isotherms approach the hyperbolae characteristic for ideal gases.

2.4.6 3D phase diagram

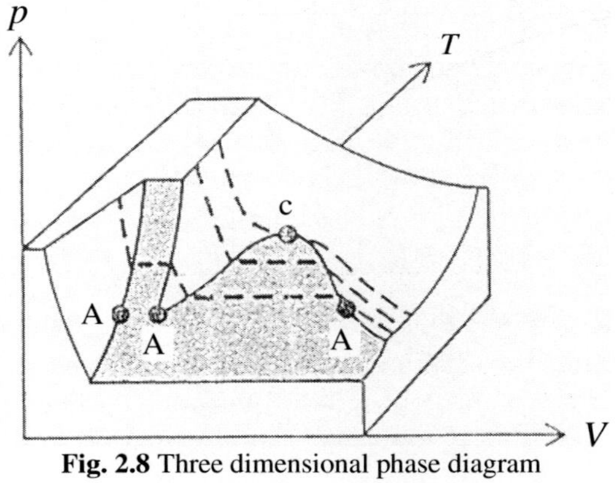

Fig. 2.8 Three dimensional phase diagram

It is instructive to plot a three-dimensional (p,v,T)-diagram. This provides an integrated view of the possibly confusing amount of information represented by the diagrams of Figs. 2.5 through 2.7. Fig. 2.8 shows such a three-dimensional plot. It identifies the critical point C, the triple line AAA, and some isotherms indicated by dashed lines. Viewing this plot in the T-direction we recognize the melting region, the sublimation region and the wet region; these regions are shaded. In the view against the υ-axis all of these regions are projected onto lines in a (p,T)-diagram, since the two-phase regions are cylindrical.

2.4.7 *Heat of evaporation and (h,T)–diagram of water*

The heat of evaporation of water – and of all other liquids – decreases with increasing temperature. For water it starts with the value $2501.6\,\text{kJ/kg}$ at the triple point and goes to zero at the critical point. At T=100°C, *i.e.* $p = 1\,\text{atm}$, it has the value $2256.9\,\text{kJ/kg}$. These values and others may be read off from Table 2.4, and Fig. 2.9 shows the $r(T)$-dependence graphically.

The heat of melting and the heat of sublimation do not significantly depend upon temperature, at least not in the technically relevant range of temperatures. We note their values as

$$r^{\text{melting}} = 333.4\tfrac{\text{kJ}}{\text{kg}} \quad \text{and} \quad r^{\text{sublimation}} = 2835\tfrac{\text{kJ}}{\text{kg}},$$

so that $r^{\text{melting}} + r(T_{\text{Tr}}) = r^{\text{sublimation}}$ holds, as it must.

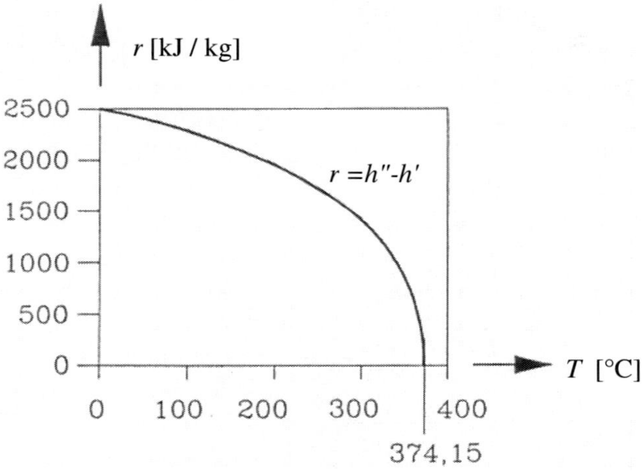

Fig. 2.9 Heat of evaporation as a function of T

The functions $h'(T)$ for a boiling liquid and $h''(T)$ for saturated steam define the boiling curve and the dew curve in the (h,T)-diagram. In-between lies the wet steam region. Fig. 2.10 a shows all two-phase regions schematically and the important curves between them. The inevitable additive constant in h is chosen here such that $h'(T_{\text{Tr}}) = 0$ holds.

Since $r = h'' - h'$ holds, the heat of evaporation may be read off from Fig. 2.10 as the vertical distance of the dew line and the boiling line. We also see that all isobars in the liquid region, up to about $250\,\text{bar}$, practically coincide with the boiling line. For large values of T the specific enthalpy becomes nearly independent of p and approaches a linear function of T. That behavior corresponds to the limiting case of an ideal gas.

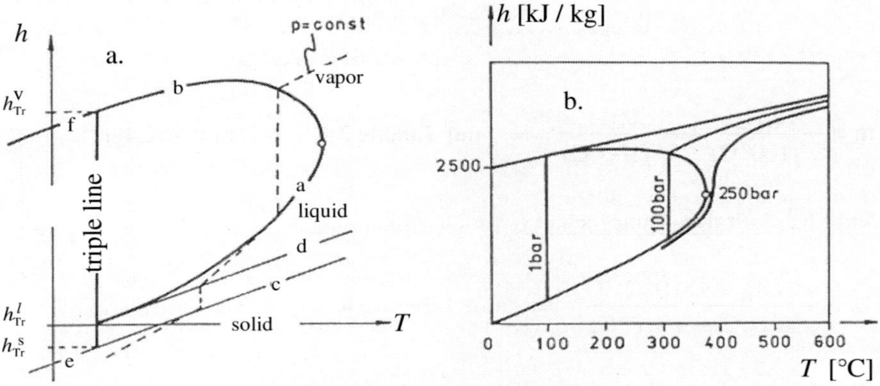

Fig. 2.10 a. Two phase regions (schematic)
Notation as in Fig. 2.7 a
b. (h,T)-diagram of water.

2.4.8 *Applications of saturated steam*

- The preservation jar

In former times, before the advent of refrigeration and deep freeze, housewives preserved fruit and vegetables in partially evacuated jars. Fruits and water are filled into a jar which is loosely closed with a lid. A sealing ring is inserted between the jar and the lid, and all of this is nearly fully immersed in a water bath which is heated so that the water in the jar starts to boil. The air on top of the fruit is thus driven out and replaced by saturated steam of 100°C. Afterwards, when the jar and its contents cools down, the steam condenses and a TORRICELLI vacuum forms below the lid. The pressure equals the small vapor pressure appropriate to room temperature. And the outer atmospheric pressure presses the lid onto the sealing ring so that the jar is tightly closed. The contents does not spoil, since all germs are killed by the boiling and no new ones can get into the jar.

For a rough quantitative study we consider the fruit and the water in a jar of $V = 1\,l$ as pure water and let the water fill 90% of the jar. When the water boils the remaining 10% of the space in the jar is filled by saturated vapor after the air has been driven out. The lid weighs $2\,\text{N}$ and has an area of 100cm^2. The temperature of the kitchen is 10°C.

We ask two questions: • What is the force which the lid exerts on the sealing ring, if the outer pressure equals $1\,\text{bar}$? • How large is the steam content?

The pressure below the lid may be read off from Table 2.4 as the pressure corresponding to 10°C. It amounts to $12.27\,\text{mbar}$ or $12.27\,\text{hPa}$. The pressure difference between the two sides of the top is therefore equal to 987.73hPa.

Considering the weight of the lid we obtain a sizable amount of 989.73N as the force acting on the seal. The mass in the jar is

$$m = \frac{0.91}{\upsilon'(100°C)} + \frac{0.11}{\upsilon''(100°C)} \quad \text{hence, by Table 2.4:} \quad m = 862.4\,\text{g}.$$

It follows that the steam content equals

$$x = \frac{V/m - \upsilon'(10°C)}{\upsilon''(10°C) - \upsilon'(10°C)} \quad \text{hence, again by Table 2.4:} \quad x = 1.5 \cdot 10^{-6}.$$

For people in a hurry there is a chemical variant of this way of preservation. A few drops of "preservation alcohol" are sprayed on the fruit – now without water – in the jar. The alcohol is ignited and, while it burns, the lid is placed on the sealing ring. The oxygen in the jar is consumed by combustion according to the reaction

$$C_2H_5OH + 3O_2 \rightarrow 2CO_2 + 3H_2O$$

and the pressure drops by one third, since the emerging water is liquid at room temperature. Thus the jar is sealed again, and no oxygen is left in it so that germs cannot enter nor multiply.

- Pressure cooker

We consider the evaporation of a mass m of water below a movable piston of cross-section A, suspended on a linear-elastic spring with the stiffness λ. The initial pressure is p_0 and the initial temperature is the corresponding boiling temperature. A safety opening is located at height L, *cf.* Fig. 2.11.

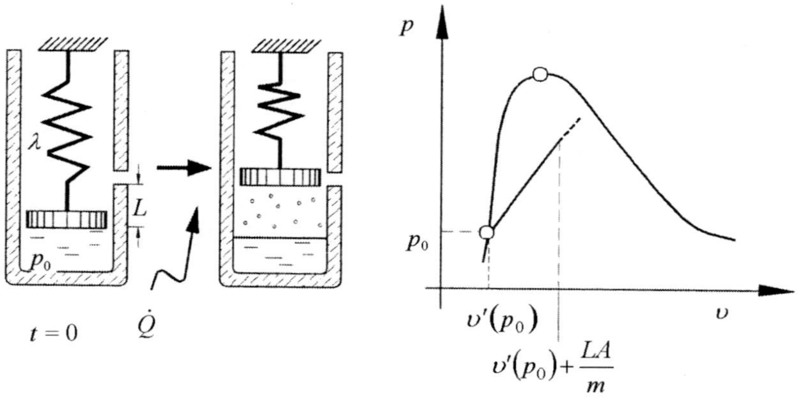

Fig. 2.11 Evaporation under increasing pressure

As soon as the evaporation begins the piston is pushed upwards by the saturated steam and the spring is compressed. The pressure increases according to the equation

$$p = p_0 + \frac{\lambda(V - V_0)}{A^2} \quad \text{or, with } V_0 = m\upsilon'(p_0)\,,\ V = m\upsilon$$

$$p = p_0 + \frac{m\lambda}{A^2}\left[v - v'(p_0)\right] .$$

It follows that the pressure is a linear function of v as indicated in the (p,v)-diagram of Fig. 2.11. We ask how far the temperature below the piston can rise. We put

$$p_0 = 1\,\text{bar} \ , \quad m = 1\,\text{kg} \, , \ \lambda = 10^5 \frac{\text{N}}{\text{m}^2} \ , \quad A = 10^2\,\text{cm}^2 \, , \ L = 3\,\text{cm}$$

and obtain

$$v_{\max} - v'(p_0) = 3 \cdot 10^{-4} \frac{\text{m}^3}{\text{kg}} \quad \text{and} \quad p = 4 \cdot 10^5 \frac{\text{N}}{\text{m}^2} .$$

By Table 2.4 this pressure corresponds to a boiling temperature of 143°C.

The example is a simplified model of the well-known pressure cooker. It is clear that meat becomes well-cooked more rapidly at 143°C than at 100°C.

The pressure cooker was a scientific sensation in the 17th century. Its inventor, the French physicist Denis PAPIN (1647-1712) was rewarded for it with a membership of the Royal Society of London, and he cooked an impressive meal for King CHARLES II with what he called his "digester."

2.4.9 *Van der Waals equation*

Johannes Diderik VAN DER WAALS (1837-1923) improved the thermal equation of state of ideal gases in the hope of making it applicable for steam, or vapors.

The basis for his analysis was the assumption that in a real gas – as opposed to an ideal gas – the atoms have so small a (mean) distance that

the actual volume of the atoms, and

the attractive force between atoms

cannot be neglected. VAN DER WAALS postulated a potential energy between two atoms which has the form shown quantitatively in Fig. 2.12 a: strongly repulsive at small atomic distances – where the atoms nearly touch each other – and weakly attractive at larger distances.

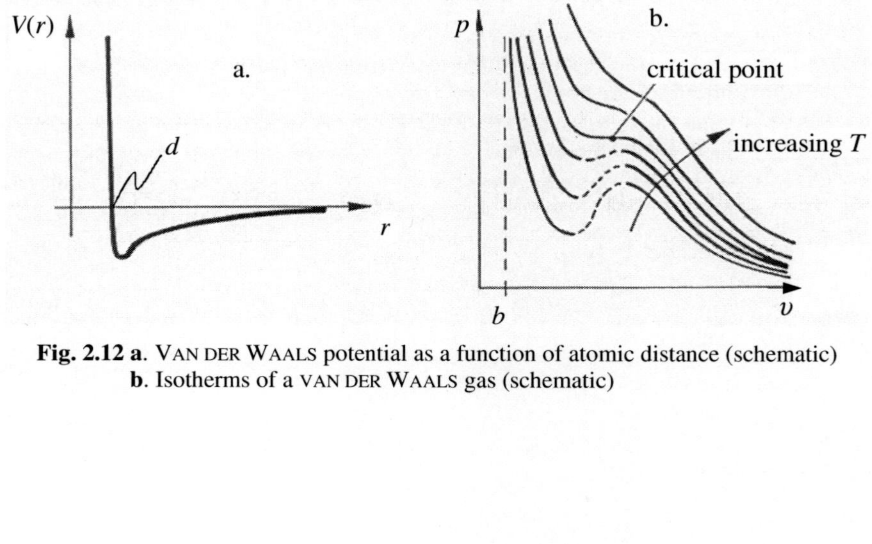

Fig. 2.12 a. VAN DER WAALS potential as a function of atomic distance (schematic)
b. Isotherms of a VAN DER WAALS gas (schematic)

In an ingenious analysis based on statistical mechanics VAN DER WAALS was able to derive a thermal equation of state of the form

$$p = \frac{\frac{R}{M}T}{\upsilon - b} - \frac{a}{\upsilon^2} \quad \text{with} \quad \begin{aligned} a &= -\frac{1}{2m^2}\int_d^\infty V(r)\, 4\pi\, r^2\, \mathrm{d}r \\ b &= \frac{1}{2m}\frac{4\pi}{3}d^3. \end{aligned} \tag{2.33}$$

The equation contains two new constants – a and b – both positive and both representing different aspects of the VAN DER WAALS potential $V(r)$. The constant a represents the attraction between atoms. It reduces the pressure of a real gas on a wall – when compared to the pressure of an ideal gas – because an atom approaching the wall is pulled back "from behind." The constant b represents the actual volume of the atoms. This volume reduces the total space available to the atoms of the gas.

VAN DER WAALS first became a school teacher for mathematics and physics, because academic studies were closed to him, who had no knowledge of Latin or Greek. Later that prohibition was relaxed and so VAN DER WAALS could present a doctoral thesis – in his native Dutch – *Over de Continuïteit van den Gas - en Vloeistoftoestand.* In this work he derived the van der Waals equation using ingenious statistical arguments which immediately put him in the forefront of molecular science. BOLTZMANN was so impressed that he called VAN DER WAALS the NEWTON of real gases.

Later VAN DER WAALS became a professor of physics in the newly created Univeristy of Amsterdam – formely the Athenaeum Illustre – and in 1910 he received the Nobel Prize in physics.

Some isotherms of a VAN DER WAALS gas are shown in Fig. 2.12 b. The most important observation is that in a certain range of temperatures the isotherms are non-monotonic. The branches with a positive slope – shown as dashed lines in the figure – are to be ignored; they are obviously unrealistic since they indicate a volume increase due to an increase of pressure.* Still, there remain two specific volumes at the same pressure and the same temperature, and we identify these as $\upsilon'(p)$ and $\upsilon''(p)$, the specific volumes of the boiling liquid and of the saturated vapor. Thus the idea arises that the VAN DER WAALS equation can represent the transition from a liquid to its vapor.

However, note that the temperature does not uniquely determine the pressure of the transition. Indeed, there is a whole *range* of pressures for which an isobar intersects an isotherm in two – or actually three – points. We shall come back to this issue in Paragraph 4.2.5. This point, however, has not confused VAN DER WAALS. He was mostly interested in the fact that at some high temperature the isotherms become monotonic. After that there is no possibility for a phase transition anymore. This possibility vanishes where the maximum and the minimum of an isotherm coalesce to form a horizontal point of inflection. VAN DER WAALS identified

---

* Such branches are unstable. The fact that they appear at all represents an artifact in the statistical mechanics derivation of the equation. We shall say more about this feature later, *cf.* Paragraphs 4.2.5 and 4.2.8.

this point with the critical point in the (p,v)-diagram. From (2.33) the coordinates of this point can easily be determined in terms of the coefficients a and b. They are

$$p_C = \frac{1}{27}\frac{a}{b^2}\ ,\quad v_C = 3b\ ,\quad \frac{R}{M}T_C = \frac{8}{27}\frac{a}{b}\,. \tag{2.34}$$

One may use these critical values to define a non-dimensional pressure π, volume ν, and temperature τ by

$$\pi = \frac{p}{p_C}\ ,\quad \nu = \frac{v}{v_C}\ ,\quad \tau = \frac{T}{T_C}\,. \tag{2.35}$$

If these quantities are introduced into the VAN DER WAALS equation (2.33), one obtains the universal equation

$$\pi = \frac{8\tau}{3\nu - 1} - \frac{3}{\nu^2}\,, \tag{2.36}$$

which is formally independent of the nature of the gas. This observation shows that the van der Waals equation can represent the universal features of the saturated vapor curve and of the wet region which we have commented on before.

According to VAN DER WAALS the equations of state of all real gases are equal, if only we measure pressure, volume, and temperature in units of their critical values. The statement is known as the *correspondence principle*: States with equal values of π, ν, and τ *correspond to each other*. The reason for this universal feature is, of course, the fairly simple form of the interaction potential $V(r)$, shown in Fig. 2.12 a, which is equal – in its broad features – for all gases.

2.4.10 *On the history of liquefying gases and solidifying liquids*

It was not easy in the 19th century to liquefy gases or vapors. Of course, the phenomenon was known from water, so it was clear that cooling and an increase of pressure will promote condensation. The latter was particularly plausible to the scientists: When the molecules of a vapor are compressed into a small volume, the vapor should surely become a liquid, or so they thought.

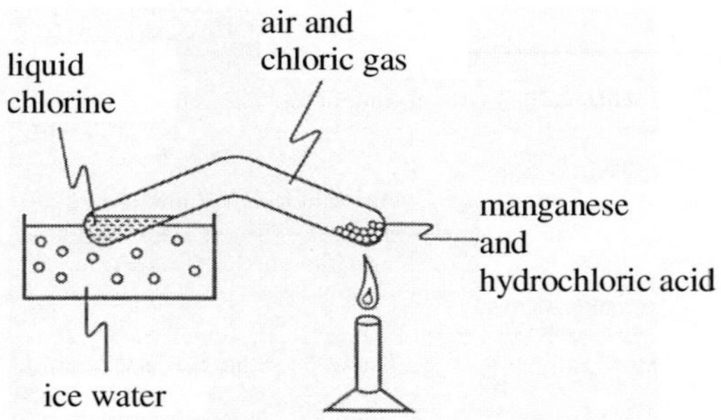

Fig. 2.13 MICHAEL FARADAY
FARADAY's liquefaction of chlorine

The eminent physicist Michael FARADAY (1791-1867) cleverly combined cooling and pressurizing in a manner illustrated in Fig. 2.13. He used a glass tube formed like a boomerang and placed some manganese dioxide with a little hydrochloric acid in one end. Gentle heating produced gaseous chlorine. The other end was sealed and placed in ice water. The emerging gas chlorine increased the pressure in the tube, and the cooling produced a puddle of liquid chlorine on the cold side.

The chemist C.S.A. THILORIER (1771-1833) used FARADAY's method to liquefy a sizable amount of CO_2 under high pressure; he used a strong steel boomerang instead of glass. Afterwards he isolated the tube adiabatically and reduced the pressure. At the lower pressure a part of the liquid CO_2 evaporated; because of the isolation the heat of evaporation was taken from the liquid itself, and so some of the remaining liquid turned solid. Thus for the first time there appeared "dry ice," which under 1 bar sublimates at -78.5°C. THILORIER achieved temperatures as low as -110°C by mixing highly volatile liquid diethyl ether with solid CO_2. The ether evaporates and cools the solid CO_2.

FARADAY adopted THILORIER's method and used the cold mixture of CO_2 and ether – instead of his former ice water – to liquefy all kinds of gases. However, he did not succeed with

hydrogen, oxygen, nitrogen, CO , NO , and methane,

although he employed very high pressure. These gases he called "permanent" and he could have added helium, neon, and argon to the list, had he known these noble gases.

Of course, we know why FARADAY could not succeed. His cooling temperature of $-110°C$ lies above most of the critical temperatures for the "permanent gases" which are, respectively

$$-240°C\ ,\ -118°C\ ,\ -147°C\ ,\ -140°C\ ,\ -93°C\ ,\ -82°C\ .$$

After all, water vapor of $400°C$ cannot be liquefied either by pressure, since the $400°C$ isotherm does not pass the wet region, see Fig. 2.7 b.

But then, FARADAY knew nothing about the critical temperature when he made his experiments. This concept was introduced by the Irish chemist Thomas ANDREWS (1813-1885). ANDREWS worked with CO_2 which can be liquefied by pressure, even at room temperature; its critical temperature lies at 31°C. He suspected correctly that all gases have a critical temperature. This suggestion led to increased efforts in the attempts to reach lower temperatures and eventually all of FARADAY'S permanent gases were liquefied, the last one, hydrogen, in the year 1898.

3 Reversible processes and cycles. "*p* d*V* thermodynamics" for the calculation of thermodynamic engines

3.1 Work and heat for reversible processes

For a heuristically important and qualitatively correct treatment of thermodynamic processes one usually ignores shear stresses, heat conduction, and temperature- and pressure gradients. We have discussed the working $\dot{W}$ of such idealized – reversible – processes in Paragraph 1.5.6. The stress work –or internal work* – done in the time $\mathrm{d}t$ is given by

$$\dot{W}_{\text{stress}}\mathrm{d}t = -p\,\mathrm{d}V\ . \tag{3.1}$$

A reversible process is characterized by a pair of time-dependent functions

$$\text{either } T(t),\ V(t)\ ,\ \text{or}\ \ p(t),\ V(t)\ ,\ \text{or}\ \ T(t),\ p(t), \tag{3.2}$$

and it may therefore be represented by a curve in a (T,V)-, or (p,V)-, or (T,p)-diagram, *cf.* Fig. 3.1. If the curve begins at time t_{B} in the point B with p_{B}, V_{B} (say) and ends at time t_{E} in the point E with p_{E}, V_{E}, the transmitted work is obviously

$$W_{\mathrm{BE}} = \int_{t_{\mathrm{B}}}^{t_{\mathrm{E}}} \dot{W}\mathrm{d}t = -\int_{t_{\mathrm{B}}}^{t_{\mathrm{E}}} p\frac{\mathrm{d}V}{\mathrm{d}t}\mathrm{d}t = -\int_{\mathrm{B}}^{\mathrm{E}} p\,\mathrm{d}V\ , \tag{3.3}$$

where the integral is a line integral along the representative curve $p(t)$, $V(t)$ in the (p,V)-diagram. We therefore conclude that the work is graphically represented by the area below the process curve in a (p,V)-diagram as indicated in Fig. 3.1. There is no such easy visualization of the work in the (T,V)- or in the (T,p)-diagram.

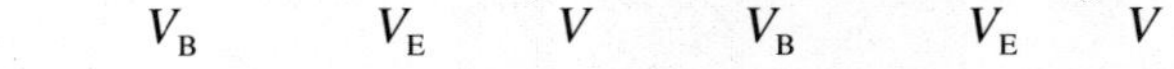

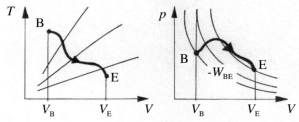

Fig. 3.1 Reversible process in a (T,V)- and a (p,V)-diagram

---

* Both are equal in reversible processes, *cf.* Paragraph 1.5.6.

The transmitted heat cannot be graphically interpreted – at least not in a (p,V,T)-diagram.* But the heat may also be represented by a line integral at least for a closed system of mass m, where we have

$$\dot{Q}\,\mathrm{d}t = \mathrm{d}U + p\,\mathrm{d}V \qquad \text{or} \qquad \dot{Q}\,\mathrm{d}t = \mathrm{d}H - V\,\mathrm{d}p \tag{3.4}$$

$$\dot{Q}\,\mathrm{d}t = \left(\frac{\partial U}{\partial T}\right)_V \mathrm{d}T + \left(\left(\frac{\partial U}{\partial V}\right)_T + p\right)\mathrm{d}V \quad \text{or} \quad \dot{Q}\,\mathrm{d}t = \left(\frac{\partial H}{\partial T}\right)_p \mathrm{d}T + \left(\left(\frac{\partial H}{\partial p}\right)_T - V\right)\mathrm{d}p\,. \tag{3.5}$$

Thus by integration from t_B to t_E we obtain

$$Q_{\mathrm{BE}} = m\left[\int_B^E c_\upsilon \mathrm{d}T + \int_B^E\left(\left(\frac{\partial u}{\partial \upsilon}\right)_\upsilon + p\right)\mathrm{d}\upsilon\right] \quad \text{or} \quad Q_{\mathrm{BE}} = m\left[\int_B^E c_p \mathrm{d}T + \int_B^E\left(\left(\frac{\partial h}{\partial p}\right)_T - \upsilon\right)\mathrm{d}p\right], \tag{3.6}$$

where c_υ and c_p are the specific heats, *cf.* Paragraph 2.3.3.

3.2 Compressor and pneumatic machine. The hot air engine

3.2.1 *Work needed for the operation of a compressor*

The compressor process consists of four steps, *cf.* Fig. 3.2. For simplicity we assume that the pressure behind the moving piston is equal to zero; this may be done without essential loss of generality.

1-2: By the backwards movement of the piston air is drawn into the cylinder by suction through the open entrance valve. We ignore the slight depression of the air caused by the retreating piston and assume that this step occurs at the constant pressure p_{L} of the entrance duct.

2-3: Compression of the air by the forward motion of the piston, while both valves are closed.

3-4: When the pressure reaches the desired value p_{H}, the exit valve opens and the compressed air is pushed into the high-pressure duct at constant pressure.

4-1: When the piston changes direction, the exit valve closes and the entrance valve opens so that the pressure drops from p_{H} to p_{L}, ideally at zero volume.

The pressure during these steps is shown as a function of *V* on the right hand side of Fig. 3.2. Note that along the branches 1-2 and 3-4 the cylinder represents an open system, since the mass changes in time.

According to (3.3) the work done to the piston reads:

$$W_{12} = -p_{\mathrm{L}}V_2\,, \quad W_{23} = -\int_2^3 p\,\mathrm{d}V\,, \quad W_{34} = +p_{\mathrm{H}}V_3\,, \quad W_{41} = 0\,. \tag{3.7}$$

* Later we shall learn about a (temperature, entropy)-diagram in which the heat is represented by the area below the process curve.

Thus obviously the total work is equal to the area within the process curve. The work done along the branch 2-3 depends on the shape of that branch. For the extreme cases of adiabatic and isothermal compression one obtains

$$\begin{array}{ll} \text{adiabatic with} & p_L V_2^{\kappa} = pV^{\kappa} \\ \text{isothermal with} & p_L V_2 = pV \end{array} \quad W_{23} = \begin{cases} \frac{1}{\kappa-1} m \frac{R}{M} T_2 \left[\left(\frac{p_H}{p_L} \right)^{\frac{\kappa-1}{\kappa}} - 1 \right] \\ m \frac{R}{M} T_2 \ln \frac{p_H}{p_L} \end{cases} \tag{3.8}$$

Thus the total work for the two cases comes out as

$$\begin{aligned} W^{\text{adiabatic}} &= \frac{\kappa}{\kappa-1} m \frac{R}{M} T_2 \left[\left(\frac{p_H}{p_L} \right)^{\frac{\kappa-1}{\kappa}} - 1 \right] \\ W^{\text{isothermal}} &= m \frac{R}{M} T_2 \ln \frac{p_H}{p_L} . \end{aligned} \tag{3.9}$$

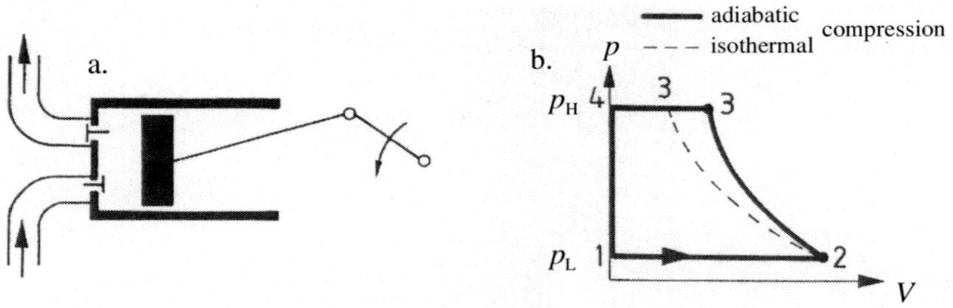

Fig. 3.2 a. Schematic view of a compressor
b. Process curve in a (p,V)-diagram

In the adiabatic case the work required by the compressor is larger than in the isothermal one. This fact is most convincingly confirmed by Fig. 3.2b: Indeed the area on the left of the dashed isothermal line is smaller than the one on the left of the adiabate, since adiabates are steeper than isotherms, *cf.* Paragraph 1.5.8. Actual compressors run so quickly – at several hundred revolutions per minute – that an effective cooling of the air during the compression phase is impossible. One can cool the *wall* of the cylinder. However, this does not produce much of a cooling effect in the interior. Therefore the adiabate of Fig. 3.2 represents the realistic version of the compression although it is undesirable.

It is noteworthy, perhaps, that – by (3.9) – the work required by a compressor depends only on the *ratio* of the pressures p_H and p_L, and *not* on their difference. As a result the compression of air from 1 bar to 10 bar requires just as much work as the compression from 10 bar to 100 bar, although the practical realization of the latter case is more demanding.

3.2.2 *Two-stage compressor*

Now, if we cannot perform an isothermal compression – because of the high number of revolutions – it is still possible to approximate isothermal conditions by building two-stage compressors, *i.e.* two compressors in series. The first one produces an intermediate pressure ratio p_M / p_L adiabatically, and pushes the air at p_M into a cooling duct, where it is cooled back to its initial temperature. The air is then fed into a second compressor, where it reaches the desired pressure p_H, again adiabatically.

Fig. 3.3 shows a (p,V)-diagram of the two-stage process. In this case the work required is, by $(3.9)_1$

$$W = \frac{\kappa}{\kappa-1} m \frac{R}{M} T \left[\left(\frac{p_M}{p_L} \right)^{\frac{\kappa-1}{\kappa}} + \left(\frac{p_H}{p_M} \right)^{\frac{\kappa-1}{\kappa}} - 2 \right]. \tag{3.10}$$

It is smaller than the one-stage process by the shaded area so that there is some profit. This profit is maximal when W as a function of p_M has a maximum and this occurs for

$$p_M = \sqrt{p_H p_L} \; . \tag{3.11}$$

as one can easily check. In this case both compressors require the same work.

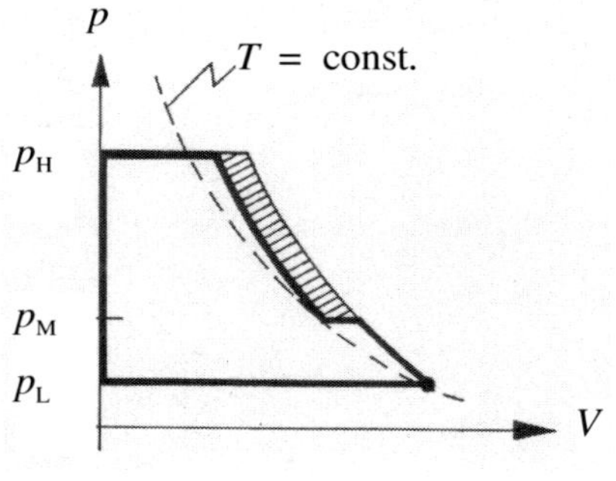

Fig. 3.3 Two-stage compressor

With more than two stages the effective compression curve is a zig-zag line that may approach the desired isotherm in the limit of very many stages. We know of compressors with up to ten stages, and often the cooling ducts are led through the groundwater.

The foregoing discussion about the two-stage compressor is a good example for the heuristic conclusions that may be derived from the consideration of reversible processes.

3.2.3 *Pneumatic machine*

Air of high pressure has many uses. High pressure chisels and high pressure hammers are common applications, *cf.* Fig. 3.4. The process in such a machine is essentially the reverse of the compressor process and in a (p,V)-diagram there are again four branches. Referring to the figure we may characterize these branches as follows

1-2: High pressure air is drawn into a cylinder.
2-3: Air is decompressed at closed valves.
3-4: Decompressed air is pushed out of the cylinder.
4-1: Change of pressure by closing exit valve and opening high pressure valve.

Because of the speed of the motion of the piston the decompression occurs adiabatically. And, once again, it would be better, if the process ran isothermally, because more work could be gained. In Fig. 3.4 that additional work is represented by the shaded triangular strip.

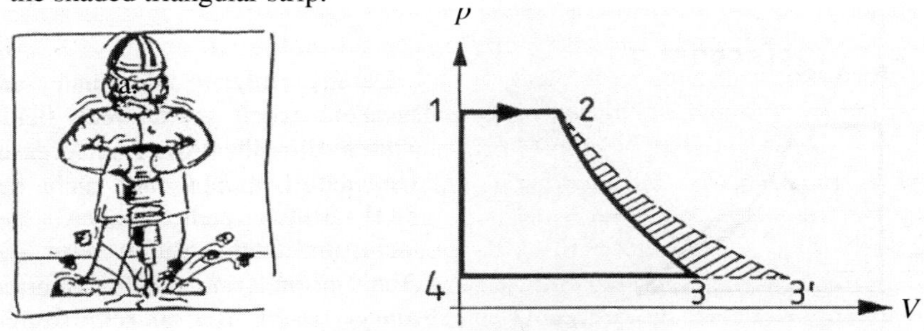

Fig. 3.4 Pneumatic machine

3.2.4 *Hot air engine*

We may consider a situation in which the work of a pneumatic engine is used to drive the compressor which furnishes the compressed air. If both machines ran truly reversibly, such a combination would be possible, albeit as a useless toy that could drive only itself. However, with a small alteration such a coupling of a compressor and a pneumatic machine becomes a useful *heat engine*: The compressed air furnished by the compressor is heated in a heat exchanger. When this is done, the volume increases – at constant pressure – and the adiabatic branch of the process curve of the pneumatic machine is shifted to the right, *cf.* Fig. 3.5. For one revolution of the crank, the work gained is equal to the difference of the areas inside the two process curves.

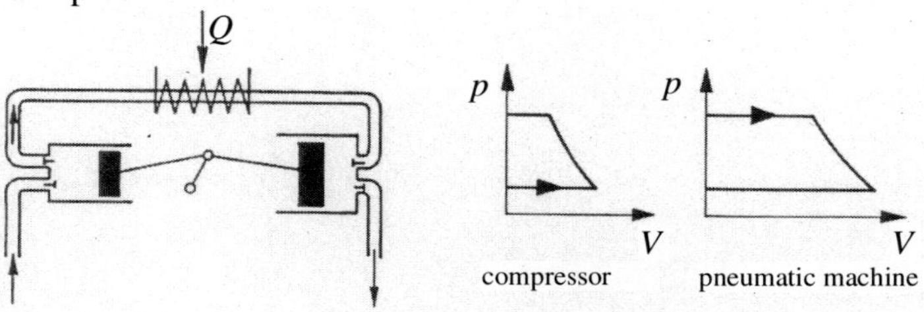

Fig. 3.5 Coupling of a compressor and of a pneumatic machine

This coupled process becomes the prototypical *cycle* of a heat engine when the air exiting the pneumatic machine is cooled and fed back into the compressor, *cf.* Fig. 3.6. In this manner we may construct a hot air engine, the prototype of all heat engines. The cycle consists of two isobars and two adiabates and it is known as the Joule process.

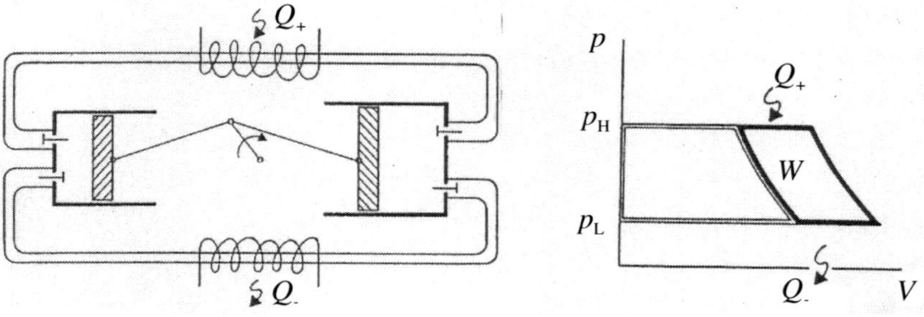

Fig. 3.6 Hot air engine
Q_+ and Q_- denote the heats exchanged in a heater and in a cooler, respectively. The area W represents the work gained.

3.3 Work and heat for reversible processes in ideal gases. "Iso-processes" and adiabatic processes

Many thermodynamic engines perform cycles whose individual branches consist of isotherms, or isochors, or isobars, or which are adiabatic – at least approximately. It is therefore helpful for later reference to make a list of the expressions for work and heat in such "iso-processes." The entries in the list are special cases of the expressions (3.3) and (3.6) for ideal gases, the only case for which we have analytic equations of state.

In an isothermal process the specific work reads

$$w_{\mathrm{BE}} = -\int_{\mathrm{B}}^{\mathrm{E}} p\,\mathrm{d}v \quad \Rightarrow \quad \text{by } p = \frac{1}{v}\frac{R}{M}T: \quad w_{\mathrm{BE}} = -\frac{R}{M}T\ln\frac{v_{\mathrm{E}}}{v_{\mathrm{B}}}.$$

The specific value q_{BE} of the heat is equal to $-w_{BE}$ by $(3.6)_1$, since u is independent of v in an ideal gas.

In an isobaric process we have

$$w_{\mathrm{BE}} = -\int_{\mathrm{B}}^{\mathrm{E}} p\,\mathrm{d}v = -p(v_{\mathrm{E}} - v_{\mathrm{B}}) = -\frac{R}{M}(T_{\mathrm{E}} - T_{\mathrm{B}}),$$

and the heat results from $(3.6)_2$ as

$$q_{\mathrm{BE}} = -\int_{\mathrm{B}}^{\mathrm{E}} c_p\,\mathrm{d}T \quad \Rightarrow \quad \text{by } c_p = \frac{\kappa}{\kappa-1}\frac{R}{M}: \quad q_{\mathrm{BE}} = \frac{\kappa}{\kappa-1}\frac{R}{M}(T_{\mathrm{E}} - T_{\mathrm{B}});$$

as always $z = \frac{1}{\kappa-1}$ holds, where κ is the ratio of specific heats.

In an isochoric process the work w_{BE} is obviously zero, while the heat, by $(3.6)_1$, comes out as

$$q_{\mathrm{BE}} = -\int_{\mathrm{B}}^{\mathrm{E}} c_v \,\mathrm{d}T \quad \Rightarrow \quad \text{by } c_v = \frac{1}{\kappa-1}\frac{R}{M}: \; q_{\mathrm{BE}} = \frac{1}{\kappa-1}\frac{R}{M}\left(T_{\mathrm{E}} - T_{\mathrm{B}}\right).$$

Finally we obviously have $q_{\mathrm{BE}} = 0$ in an adiabatic process and therefore the work is given by

$$w_{\mathrm{BE}} = -\int_{\mathrm{B}}^{\mathrm{E}} p \,\mathrm{d}v = u(T_{\mathrm{E}}) - u(T_{\mathrm{B}}) \text{ hence } w_{\mathrm{BE}} = \frac{1}{\kappa-1}\frac{R}{M}\left(T_{\mathrm{E}} - T_{\mathrm{B}}\right).$$

All these results are summarized in Table 3.1.

Table 3.1 Work and heat for ideal gases in special processes

	isothermal	isobaric	isochoric	adiabatic
work	$-\frac{R}{M}T\ln\frac{v_{\mathrm{E}}}{v_{\mathrm{B}}}$	$-\frac{R}{M}\left(T_{\mathrm{E}} - T_{\mathrm{B}}\right)$	0	$\frac{1}{\kappa-1}\frac{R}{M}\left(T_{\mathrm{E}} - T_{\mathrm{B}}\right)$
heat	$\frac{R}{M}T\ln\frac{v_{\mathrm{E}}}{v_{\mathrm{B}}}$	$\frac{\kappa}{\kappa-1}\frac{R}{M}\left(T_{\mathrm{E}} - T_{\mathrm{B}}\right)$	$\frac{1}{\kappa-1}\frac{R}{M}\left(T_{\mathrm{E}} - T_{\mathrm{B}}\right)$	0

3.4 Cycles

3.4.1 *Efficiency in the conversion of heat to work*

In a cycle the state of a fixed mass of the working agent changes periodically. In the hot air engine, *cf.* Paragraph 3.2.4, this happened to air and in the steam engine it happens to water. It is then appropriate to apply all thermodynamic calculations to a fixed mass and to write the First Law for closed systems in its specific form as

$$\dot{q}\,\mathrm{d}t = \mathrm{d}u - \dot{w}\,\mathrm{d}t\,. \tag{3.12}$$

Integration over a period – or a complete cycle – gives $\oint \mathrm{d}u = 0$, since the states at the beginning and at the end are equal. Thus (3.12) provides

$$q_{\circ} = -w_{\circ} \quad \text{read: } q_{\text{cycle}} = -w_{\text{cycle}}\,. \tag{3.13}$$

In a heat engine $w_{\circ}$ is negative, since work is gained and, consequently, $q_{\circ}$ is positive. Experience has shown that the total heat $q_{\circ}$ contains positive and negative parts, *i.e.* we have

$$q_{\circ} = q_{+} + q_{-} = q_{+} - \left|q_{-}\right|. \tag{3.14}$$

Therefore we may express (3.13) by saying that the difference between the provided heat and the heat withdrawn is equal to the work.

It is only the provided heat that needs to be paid for in the operation of the heat engine and, of course, it is the work that can be sold. Therefore it makes sense to define the efficiency e of the engine as the ratio of $\left|w_{\circ}\right|$ and q_{+}

$$e = \frac{|w_\circ|}{q_+} = 1 - \frac{|q_-|}{q_+}. \tag{3.15}$$

e would be equal to one, if q_- were zero. We shall later see that this is impossible; it contradicts the Second Law, see Chap. 4.

In a *reversible* cycle we have $\dot{w}\,\mathrm{d}t = -p\,\mathrm{d}\upsilon$ and then the cycle may be represented as a closed loop in a (T,υ)-diagram or a (p,υ)-diagram. In the (p,υ)-diagram the area enclosed by the loop represents the work $w_\circ$ of the reversible engine, because the area is equal to $\oint p\,\mathrm{d}\upsilon = 0$, *cf.* Fig. 3.7. In a heat engine the states of the working agent move along the closed curve clockwise in the (p,υ)-diagram. And this is usually also the case in a (T,υ)-diagram although there are exceptions: Indeed, if $\left.\frac{\partial p}{\partial T}\right|_\upsilon < 0$ holds the states of a heat engine move counter-clockwise along the (T,υ)-curve. This is the case, for instance, for water below 4°C where water behaves “anomalously,” see above, Paragraph 2.4.4.

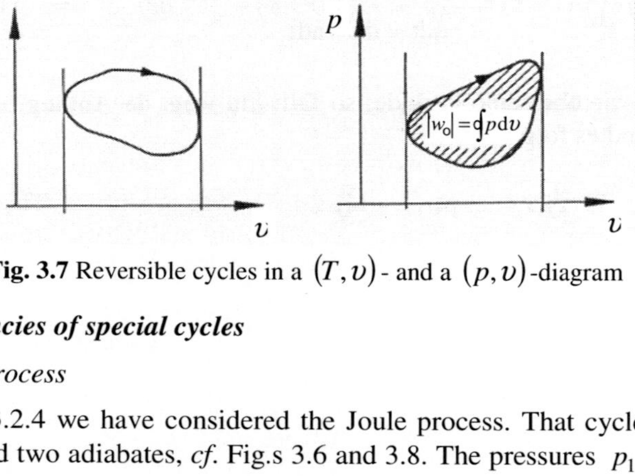

Fig. 3.7 Reversible cycles in a (T,υ)- and a (p,υ)-diagram

3.4.2 *Efficiencies of special cycles*

- *Joule process*

In Paragraph 3.2.4 we have considered the Joule process. That cycle consists of two isobars and two adiabates, *cf.* Fig.s 3.6 and 3.8. The pressures p_1 and p_3 are prescribed.

If the working agent is an ideal gas, the heats and works on the individual branches follow from Table 3.1

$$q_{12} = \frac{\kappa}{\kappa-1}\frac{R}{M}(T_2 - T_1) > 0 \qquad w_{12} = -\frac{R}{M}(T_2 - T_1)$$

$$q_{23} = 0 \qquad w_{23} = \frac{1}{\kappa-1}\frac{R}{M}(T_3 - T_2)$$

$$q_{34} = \frac{\kappa}{\kappa-1}\frac{R}{M}(T_4 - T_3) < 0 \qquad w_{34} = -\frac{R}{M}(T_4 - T_3)$$

$$q_{41} = 0 \qquad w_{41} = \frac{1}{\kappa-1}\frac{R}{M}(T_1 - T_4).$$

$w_\circ$ results from summing up all works

$$w_{\circ} = \frac{\kappa}{\kappa - 1} \frac{R}{M} \left[(T_3 - T_4) - (T_2 - T_1) \right]$$

and, since q_{12} is the only positive contribution to $q_{\circ}$, we obtain for the efficiency

$$e = 1 - \frac{T_3 - T_4}{T_2 - T_1} = 1 - \frac{T_3}{T_2} \frac{1 - \frac{T_4}{T_3}}{1 - \frac{T_1}{T_2}}.$$

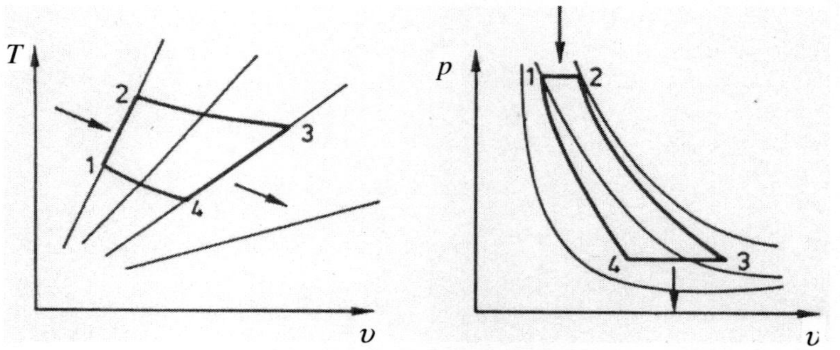

Fig. 3.8 Reversible cycle of the hot air engine

The points 2 and 3 and the points 1 and 4 are connected by the adiabatic equations of state (1.62) and the points 1 and 2, and 3 and 4 lie on isobars. Therefore we have

$$\frac{T_3}{T_2} = \left(\frac{p_3}{p_1} \right)^{\frac{\kappa - 1}{\kappa}} \quad \text{and} \quad \frac{T_4}{T_1} = \left(\frac{p_3}{p_1} \right)^{\frac{\kappa - 1}{\kappa}} \quad \Rightarrow \quad \frac{T_4}{T_3} = \frac{T_1}{T_2},$$

and the efficiency may be written as

$$e = 1 - \frac{T_3}{T_2}.$$

This is a correct expression; however, it is not useful, because the corner temperatures T_2 and T_3 are unknown to begin with. What is prescribed, though, are the pressures p_1 and p_3. In terms of these the efficiency reads

$$e = 1 - \left(\frac{p_3}{p_1} \right)^{\frac{\kappa - 1}{\kappa}}; \tag{3.16}$$

it depends on the ratio of the pressures and it is different for different gases, if their κ-values are different. For air with $\kappa = 1.4$ we obtain $e = 0.48$ for a pressure ratio of 10:1.

- *Carnot cycle*

The Carnot engine exchanges heat at two prescribed temperatures only, *i.e.* its cycle consists of two isotherms and two adiabates, *cf.* Fig. 3.9.

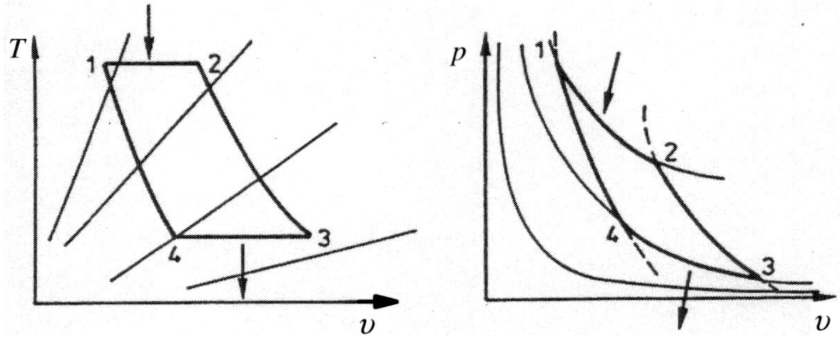

Fig. 3.9 Reversible cycle in a Carnot engine

For an ideal gas heats and works of the branches are given by, *cf.* Table 3.1

$$q_{12} = \frac{R}{M} T_1 \ln \frac{v_2}{v_1} > 0 \qquad\qquad w_{12} = -\frac{R}{M} T_1 \ln \frac{v_2}{v_1}$$

$$q_{23} = 0 \qquad\qquad w_{23} = \frac{1}{\kappa - 1} \frac{R}{M} (T_3 - T_1)$$

$$q_{34} = \frac{R}{M} T_2 \ln \frac{v_4}{v_3} < 0 \qquad\qquad w_{34} = -\frac{R}{M} T_2 \ln \frac{v_4}{v_3}$$

The total work is therefore

$$w_{\circ} - \frac{R}{M} T_1 \ln \frac{v_2}{v_1} - \frac{R}{M} T_3 \ln \frac{v_4}{v_3}$$

and the efficiency reads

$$e = 1 + \frac{T_3}{T_1} \frac{\ln \frac{v_4}{v_3}}{\ln \frac{v_2}{v_1}} \quad \text{or, by} \quad \frac{v_4}{v_3} = \left(\frac{T_1}{T_3} \right)^{\frac{1}{\kappa - 1}} \quad \text{and} \quad \frac{v_3}{v_2} = \left(\frac{T_1}{T_3} \right)^{\frac{1}{\kappa - 1}}$$

$$e = 1 - \frac{T_3}{T_1} . \tag{3.17}$$

The Carnot process and the Carnot efficiency play an important role in the formulation of the Second Law of thermodynamics. Once we have formulated this law we shall see that the Carnot process has a universal efficiency – independent of the working agent – which is maximal among the efficiencies of all cycles with the maximal temperature T_1 and a minimal temperature T_3. Actually, this important result for the Carnot process is forecast by the present result. We can state at this time that the efficiency depends on the prescribed temperature ratio and is independent of the gas. For a given cooling temperature T_3 the Carnot efficiency grows with increasing heating temperature T_1.

- *Modified Carnot cycle*

The heating in reversible processes is always a conceptually difficult phenomenon, because – strictly speaking – it has to occur between a heat reservoir and an engine (say) of the same temperature. Customary explanations about how this can be realized involve an infinitesimal temperature difference between the reservoir and the engine and infinitely long times for the heating to occur. That is neither practical nor conceptually very satisfactory.

Therefore one may conceive of the heating and cooling phases in a cycle as a realistic problem of heat conduction through auxiliary heat conductors. Thus for instance in a reversible Carnot cycle between T_3 and T_1, *cf.* Fig 3.9, – with the efficiency

$$e = \frac{|w_\circ|}{q_+} = \frac{q_+ + q_-}{q_+} = 1 - \frac{|q_-|}{q_+} = 1 - \frac{T_3}{T_1}, \tag{3.18}$$

according to (3.17) – the heating may be realized by a conductor with temperatures τ_1 on the hot side and T_1 on the cold side. And the cooling may be realized by a conductor with T_3 on its hot side and τ_3 on the cold side. This means that we have

$$q_+ = C_+(\tau_1 - T_1) \text{ and } q_- = C_-(T_3 - \tau_3), \tag{3.19}$$

where $C_\pm$ are constants depending on the properties of the conductors, *e.g.* the thermal conductivities and the widths of the conductors.

In that case τ_1 and τ_3 are the highest and the lowest temperatures of the engine. Let those be prescribed and let us ask whether the efficiency can be expressed in terms of those temperatures and, if so, what its value is.

Since the efficiency is always equal to

$$e = 1 - \frac{T_3}{T_1} \text{ with } \frac{|q_-|}{q_+} = \frac{T_3}{T_1}, \tag{3.20}$$

we need to express T_1 and T_3 in terms of τ_1 and τ_3. Obviously, by (3.19) and (3.20) we have

$$\frac{T_3}{T_1} = \frac{C_-}{C_+}\frac{T_3 - \tau_3}{\tau_1 - T_1}, \text{ hence } T_3 = \frac{C_- T_1 \tau_3}{(C_+ + C_-)T_1 - C_+\tau_1}. \tag{3.21}$$

The efficiency is therefore equal to

$$e = 1 - \frac{C_- \tau_3}{(C_+ + C_-)T_1 - C_+\tau_1} \tag{3.22}$$

and that is all we can say unless we add another equation in order to determine T_1 as a function of τ_1, τ_3.

We let that additional equation come from the requirement that T_1 be such as to make $|w_\circ|$ maximal. We have

$$\begin{aligned} |w_\circ| &= q_+ - |q_-| && \text{and by (3.19)} \\ &= C_+(\tau_1 - T_1) - C_-(T_3 - \tau_3) && \text{and by } (3.21)_2 \\ &= C_+(\tau_1 - T_1) - C_-\left[\frac{C_- T_1 \tau_3}{(C_+ + C_-)T_1 - C_+\tau_1} - \tau_3\right] \end{aligned}$$

which is maximal as a function of T_1 for

$$T_1 = \frac{1}{C_+ + C_-}\left(C_+\tau_1 \pm C_-\sqrt{\tau_1\tau_3}\right). \tag{3.23}$$

This is the required function $T_1 = T_1(\tau_1, \tau_3)$.

Insertion of (3.23) into $(3.21)_2$ gives

$$T_3 = \frac{1}{C_+ + C_-}\left(C_-\tau_3 \pm C_-\sqrt{\tau_1\tau_3}\right) \tag{3.24}$$

and hence follows the efficiency by insertion of (3.23) and (3.24) into $(3.18)_4$ and by some rearrangement

$$e = 1 - \sqrt{\frac{\tau_3}{\tau_1}}. \tag{3.25}$$

The lower sign in (3.23) and (3.24) is irrelevant, since T_1 and T_3 must both be positive.

This is a neat result: The efficiency of the modified Carnot cycle – with realistic heating and cooling – is still given by the ratio of the smallest and highest temperatures. However, it is the *square root* of that ratio that enters the formula. Thus, in a manner of speaking, the efficiency has been decreased by the employment of heat conduction on the isothermal branches of the Carnot process.

- *Ericson cycle*

In the Ericson engine the working agent is carried through a cycle consisting of two isotherms and two isobars, *cf.* Fig. 3.10.

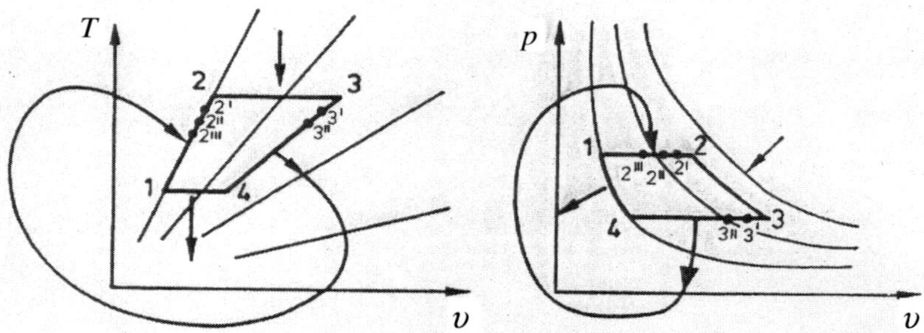

Fig. 3.10 Reversible cycle of an Ericson engine

Works and heats follow from Table 3.1

$$q_{12} = \frac{\kappa}{\kappa-1}\frac{R}{M}(T_2 - T_1) > 0 \qquad w_{12} = -\frac{R}{M}(T_2 - T_1)$$

$$q_{23} = \frac{R}{M}T_2 \ln\frac{\upsilon_3}{\upsilon_2} > 0 \qquad w_{23} = -\frac{R}{M}T_2 \ln\frac{\upsilon_3}{\upsilon_2}$$

$$q_{34} = \frac{\kappa}{\kappa-1}\frac{R}{M}(T_1 - T_2) < 0 \qquad w_{34} = -\frac{R}{M}(T_1 - T_2)$$

$$q_{41} = \frac{R}{M}T_1 \ln\frac{\upsilon_1}{\upsilon_4} < 0 \qquad w_{41} = -\frac{R}{M}T_1 \ln\frac{\upsilon_1}{\upsilon_4}.$$

The works w_{12} and w_{34} compensate each other and we have $\frac{\upsilon_1}{\upsilon_2} = \frac{\upsilon_4}{\upsilon_3}$, because the points 2 and 3, and 1 and 4 lie on isotherms and because 1 and 2, and 3 and 4 lie on isobars. Therefore the total work is given by

$$w_\circ = \frac{R}{M}(T_2 - T_1)\ln\frac{\upsilon_1}{\upsilon_4}.$$

The heats provided to the process are q_{12} and q_{23}. However, a clever process management will take the heat q_{12} from the process itself, to wit from the heat q_{34} which is equal and opposite to q_{12}, *cf.* Figs. 3.10 and 3.11. If this is done, the only expenditure concerns q_{23} and therefore the efficiency reads

$$e = \frac{|w_\circ|}{q_{23}} = 1 - \frac{T_1}{T_2}, \tag{3.26}$$

so that it is equal to the Carnot efficiency. It is true that in the Ericson process – in contrast to the Carnot process – heats are exchanged on the non-isothermal branches, however, these heats are transferred *within the process*. Such an internal transfer of heat is often called *regeneration*.

Two remarks are appropriate in this context: First of all, an isothermal process is technically unrealistic, because the effective cooling of a fast running engine is impossible. We have discussed this difficulty in Paragraphs 3.2.1 and 3.2.2 and it applies to the Carnot and to the Ericson processes as well. Both represent essentially theoretical possibilities for a conversion of heat into work.

Second, it is all very well to say that the heat q_{34}, which was withdrawn from the process, is supplied to the branch 1-2. Such a shift of heats, however, can only occur from the warmer to a colder body. If the rearrangement of heats is to be realized, it must occur in very small steps: The portion that is released between 3 and 3′ – *cf.* Fig. 3.10 – must be supplied to the process between 2′ and 2″; and what is released between 3′ and 3″ must be supplied from 2″ and 2‴, *etc.* One way to realize this transfer – at least approximately – is the application of a heat exchanger with flows in opposite directions as indicated in Fig. 3.11.

Compression and decompression are conducted isothermally so that they require cooling and heating, respectively.

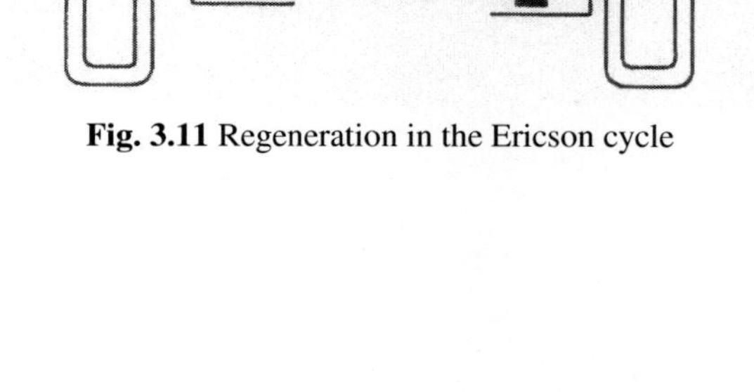

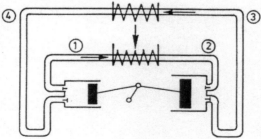

Fig. 3.11 Regeneration in the Ericson cycle

- *Stirling cycle*

In the Stirling engine the cycle consists of two isotherms and two isochors, *cf.* Fig. 3.12.

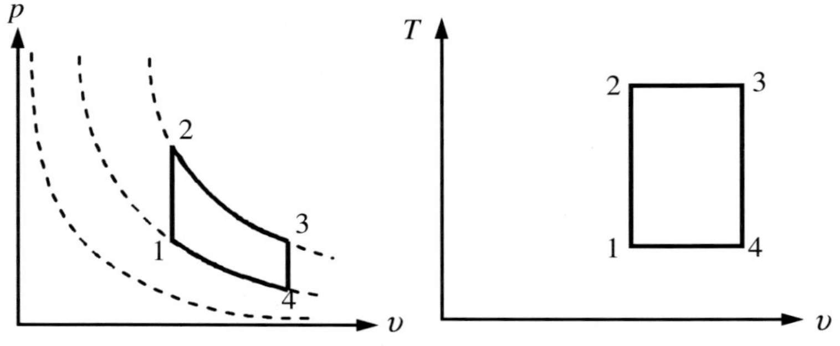

Fig. 3.12 Reversible Stirling cycle

The works and heats on the individual branches are

$$q_{12} = \frac{1}{\kappa-1}\frac{R}{M}(T_2 - T_1) > 0 \qquad w_{12} = 0$$

$$q_{23} = \frac{R}{M}T_2 \ln\frac{\upsilon_3}{\upsilon_2} > 0 \qquad w_{23} = -\frac{R}{M}T_2 \ln\frac{\upsilon_3}{\upsilon_2}$$

$$q_{34} = \frac{1}{\kappa-1}\frac{R}{M}(T_1 - T_2) < 0 \qquad w_{34} = 0$$

$$q_{41} = \frac{R}{M}T_1 \ln\frac{\upsilon_1}{\upsilon_4} < 0 \qquad w_{41} = -\frac{R}{M}T_1 \ln\frac{\upsilon_1}{\upsilon_4}.$$

We face much of the same situation as for the Ericson cycle with the possibility of regeneration: The heating during step 1-2 may be taken from the cooling during step 3-4. And obviously $\upsilon_2 = \upsilon_1$ and $\upsilon_3 = \upsilon_4$ holds so that the efficiency reads

$$e = 1 - \frac{T_1}{T_2}, \tag{3.27}$$

just like for the Carnot process. With an efficient regeneration the Stirling engine run with an ideal gas may thus achieve maximal efficiency.

3.5 Internal combustion cycles

3.5.1 *Otto cycle*

We know that the number of revolutions per minute of the crankshaft of internal combustion engines may reach many thousands, – up to 19,000 –, and that during ignition and during the exhaust process large in-homogeneities occur in the pressure and temperature fields inside the cylinder. And yet we obtain important heuristic results when we treat the process as a reversible one.

Also the working agent is considered as air, although the combustible mixture contains finely dispersed droplets of petrol, and although water vapor and soot particles constitute a good part of the combustion products.

The process in the internal combustion engine is *not* a cycle, because the working agent is regularly exchanged during two out of the four strokes. Indeed, in the Otto engine – or four stroke engine – the following partial processes occur, *cf.* Fig. 3.13 a

0-1: Intake of the combustible mixture (1st stroke).
1-2: Compression of the mixture (2nd stroke).
2-3: Combustion after ignition by a spark plug.
3-4: Expansion (3rd stroke, working stroke).
4-1′: Exhaust after opening of exit valve.
1′-0′: Push-out of the combustion products (4th stroke).

Combustion and exhaust occur so quickly that the motion of the piston may be neglected during these processes; therefore they are considered as isochoric. Compression and expansion are approximately adiabatic processes. The intake and the push-out of the mixture occur at only slightly different pressures. Therefore the corresponding lines are nearly on top of each other so that their contributions cancel each other in the work balance.

With all this in mind one arrives at a substitute process as shown in Fig. 3.13 b. The exhaust process 4-1 is replaced by an isochoric cooling and the intake and push-out parts of the process are missing altogether. The heating along the branch 2-3 is, of course, due to the heat of combustion of the fuel. In this manner the Otto process has been replaced by a fictional cycle of two isochors and two adiabates and this is used for thermodynamic calculations, in particular for the calculation of the efficiency.

The works and heats on the branches of the substitute process are taken from Table 3.1

$$q_{12}=0 \qquad w_{12}=\frac{1}{\kappa-1}\frac{R}{M}\left(T_2-T_1\right)$$

$$q_{23}=\frac{1}{\kappa-1}\frac{R}{M}\left(T_3-T_2\right)>0 \qquad w_{23}=0$$

$$q_{34}=0 \qquad w_{34}=\frac{1}{\kappa-1}\frac{R}{M}\left(T_4-T_3\right)$$

$$q_{41}=\frac{1}{\kappa-1}\frac{R}{M}\left(T_1-T_4\right)<0 \qquad w_{41}=0\,.$$

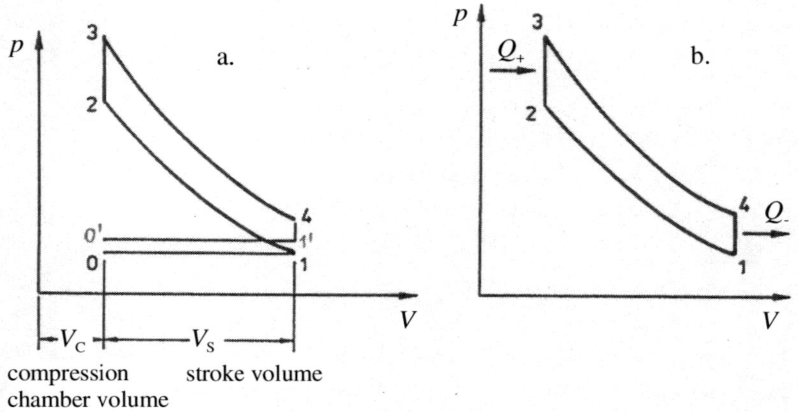

Fig. 3.13 a. 4-stroke process
b. Otto cycle

Therefore the efficiency is given by

$$e = \frac{T_1 - T_2 + T_3 - T_4}{T_3 - T_2} = 1 - \frac{T_4 - T_1}{T_3 - T_2} = 1 - \frac{T_1}{T_2}\frac{1 - \frac{T_4}{T_1}}{1 - \frac{T_3}{T_2}}.$$

Since the points 1 and 2, and the points 3 and 4 are connected by adiabates, we have

$$\frac{T_2}{T_1} = \left(\frac{V_1}{V_2}\right)^{\kappa-1} \text{ and } \frac{T_3}{T_4} = \left(\frac{V_4}{V_3}\right)^{\kappa-1} = \left(\frac{V_1}{V_2}\right)^{\kappa-1} \Rightarrow \frac{T_4}{T_1} = \frac{T_3}{T_2},$$

so that the efficiency comes out as

$$e = 1 - \frac{T_1}{T_2} = 1 - \left(\frac{V_2}{V_1}\right)^{\kappa-1}.$$

Thus the efficiency is determined by the increase of temperature during adiabatic compression. This increase, in turn, is determined by the stroke volume $V_S = V_1 - V_2$ and by the volume $V_C = V_2$ of the compression chamber. If we introduce the compression ratio

$$\varepsilon = \frac{V_1}{V_2} = \frac{V_C + V_S}{V_C},$$

we may write the efficiency in the form

$$e = 1 - \frac{1}{\varepsilon^{\kappa-1}} \tag{3.28}$$

so that it grows with increasing compression ratio.

During the construction of the engine on the drawing board the designer may therefore be tempted to increase ε indefinitely in order to obtain a better efficiency. In practice, however, ε must be smaller than 10, because for a higher compression – and the concomitant higher temperature – the combustible mixture would ignite prematurely, *i.e.* before the upper dead-center is reached. Obviously this would disturb the proper operation of the engine and, in fact, damage the engine.

Still with $\varepsilon = 8$ (say) and $\kappa = 1.4$ we obtain the efficiency $e = 0.56$. This is not a bad value at all. However, unfortunately, the losses in the actual process through irreversibility and the mechanical losses by friction both reduce the available work by the factor 0.7. Therefore the 4-stroke engine makes use of only about 25% of the energy contained in the fuel. It is difficult to *calculate* these losses reliably, but they may be measured, of course.

3.5.2 *Diesel cycle*

Despite the discrepancy between the calculated and the measured values of the efficiency of the Otto cycle, the formula (3.28) has considerable heuristic value. Indeed, it helps the engineer to realize that he should strive for a higher compression ratio, if he wishes to improve performance. Thus, if a premature ignition of the combustible mixture prevents a higher ratio, – as it does in the Otto cycle the engineer may wish to compress pure air, and add the fuel by injection *after* the compression. This has the additional advantage that the air is so hot after the drastic compression that the fuel – when injected – ignites all by itself, *i.e.* without the help of a spark plug. This new process was invented by Rudolf DIESEL (1858-1913), and it is called the Diesel process; it is a variant of the Otto process and also uses four strokes.

The variation is mostly due to the fact that the injection of the fuel takes time and during that time the piston moves backwards and increases the volume of the cylinder. The pressure increase expected from the combustion of the fuel is reduced by the backward motion of the piston and therefore the injection and combustion occur under nearly isobaric conditions. The substitute process is shown in Fig. 3.14. It has an isobaric branch of length V_{I}, the injection volume.

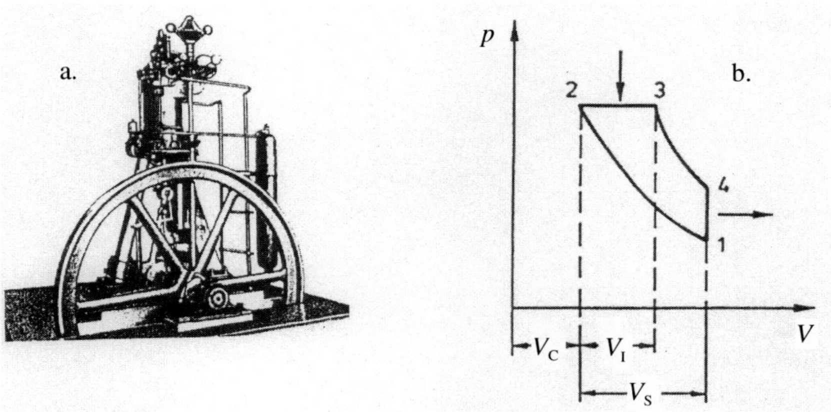

Fig. 3.14 a. First Diesel engine of 1897
b. Substitute cycle for the 4-stroke Diesel process

The heats and works along the different branches may be read off from Table 3.1

$$q_{12}=0 \qquad w_{12}=\frac{1}{\kappa-1}\frac{R}{M}(T_2-T_1)$$

$$q_{23}=\frac{1}{\kappa-1}\frac{R}{M}(T_3-T_2)>0 \qquad w_{23}=\frac{R}{M}(T_2-T_3)$$

$$q_{34}=0 \qquad w_{34}=\frac{1}{\kappa-1}\frac{R}{M}(T_4-T_3)$$

$$q_{41}=\frac{1}{\kappa-1}\frac{R}{M}(T_1-T_4)<0 \qquad w_{41}=0\,.$$

Hence follows

$$q_+=\frac{\kappa}{\kappa-1}\frac{R}{M}(T_3-T_2) \text{ and } w_+=-\frac{\kappa}{\kappa-1}\frac{R}{M}(T_3-T_2)\left(1-\frac{1}{\kappa}\frac{T_1-T_4}{T_2-T_3}\right)$$

and for the efficiency

$$e=1-\frac{1}{\kappa}\frac{T_1-T_4}{T_2-T_3}=1-\frac{1}{\kappa}\frac{T_1}{T_2}\frac{1-\frac{T_4}{T_1}}{1-\frac{T_3}{T_2}}\,.$$

On the individual branches we have

$$\text{1-2 (adiabatic): } T_1V_1^{\kappa-1}=T_2V_2^{\kappa-1}\Rightarrow \frac{T_1}{T_2}=\left(\frac{V_2}{V_1}\right)^{\kappa-1}$$

$$\text{2-3 (isobaric): } \frac{T_2}{V_2}=\frac{T_3}{V_3} \Rightarrow \frac{T_3}{T_2}=\frac{V_3}{V_2}$$

$$\text{3-4 (adiabatic): } p_4V_4^{\kappa}=p_3V_3^{\kappa}\Rightarrow \frac{p_4}{p_3}=\left(\frac{V_3}{V_1}\right)^{\kappa}$$

$$\text{4-1 (isochoric): } \frac{T_1}{p_1}=\frac{T_4}{p_4}\,.$$

Hence follows

$$\frac{T_4}{T_1}=\frac{p_4}{p_1}=\frac{p_3}{p_1}\left(\frac{V_3}{V_1}\right)^{\kappa}=\frac{p_2}{p_1}\left(\frac{V_3}{V_1}\right)^{\kappa}=\left(\frac{V_1}{V_2}\right)^{\kappa}\left(\frac{V_3}{V_1}\right)^{\kappa}=\left(\frac{V_3}{V_2}\right)^{\kappa},$$

and the efficiency becomes

$$e=1-\frac{1}{\kappa}\frac{1}{\left(\frac{V_1}{V_2}\right)^{\kappa-1}}\frac{\left(\frac{V_3}{V_2}\right)^{\kappa}-1}{\frac{V_3}{V_2}-1}\,.$$

We define the compression ratio ε – as before – and in addition the injection ratio φ as

$$\varepsilon=\frac{V_1}{V_2} \text{ and } \varphi=\frac{V_3}{V_2}$$

so that

$$e = 1 - \frac{1}{\kappa} \frac{1}{\varepsilon^{\kappa-1}} \frac{\varphi^{\kappa} - 1}{\varphi - 1} . \tag{3.29}$$

For the reasons described above, the Diesel engine tolerates compression ratios that are two to three times larger than those of the Otto engine. If we choose $\varepsilon = 20$ and $\varphi = 2.5$ along with $\kappa = 1.4$, we obtain an efficiency of 0.63. This must be compared with the value 0.56 for the Otto process and we conclude that the Diesel efficiency is higher. In addition the thermodynamic losses reduce the efficiency only by 15%, – rather than 30% for the Otto process –, while the mechanical losses amount to 30% for both. Thus the Diesel process makes use of about 35% of energy contained in the fuel and that is considerably more than the 25% of the Otto engine.

3.5.3 *On the history of the internal combustion engine*

The steam engine, where the heating occurs outside of the working cylinder, preceded the internal combustion engine by more than 100 years. The difficulty was that the fuel had to be a gas or a volatile liquid, and such agents were not available in quantity before petroleum, *i.e.* mineral oil was discovered and exploited in the second half of the 19th century.

The inventor Jean Joseph Etienne LENOIR (1822-1900) used coal gas, *i.e.* the household gas still used in many homes for cooking purposes. This gas is extracted from coal by partial burning. He built the first internal combustion engine and mounted it on a carriage in 1860 thus producing the first "automobile."* LENOIR's engine was quite wasteful of fuel and this made it impractical; so, although Lenoir laid the foundation for an immensely prosperous industry, and although his contribution was recognized, he died a poor man.

Nikolaus August OTTO (1832-1891), a traveling agent, read about LENOIR's engine and improved it by using the four-stroke cycle described above which, in his honor, is often called the Otto process. Otto received a patent for the process in 1877 and founded a firm in Cologne which sold 35,000 engines within a few years.

Otto's assistant Gottlieb Wilhelm DAIMLER (1834-1900) set up a business himself in 1883 and endeavored to build light-weight engines with a good efficiency. He used a *carburetor* which dispersed gasoline into droplets in air and thus produced a combustible mixture. DAIMLER mounted his engine first on a boat, then on a bicycle, and – finally – on a four-wheel car thus producing the first automobile of modern kind. He founded the Daimler motor company in 1890 which produces the Mercedes cars to this very day. They were named after the daughter of DAIMLER's Austrian agent Emil JELLINEK, who had impressed DAIMLER with her beauty and liveliness. Gasoline in those days was bought at drug stores or in pharmacies.

Eventually the automobile became an overwhelming success, as we all know. This was essentially the merit of Henry FORD (1863-1947). In 1908 he invented the "assembly line" for mass production at cheap prizes. His Model T – the Tin Lizzy – was available at US$ 750 and it started modern life as we now know it. Millions of cars with standardized parts were turned out.

Rudolf DIESEL (1858-1913) studied engineering at the Technical High School in Munich. He passed exams with the best grades in Mechanical Engineering ever achieved. Afterwards he worked in the ice factory of the eminent inventor Karl VON LINDE (1842-1934). DIESEL improved the Otto process by fuel injection – doing away with the carburetor and the spark plugs – and his engine had the additional advantage of using cheaper fuel:

* Well, not quite: There had been steam automobiles before.

kerosene instead of gasoline. The first Diesel engine was built in St. Louis in 1897, financed by a brewer of that city. Between World Wars I and II the Diesel engine largely replaced steam engines on ships and locomotives. For a long time the large size and heavy weight of the engine prevented its use in trucks and passenger cars, but nowadays engineers have succeeded to produce lightweight engines that can be used in cars as well. The fuel is cheaper and the consumption is less. However, taxes make these advantages largely irrelevant to the car owner.

DIESEL was much in demand by Navy engineers as a consultant. He was returning from a consultation of the British Navy when he fell from the Channel ferry and drowned. There were wild rumors that the British secret service had a hand in the accident.

3.6 Gas turbine

3.6.1 *Brayton process*

In the Brayton process three essential steps of a heat engine, namely compression, heating and expansion are combined in a single rotating machinery, the gas turbine. The fourth step, cooling, is absent, because the gas turbine uses fresh air from the environment and releases the exhaust gases into the atmosphere, – just like other internal combustion engines.

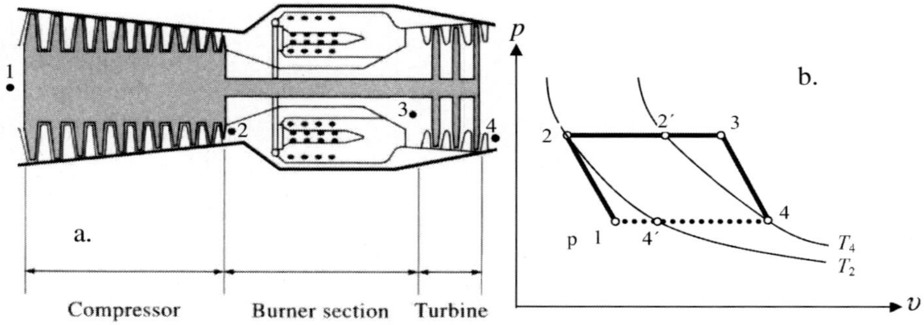

Fig. 3.15 a. Schematic picture of a gas turbine
b. Brayton cycle in a (p, v)-diagram. (The dashed line represents the closure of the substitute cycle)

Fig. 3.15 a shows a schematic picture. The numbers characterize the individual branches as follows.

1-2: adiabatic compression
2-3: isobaric heating by burning the fuel injected into the burner.
3-4: adiabatic expansion in the turbine.

For an efficiency calculation we consider this open process closed into a cycle by replacing the heat loss of the exhaust gases by an isobaric heat exchange that leads back from 4 to 1. The working agent is assumed to be air so that the mass and the properties of the burned fuel are neglected. This is no different from the procedure employed for the internal combustion engines. The resulting substitute process consists of two adiabates and two isobars, *cf.* Fig. 3.15 b, *i.e.* it is a Joule process and has the efficiency (3.16) of the Joule process, *viz.*

$$e = 1 - \frac{T_4}{T_3} = 1 - \left(\frac{p_1}{p_2}\right)^{\frac{\kappa-1}{\kappa}} . \tag{3.30}$$

When a gas turbine is used, the Joule process is known as the Brayton cycle after George BRAYTON, a pioneer of oil burning engines in the 19th century.

There is the possibility for regeneration because the heat withdrawn from the exhaust gases on the branch 4-4´ may be used to preheat the air entering the combustion chamber on the branch 2-2´. In this manner, ideally, it is only the heat $q_{2'3}$ that is really paid for and enters the efficiency. Thus we obtain in a manner often employed before

$$e = \frac{|w_\circ|}{q_{2'3}} = 1 - \frac{T_2}{T_3} = 1 - \frac{T_1}{T_3}\left(\frac{p_2}{p_1}\right)^{\frac{\kappa-1}{\kappa}} . \tag{3.31}$$

Inspection shows that this is significantly larger than the efficiency (3.30) so that regeneration is a useful measure.

3.6.2 *Jet propulsion process*

The most conspicuous use of turbines is for the propulsion of aircraft. In that application the turbine does not serve for the complete expansion – down to atmospheric pressure – of the compressed and heated gas in the burner. Rather the turbine furnishes only the working of the compressor. Thus the gas leaving the turbine is still under high pressure and it expands through a nozzle, – usually supersonically –, in the manner that was discussed in Paragraph 1.5.9. A schematic picture of a turbojet engine is shown in Fig. 3.16 a. Obviously it differs from the gas turbine by the nozzle at the end and also by the conical converging inlet which increases the pressure of the air in front of the turbine. Fig. 3.16 b shows the process in a (p, v)-diagram, and it emphasizes the internal transfer of work from the turbine to the compressor.

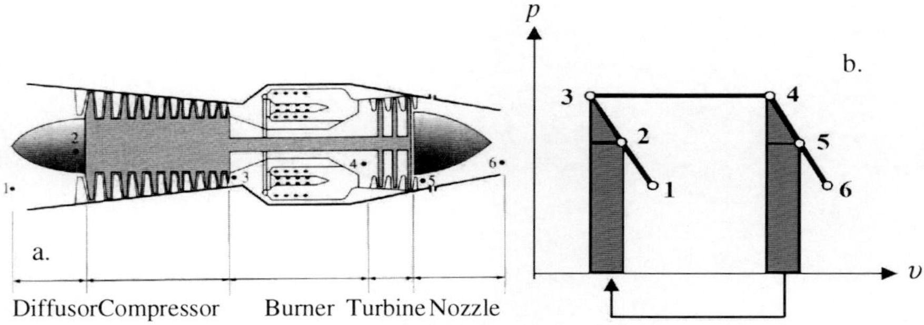

Fig. 3.16 a. Schematic picture of a jet propulsion engine
b. (p, v)-diagram of the jet propulsion process
[adapted from Y.A. Çengel, M.A. Boles, Thermodynamics – An Engineering Approach, 3rd ed., 1998]

If V is the speed of the aircraft and a the speed of the exhaust gas with respect to the craft, the thrust is determined by (1.30) and we have

$$F = \dot{m}(V - a) . \tag{3.32}$$

The working, or power, of the thrust is then obviously equal to $\dot{W} = \dot{m}(V - a)V$ so that we may define the *propulsive efficiency* by the ratio

$$\eta = \frac{\dot{W}}{\dot{Q}} = \frac{\dot{m}(V - a)V}{\dot{Q}} , \tag{3.33}$$

where $\dot{Q}$ is the heating expended in the burner.

3.6.3 *Turbofan engine*

When jet propulsion was first used in aircraft during WWII, the engines were long and slim, whereas nowadays they appear fat, and even a little stubby. The reason is that the central part of the engine is surrounded by a cylindrical duct, or cowl, or bypass, through which air is propelled by a fan, *cf.* Fig. 3.17. The fan itself is usually driven by a second turbine, a low pressure turbine. In this manner the thrust is increased, because there is a greater accelerated mass $\dot{m}$. Actually there are now two mass rates $\dot{m}_{\text{nozzle}}$ and $\dot{m}_{\text{fan}}$. It is true that the fan does not accelerate the air to supersonic speeds, of course, as the nozzle expansion does, but the bypass ratio $\frac{\dot{m}_{\text{fan}}}{\dot{m}_{\text{nozzle}}}$ may be large, – up to 10 – so that the fan contributes significantly to the overall thrust.

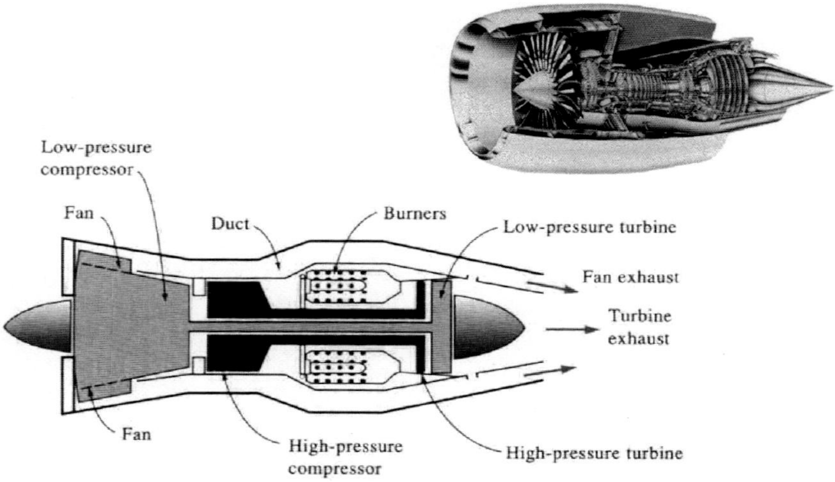

Fig. 3.17 Fanjet engine schematic, and GE GE90-115B high bypass turbofan [adapted from Y.A. Çengel, M.A. Boles, Thermodynamics – An Engineering Approach, 3rd ed., 1998]

Yet another variant of jet propulsion is the *propjet*, where the cowl is eliminated and the fan is replaced by a conventional propeller.

4 Entropy

4.1 The Second Law of thermodynamics

4.1.1 Formulation and exploitation

- *Formulation*

Rudolf Julius CLAUSIUS (1822-1888) has drawn conclusions from the following experience

Heat cannot pass by itself pass from a colder to a warmer body

or, in a later version

Heat cannot pass from a colder to a warmer body without compensation.

These are two formulations of the Second Law of thermodynamics.

We have already mentioned – as an assumption – that there are always positive *and* negative parts of the heat exchanged with a heat engine. Indeed, this is a consequence of the Second law. If it were different, we should be able to convert the heat of a part of the cold sea fully into work, and then convert the work back into heat by stirring a hot liquid. In this manner, in effect, heat would have passed from the cold sea to the hot liquid and this contradicts the Second Law. William THOMSON (Lord KELVIN) has used this argument to express the Second Law in an alternative form, *viz.*

It is impossible to gain work in a heat engine by just cooling a body.

All of these statements are open to the criticism that they are verbally expressed and lack the stringency of mathematical formulae. It is a somewhat idle effort, however, to try and make these suggestive formulations strict, because in the end, when we have gone through CLAUSIUS's argument, there is a mathematical formula, an inequality, and this is the proper mathematical form of the Second Law – and that *is* strict.

- *Universal efficiency of the Carnot process*

CLAUSIUS exploits his Second Law by attempting to contradict it with the operation of two Carnot engines I and II of which one is a heat engine, while the other one is a heat pump or a refrigerator. The engines may employ different working agents in different ranges of density, but they do operate between the same temperatures, denoted by T_L and $T_H > T_L$. The work produced by one engine is consumed by the other one. The heat engine absorbs the heat Q_H at the temperature T_H and gives off the heat $|Q_L|$ at T_L. If it were different, we should have a transition from cold to hot and work produced, and that contradicts the Second Law.

In a first step – always following CLAUSIUS – we consider reversible Carnot engines, *cf.* Fig. 4.1. The arrows represent the direction of the cyclic processes. The strategy of the argument is to let both engines run through an equal number of cycles, so that no work is left over, and then compare the heats exchanged at T_L and T_H. By (3.13) we have

$$-W_\circ^{I} = Q_L^{I} - \left|Q_H^{I}\right| < 0 \quad \text{and} \quad -W_\circ^{II} = Q_H^{II} - \left|Q_L^{II}\right| > 0 . \tag{4.1}$$

Hence follows from $-W_\circ^{I} = W_\circ^{II}$

$$Q_L^{I} - \left|Q_H^{I}\right| = -\left(Q_H^{II} - \left|Q_L^{II}\right|\right) \Rightarrow Q_L^{I} - \left|Q_L^{II}\right| = \left|Q_H^{I}\right| - Q_H^{II} . \tag{4.2}$$

Now let us assume that $Q_L^{I} - \left|Q_L^{II}\right| > 0$ holds so that engine I absorbs more heat at T_L than engine II delivers there. In this case (4.2) implies that engine I delivers more heat at T_H that engine II absorbs there. In effect – after both engines have run through equally many cycles – the result is a transition of heat from T_L to T_H and that is forbidden. But the alternative $Q_L^{I} - \left|Q_L^{II}\right| < 0$ is also forbidden. To be sure: If engine I absorbs less heat at T_L than engine II delivers there, the two engines would effect a transition of heat from high to low temperature which is allowed. However, the engines are reversible and when we reverse their operation, the heats are reversed and we are brought back to the former impossible situation.

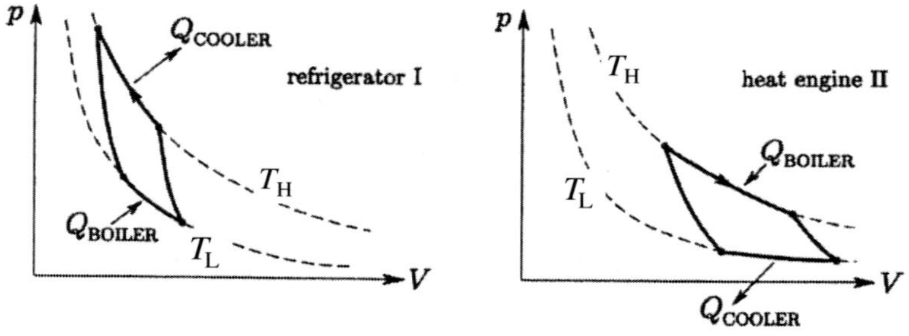

Fig. 4.1 CLAUSIUS's competing Carnot engines.

Therefore the only possible case is

$$Q_L^{I} = \left|Q_L^{II}\right| \quad \text{and, by (4.2)} \quad \left|Q_H^{I}\right| = Q_H^{II} . \tag{4.3}$$

It follows that the efficiencies of both engines are equal, because we have

$$e_C^{I} = \frac{W_\circ^{I}}{\left|Q_H^{I}\right|} \quad \text{and} \quad e_C^{II} = \frac{-W_\circ^{II}}{Q_H^{II}} \quad \text{or, by (4.3)} \quad e_C^{I} = e_C^{II} . \tag{4.4}$$

Since both engines may be working with different agents, (4.4) implies that the efficiency of a Carnot engine is universal, *i.e.* independent of the agent, be it a gas, a vapor, a liquid, or a solid. Since this is so, we may identify the efficiency with the one of an ideal gas. Therefore (3.17) holds for all agents and we write

$$e_C = 1 - \frac{T_L}{T_H} . \tag{4.5}$$

- *Absolute temperature as an integrating factor*

By (4.4) and (4.5) we conclude for the reversible Carnot process of a heat engine

$$e_{\mathrm{C}} = \frac{-W_{\circ}}{Q_{\mathrm{H}}} = \frac{Q_{\mathrm{H}} - |Q_{\mathrm{L}}|}{Q_{\mathrm{H}}} = 1 - \frac{|Q_{\mathrm{L}}|}{Q_{\mathrm{H}}} = 1 - \frac{T_{\mathrm{L}}}{T_{\mathrm{H}}} .$$

The last equation implies

$$\frac{Q_{\mathrm{H}}}{T_{\mathrm{H}}} + \frac{Q_{\mathrm{L}}}{T_{\mathrm{L}}} = 0 . \tag{4.6}$$

Thus the quantity Q/T enters the heat engine at high temperature and leaves it – unchanged in value – at low temperature. CLAUSIUS calls this quantity the *entropy* and denotes it by *S*. We may write $S_{\mathrm{H}} = S_{\mathrm{L}}$ and conclude that the entropy passes through the engine from hot to cold, unchanged in amount. It follows that at the low temperature the heat delivered is smaller than the heat absorbed at the high temperature. The heat difference is, of course, the work.

Up to this point CLAUSIUS's reasoning concerned Carnot cycles and their efficiency. But now starts a chain of arguments which extrapolates the equation (4.6) so that in the end we shall arrive at an equation for an arbitrary cycle. Hence follows the generalization of the concept of entropy and some properties of this quantity. We develop the argument in two steps.

The statement (4.6) may easily be generalized to reversible cycles that consist of isotherms and adiabates, *cf.* Fig. 4.2 a. and the octagone emphasized there by bold lines. The two dashed lines show how the process may be decomposed into three Carnot cycles *a*, *b*, and *c*. For each one we have the equivalent of (4.6), *viz.*

$$\frac{Q_{\mathrm{H}}^{a}}{T_{\mathrm{H}}^{a}} + \frac{Q_{\mathrm{L}}^{a}}{T_{\mathrm{L}}^{a}} = 0 , \quad \frac{Q_{\mathrm{H}}^{b}}{T_{\mathrm{H}}^{b}} + \frac{Q_{\mathrm{L}}^{b}}{T_{\mathrm{L}}^{b}} = 0 , \quad \frac{Q_{\mathrm{H}}^{c}}{T_{\mathrm{H}}^{c}} + \frac{Q_{\mathrm{L}}^{c}}{T_{\mathrm{L}}^{c}} = 0 . \tag{4.7}$$

From Fig. 4.2 a. we identify

$$T_{\mathrm{H}}^{a} = T_{\mathrm{H}}^{b} = T_1 , \quad T_{\mathrm{H}}^{c} = T_2 , \quad T_{\mathrm{L}}^{b} = T_{\mathrm{L}}^{c} = T_4 , \quad T_{\mathrm{L}}^{a} = T_3 .$$

Therefore the sum of the three equations (4.7) gives

$$\frac{Q_{\mathrm{H}}^{a} + Q_{\mathrm{H}}^{b}}{T_1} + \frac{Q_{\mathrm{H}}^{c}}{T_2} + \frac{Q_{\mathrm{L}}^{a}}{T_3} + \frac{Q_{\mathrm{L}}^{c} + Q_{\mathrm{L}}^{b}}{T_4} = 0 ,$$

or with Q_i for the heat exchanged at temperature T_i

$$\sum_{i=1}^{4} \frac{Q_i}{T_i} = 0 . \tag{4.8}$$

The process curve of any reversible cycle may be approximated by a zig-zag line of isotherms and adiabates in the manner indicated in Fig. 4.2 b. When this zig-zag line is constructed, one must take care that the heat exchanged on each of the infinitesimal isothermal branches is equal to the heat exchange on the smooth curve; the blow-up in Fig. 4.2 b. shows how this can be achieved: The shaded areas in each step must be equal. By this construction we also ensure that the work of the smooth cycle is equal to that of the zig-zag curve, since the areas inside the curves are equal. As far as heat and work are concerned, the process is thus equivalent to a combined process of many infinitesimal Carnot cycles.

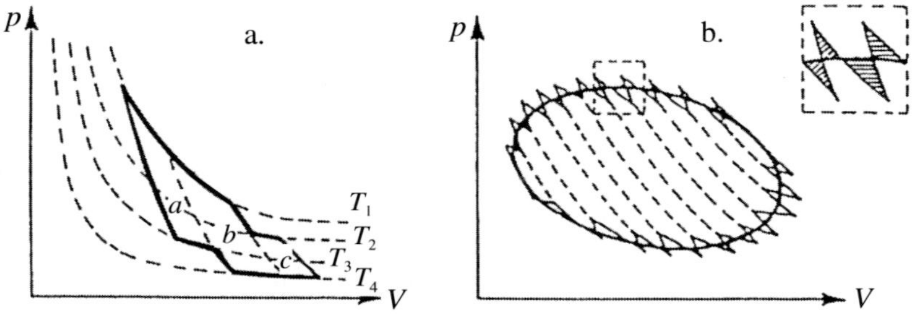

Fig. 4.2 a. A reversible cycle of isotherms and adiabates
b. An arbitrary reversible cycle decomposed into isotherms and adiabates.

Given this we may now apply the argument, which has led to (4.8), to the complex cycle of Fig. 4.2 b. The heat exchanged on an infinitesimal isothermal step is $\dot{Q}\,\mathrm{d}t$, and the entropy exchange is

$$\mathrm{d}S = \frac{\dot{Q}\mathrm{d}t}{T}. \tag{4.9}$$

Integration over the cycle between the initial and final times gives

$$\int_{t_i}^{t_f} \frac{\dot{Q}\mathrm{d}t}{T} = 0, \tag{4.10}$$

since the entropy exchange $\mathrm{d}S$ – or $\frac{\dot{Q}\mathrm{d}t}{T}$ – goes in and out of the infinitesimal Carnot cycles unchanged in amount. By (3.4) the equation (4.9) may be written in the form

$$\mathrm{d}S = \frac{1}{T}(\mathrm{d}U + p\mathrm{d}V). \tag{4.11}$$

Obviously the entropy S is a function of U and V, and we may express (4.11) by saying that the absolute temperature is an integrating denominator of the differential form $(\mathrm{d}U + p\mathrm{d}V)$.

- *Growth of entropy*

We continue to pursue CLAUSIUS's argument on the competition of Carnot engines that work between the temperatures T_L and T_H and of which engine II is a heat engine whose work drives the heat pump or refrigerator I. Now, however, we assume that engine II does not work *reversibly*, such that the heats exchanged at T_L and T_H are not reversed when the operation of the engine is reversed.

The relation (4.2) continues to hold and we assume that engine I absorbs more heat at temperature T_L than engine II delivers there. The net result is that heat has moved from cold to hot and this is forbidden by the second law. We consider the alternative: Engine I absorbs less heat at T_L than engine II delivers there. In this situation heat passes from hot to cold, which is allowed. Nor can this possibility be excluded – as it was previously – by reversing the direction of operation of the engines. After all, engine II is not reversible.

Therefore we conclude from (4.2) that we must have

$$Q_L^I - \left|Q_L^{II}\right| = \left|Q_H^I\right| - Q_H^{II} < 0,$$

i.e. the absorbed heat Q_H^I of the reversible *heat engine* I is smaller than the absorbed heat Q_H^{II} of the irreversible heat engine II, or else we have

$$\frac{-W_\circ^I}{Q_H^I} > \frac{-W_\circ^{II}}{Q_H^{II}} \quad \text{or} \quad e_C^I > e_C^{II}. \tag{4.12}$$

Thus the efficiency of the reversible engine is greater than the efficiency of the irreversible one. But, since engine I is reversible, we have $e_C^I = 1 - T_L/T_H$ and therefore by $(4.12)_1$ and with $-W_\circ^{II} = Q_H^{II} - \left|Q_L^{II}\right|$

$$e_C^I = 1 - \frac{T_L}{T_H} > 1 + \frac{Q_L^{II}}{Q_H^{II}} \quad \text{or} \quad \frac{Q_H^{II}}{T_H} + \frac{Q_L^{II}}{T_L} < 0. \tag{4.13}$$

The inequality (4.13) now replaces the equation (4.6). The latter one led to (4.10) in a sequence of steps which we may repeat for the present case. Thus we arrive at the inequality

$$\int_{t_i}^{t_f} \frac{\dot{Q}\,dt}{T} < 0 \tag{4.14}$$

which must hold for an arbitrary irreversible cycle.

In order to proceed further we decompose the irreversible cycle between the times t_i and t_f into an irreversible branch between t_i and t_m and a reversible branch between t_m and t_f. We obtain

$$\int_{t_i}^{t_m} \frac{\dot{Q}\,dt}{T} + \int_{t_m}^{t_f} \frac{\dot{Q}\,dt}{T} < 0. \tag{4.15}$$

By (4.9) the second integral may be evaluated; it turns out to be equal to $S_f - S_m$ or $S_i - S_m$, since $S_f = S_i$. Hence we obtain from (4.15)

$$S_m - S_i > \int_{t_i}^{t_m} \frac{\dot{Q}\,dt}{T} \tag{4.16}$$

In words: The entropy change in an irreversible process is greater than the accumulated heating, divided by the temperature at which the heating occurs.

- (T, S)*-diagram and maximal efficiency of the Carnot process*

When the thermal and caloric equation of state are given, the equation (4.11) may be used to calculate $S = S(T, V)$ by integration to within an additive constant. After that has been done, a reversible cycle of a heat engine may be represented by a closed curve in a (T, S)-diagram, *cf.* Fig. 4.3 a. Such a diagram has several advantages:

- From (4.9) it follows that $Q_+ = \int T\,\mathrm{d}S$ holds when the integration is carried out over the upper part of the curve. Therefore the area below the upper part of the curve is equal to Q_+.
- Likewise $Q_- = \int T\,\mathrm{d}S$ holds by integration over the lower part of the curve. Therefore the area below the cycle is equal to $|Q_-|$.
- Furthermore we have $-W_\circ = Q_+ - |Q_-|$ so that $|W_\circ|$ is equal to the area inside the curve. This property is shared by the (T, S)-diagram and by the (p, V)-diagram, *cf.* Section 3.1.
- A reversible adiabatic process is an *isentropic* process, because we have $\mathrm{d}S = \dot{Q}\,\mathrm{d}t = 0$. This means that the process is represented by a vertical line in the (T, S)-diagram.

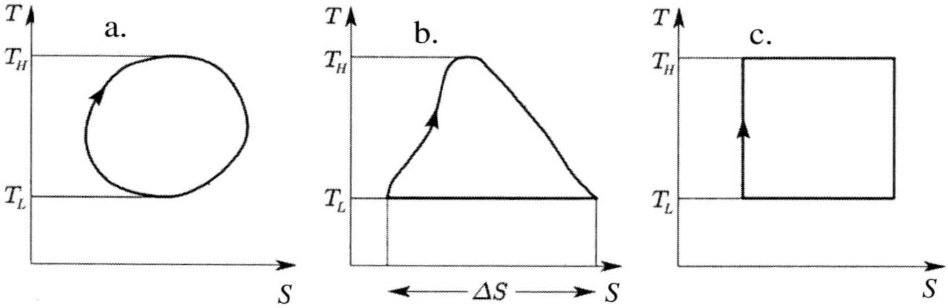

Fig. 4.3 a. Reversible cycle in the (T, S)-diagram
b. Cycle with cooling at T_L
c. Carnot cycle.

By use of the (T, S)-diagram we may find out – purely geometrically – which reversible cycle, running between an upper temperature T_H and a lower one T_L, has the largest efficiency. Fig. 4.3 shows three cycles of that type, all with the same value Q_+, the area below the upper part of the process curve. We may increase the efficiency

$$e = 1 - \frac{|Q_-|}{Q_+}$$

at constant Q_+ by making $|Q_-|$, the integral $\int T\,\mathrm{d}S$ along the lower part of the curve, as small as possible. This is achieved by releasing all heat at the lowest temperature T_L, see Fig. 4.3 b. In this case $|Q_-| = T_\mathrm{L}\Delta S$ and $|Q_-|$ becomes minimal, when ΔS is made as small as possible. Now, a smaller ΔS – for fixed Q_+ – is realized when the temperature of the added heat is increased. And ΔS becomes minimal, when all contributions to Q_+ are added at the highest temperature T_H. Thus in the (T, S)-diagram a rectangle appears, defined by two horizontal isotherms and two vertical isentropes. This process curve is the one of the Carnot process.

The argument confirms what has merely been stated before, namely that the Carnot process has the largest efficiency among reversible cycles between the temperatures T_H and T_L. To be sure there are other cycles with the same efficiency, *e.g.* the Erickson cycles and the Stirling cycle in ideal gases, – *cf.* Paragraph 3.4.2 – but those require regeneration. Also their efficiency is not universal.

4.1.2 *Summary*

For reversible processes in closed systems for which, by (1.57), the First Law reads

$$\dot{Q}\,\mathrm{d}t = \mathrm{d}U + p\,\mathrm{d}V\,, \tag{4.17}$$

the Second Law implies that $\dot{Q}\,\mathrm{d}t$, the exchanged heat, changes a quantity S, the entropy which depends on the states (U, V), or (T, V), or (p, V), *etc.* The change of entropy amounts to $\dot{Q}\,\mathrm{d}t$ divided by the absolute temperature, so that we have

$$\mathrm{d}S = \frac{\dot{Q}\,\mathrm{d}t}{T} \quad \text{or by (4.17)} \tag{4.18}$$

$$\mathrm{d}S = \frac{1}{T}\left(\mathrm{d}U + p\,\mathrm{d}V\right) \quad \text{or with } U = U(T, V) \tag{4.19}$$

$$\mathrm{d}S = \frac{1}{T}\left(\frac{\partial U}{\partial T}\right)_V \mathrm{d}T + \frac{1}{T}\left[\left(\frac{\partial U}{\partial V}\right)_T + p\right]\mathrm{d}V\,. \tag{4.20}$$

The equations (4.19) or (4.20) are known as *Gibbs equation*. In words the equation may be expressed by saying that the absolute temperature serves as an *integrating denominator* of the expression $\mathrm{d}U + p\,\mathrm{d}V$.

For irreversible processes the Second Law implies that the change of entropy between the initial time t_i and the final time t_f satisfies the inequality

$$S_\mathrm{f} - S_\mathrm{i} > \int_{t_\mathrm{i}}^{t_\mathrm{f}} \frac{\dot{Q}\,\mathrm{d}t}{T}\,. \tag{4.21}$$

The relations (4.18) and (4.21) may be summarized by the *Clausius inequality* in the form

$$S_{\mathrm{f}} - S_{\mathrm{i}} \geq \int_{t_{\mathrm{i}}}^{t_{\mathrm{f}}} \frac{\dot{Q}\,\mathrm{d}t}{T}, \text{ or for short processes } \frac{\mathrm{d}S}{\mathrm{d}t} \geq \frac{\dot{Q}}{T}, \tag{4.22}$$

where the equality holds for irreversible processes.

In particular (4.22) implies that the entropy cannot decrease in an adiabatic process; in equilibrium, at the end of the process, the entropy is maximal.

Of course, the usefulness of the Second Law must not exhaust itself in the introduction and characterization of a quantity, the entropy, which we have not missed up to now. The knowledge to be gained from entropy, Gibbs equation, and Clausius inequality will be discussed in Sect. 4.2 and in later chapters.

One point which we may now understand concerns the occurrence of $|Q_-| \neq 0$ in every cycle. Indeed a cycle has always parts with Q_+ and Q_-. This is made obvious by Fig. 4.3 and the discussion of the (T, S)-diagram. It is therefore impossible to run a working cycle by only absorbing heat from a reservoir without releasing heat during certain parts of the cycle. This is the statement which Kelvin chose as his version of the Second Law, see Paragraph 4.1.1.

The Gibbs equation (4.19) makes a statement about the function $S(U,V)$. Equally well we may write it – for a fluid in a *reversible process* characterized by $U(t)$ and $V(t)$ – in the form

$$\frac{\mathrm{d}S}{\mathrm{d}t} = \frac{1}{T}\left(\frac{\mathrm{d}U}{\mathrm{d}t} + p\frac{\mathrm{d}V}{\mathrm{d}t}\right).$$

This gives the rate of change of S in time when U and V, hence $p=p(U,V)$ and $T=T(U,V)$, change in time.

On the other hand in an *irreversible process* we have the Clausius inequality (4.22) or, by $(1.53)_1$

$$\frac{\mathrm{d}S}{\mathrm{d}t} > \frac{1}{T}\left(\frac{\mathrm{d}(U+K)}{\mathrm{d}t} - \dot{W}\right) \quad \text{or, by } (1.53)_2$$

$$\frac{\mathrm{d}S}{\mathrm{d}t} > \frac{1}{T}\left(\frac{\mathrm{d}(U+E_{\mathrm{pot}}+K)}{\mathrm{d}t} - \dot{W}_{\mathrm{stress}}\right)$$

There is no Gibbs equation in this case;* the inequality provides a lower bound on the rate of change of entropy in terms of the fields $\rho(x_j,t)$, $w_i(x_j,t)$, and $T(x_j,t)$ inside the system, and of the working of the stress on the boundary.

Note also that T in the Clausius inequality is always the homogeneous temperature on the boundary, or at least on that part of the boundary across which heat is exchanged. In a reversible process this boundary temperature is equal to the homogeneous temperature throughout the system, but not in an irreversible process,

* Except perhaps locally, *cf.* Chap. 12

where generally we have arbitrary fields inside the system, including an arbitrary temperature field.

4.2 Exploitation of the Second Law

4.2.1 *Integrability condition*

We start with the Gibbs equation in the form (4.20). Mathematicians call an expression like that a total differential. The relation is a short form of expressing two differential equations, namely

$$\left(\frac{\partial s}{\partial T}\right)_{\upsilon} = \frac{1}{T}\left(\frac{\partial u}{\partial T}\right)_{\upsilon} \quad \text{and} \quad \left(\frac{\partial s}{\partial \upsilon}\right)_{T} = \frac{1}{T}\left[\left(\frac{\partial u}{\partial \upsilon}\right)_{T} + p\right], \tag{4.23}$$

where we have switched to specific values by dividing (4.20) by the mass of the body.

Mixed second derivatives can be interchanged. Therefore we must have

$$\left(\frac{\partial}{\partial \upsilon}\left[\frac{1}{T}\left(\frac{\partial u}{\partial T}\right)_{\upsilon}\right]\right)_{T} = \left(\frac{\partial}{\partial T}\left[\frac{1}{T}\left[\left(\frac{\partial u}{\partial \upsilon}\right)_{T} + p\right]\right]\right)_{\upsilon}.$$

The mixed second derivatives of u cancel and a short calculation will then show that

$$\left(\frac{\partial u}{\partial \upsilon}\right)_{T} = -p + T\left(\frac{\partial p}{\partial T}\right)_{\upsilon}. \tag{4.24}$$

The significance of this equation can hardly be overestimated. It saves infinitely much money and labor, because it relates the caloric equation of state to the thermal equation of state. Let us consider this:

We recall that the thermal equation of state is relatively easy to determine, while the determination of the caloric equation of state requires many measurements of c_p and c_υ, the specific heats. And those, being caloric measurements, are difficult, expensive, and unreliable. The integrability condition (4.24) helps us to reduce the number of caloric measurements *drastically* and therein lies its importance.

Indeed, in Paragraph 2.3.3 we have seen that the determination of $\left(\frac{\partial u}{\partial T}\right)_{\upsilon}$ and $\left(\frac{\partial u}{\partial \upsilon}\right)_{T}$ makes it necessary to measure c_p and c_υ for all pairs of (υ, T). Now, since by the integrability condition (4.24) the derivative $\left(\frac{\partial u}{\partial \upsilon}\right)_{T}$ may be calculated from the thermal equation of state, *we do not need to measure* c_p *anymore*. And this is not all yet: If we differentiate (4.24), left and right, with respect to T, we obtain

$$\frac{\partial^2 u}{\partial T \partial \upsilon} = T\left(\frac{\partial^2 p}{\partial T^2}\right)_{\upsilon} \quad \text{or, with } c_\upsilon = \left(\frac{\partial u}{\partial T}\right)_{\upsilon}: \quad \left(\frac{\partial c_\upsilon}{\partial \upsilon}\right)_{T} = T\left(\frac{\partial^2 p}{\partial T^2}\right)_{\upsilon}. \tag{4.25}$$

It follows that the υ-dependence of c_υ follows from the thermal equation of state as well. All that remains to be measured is c_υ for *one* choice of υ as a function of T. Thus the number of required caloric measurements has indeed been reduced most severely.

4.2.2 *Internal energy and entropy of a van der Waals gas and of an ideal gas*

An example for the partial determination of the caloric equation $u(\upsilon,T)$ from the thermal equation $p(\upsilon,T)$ is furnished by the van der Waals gas, *cf.* Paragraph 2.4.9. Here we know only the thermal equation of state (2.33). However, the integrability condition (4.24) allows us to determine the υ-dependence of $u(\upsilon,T)$, *viz.*

$$\left(\frac{\partial u}{\partial \upsilon}\right)_T = +\frac{a}{\upsilon^2}, \text{ hence } u(\upsilon,T) = -\frac{a}{\upsilon} + F(T), \tag{4.26}$$

where $F(T)$ is a constant of integration, namely a constant with respect to υ.

Furthermore, by (4.25), c_υ is independent of υ in a van der Waals gas, because p is a linear function of T. The constant of integration $F(T)$ may therefore be written as $\int c_\upsilon(T)\,\mathrm{d}T$, because, by (4.26), we have $c_\upsilon = F'(T)$. Therefore the generic form of the caloric equation of state for a van der Waals gas reads

$$u(\upsilon,T) = -\frac{a}{\upsilon} + \int c_\upsilon(T)\,\mathrm{d}T\,. \tag{4.27}$$

In order to determine it fully we must measure $c_\upsilon(T)$. Note that the van der Waals case is a little too simple to be typical, because in general c_υ will not only depend on T but also on υ.

By use of the Gibbs equation (4.20) we may now calculate the specific entropy of a van der Waals gas. By (4.20), (4.24), and (4.27) we obtain

$$\mathrm{d}s = \frac{1}{T}c_\upsilon(T)\,\mathrm{d}T + \left(\frac{\partial p}{\partial T}\right)_\upsilon \mathrm{d}\upsilon \quad \text{or, by (2.33)}$$

$$\mathrm{d}s = \frac{1}{T}c_\upsilon(T)\,\mathrm{d}T + \frac{\frac{R}{M}}{\upsilon - b}\mathrm{d}\upsilon \quad \text{or, by integration} \tag{4.28}$$

$$s = \int \frac{c_\upsilon(T)}{T}\,\mathrm{d}T + \tfrac{R}{M}\ln(\upsilon - b)\,. \tag{4.29}$$

The integral term, of course, contains an arbitrary constant.

The entropy of an ideal gas follows from (4.29) as the special case, where c_υ is constant and b is zero. We have

$$s(\upsilon,T) = c_\upsilon \ln T + \frac{R}{M}\ln \upsilon + \beta'\,, \tag{4.30}$$

where β' is a constant of integration. Often s is expressed as a function of T and p, rather than T and υ. Since $p\upsilon = \frac{R}{M}T$ and $c_p - c_\upsilon = \frac{R}{M}$ hold, we obtain

$$s(T,p) = c_p \ln T - \frac{R}{M}\ln p + \beta\,, \tag{4.31}$$

where $\beta = \beta' - \frac{R}{M} \ln \frac{R}{M}$.

The value of the additive constants in energy and entropy are of no interest until we come to consider chemical reactions, *cf.* Chap. 8. For all other processes they drop out of the relevant equations.

We know now that a reversible adiabatic process is isentropic so that s = const holds in such a process. Thus (4.30) with $c_p - c_v = \frac{R}{M}$ and $\kappa = \frac{c_p}{c_v}$ implies $\frac{p^{\frac{\kappa-1}{\kappa}}}{T} = \text{const}$, a relation which we have previously called an "adiabatic equation of state", *cf.* (1.62). Of course, this only means that our concepts and calculations are consistent, at least with respect to isentropy.

4.2.3 *Alternatives of the Gibbs equation and its integrability conditions*

By easy manipulations the Gibbs equation (4.19) may be written in the alternative forms

$$\begin{aligned}
&\mathrm{d}u = T\,\mathrm{d}s - p\,\mathrm{d}v & & \left(\frac{\partial T}{\partial v}\right)_s = -\left(\frac{\partial p}{\partial s}\right)_v \\
&\mathrm{d}(u - Ts) = -s\,\mathrm{d}T - p\,\mathrm{d}v & & \left(\frac{\partial s}{\partial v}\right)_T = \left(\frac{\partial p}{\partial T}\right)_v \\
&\mathrm{d}(u + pv) = T\,\mathrm{d}s + v\,\mathrm{d}p & \text{hence} \quad & \left(\frac{\partial T}{\partial p}\right)_s = \left(\frac{\partial v}{\partial s}\right)_p \\
&\mathrm{d}(u + pv - Ts) = -s\,\mathrm{d}T + v\,\mathrm{d}p & & \left(\frac{\partial s}{\partial p}\right)_T = -\left(\frac{\partial v}{\partial T}\right)_p .
\end{aligned} \tag{4.32}$$

The quantities appearing in (4.32) are called

$$\begin{aligned}
&u && \text{- internal energy} \\
&f = u - Ts && \text{- Helmholtz free energy, or free energy} \\
&h = u + pv && \text{- enthalpy} \\
&g = u + pv - Ts && \text{- Gibbs free energy, or free enthalpy.}
\end{aligned} \tag{4.33}$$

They are also known as *thermodynamic potentials*, because their derivatives are simple variables or state functions, namely p, v, T, and s.

The equations on the right hand side of (4.32) are the integrability conditions, which follow from the alternative forms of the Gibbs equations by inspection. We shall demonstrate the usefulness of these integrability conditions by showing that the ratio of the isothermal compressibility κ_T and the adiabatic compressibility κ_s is equal to the ratio of the specific heats c_p and c_v.

The adiabatic – or isentropic – compressibility κ_s is defined as

$$\kappa_s = -\frac{1}{v}\left(\frac{\partial v}{\partial p}\right)_s$$

$$= -\frac{1}{\upsilon}\left(\frac{\partial \upsilon}{\partial T}\right)_s \left(\frac{\partial T}{\partial p}\right)_s \qquad \text{or, by (2.13):} \left(\frac{\partial \upsilon}{\partial T}\right)_s = -\left(\frac{\partial \upsilon}{\partial s}\right)_T \left(\frac{\partial s}{\partial T}\right)_\upsilon$$

$$= \frac{1}{\upsilon}\left(\frac{\partial \upsilon}{\partial s}\right)_T \left(\frac{\partial s}{\partial T}\right)_\upsilon \left(\frac{\partial T}{\partial p}\right)_s \qquad \text{or, by (4.32)}_2\text{:} \left(\frac{\partial \upsilon}{\partial s}\right)_T = \left(\frac{\partial T}{\partial p}\right)_\upsilon \text{ and with (2.13)}$$

$$= -\left(\frac{\partial T}{\partial \upsilon}\right)_p \left(\frac{\partial \upsilon}{\partial p}\right)_T$$

$$= -\frac{1}{\upsilon}\left(\frac{\partial \upsilon}{\partial p}\right)_T \left(\frac{\partial s}{\partial T}\right)_\upsilon \left(\frac{\partial T}{\partial \upsilon}\right)_p \left(\frac{\partial T}{\partial p}\right)_s \qquad \text{or, by: } \kappa_T = -\frac{1}{\upsilon}\left(\frac{\partial \upsilon}{\partial p}\right)_T \text{ and } c_\upsilon = T\left(\frac{\partial s}{\partial T}\right)_\upsilon$$

$$= \frac{1}{T}\kappa_T c_\upsilon \left(\frac{\partial T}{\partial \upsilon}\right)_p \left(\frac{\partial T}{\partial p}\right)_s \qquad \text{or, by (4.31)}_3 \text{ with } \left(\frac{\partial T}{\partial p}\right)_s = \left(\frac{\partial \upsilon}{\partial s}\right)_p$$

$$= \frac{1}{T}\kappa_T c_\upsilon \left(\frac{\partial T}{\partial s}\right)_p \qquad \text{or, by: } c_p = T\left(\frac{\partial s}{\partial T}\right)_p$$

$$= \kappa_T \frac{c_\upsilon}{c_p}. \tag{4.34}$$

This chain of equations is typical for the calculation of relations between material coefficients. It is sometimes difficult to find the right path, because in such calculations it is easy to write down infinitely many correct equations and yet move away from the desired result.

In the above derivation we have used expressions for the specific heats c_p and c_υ which were not listed before, *viz.*

$$c_\upsilon = T\left(\frac{\partial s}{\partial T}\right)_\upsilon \quad \text{and} \quad c_p = T\left(\frac{\partial s}{\partial T}\right)_p. \tag{4.35}$$

These are straightforward consequences from the previous definitions (2.20) and (2.25) when $(4.32)_{1,3}$ are used.

Another example for the usefulness of (4.32) is the determination of the difference $c_p - c_\upsilon$ from the thermal equation of state. We have by (2.22)

$$c_p - c_\upsilon = \left(\frac{\partial \upsilon}{\partial T}\right)_p \left[\left(\frac{\partial u}{\partial \upsilon}\right)_T + p\right] \text{ or, with } \alpha = \frac{1}{\upsilon}\left(\frac{\partial \upsilon}{\partial T}\right)_p \text{ and (4.24)}$$

$$= \upsilon T \alpha \left(\frac{\partial p}{\partial T}\right)_\upsilon \qquad \text{or, with } \beta = \frac{1}{p}\left(\frac{\partial p}{\partial T}\right)_\upsilon$$

$$= p\upsilon T\alpha\beta \qquad \text{or, with } \alpha = p\kappa_T\beta \text{ by (2.14)}$$

$$c_p - c_\upsilon = p^2 \upsilon T \kappa_T \beta^2. \tag{4.36}$$

In particular it follows that in an incompressible body the specific heats are equal.

4.2.4 *Phase equilibrium. Clausius-Clapeyron equation*

In Paragraph 2.4.1 we have shown that at the phase boundary between liquid and vapor we have, *cf.* (2.30)

$$r = h'' - h' \qquad \text{or, by } h = u + pv$$

$$r = u'' + pv'' - (u' + pv'), \tag{4.37}$$

since the pressure is continuous on the boundary. r is the specific heat of evaporation. If the evaporation occurs at constant temperature T we have, by (4.18)

$$r = T(s'' - s'). \tag{4.38}$$

Note that T is also continuous at the phase boundary. Elimination of r between (4.37) and (4.38) provides

$$(u + pv - Ts)'' = (u + pv - Ts)', \tag{4.39}$$

so that *the specific Gibbs free energy is continuous at the phase boundary*, – along with T and p. We may write

$$g''(T,p) = g'(T,p), \tag{4.40}$$

and this is an implicit form of the equation $p = p(T)$ for saturated steam and boiling liquid. Unfortunately we do not know the functions $g'(T,p)$ and $g''(T,p)$ analytically, therefore we cannot have an explicit analytic form of $p(T)$.

And yet, we can make an interesting statement about $p(T)$ by writing (4.39) as

$$p(T) = -\frac{u'' - Ts'' - (u' - Ts')}{v'' - v'}, \tag{4.41}$$

where all the quantities on the right hand side are functions of T. Differentiation with respect to T provides*

$$\frac{\mathrm{d}p}{\mathrm{d}T} = \frac{1}{T}\frac{r}{v'' - v'}. \tag{4.42}$$

This is the *Clausius-Clapeyron equation.* We may call $p = p(T)$ and $r = r(T)$ the thermal and the caloric equations of state of saturated vapor, and then (4.42) represents a relation between these state equations, just as (4.24) implies a relation between the two equations of state of a one-phase fluid. Indeed, the Clausius-Clapeyron equation with $r = h'' - h'$ and $h = u + pv$ may be rewritten in the form

$$\frac{u'' - u'}{v'' - v'} = -p + T\frac{\mathrm{d}p}{\mathrm{d}T}, \tag{4.43}$$

which is clearly the analogue of the integrability condition (4.24) applied to evaporation.

---

* One might suspect that there must be an additional term in (4.42) due to $\frac{\mathrm{d}u''}{\mathrm{d}T}, \frac{\mathrm{d}s''}{\mathrm{d}T}, \frac{\mathrm{d}v''}{\mathrm{d}T}$ and $\frac{\mathrm{d}u'}{\mathrm{d}T}, \frac{\mathrm{d}s'}{\mathrm{d}T}, \frac{\mathrm{d}v'}{\mathrm{d}T}$. However, that term vanishes because of the Gibbs equation applied to the saturated vapor and the boiling liquid.

With some simplifying assumptions the Clausius-Clapeyron equation may be integrated to provide an analytic form for the saturated vapor curve $p = p(T)$. We assume

$$\begin{aligned} &\text{i.)} \quad r \text{ is independent of } T, \\ &\text{ii.)} \quad \upsilon' << \upsilon'', \\ &\text{iii.)} \quad \upsilon' = \frac{1}{p}\frac{R}{M}T \text{ , i.e. vapor is considered an ideal gas,} \end{aligned} \tag{4.44}$$

and write (4.42) as

$$\frac{\mathrm{d}p}{p} = \frac{r}{\frac{R}{M}}\frac{\mathrm{d}T}{T^2} \text{ , hence } p = p(T_0)\exp\left[-\frac{r}{\frac{R}{M}}\left(\frac{1}{T}-\frac{1}{T_0}\right)\right], \tag{4.45}$$

where T_0 is some reference temperature. According to this equation the saturated vapor curves must be straight lines in a $(\ln p, 1/T)$-diagram. Fig. 4.4 shows experimental curves which confirm this linearity *in the whole range of temperature between the triple point and the critical point*. This agreement is most surprising, because none of the assumptions (4.44) is very trustworthy: all are bad in the neighborhood of the critical point. We must conclude that several mistakes in the assumptions compensate each other. Anyway we are thus left with a reliable analytic form of the saturated vapor curve.

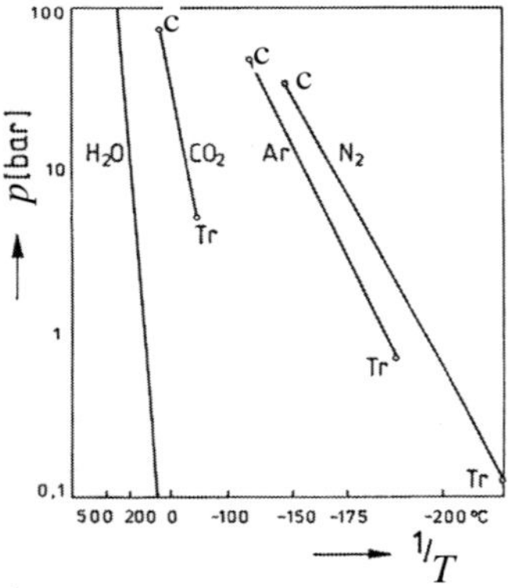

Fig. 4.4 Benoît Pierre Émile CLAPEYRON (1799-1864)
Vapor-pressure-curves in a $\left(\ln p, \frac{1}{T}\right)$-diagram

CLAPEYRON was a professor at the École Polytechnique in Paris. He did much to clarify CARNOT's work but he could not overcome the caloric theory. Therefore he could not identify the factor $\frac{1}{T}$ in the Clausius-Clapeyron equation; this was left to do for CLAUSIUS. CLAPEYRON, however, made an everlasting contribution to $p\,\mathrm{d}V$-thermodynamics by representing reversible processes as lines in (p,V)-diagram.

4.2.5 *Phase equilibrium in a van der Waals gas*

In Paragraph 2.4.9 we have discussed the isotherms of a van der Waals gas. A typical isotherm has the form shown in Fig. 4.5 a. The ascending branch, where the volume grows with increasing pressure, is physically unrealistic; later – in Paragraph 4.2.8 – we shall recognize that branch as unstable.

But even without this peculiarity the van der Waals equation provides the possibility to have two different volumes for the same pressure and the same temperature. Therefore we say that the equation can describe a fluid in the range of the phase transition liquid-vapor. The greater specific volume may then be called $\upsilon''(T)$ and the smaller one $\upsilon'(T)$.

This sounds promising. However, for a fixed temperature there is a whole *range* of pressures for which liquid and vapor may coexist; three such pressures are shown in Fig. 4.5 a. On the other hand we already know from the section on wet steam that there is only *one* $p(T)$, and so we must ask how we can identify this single pressure.

The answer is best given graphically by a construction which goes back to MAXWELL. We recall (4.41) and write, using $f = u - Ts$

$$p(T)(\upsilon'' - \upsilon') + f'' - f' = 0 \quad \text{or, with} \quad f = -\int p \, \mathrm{d}\upsilon \,, \; cf. \; (4.32) \tag{4.46}$$

$$p(T)(\upsilon'' - \upsilon') - \int_{\upsilon'}^{\upsilon''} p(\upsilon, T) \, \mathrm{d}\upsilon = 0 \,, \tag{4.47}$$

where the integral has to be taken along the isotherm; *i.e.* the integral represents the area below the isotherm between υ' and υ''. By (4.47) this area must be equal to the area of the rectangle $p(T)(\upsilon'' - \upsilon')$. Therefore we find $p(T)$ as *the* isobar – the Maxwell line – with equal areas between the isotherm above and below, *cf.* Fig. 4.5 b. Afterwards υ' and υ'' come out as shown in the figure.

It is instructive to consider this case from another point of view by use of the free energy. We have $p = -\left(\frac{\partial f}{\partial \upsilon}\right)_T$, so that the non-monotonic isotherm in the (p, υ)-diagram corresponds to a non-convex isotherm in the (f, υ)-diagram shown in Fig. 4.5 c. For the wet steam we write, as in Paragraph 2.4.1

$$\begin{aligned} f &= (1 - x) f' + x f'' \\ \upsilon &= (1 - x) \upsilon' + x \upsilon'' \end{aligned} \quad \text{or, by (4.46):} \; f = f' - p(T)(\upsilon - \upsilon') \,. \tag{4.48}$$

It follows that the free energy of the wet steam is a linear function of υ with the slope $-p(T)$. Graphically this function is equal to the common tangent of the two convex parts of $f(\upsilon, T)$, *cf.* Fig. 4.5 c. In this manner it is obvious that the free energy of wet steam is lower than the free energy of either homogeneous phase.

The graphical methods described above are known in the literature as the "equal area rule" and as the "common tangent rule." Mathematicians speak of the *convexification* of the free energy function.

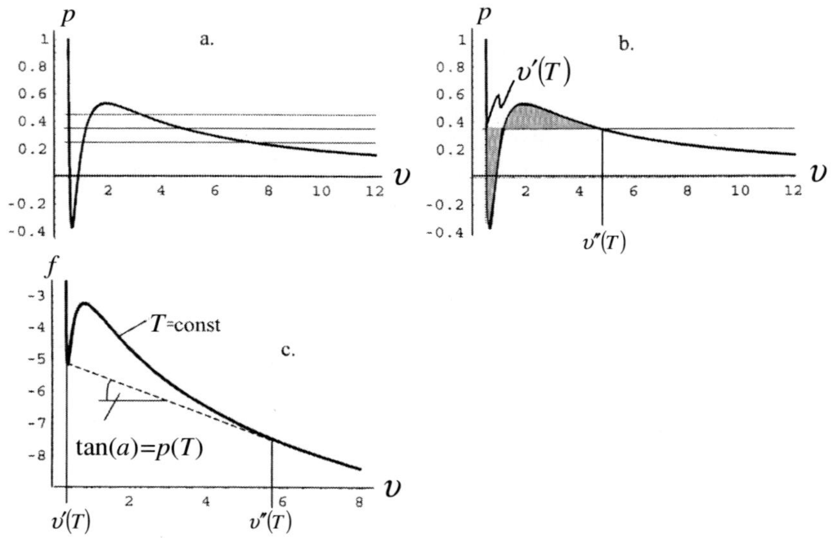

Fig. 4.5 a. Isotherm of a van der Waals gas
b. Construction of the Maxwell line $p(T)$
c. Free energy of a van der Waals gas, and of wet steam. (For a clear graphical representation the temperatures in a. and b. are different from the temperature in c.)

4.2.6 *Temperature change during adiabatic throttling Example: Van der Waals gas*

Throttling is a means for decreasing the pressure of a fluid and in Paragraph 1.5.9 we have seen that during adiabatic throttling the specific enthalpy is equal in front and far behind the throttle valve. Thus we have, at least approximately

$$\mathrm{d}h = \left(\frac{\partial h}{\partial T}\right)_p \mathrm{d}T + \left(\frac{\partial h}{\partial p}\right)_T \mathrm{d}p = 0 \quad \text{and therefore , with} \quad c_p = \left(\frac{\partial h}{\partial T}\right)_p$$

$$\left(\frac{\partial T}{\partial p}\right)_h = \frac{1}{c_p}\left(\frac{\partial h}{\partial p}\right)_T . \tag{4.49}$$

This equation determines the temperature change that accompanies the pressure decrease. We proceed to replace $\left(\frac{\partial h}{\partial p}\right)_T$ by quantities that follow from the thermal equation of state. For this purpose we employ the Gibbs equation (4.19)

$$\mathrm{d}s = \frac{1}{T}(\mathrm{d}u + p\,\mathrm{d}v) \quad \text{with} \quad p\,\mathrm{d}v = \mathrm{d}(pv) - v\,\mathrm{d}p \quad \text{and} \quad h = u + pv$$

$$\mathrm{d}s = \frac{1}{T}(\mathrm{d}h - v\,\mathrm{d}p) \quad \text{with} \quad h = h(T, p):$$

$$\mathrm{d}s = \frac{1}{T}\left(\frac{\partial h}{\partial T}\right)_p \mathrm{d}T + \frac{1}{T}\left[\left(\frac{\partial h}{\partial p}\right)_T - \upsilon\right]\mathrm{d}p .$$

The implied integrability condition reads

$$\left(\frac{\partial}{\partial p}\left[\frac{1}{T}\left(\frac{\partial h}{\partial T}\right)_p\right]\right)_T = \left(\frac{\partial}{\partial T}\left[\frac{1}{T}\left[\left(\frac{\partial h}{\partial p}\right)_T - \upsilon\right]\right]\right)_p \Rightarrow \left(\frac{\partial h}{\partial p}\right)_T = \upsilon - T\left(\frac{\partial \upsilon}{\partial T}\right)_p . \tag{4.50}$$

Insertion into (4.49) gives

$$\left(\frac{\partial T}{\partial p}\right)_h = \frac{\upsilon T}{c_p}\left(\alpha - \frac{1}{T}\right), \text{ where } \alpha = \frac{1}{\upsilon}\left(\frac{\partial \upsilon}{\partial T}\right)_p \tag{4.51}$$

is the thermal expansion coefficient, *cf.* (2.12).

For an ideal gas, where $\alpha = 1/T$ holds, there is no temperature drop during throttling. Later, *cf.* Paragraph 4.2.8, we shall show that c_p is positive and therefore (4.51) implies that in throttling we have

$$\begin{matrix}\text{heating} \\ \text{cooling}\end{matrix} \text{ if } \alpha \begin{matrix} < \\ > \end{matrix} \frac{1}{T} . \tag{4.52}$$

As an instructive example we investigate the throttling of a van der Waals gas. The thermal equation of state is given by (2.33) so that by differentiation we have

$$\mathrm{d}p = \frac{\frac{R}{M}}{\upsilon - b}\mathrm{d}T - \left(\frac{\frac{R}{M}T}{(\upsilon - b)^2} - \frac{2a}{\upsilon^3}\right)\mathrm{d}\upsilon$$

from which we determine the expansion coefficient

$$\alpha = \frac{\frac{R}{M}\frac{\upsilon - b}{\upsilon}}{\frac{R}{M}T - \frac{2a}{\upsilon}\left(\frac{\upsilon - b}{\upsilon}\right)^2}$$

and the quantity $\alpha - \frac{1}{T}$ relevant for the temperature change

$$\alpha - \frac{1}{T} = \frac{1}{\upsilon T}\frac{-b + \frac{2a}{\frac{R}{M}T}\left(\frac{\upsilon - b}{\upsilon}\right)^2}{1 - \frac{2a}{\upsilon\frac{R}{M}T}\left(\frac{\upsilon - b}{\upsilon}\right)^2} . \tag{4.53}$$

Loosely speaking the terms with a and b are small and therefore the sign of $\alpha - 1/T$ is determined by the numerator of this expression. The sign of the numerator depends on the pair (υ, T) before the throttle valve (say). We have

$$\begin{matrix}\text{heating} \\ \text{cooling}\end{matrix} \text{ for } \frac{\upsilon - b}{\upsilon} \begin{matrix} > \\ < \end{matrix} \sqrt{\frac{\frac{R}{M}T}{2\frac{a}{b}}} \text{ or } v \begin{matrix} > \\ < \end{matrix} \frac{1}{3 - 2\sqrt{\frac{1}{3}\gamma}} . \tag{4.54}$$

In $(4.54)_2$ we have used dimensionless variables in order to emphasize the universal properties of the van der Waals equation, *cf.* (2.34),(2.35). We conclude that the curve

$$v = \frac{1}{3-2\sqrt{\frac{1}{3}\gamma}}, \text{ or } \gamma = \frac{3}{4}\left(3-\frac{1}{v}\right)^2. \tag{4.55}$$

divides the first quadrant of the (v,γ)-space into two regions: one where cooling occurs and one for heating. This curve is called the inversion curve; it may also be calculated in a (π,γ)- or a (π,v)-diagram by use of the van der Waals equation of state (2.36). We obtain

$$\pi = 24\sqrt{3}\sqrt{\gamma} - 12\gamma - 27 \quad \text{or} \quad \pi = \frac{18}{v} - \frac{9}{v^2}. \tag{4.56}$$

Fig. 4.6 shows these graphs and also – as a thin line – the critical isochor and the critical isotherm, respectively. The regions of heating and cooling are indicated.

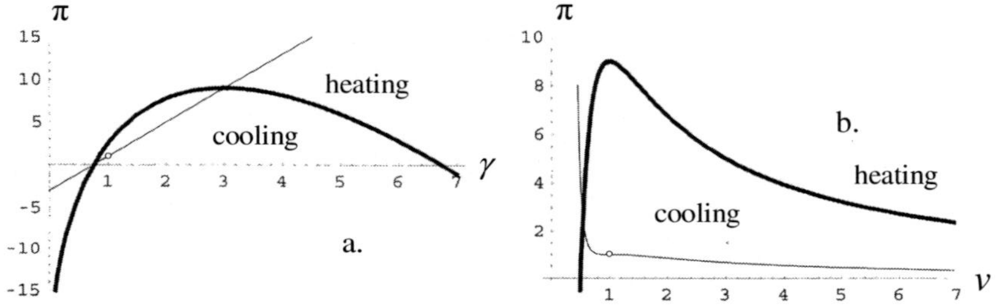

Fig. 4.6 a. Inversion curve and critical isochor $\pi = 4\gamma - 3$

b. Inversion curve and critical isotherm $\pi = \frac{8}{3v-1} - \frac{3}{v^2}$.

For nitrogen, oxygen, and hydrogen we have listed critical data in Table 4.1. It follows from this data that those gases have the following dimensionless pressures and temperatures at 1 bar and 300 K

$$\pi_{N_2} = 0.03 \qquad \pi_{O_2} = 0.02 \qquad \pi_{H_2} = 0.08$$
$$\gamma_{N_2} = 2.4 \qquad \gamma_{O_2} = 1.9 \qquad \gamma_{H_2} = 9.1.$$

For hydrogen these data place the state (1bar, 300K) firmly outside the region of cooling, so that hydrogen heats up by throttling from this state.

Therefore hydrogen-carrying pipes must be carefully sealed, because escaping – throttled – hydrogen forms a highly explosive mixture with air and can easily be ignited at the leak where the gas heats up. Most other gases are cooled at a leak, – so also O_2 and N_2.

JOULE and THOMSON (LORD KELVIN) have first observed the cooling of gases by throttling. From the molecular point of view the phenomenon is easy to understand. By reducing the pressure we increase the specific volume and therefore the mean distance of the molecules. According to Fig. 2.12 a this means that the potential energy of the molecular interaction grows, and – since the process is adiabatic – the kinetic energy, hence the temperature must drop. The temperature drop is bigger, if the gas was colder before the throttle valve.

Table 4.1 Critical values of N_2, O_2, and H_2

	p_C [bar]	T_C [K]
N_2	34.5	126
O_2	51.8	155
H_2	13.2	33

Louis Paul CAILLETET (1832-1913) used this observation to liquefy small amounts of oxygen by throttling and Carl Ritter VON LINDE (1842-1934) invented a clever repetitive process in order to liquefy large amounts of air for technical purposes. He used the throttled air for pre-cooling newly supplied air before throttling it, and then he used that throttled air for pre-cooling yet more air, *etc.* LINDE succeeded to construct a continuous process upon this principle. Thus liquid air became a commodity when before it had been a scientific curiosity.

Louis Paul Cailletet

Carl Ritter von Linde

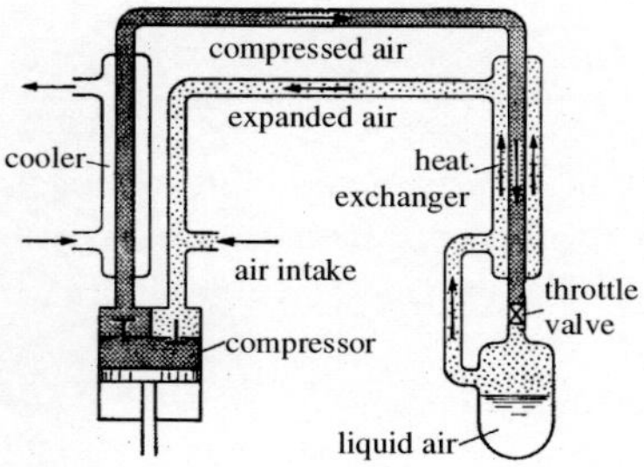

Linde's liquefying apparatus

4.2.7 *Available free energies*

We have seen that the entropy in an adiabatic system cannot decrease, so that in equilibrium – when all rates of change have come to an end – the entropy assumes its maximal value. We write

$$S \Rightarrow \text{ maximum in an adiabatic body.} \tag{4.57}$$

This is the prototype of all thermodynamic stability conditions. There are alternatives for non-adiabatic bodies, and we proceed to derive those.

We rely on the first and second laws for this purpose and write, – *cf.* Paragraphs 1.5.2 through 1.5.5 and 4.1.2 – ignoring radiation

$$\frac{\mathrm{d}(U + E_{\text{pot}} + K)}{\mathrm{d}t} = \dot{Q} + \dot{W}_{\text{stress}}\,, \quad \text{where } \dot{W}_{\text{stress}} = \int_{\partial V} t_{ij} w_i n_j \mathrm{d}A$$

$$\frac{\mathrm{d}S}{\mathrm{d}t} \geq \frac{\dot{Q}}{T}\,, \tag{4.58}$$

where T is a homogeneous temperature on ∂V. Elimination of $\dot{Q}$ provides

$$\frac{\mathrm{d}(U + E_{\mathrm{pot}} + K - TS)}{\mathrm{d}t} \leq -S\frac{\mathrm{d}T}{\mathrm{d}t} + \int\limits_{\partial V} t_{ij} w_i n_j \mathrm{d}A \,. \tag{4.59}$$

If the surface temperature is constant – as well as homogeneous – and if the surface does not move, so that its velocity is zero, we have the stability condition

$$U + E_{\mathrm{pot}} + K - TS \Rightarrow \text{ minimum for } T\text{=const and } w_i = 0 \quad \text{on } \partial V \,. \tag{4.60}$$

Note that inside V anything and everything may occur initially: turbulent motion, friction, heat conduction, phase changes, chemical reactions, *etc.* As long as the surface is at rest and has a constant homogeneous temperature, the expression $U + E_{\mathrm{pot}} + K - TS$ tends to a minimum as equilibrium is approached. Consequently we conclude that a *decrease of energy* is conducive to equilibrium and so is an *increase of entropy*. In a manner of speaking we might say that the energy *wants* to reach a minimum, and that the entropy *wants* to reach a maximum.

There is another interpretation of (4.59) when ∂V contains a movable part on which $t_{ij} = -p\delta_{ij}$ holds with a homogeneous pressure p, prescribed and maintained there. In that case $\int\limits_{\partial V} t_{ij} w_i n_j \mathrm{d}A = -p\frac{\mathrm{d}V}{\mathrm{d}t}$ holds and we may write

$$\frac{\mathrm{d}(U + E_{\mathrm{pot}} + K - TS)}{\mathrm{d}t} \leq -S\frac{\mathrm{d}T}{\mathrm{d}T} - p\frac{\mathrm{d}V}{\mathrm{d}t} \,.$$

So, if the surface temperature is constant and homogeneous on ∂V, and if p is constant and homogeneous on its movable part, we have the stability condition

$$U+E_{\mathrm{pot}}+K+pV-TS \Rightarrow \text{minimum, for } T\text{=const on } \partial V \text{ and for } p\text{=const on its movable part.} \tag{4.61}$$

These arguments illustrate that the thermodynamic stability conditions – just as mechanical ones – depend on the system and on its thermo-mechanical loading. The foregoing considerations do not in any way exhaust the possibilities: Indeed, the quantity that must be minimized, or maximized, may be different for different problems and must be carefully identified by the exploitation of the two relations (4.58) in each case. Generically we may call that quantity the *available free energy* and we shall usually denote it by the letter $\mathcal{A}$.

The above considerations indicate that, if the surface temperature T is small, $\mathcal{A}$ tends to a minimum, because its energetic part does. If, on the other hand, T is large, $\mathcal{A}$ tends to a minimum, because the entropy tends to a maximum. In general, however, – for intermediate values of T – both tendencies, the energetic one and the entropic one, have to compromise. We shall later treat some illustrative examples of this competition.

Note that the T in (4.58) through (4.61) is the homogeneous temperature of *the surface* ∂V and that p in (4.61) is the homogeneous pressure *on the movable part of* ∂V. Therefore the available free energies in (4.60) and (4.61) are not the Helmholtz free energy and Gibbs free energy, – not even without the kinetic energy K. How could they? After all, there is generally no single T, nor a single p inside V. Having said this, we must qualify: It is often the case that in the approach to equilibrium the temperature and the pressure are first to assume homogeneous values inside V, and the kinetic energy K becomes homogeneously zero, while other processes – like phase transitions and chemical reactions – are still approaching equilibrium. In such cases the availabilities $U + E_{\text{pot}} + K - TS$ in (4.60) and $U + E_{\text{pot}} + pV - TS$ in (4.61) are indeed the Helmholtz and Gibbs free energies, respectively, – plus the potential energy –, and they approach a minimum after the velocity and the gradients of temperature and pressure are "relaxed" to zero.

4.2.8 *Stability conditions*

We consider the situation, where in an adiabatic container of volume V a fixed piston divides a fluid into equal masses m with the states U_1 ,V_1 and U_2 ,V_2. The entropy of both parts together is the sum of the entropies of the individual parts

$$S_{\text{tot}}(U_1,V_1,U_2,V_2) = S(U_1,V_1) + S(U_2,V_2) \tag{4.62}$$

S_{tot} depends on four variables, but only two of them are independent, since we have $V=V_1+V_2$ and $U=U_1+U_2$ and both are constant. Let (U_1,V_1) be the independent pair.

If the piston is now allowed to move, the entropy S_{tot} tends to a maximum – by (4.57) – and we obtain as sufficient condition for the eventual equilibrium

$$\frac{\partial S_{\text{tot}}}{\partial U_1} = 0, \frac{\partial S_{\text{tot}}}{\partial V_1} = 0, \frac{\partial^2 S_{\text{tot}}}{\partial U_1 \partial U_1}\delta U_1^2 + 2\frac{\partial^2 S_{\text{tot}}}{\partial U_1 \partial V_1}\delta U_1 \delta V_1 + \frac{\partial^2 S_{\text{tot}}}{\partial V_1 \partial V_1}\delta V_1^2 < 0 \tag{4.63}$$

for arbitrary small values of δU_1, δV_1 . By use of (4.62) we obtain

$$\left(\frac{\partial S}{\partial U_1}\right)_{V_1} = \left(\frac{\partial S}{\partial U_2}\right)_{V_2}, \quad \text{and} \quad \left(\frac{\partial S}{\partial V_1}\right)_{U_1} = \left(\frac{\partial S}{\partial V_2}\right)_{U_2} \quad \text{and} \tag{4.64}$$

$$\frac{\partial^2 S}{\partial U_1^2}\delta U_1^2 + 2\frac{\partial^2 S}{\partial U_1 \partial V_1}\delta U_1 \delta V_1 + \frac{\partial^2 S}{\partial V_1^2}\delta V_1^2 < 0\,. \tag{4.65}$$

By the Gibbs equation (4.19) the equations (4.64) imply $T_1 = T_2$ and $p_1 = p_2$ – not unexpectedly for an equilibrium condition. The inequality (4.65) indicates that the entropy is a concave function of U_1 and V_1. For simplicity we drop the index 1 and write (4.65) in the form

$$\left[\frac{\partial}{\partial U}\left(\frac{\partial S}{\partial U}\right)\delta U + \frac{\partial}{\partial V}\left(\frac{\partial S}{\partial U}\right)\delta V\right]\delta U + \left[\frac{\partial}{\partial U}\left(\frac{\partial S}{\partial V}\right)\delta U + \frac{\partial}{\partial V}\left(\frac{\partial S}{\partial V}\right)\delta V\right]\delta V < 0.$$

By $\dfrac{\partial S}{\partial U} = \dfrac{1}{T}$ and $\dfrac{\partial S}{\partial V} = \dfrac{p}{T}$ we obtain

$$\delta\left(\frac{1}{T}\right)\delta U + \delta\left(\frac{p}{T}\right)\delta V < 0\,.$$

We insert

$$\delta U = \left(\frac{\partial U}{\partial T}\right)_V \delta T + \left(\frac{\partial U}{\partial V}\right)_T \delta V = m c_v \delta T + \left(-p + T\left(\frac{\partial p}{\partial T}\right)_V\right)\delta V \quad \text{and}$$

$$\delta\left(\frac{p}{T}\right) = -\frac{p}{T^2}\delta T + \frac{1}{T}\left(\frac{\partial p}{\partial T}\right)_V \delta T + \frac{1}{T}\left(\frac{\partial p}{\partial V}\right)_T \delta V$$

so that a quadratic form follows without a mixed term, *viz.*

$$-\frac{1}{T^2}c_v(\delta T)^2 + \frac{1}{T}\left(\frac{\partial p}{\partial V}\right)_T (\delta V)^2 < 0 . \tag{4.66}$$

Since this must hold for arbitrary small values of δT and δV, the inequality implies, *cf.* (2.12)

$$c_v > 0 \quad \text{and} \quad \kappa_T > 0 . \tag{4.67}$$

The specific heat c_v *and the compressibility* κ_T *are positive.*

We conclude that isochoric heating will always raise the temperature and that an isothermal increase of pressure will always decrease the volume. Both phenomena correspond so clearly to everyday experience, that any other result would be absurd. However, it is interesting to note that our experience conforms to the second law, since the present result is a direct consequence of that law.

The inequalities (4.67) have corollaries that are not quite as obvious. Indeed, by (4.34) and (4.36), they imply

$$\kappa_T > \kappa_S \quad \text{and} \quad c_p > c_v . \tag{4.68}$$

These relations are therefore also consequences of the Second Law. Nor are they the only corollaries of (4.67). A particularly interesting one results from (4.36) (2.14) and (4.68). It concerns the thermal expansion coefficient α^2 and it reads

$$\alpha^2 < \frac{1}{vT}c_p\kappa_T . \tag{4.69}$$

It follows that a vanishing compressibility implies that there is no thermal expansion either, *i.e.* the density ρ is strictly constant in an incompressible body; it depends neither on p nor T.

The conditions (4.67) through (4.69) are called thermodynamic stability conditions. They are universal restrictions on constitutive properties. If there existed a material, in which the restrictions did not apply, the Second Law would be violated by processes in such materials.

4.2.9 *Specific heat cp is singular at the critical point*

It is clear from $(4.67)_2$ that the van der Waals isotherms for $T<T_C$ violate thermodynamic stability in a range of states where $\frac{\partial p}{\partial v} > 0$ holds. Such states are unwanted byproducts of the statistical mechanical derivation of the van der Waals equation, and under most circumstances they must be ignored.* This is effectively done by the convexification procedure described in Paragraph 4.2.5.

The lowest monotonically decreasing isotherm is the critical one which, however, also violates thermodynamic stability, albeit in a single point, namely the critical point, where $\frac{\partial p}{\partial v} = 0$ holds. The caloric equation of state also exhibits a singular behavior in that point. In order to see that we consider the specific heats.

* The unstable states play a certain role in the investigation of hysteretic phenomena, *cf.* Sect. 12.5.

We have already seen, *cf.* (4.27) that c_υ in a van der Waals gas depends on T only and for the present purposes we take it to be constant.* On the other hand c_p is given by (4.36) as

$$c_p - c_\upsilon = \upsilon T \kappa_T \left(\frac{\partial p}{\partial T} \right)_\upsilon^2 .$$

κ_T is infinite at the critical point, while $\frac{\partial p}{\partial T}$ is finite. Therefore, as we approach the critical point, c_p tends to infinity. Fig. 4.7 shows graphs of c_p for superheated water vapor which illustrate the emerging singularity as the critical point is approached.

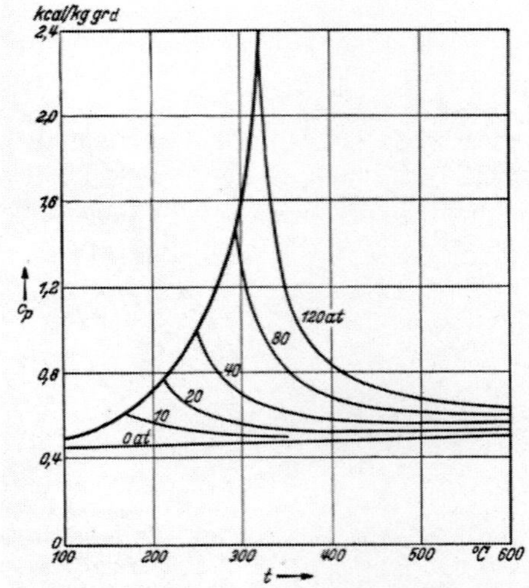

Fig. 4.7 c_p for superheated water vapor as a function of temperature. Points on the concave limiting curve correspond to saturated vapor. [The graph is taken from E.Schmidt, Thermodynamik. Springer, published in 1956. Hence the archaic units]

4.3 A layer of liquid heated from below – onset of convection

We consider a horizontal fluid layer of thickness D. The fluid is at rest and in a stationary state with respect to the density ρ and temperature T; it is heated from below so that the temperatures on top and at the bottom are T_D and T_0, respectively, with $T_0 > T_D$. The mass balance is identically satisfied and the balance of momentum and energy read

$$\text{momentum balance} \quad \frac{\mathrm{d}p}{\mathrm{d}x} = -\rho\, g \quad \text{with } p = p(\rho, T)$$

* The reference to the van der Waals equation is only convenient, not essential; the result of this paragraph holds for all fluids.

$$\text{energy balance}\quad \frac{\mathrm{d}q}{\mathrm{d}x}=0 \quad\text{with}\quad q=-\kappa\frac{\mathrm{d}T}{\mathrm{d}x}.$$

Hence follows for a constant thermal conductivity κ

$$T(x)=T_0+\frac{T_D - T_0}{D}x\text{ , and}$$

$$\underbrace{\left(\frac{\partial p}{\partial \rho}\right)_T}_{\frac{1}{\rho}\frac{1}{\kappa_T}}\frac{\mathrm{d}\rho}{\mathrm{d}x}+\underbrace{\left(\frac{\partial p}{\partial T}\right)_\rho}_{\frac{\alpha}{\kappa_T}}+\frac{T_D-T_0}{D}=-\rho g\ , \tag{4.70}$$

where the isothermal compressibility κ_T and the thermal expansion coefficient α may be introduced to give

$$\left.\frac{\mathrm{d}\rho}{\mathrm{d}x}\right|_{\text{rest}}=\alpha\rho\frac{T_0-T_D}{D}-\rho^2\kappa_T g\ . \tag{4.71}$$

This is the density gradient in the layer at rest. We conclude that the gradient is bigger than in an isothermal layer,* although generally still negative.** Thus for a given temperature T_D the fluid is lighter at the bottom than in the isothermal case. This may be suspected to be an unstable situation.

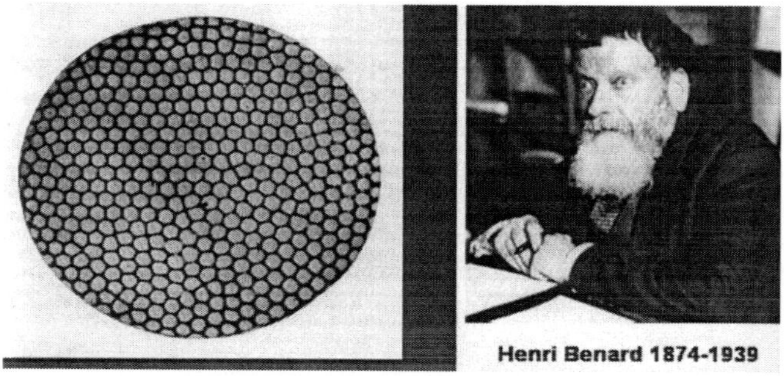

Fig. 4.8 Hexagonal convection cells. Henri Bénard, a pioneer of convection processes

Namely, the relatively light fluid at the bottom – relative to the isothermal case – may have a tendency to rise; or else, the relatively heavy fluid at the top has a tendency to sink. Observations show that this is indeed the case: The rising and sinking occurs in "convection bubbles" such that a light "bubble" rises and pushes the heavier fluid at the top sideways; subsequently that heavier fluid is sucked down into the space that opens below it when the ascending fluid is replaced by a horizontal flow at the bottom. In this way – after some time – a stationary pattern appears with fluid moving upwards, sideways and downwards. If the boundaries

---

* As long as $\alpha > 0$ is assumed which is usually the case, barring the anomaly of water between 0°C and 4°C.

** Of course, if T_0 is sufficiently bigger than T_D, the density may actually grow with height.

of the fluid permit it, the pattern is geometrically regular, and it may consist of hexagonal *convection cells* as shown in Fig. 4.8.

The rising and sinking bubbles are essentially adiabatic, because the thermal conductivity is too small to allow for an effective exchange of heat between the bubbles and the surrounding fluid. We use the adiabaticity to calculate the density gradient in the center of a cell, where the fluid moves vertically upwards. This gradient is called the *adiabatic density gradient*. We have

$$\dot{q}\mathrm{d}t = \mathrm{d}u + p\,\mathrm{d}\upsilon = 0 \quad \text{or}$$

$$\mathrm{d}h - \upsilon\mathrm{d}p = 0$$

$$\underbrace{\left(\frac{\partial h}{\partial T}\right)_p}_{c_p}\frac{\mathrm{d}T}{\mathrm{d}x} + \underbrace{\left(\left(\frac{\partial h}{\partial p}\right)_T - \upsilon\right)}_{-T\left(\frac{\partial \upsilon}{\partial T}\right)_p}\frac{\mathrm{d}p}{\mathrm{d}x} = 0$$

$$c_p\left(\frac{\partial T}{\partial \rho}\right)_p\frac{\mathrm{d}\rho}{\mathrm{d}x} + \left(c_p\left(\frac{\partial T}{\partial p}\right)_\rho - T\left(\frac{\partial \upsilon}{\partial T}\right)_p\right)\underbrace{\frac{\mathrm{d}p}{\mathrm{d}x}}_{-\rho g} = 0$$

$$c_p\frac{\mathrm{d}\rho}{\mathrm{d}x} + \left(c_p\left(\frac{\partial \rho}{\partial p}\right)_T - \frac{T}{\rho^2}\left(\frac{\partial \rho}{\partial T}\right)^2\right)\rho g = 0$$

Therefore we obtain

$$\left.\frac{\mathrm{d}\rho}{\mathrm{d}x}\right|_{\text{convection}} = \frac{g\alpha^2\rho T}{c_p} - \rho^2\kappa_T g\,. \tag{4.72}$$

Thus both gradients, the one for the fluid at rest and the convective – adiabatic – one, are bigger than the isothermal gradients, so that both situations are possibly unstable. It is now a question which of the situations is "more unstable" and concerning this we assume that the fluid at rest is stable, if

$$\left.\frac{\mathrm{d}\rho}{\mathrm{d}x}\right|_{\text{rest}} \le \left.\frac{\mathrm{d}\rho}{\mathrm{d}x}\right|_{\text{convection}}\,. \tag{4.73}$$

Otherwise the stationary convection is stable. The idea is this: When a bubble starts to rise and its density becomes larger than the density of the surrounding fluid at rest, the bubble will sink back and stability of the rest state is maintained. By (4.71), (4.72) the criterion (4.73) places a condition on the maximal value of T_0-T_D for which the fluid can be at rest, namely

$$\frac{T_0 - T_D}{D} \le \frac{g\alpha T}{c_p} \quad \text{(stability of the state of rest).} \tag{4.74}$$

By (4.72) – after an easy calculation – we see that the condition (4.74) is tantamount to saying that, for convection to occur, the temperature gradient of the rest-solution exceeds the adiabatic temperature gradient.

If the criterion (4.73) is believed, this inequality provides a necessary condition for stability of the rest state. We have ignored viscosity in the argument which, however, may serve to keep the fluid at rest up to a bigger temperature gradient T_0-T_D than the one implied by (4.74). Viscosity, *i.e.* internal friction, may delay the onset of motion.*

So far, entropy was not mentioned in the argument. However, if we introduce entropy, we obtain yet another suggestive form of the stability criterion for the state of rest. Since the fluid in the rising bubble undergoes a reversible adiabatic change, we have $\left.\frac{ds}{dx}\right|_{\text{convection}} = 0$. On the other hand, the entropy gradient for the fluid at rest reads

$$\left.\frac{ds}{dx}\right|_{\text{rest}} = \underbrace{\left(\frac{\partial s}{\partial p}\right)_T}_{-\alpha v} \underbrace{\frac{dp}{dx}}_{\rho g} + \underbrace{\left(\frac{\partial s}{\partial T}\right)_p}_{\frac{1}{T}c_p} \underbrace{\frac{dT}{dx}}_{\frac{T_D - T_0}{D}}$$

$$\left.\frac{ds}{dx}\right|_{\text{rest}} = \frac{c_p}{T}\left(\frac{T_D - T_0}{D} + \frac{\alpha g T}{c_p}\right). \tag{4.75}$$

Comparison with (4.74) shows that the rest state is stable as long as $\left.\frac{ds}{dx}\right|_{\text{rest}} \geq 0$, *i.e.* as long as the entropy grows with height.

Convection plays an important role in the heating of our houses. Indeed, the warm air on top of the heater, – sometimes called "radiator" – rises and pushes the cold air near the ceiling down into contact with the heater. Thus a convection cell develops in our rooms and that is responsible for efficient, more or less homogeneous heating. If the heating of a room had to rely on conduction, it would take a very long time indeed.

On a larger scale convection of air, and thermal updrafts, are responsible for the mixing of air in the lower atmosphere. If the air contains moisture, the upward convection – and the concomitant adiabatic cooling – creates cumulus clouds, the "good weather clouds" on a fine summer day. We describe that phenomenon in Paragraph 10.4.

On the global scale convection creates the earth-spanning convections cells: The Hadley cells on both sides of the equator, the Ferell cells in the temperate zones of the earth, and the polar cells. To a very large extent these convection cells are responsible for the moderate temperatures on much of the earth's surface.

We have already discussed the convection in water and the seas, *cf.* Paragraph 2.4.4. In that case the anomaly of water – with $\alpha < 0$ – adds a complication according to (4.71). But the complication is benign; it helps to make the earth habitable as we have argued in Chapter 2.

Finally convection plays a very important part in stars like the sun, because it is essential for the transport of the nuclear heat in the inner core of the star to the surface.

For all this confirmed importance of convection we must realize that the stability criterion (4.74) is *ad hoc*. It may be plausible but it does not follow from the Second Law – or not in any way that we are aware of.

---

* A proper mathematical treatment of this problem does not use the condition (4.73). It proceeds by a linear stability analysis of the field equations, – including viscous terms –, under small disturbances. That treatment requires numerical methods and it is outside the scope of this book.

4.4 On the history of the Second Law

Apart from Clausius, whose ideas we have discussed in detail, we must mention Nicolas Léonard Sadi CARNOT (1796-1832) as one of the forerunners of the Second Law. CARNOT´s work *Reflexions sur la Puissance Motrice du Feu et sur les Machines Propres a développer cette Puissance* appeared in 1824. At that time there were already thousands of steam engines at work – most of them in England – and their efficiency was less than 10%.

CARNOT posed the question, whether and how the efficiency might be increased. Everything seemed possible: It seemed conceivable that the efficiency could be improved by changing the process of the steam engines, or by replacing water as the working agent by some other material, *e.g.* mercury or sulphur. CARNOT obtained two results in this matter.

- The maximal efficiency of a heat engine working between a highest temperature T_H and a lowest temperature T_L occurs when the engine exchanges heat only at those temperatures. That process is now called a Carnot process, and CARNOT writes that it is

  … le plus avantageux possible, car il ne s´est fait aucun rétablissement inutile d´équilibre dans le calorique.

- The efficiency of a Carnot engine is a universal function of T_H and T_L. It is independent of the working agent and does not depend either one the volume or pressure range that is covered by the process. In CARNOT´s words:

  Le maximum de puissance motrice résultant de l´emploi de la vapeur est aussi le maximum de puissance motrice réalisable par quelque moyen que ce soit.

In the shortest possible form this means that

The Carnot efficiency is maximal and universal.*

That CARNOT could arrive at these correct conclusions, shows the power of his intuition rather than his physical insight. Indeed, CARNOT's argument is largely based on the then prevalent caloric theory, by which heat was considered a weightless fluid.

The caloric theory assumed that the caloric – the heat stuff – is added to an engine in the boiler and comes out of the cooler, *unchanged in amount*. In the transition from boiler to cooler heat was supposed to perform work much as water does when it falls from a height. That is a plausible idea, perhaps, but it turned out to be wrong.

It was left to CLAUSIUS to correct these ideas. He did this in 1854 in his work: *Über eine veränderte Form des zweiten Hauptsatzes der mechanischen Wärmetheorie*. In that work he calculates the efficiency of a Carnot engine and he confirms CARNOT´s statements about the efficiency. CLAUSIUS comes to the conclusion that it is Q/T – not Q itself – that passes through a heat engine from boiler to cooler unchanged in amount and that quantity he calls the *entropy*. The motivation for inventing this word is a little vague: CLAUSIUS envisages changes of heat to work and of heat of high temperature to heat of low temperature, and he sees entropy as a measure for the capacity of such changes. He says:

Ich habe den Vorschlag gemacht, diese Verwandlung nach dem griechischen Wort τροπη die Entropie des Körpers zu nennen.

* CARNOT did not have the form $1-\frac{T_L}{T_H}$ for the efficiency. That came 30 years later with CLAUSIUS in 1854.

... on peut comparer avec assez de justesse la puissance motrice de la chaleur à celle d´une chute d´eau : toutes deux ont une maximum que l´on ne peut pas dépasser, quelle que soit d´une part la machine employée à recevoir l´action de l´eau, et quelle que soit de l´autre la substance employées à recevoir l´action de la chaleur. La puissance motrice d´une chute d´eau dépend de sa hauteur et de la quantité du liquide ; la puissance motrice de la chaleur dépend aussi de la quantité de calorique employé, et de ce ... que nous appellerons en effet la hauteur de sa chute, c'est á dire de la différence de température.

From the beginning the entropy met with great interest in Western science and it provoked opposition. The criticism was directed primarily against the teleological tendency expressed by the entropy *inequality*, *e.g.* (4.22). The monotonic trend to a featureless homogeneous equilibrium ran counter to the idealistic idea of progress to an ever more complex world. And it was not only natural scientists who concerned themselves with entropy, but also philosophers, sociologists, and historians.

CLAUSIUS himself had provided the catchword: He pointed out that the entropy of the universe – presumably an adiabatic system – could only grow until, in a final equilibrium, the universe would die the *heat death*.

It is often said – remarks CLAUSIUS – that the world runs in a cycle. The second law of the mechanical theory of heat contradicts that idea most decisively. Indeed, one must conclude that in all natural phenomena the total value of entropy can never decrease. The entropy tends to a maximum.

The closer the world approaches that limiting value – the maximum of entropy – the fainter will be the causes for a further change and eventually, when the maximum is reached, no change can occur anymore. The world is then in a dead stagnant state.

Johann Joseph LOSCHMIDT (1821-1895) deplored *the terroristic nimbus of the second law which lets it appear as a destructive principle of all life in the universe.*

The American historian Henry ADAMS (1838-1918) cites CLAUSIUS´s formulation of the second law and has this to say:
For the ordinary and unsophisticated historian this only means that the ash heap grows ever bigger.

It should be noted that ADAMS was an inveterate pessimist; he considered optimism as a sure sign of idiocy.

Ostwald SPENGLER (1880-1936) an illustrious historian and philosopher – author of the book *The Decline of the West* – philosophized on entropy as follows:
The end of the world as the completion of an intrinsically necessary development – that is the twilight of the gods [of Germanic mythology]*; therefore the doctrine of entropy is the last, non-religious version of the myth.*

With such extrapolations of thermodynamics it is not always entirely clear what the authors think when they speculate about the First and Second Law. Let us therefore also quote a knowledgeable American jester who succinctly formulated the laws with respect to heat and work and efficiency thus:

1st Law: *You can´t win.*
2nd Law: *You cannot even break even.*

5 Entropy as *S=k* ln*W*

5.1 Molecular interpretation of entropy

Mass and momentum of a gas or a fluid are easy to interpret in terms of atoms and molecules: They are simply the sums of the masses and momenta of the constituent molecules. Pressure and temperature may also convincingly be interpreted by molecular processes and properties, *e.g.* see Prologue 4. Energy is more subtle: It is true that the kinetic energy of a gas is the sum of the kinetic energies of the molecules, but energy also contains the potential energy of the molecular interaction between at least pairs of two particles; often clusters of more than two.

Much more difficult is the molecular interpretation of entropy. Entropy represents a property of the ensemble of all molecules and depends on their spatial distribution and on the distribution of momenta in the gas. This interpretation was found by Ludwig Eduard BOLTZMANN (1844-1906) who – along with James Clerk MAXWELL (1831-1879) – developed the kinetic theory of gases. Here we cannot go into that theory in any detail, and therefore we cannot explain the derivation of the molecular interpretation of entropy.

However, we can provide the relevant simple formula and we can apply it, and make it plausible.

According to BOLTZMANN we have

$$S = k \ln W \,, \tag{5.1}$$

where k is the BOLTZMANN constant, *cf.* (P.6). W is the number of possibilities to realize a state of a gas. BOLTZMANN succeeded to show that W – and $k \ln W$ – cannot decrease in an adiabatic gas and that its maximal value equals the equilibrium entropy (4.30) of an ideal gas.

BOLTZMANN's formula (5.1) may be extrapolated away from gases and it finds applications in the calculation of entropy for arbitrary bodies. In order to illustrate this universal character of the formula we shall proceed to use it – in juxtaposition – for the calculation of the entropy of a gas and the entropy of a polymer molecule.

5.2 Entropy of a gas and of a polymer molecule

We use the BOLTZMANN formula in order to calculate the entropy of

a gas at rest consisting of N atoms with independent positions and velocities. The internal energy is given by U.	a polymer molecule modeled by a long chain of N links with independent orientations. The end-to-end distance is given by r.

Fig. 5.1 illustrates these models. The volume of the gas is V, and the length of a link in the polymer chain is b.

The states of the two systems are characterized as follows

$N_{\underline{x}\underline{c}}$ atoms with the velocity $\underline{c}$ lie at the position $\underline{x}$. We have

$$\sum_{\underline{x},\underline{c}} N_{\underline{x}\underline{c}} = N\,, \quad \sum_{\underline{x},\underline{c}} \frac{m}{2} c^2 N_{\underline{x}\underline{c}} = U$$

$N_{\vartheta\varphi}$ links point into the direction ϑ, φ. ϑ denotes the polar angle between a link and the direction of the end-to-end distance. φ is the corresponding azimuthal angle. We have

$$\sum_{\vartheta,\varphi} N_{\vartheta\varphi} = N\,, \quad \sum_{\vartheta,\varphi} b \cos\vartheta\, N_{\vartheta\varphi} = r \quad (5.2)$$

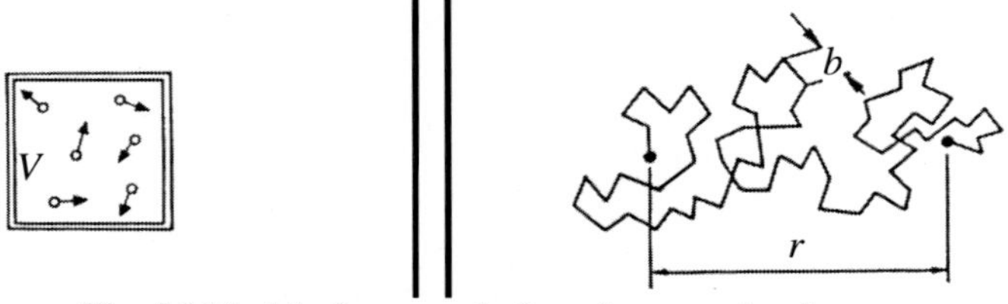

Fig. 5.1 Model of a gas and of a polymer molecule

Thus the states are given by the distributions $\{N_{\underline{x}\underline{c}}\}$ over all $\underline{x}$, $\underline{c}$, and $\{N_{\vartheta\varphi}\}$ over all ϑ, φ, respectively, which satisfy the constraints (5.2).

A typical realization of these states is given by

atoms number
5, 13, …, N-7 lie at the 1st point $\underline{x}$, $\underline{c}$
15,29, …, N-23 lie at the 2nd point $\underline{x}$, $\underline{c}$,
…
no atom lies at the i th point $\underline{x}$, $\underline{c}$,
…
53, 104, …, N-5 lie at the last point $\underline{x}$, $\underline{c}$.

links number
3, 14, …, N-5 point in 1st direction ϑ, φ,
4, 5, …, N-15 point in 2nd direction, ϑ, φ
…
…
…
5, 1, …, N-1 point in last direction ϑ, φ.

By the rules of combinatorics the number of realizations of a state thus reads

$$W = \frac{N!}{\prod_{\underline{x},\underline{c}} N_{\underline{x}\underline{c}}!} \qquad\qquad W = \frac{N!}{\prod_{\vartheta,\varphi} N_{\vartheta\varphi}!}\,, \qquad (5.3)$$

where the product ranges over all $\underline{x}$, $\underline{c}$ and all ϑ, φ. Insertion into (5.1) provides the entropies of the gas and of the rubber molecule in the states $\{N_{\underline{x}\underline{c}}\}$ and $\{N_{\vartheta\varphi}\}$.

We proceed to show that the Boltzmann entropy (5.1) grows to a maximum. This is best argued for the polymer molecule: One must realize that the chain-links are constantly moving, because they are subject to thermal motion. Therefore the chain will change its shape and end-to-end distance continuously at a rate of – typically – 10^{12} times per second.* Now it is assumed that each realization – often called a microstate – occurs just as frequently as any other one. This is the only possible unbiased assumption and it is therefore forced upon us. And it implies

* Particles in liquids and solids typically have 10^{12} collisions per second, while in gases, which are one thousand times more rarefied, that number is 10^9.

that a state with many microstates occurs more often in the course of thermal motion than a state with few microstates. Most frequent is *the* state which is realizable by most microstates and this is – rather obviously – a state where the links are isotropically distributed and where the end-to-end distance is zero. Thus, if we start out with a fully stretched polymer molecule with end-to-end distance Nb, the state can only be realized in one single manner; therefore we have $W=1$ and $S=0$ initially. The thermal motion will, however, quickly destroy this orderly arrangement and lead to states with kinked molecules which have more realizations and therefore higher entropy. Eventually the chain will end up in the state with most realizations and maximal entropy. That is the nature of entropic growth, and the final state is called an equilibrium.

Obviously this interpretation of entropy has a probabilistic flavor. Indeed, since all realizations occur equally frequently, there is a certain – very small – probability that the molecule departs from equilibrium and assumes a state with less realizations. This happens in a thermal fluctuation.

For the determination of the equilibrium entropies we calculate the distributions $\{N_{\underline{x}\underline{c}}\}$ and $\{N_{\vartheta\varphi}\}$ which maximize W or $S=k\ln W$. Of course, the constraints (5.2) must be observed. This requires two Lagrange multipliers – α and β – such that the expressions

$$k\ln\frac{N!}{\prod\limits_{\underline{x},\underline{c}} N_{\underline{x}\underline{c}}!}-\alpha\left(\sum_{\underline{x},\underline{c}} N_{\underline{x}\underline{c}}-N\right)-\beta\left(\sum_{\underline{x},\underline{c}}\frac{m}{2}c^2 N_{\underline{x}\underline{c}}-U\right)$$

$$k\ln\frac{N!}{\prod\limits_{\vartheta,\varphi} N_{\vartheta\varphi}!}-\alpha\left(\sum_{\vartheta\varphi} N_{\vartheta\varphi}-N\right)-\beta\left(\sum_{\vartheta,\varphi} b\cos\vartheta\, N_{\vartheta\varphi}-r\right)$$

must be maximized *without* constraints. These are functions of N_{xc} and $N_{\vartheta\varphi}$; their derivatives with respect to $N_{\underline{x}\underline{c}}$ and $N_{\vartheta\varphi}$ must vanish.

For the calculation we assume that all $N_{\underline{x}\underline{c}}$ and all $N_{\vartheta\varphi}$ are much bigger than 1. In this case we may use the Stirling approximation $\ln a! \approx a\ln a - a$ and write

$$k\ln\frac{N!}{\prod\limits_{\underline{x},\underline{c}} N_{\underline{x}\underline{c}}!}\approx N\ln N-\sum_{\underline{x},\underline{c}} N_{\underline{x}\underline{c}}\ln N_{\underline{x}\underline{c}} \qquad k\ln\frac{N!}{\prod\limits_{\vartheta,\varphi} N_{\vartheta\varphi}!}\approx N\ln N-\sum_{\vartheta,\varphi} N_{\vartheta\varphi}\ln N_{\vartheta\varphi} \tag{5.5}$$

Thus the equilibrium distributions come out as

$$N_{\underline{x}\underline{c}}^{\mathrm{E}}=\mathrm{e}^{-\left(1+\frac{\alpha}{k}\right)}\mathrm{e}^{-\frac{\beta}{k}\frac{m}{2}c^2} \qquad N_{\vartheta\varphi}^{\mathrm{E}}=\mathrm{e}^{-\left(1+\frac{\alpha}{k}\right)}\mathrm{e}^{-\frac{\beta}{k}b\cos\vartheta}. \tag{5.6}$$

For the determination of the Lagrange multipliers we insert these distributions into the constraints (5.2) and obtain

$$e^{-\left(1+\frac{\alpha}{k}\right)}=N\frac{1}{\sum\limits_{\underline{x},\underline{c}} e^{-\frac{\beta}{k}\frac{m}{2}c^2}} \qquad \Bigg\| \qquad e^{-\left(1+\frac{\alpha}{k}\right)}=N\frac{1}{\sum\limits_{\vartheta,\varphi} e^{-\frac{\beta}{k}b\cos\vartheta}} \tag{5.7}$$

$$U=N\frac{\sum\limits_{\underline{x},\underline{c}}\frac{m}{2}c^2 e^{-\frac{\beta}{k}\frac{m}{2}c^2}}{\sum\limits_{\underline{x},\underline{c}} e^{-\frac{\beta}{k}\frac{m}{2}c^2}} =-Nk\frac{\partial}{\partial\beta}\left[\ln\sum_{\underline{x},\underline{c}} e^{-\frac{\beta}{k}\frac{m}{2}c^2}\right] \qquad \Bigg\| \qquad r=N\frac{\sum\limits_{\vartheta,\varphi} b\cos\vartheta\, e^{-\frac{\beta}{k}b\cos\vartheta}}{\sum\limits_{\vartheta,\varphi} e^{-\frac{\beta}{k}b\cos\vartheta}} =-Nk\frac{\partial}{\partial\beta}\left[\ln\sum_{\vartheta,\varphi} e^{-\frac{\beta}{k}b\cos\vartheta}\right]. \tag{5.8}$$

The equations (5.7) permit the determination of α in terms of β, while β is determined by (5.8) in terms of U, or r, respectively, albeit implicitly, because the equations (5.8) cannot be solved for β analytically.

The equilibrium entropies are obtained by insertion of (5.6), (5.7) into $S=k\ln W$ with W from (5.3). With the Stirling formula we obtain for the equilibrium entropy S^{E} of the gas and for the equilibrium entropy $S^{\mathrm{E}}{}_{\mathrm{Mol}}$ of the polymer molecule

$$S=k\left(N\ln N-\sum_{\underline{x},\underline{c}} N_{\underline{x}\underline{c}}\ln N_{\underline{x}\underline{c}}\right) \qquad \Bigg\| \qquad S_{\mathrm{Mol}}=k\left(N\ln N-\sum_{\vartheta,\varphi} N_{\vartheta\varphi}\ln N_{\vartheta\varphi}\right)$$

$$S^{\mathrm{E}}=\beta U+Nk\ln\left(\sum_{\underline{x},\underline{c}} e^{-\frac{\beta}{k}\frac{m}{2}c^2}\right) \qquad \Bigg\| \qquad S^{\mathrm{E}}_{\mathrm{Mol}}=\beta r+Nk\ln\left(\sum_{\vartheta,\varphi} e^{-\frac{\beta}{k}b\cos\vartheta}\right) \tag{5.9}$$

For further evaluation of (5.8) and (5.9) we must determine the sums occurring in these equations. This requires that the sums be converted to integrals by assuming that

the number of positions between $\underline{x}$ and $\underline{x}+\mathrm{d}\underline{x}$, and velocities between $\underline{c}$ and $\underline{c}+\mathrm{d}\underline{c}$ is proportional to $\mathrm{d}\underline{x}\,\mathrm{d}\underline{c}=\mathrm{d}x_1\mathrm{d}x_2\mathrm{d}x_3\mathrm{d}c_1\mathrm{d}c_2\mathrm{d}c_3$, or equal to $Y\mathrm{d}\underline{x}\,\mathrm{d}\underline{c}$.	the number of orientations of the chain links between the angles ϑ and $\vartheta+\mathrm{d}\vartheta$, and φ and $\varphi+\mathrm{d}\varphi$ is proportional to the element of solid angle $\sin\vartheta\,\mathrm{d}\vartheta\mathrm{d}\varphi$, or equal to $Z\sin\vartheta\,\mathrm{d}\vartheta\mathrm{d}\varphi$.

The factors of proportionality Y and Z "quantize" the $(\underline{x},\underline{c})$-space and the unit sphere. Indeed, $\frac{1}{Y}$ is obviously the volume of the smallest cell $\mathrm{d}\underline{x}\,\mathrm{d}\underline{c}$, *i.e.* the cell that contains only one point $\underline{x},\underline{c}$. Analogously $\frac{1}{Z}$ determines the smallest element of solid angle, which contains only one orientation. Thus one obtains

$$\sum_{\underline{x},\underline{c}} \mathrm{e}^{-\frac{\beta}{k}\frac{m}{2}c^2} = Y \int_V \int_{-\infty}^{\infty} \sum_{\underline{x},\underline{c}} \mathrm{e}^{-\frac{\beta}{k}\frac{m}{2}c^2} \, \mathrm{d}\underline{x}\, \mathrm{d}\underline{c} = YV\sqrt{\frac{2\pi k}{\beta m}}^{3}$$

$$\sum_{\vartheta,\varphi} \mathrm{e}^{-\frac{\beta}{k} b\cos\vartheta} = Z \int_0^{\pi}\int_0^{2\pi} \mathrm{e}^{-\frac{\beta}{k} b\cos\vartheta} \sin\vartheta \, \mathrm{d}\vartheta \, \mathrm{d}\varphi = 4\pi Z \frac{\sinh\left(\frac{\beta}{k} b\right)}{\frac{\beta}{k} b}. \tag{5.10}$$

Therefore the entropies (5.9) come out as

$$S^{\mathrm{E}} = \beta U + Nk \ln\left[YV\sqrt{\frac{2\pi k}{\beta m}}^{3} \right] \qquad S^{\mathrm{E}}_{\mathrm{Mol}} = \beta r + Nk \ln\left[4\pi Z \frac{\sinh\left(\frac{\beta}{k} b\right)}{\frac{\beta}{k} b} \right].$$

In both cases the final task is the determination of β from (5.10) and (5.8). We obtain

$$\frac{U}{N} = +\frac{3}{2}\frac{k}{\beta}$$

In the gas we may therefore determine β explicitly in terms of U/N and since we know from Section P.4 that $U/N = \frac{3}{2}kT$ holds, we obtain

$$\beta = \frac{1}{T}$$

Thus the entropy of a monatomic gas reads

$$\frac{S^{\mathrm{E}}}{N\mu} = \frac{3}{2}\frac{k}{m}\ln T + \frac{k}{m}\ln \upsilon + \text{const},$$

and this expression corresponds – in its T- and υ-dependence – to the entropy (4.30) of a monatomic gas.

$$\frac{r}{Nb} = \frac{1}{\frac{\beta}{k} b} - \operatorname{ctgh}\left(\frac{\beta}{k} b\right). \tag{5.12}$$

For the polymer molecule (5.12) cannot be solved explicitly for β, except in the case of strong entanglement of the chain, which means $\frac{r}{Nb} \ll 1$. In this case the right hand side may be expanded linearly in $\frac{\beta}{k} b$ and we obtain

$$\beta = -3k\frac{r}{Nb^2}. \tag{5.13}$$

Insertion into (5.9) gives – again for strong entanglement –

$$S^{\mathrm{E}}_{\mathrm{Mol}} = Nk\left[\ln(4\pi Z) - \frac{r^2}{\frac{1}{3}N^2 b^2} \right]. \tag{5.14}$$

The equilibrium entropy of the polymer molecule is therefore biggest when the end-to-end distance r is zero. For the present case of strong entanglement it decreases quadratically with growing r.

5.3 Entropy as a measure of disorder

In the effort to give a suggestive interpretation of entropy one says that the entropy is a measure of disorder. Although this interpretation is sometimes somewhat precarious – “disorder” is not well defined – it is most easily understood for a polymer molecule: We consider the molecular model of Fig. 5.1, take N=32 and think of the chain in its fully stretched microstate, so that we have $r = 32b$. This most "orderly" state can only be realized in one way so that $W = 1$ holds and $S_{\text{Mol}} = 0$. The microstates, where just one link points "backwards" is already less orderly; it may be realized in 32 ways and therefore $S = k\ln 32 = 3.46\,k$ holds. A less trivial case is the one shown in Fig. 5.2: Three "disorderly" arrangements of 32 links in four directions with the distribution {9,9,2,12} and the large entropy $S = k\ln\frac{32!}{9!9!2!12!} = 35.27\,k$.

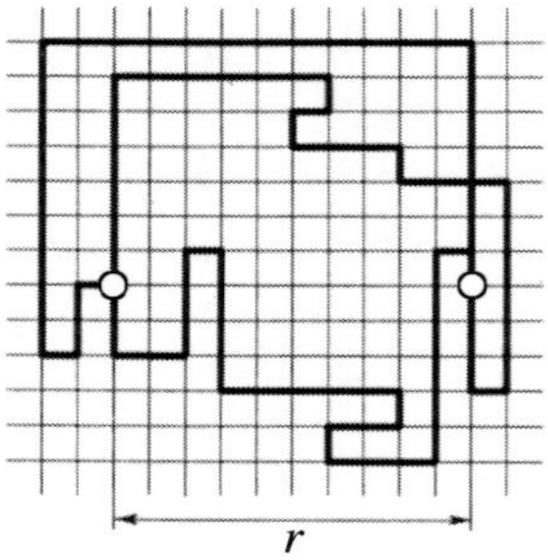

Fig. 5.2 Three out of 2.086 10^{15} microstates of the distribution {9,9,2,12}

Similar calculations make it clear that – for fixed r and fixed N – there are more and more possible realizations for more and more possible orientations. This, however, means that the entropy is bigger for a more disorderly state.

With this in mind an anonymous joker with an original turn of mind has characterized the cultural progress of our society to a higher entropy and more disorder by the following graffito:

Hamlet: to be or not to be
Camus: to be is to do
Satre: to do is to be
Sinatra: do be do be do.

Let this be a not entirely serious example for the possibility to extrapolate the concept of entropy and order.

5.4 Maxwell distribution

We briefly comment on the result $(5.6)_1$, $(5.7)_1$ for the equilibrium distribution function. We have

$$N_{\underline{x}\underline{c}} = N\frac{e^{-\frac{\beta}{k}\frac{m}{2}c^2}}{\sum\limits_{\underline{x},\underline{c}} e^{-\frac{\beta}{k}\frac{m}{2}c^2}} \quad \text{and, by } (5.10)_1\text{:}$$

$$N_{\underline{x}\underline{c}} = \frac{N}{YV}\frac{1}{\sqrt{\frac{2\pi k}{\beta m}}^{3}} e^{-\frac{mc^2}{2kT}}. \tag{5.15}$$

This is the Maxwell distribution. It implies that the particle distribution in equilibrium is homogeneous in space and a Gaussian function in the velocities. We may use this relation to calculate the mean kinetic energy of an atom and obtain

$$\frac{m}{2}\overline{c}^2 = \sum_{\underline{x},\underline{c}} \frac{m}{2} c^2 \frac{N_{\underline{xc}}}{N} = \frac{1}{YV} \frac{1}{\sqrt{\frac{2\pi k}{\beta m}}^3} \sum_{\underline{x},\underline{c}} \mathrm{e}^{-\frac{\beta}{k}\frac{m}{2}c^2} = \frac{3}{2} kT . \tag{5.16}$$

In the last step we have again replaced the sums by an integral as before in (5.10). Equation (5.16) confirms the result – anticipated in Sect. P.4 – that the temperature of a gas represents the mean kinetic energy of the atoms.

5.5 Entropy of a rubber rod

Rubber consists of long polymer chains that form a network as shown in Fig. 5.3 schematically, where the end-to-end distances of the chains are represented by lines and the joints by dots. The left figure shows a rod in its natural – *i.e.* load-free – configuration and length L_0. On the right hand side we see the rod stretched to the length L by a tensile force P. The deformation gradient in the vertical direction is $\lambda = L/L_0$ and the lateral contractions are $1/\sqrt{\lambda} = \sqrt{L_0/L}$ so that the volume of the rod is not changed by the deformation; indeed, rubber is considered as incompressible.

The entropy of the rod is equal to the sum of entropies $S_{\mathrm{Mol}}^{\mathrm{E}}$ of the rubber molecule between the joints

$$S_0 = \int S_{\mathrm{Mol}}^{\mathrm{E}}(\vartheta_1,\vartheta_2,\vartheta_3)\, z_0(\vartheta_1,\vartheta_2,\vartheta_3)\,\mathrm{d}\vartheta_1\,\mathrm{d}\vartheta_2\,\mathrm{d}\vartheta_3 \text{ in the natural state} \tag{5.17}$$

$$S = \int S_{\mathrm{Mol}}^{\mathrm{E}}(\vartheta_1,\vartheta_2,\vartheta_3)\; z(\vartheta_1,\vartheta_2,\vartheta_3)\,\mathrm{d}\vartheta_1\,\mathrm{d}\vartheta_2\,\mathrm{d}\vartheta_3 \text{ in the stretched state.}$$

ϑ_i are the components of the distance vectors between two joints, and $z_0(\vartheta_1,\vartheta_2,\vartheta_3)$ or $z(\vartheta_1,\vartheta_2,\vartheta_3)$ are the corresponding numbers of distance vectors in the range ϑ_i and $\vartheta_i + \mathrm{d}\vartheta_i$. Because of $S_{\mathrm{Mol}}^{\mathrm{E}} = k \ln W$ in the natural state that number is proportional to

$$W = \exp\left(\frac{S_{\mathrm{Mol}}^{\mathrm{E}}}{k}\right), \text{ hence } z_0(\vartheta_1,\vartheta_2,\vartheta_3) = C \exp\left(-\frac{\vartheta_1^2+\vartheta_2^2+\vartheta_3^2}{\frac{1}{3}Nb^2}\right),$$

where C is obtained by integration, because the total number of chains is n. We thus have

$$z_0(\vartheta_1,\vartheta_2,\vartheta_3) = \frac{n}{\sqrt{\frac{1}{3}\pi N b^2}^3} \exp\left(-\frac{\vartheta_1^2+\vartheta_2^2+\vartheta_3^2}{\frac{1}{3}Nb^2}\right). \tag{5.18}$$

The distribution function $z(\vartheta_1,\vartheta_2,\vartheta_3)$ in the deformed state results from $z_0(\vartheta_1,\vartheta_2,\vartheta_3)$ by the following argument: One assumes that the deformation gradient F – *cf.* Fig. 5.3 – does not only determine the total deformation of the rod but also the deformation of each distance vector $\vartheta_1, \vartheta_2, \vartheta_3$. In other words we assume

$$\vartheta_i = F_{ij}\vartheta_j^0 \quad \text{or, explicitly} \quad \begin{bmatrix} \vartheta_1 \\ \vartheta_2 \\ \vartheta_3 \end{bmatrix} = \begin{bmatrix} \lambda & 0 & 0 \\ 0 & \frac{1}{\sqrt{\lambda}} & 0 \\ 0 & 0 & \frac{1}{\sqrt{\lambda}} \end{bmatrix} \begin{bmatrix} \vartheta_1^0 \\ \vartheta_2^0 \\ \vartheta_3^0 \end{bmatrix}. \tag{5.19}$$

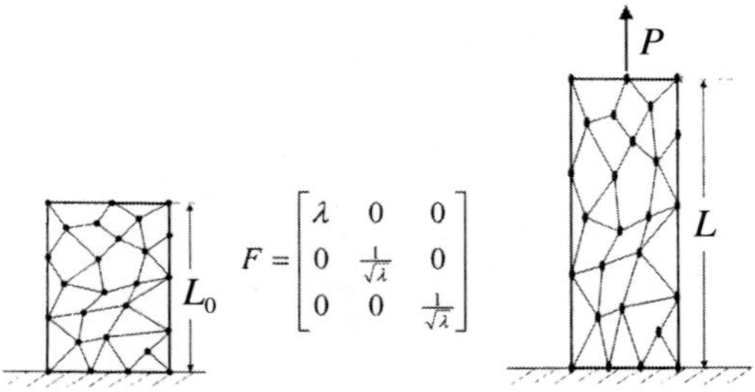

Fig. 5.3 Rubber rod in natural and stretched states

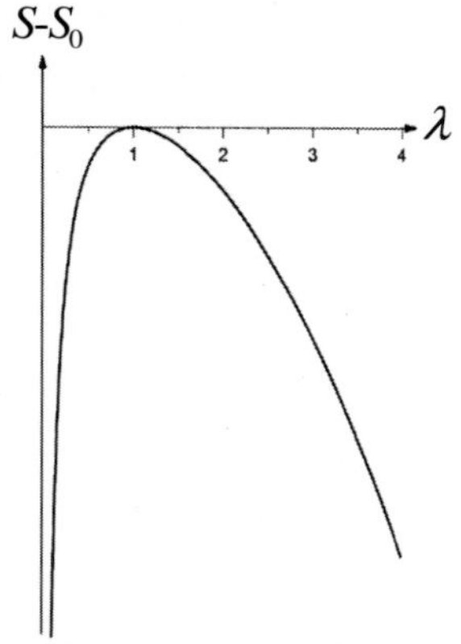

Fig. 5.4 Entropy of a rubber rod as a function of the deformation λ.

And, since during the deformation no distance vectors get lost, we must have

$$z_0\left(\vartheta_1^0, \vartheta_2^0, \vartheta_3^0\right) d\vartheta_1^0\, d\vartheta_2^0\, d\vartheta_3^0 = z(\vartheta_1, \vartheta_2, \vartheta_3)\, d\vartheta_1\, d\vartheta_2\, d\vartheta_3\,, \tag{5.20}$$

if ϑ_i and ϑ_i^0 are related by (5.19). The volume elements $d\vartheta_1\, d\vartheta_2\, d\vartheta_3$ and $d\vartheta_1^0\, d\vartheta_2^0\, d\vartheta_3^0$ are equal because of the incompressibility of rubber. Thus follows from (5.19), (5.20)

$$z(\vartheta_1, \vartheta_2, \vartheta_3)\, d\vartheta_1\, d\vartheta_2\, d\vartheta_3 = z_0\left(\frac{\vartheta_1}{\lambda}, \sqrt{\lambda}\vartheta_2, \sqrt{\lambda}\vartheta_3\right) \tag{5.21}$$

or, by (5.18)

$$z(\vartheta_1,\vartheta_2,\vartheta_3)=\frac{n}{\sqrt{\frac{1}{3}\pi Nb^2}^{\,3}}\exp\left[-\frac{\frac{1}{\lambda^2}\vartheta_1^2+\lambda\left(\vartheta_2^2+\vartheta_3^2\right)}{\frac{1}{3}Nb^2}\right]. \tag{5.22}$$

Insertion of (5.18) and (5.22) into (5.17) and a simple calculation provide

$$S_0=nk\left[N\ln(4\pi Z)-\frac{3}{2}\right] \quad\text{and}\quad S=nk\left[N\ln(4\pi Z)-\left(\frac{\lambda^2}{2}+\frac{1}{\lambda}\right)\right]. \tag{5.23}$$

Fig. 5.4 shows $S-S_0$ as a function of λ. The entropy of the rod is maximal in the natural state.

We proceed to calculate the thermal equation of state of rubber from the entropy (5.23). This equation determines the load P as a function of length L and temperature T.

5.6 Examples for entropy and Second Law. Gas and rubber

5.6.1 *Gibbs equation and integrability condition for liquids and solids*

The two laws of thermodynamics

$$\frac{\mathrm{d}(U+K)}{\mathrm{d}t}=\dot{Q}+\dot{W} \quad\text{and}\quad \frac{\mathrm{d}S}{\mathrm{d}t}\geq\frac{\dot{Q}}{T} \tag{5.24}$$

are valid universally and, in particular, for fluids and solids. Also the working*

$$\dot{W}=\int_{\partial V}t_{ij}w_i n_j\,\mathrm{d}A \tag{5.25}$$

is a universal expression. But what makes fluids and solids different, even in slow, *i.e.* reversible processes is the form of the stress tensor. In fluids we have $t_{ij}=-p\delta_{ij}$ with a homogeneous pressure, while in a solid there may be shear components t_{ij} $(i\neq j)$ and the diagonal components may differ. This leads to widely different expressions for the working $\dot{W}$ in general. However, in this chapter we minimize the difference between a fluid and a solid by considering uni-axial deformations of an incompressible elastic rod as indicated in Fig. 5.5. P is the tensile force and L is the length of the rod.

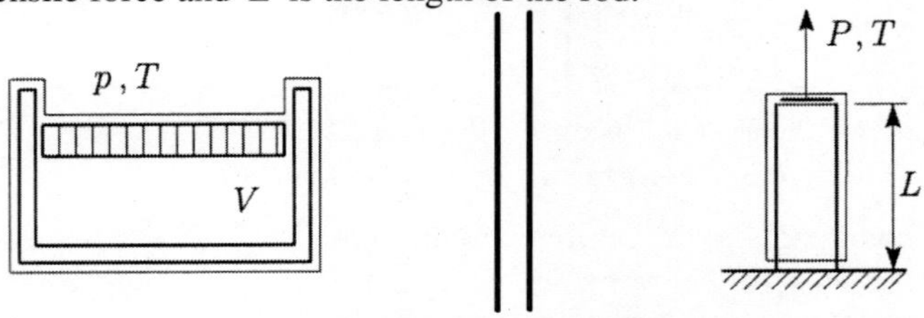

Fig. 5.5 Fluid under pressure and solid in uniaxial tension

In this case the workings are given by

* We ignore the working of gravity here.

$$\dot{W} = -p\frac{\mathrm{d}V}{\mathrm{d}t} \qquad \Big\| \qquad \dot{W} = P\frac{\mathrm{d}L}{\mathrm{d}t}. \tag{5.26}$$

Elimination of $\dot{Q}$ between the First and the Second Law (5.24) for a reversible process and neglect of the kinetic energy thus leads to the Gibbs equations

$$T\mathrm{d}S = \mathrm{d}U + p\,\mathrm{d}V \qquad \Big\| \qquad T\mathrm{d}S = \mathrm{d}U - P\,\mathrm{d}L. \tag{5.27}$$

Alternatively we have for the free energy $F = U - TS$

$$\mathrm{d}F = -S\,\mathrm{d}T - p\,\mathrm{d}V \qquad \Big\| \qquad \mathrm{d}F = -S\,\mathrm{d}T + P\,\mathrm{d}L. \tag{5.28}$$

p and P, and U and S are given by the thermal and caloric equations of state, *viz.*

$$p=p(V,T),\ U=U(V,T),\ S=S(V,T) \qquad \Big\| \qquad P=P(L,T),\ U=U(L,T),\ S=S(L,T). \tag{5.29}$$

From either (5.27) or (5.28) we obtain

$$p = -\left(\frac{\partial F}{\partial V}\right)_T = -\left(\frac{\partial U}{\partial V}\right)_T + T\left(\frac{\partial S}{\partial V}\right)_T \qquad \Big\| \qquad P = \left(\frac{\partial F}{\partial L}\right)_T = \left(\frac{\partial U}{\partial L}\right)_T - T\left(\frac{\partial S}{\partial L}\right)_T. \tag{5.30}$$

This may be interpreted by saying that p, or P have an energetic and an entropic part. Both parts may be identified in terms of the thermal equation of state. Indeed, the integrability conditions implied by (5.28) read:

$$\left(\frac{\partial S}{\partial V}\right)_T = \left(\frac{\partial p}{\partial T}\right)_V \qquad \Big\| \qquad \left(\frac{\partial S}{\partial L}\right)_T = -\left(\frac{\partial P}{\partial T}\right)_L \tag{5.31}$$

which identifies the entropic contribution of p, or P. Insertion into (5.30) gives

$$\left(\frac{\partial U}{\partial V}\right)_T = -p + T\left(\frac{\partial p}{\partial T}\right)_V \qquad \Big\| \qquad \left(\frac{\partial U}{\partial L}\right)_T = P - T\left(\frac{\partial P}{\partial T}\right)_L \tag{5.32}$$

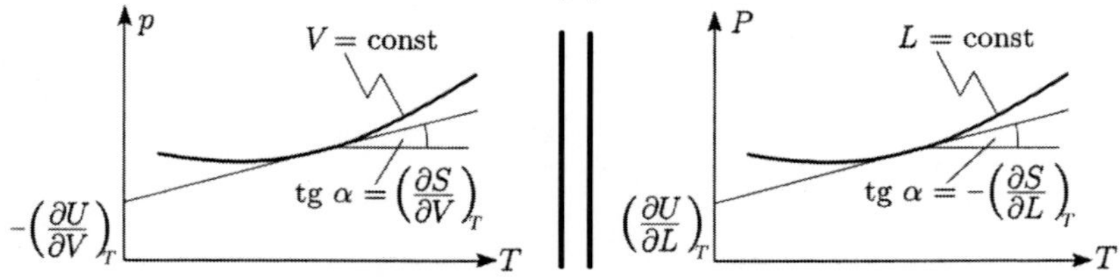

Fig. 5.6 Representation of generic curves $p = p(T)$, or $P = P(T)$. The entropic and energetic contributions to p, or P

It follows that the energetic and the entropic contribution of P may be read off from a measured (P,T)-curve: All one needs to do is to register the load needed to maintain a fixed length when the temperature changes. The entropic part of P – for one pair (L,T) – is then given by the slope of the (P,T)-curve, and the energetic part is given by the ordinate intercept of the tangent. *Mutatis mutandis* the same holds for the fluid.

5.6.2 *Examples for entropic elasticity*

Ideal gases and amorphous rubber are examples for purely *entropic* elasticity. In the case of a gas we know from the thermal equation of state – basically from the experiments by BOYLE and MARIOTTE – that p is proportional to T, *cf.* Fig. 5.7. Rubber experiments show something similar: The load is proportional to T, *i.e.* the rubber "attempts" to contract when the temperature is raised, *cf.* Fig. 5.7.* In both cases the ordinate intercept of the graphs $p(T)$, or $P(T)$ vanishes and therefore the energetic parts of p, or P are zero; U is independent of V, or L, *i.e.*

$$p = T\left(\frac{\partial S}{\partial V}\right)_T \quad \Big\| \quad P = -T\left(\frac{\partial S}{\partial L}\right)_T . \tag{5.33}$$

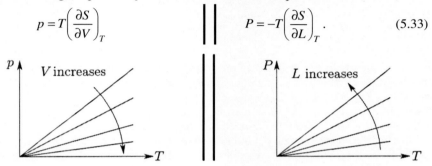

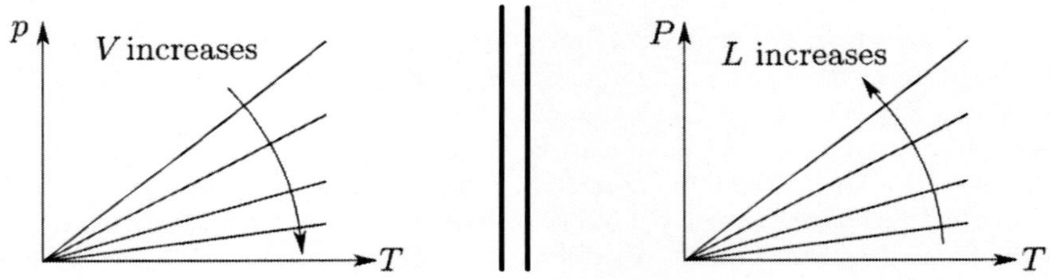

Fig. 5.7 Observed (p,T)-, or (P,T)-curves in ideal gases and rubber

In words: Pressure in ideal gases and load in rubber are "entropy induced." It follows that the knowledge of $S(V,T)$ and $S(L,T)$ implies knowledge of the thermal equations of state. And, indeed, from $(5.14)_1$ and (5.23) we derive

$$p = \frac{1}{V} NkT \quad \Big\| \quad P = nkT\left(\lambda - \frac{1}{\lambda^2}\right). \tag{5.34}$$

In the case of the gas this is the well-known thermal equation of state, mentioned many times before, so that there is nothing new here. However, for rubber we have a genuinely new result: The non-linear thermal equation of state that was not known before. It is a direct consequence of the statistical formula $S = k \ln W$ and it provides the prototype of thermal equations of state in non-linear elasticity.

Fig. 5.8 shows graphical representations of the thermal equation of state (5.34). For rubber P grows asymptotically with increasing λ toward a straight line through the origin. For $\lambda < 1$ the load is negative, *i.e.* a compressive force is needed.

* We ignore thermal expansion in this argument. According to Paragraph 4.2.8 this is consistent with the assumption of incompressibility. In reality both assumptions are idealizations.

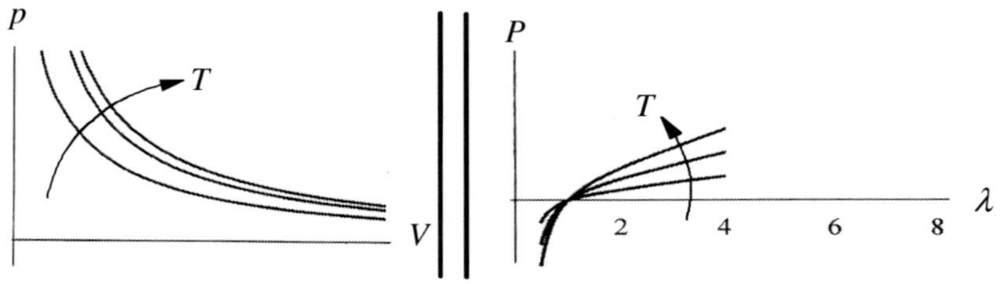

Fig. 5.8 Isotherms for ideal gases and amorphous rubber

5.6.3 *Real gases and crystallizing rubber*

When a gas is compressed, or cooled, or both, it is no longer governed by the ideal gas law. The gas becomes a "real" gas and its internal energy develops a υ-dependence. Analogously, when amorphous rubber is stretched to more than three times its natural length, the thermal equation of state (5.34) becomes invalid and the internal energy is no longer independent of L.

Rubber loses its amorphous character, because it develops longish crystallites along the direction of the load. We proceed to derive – in a very qualitative manner – the properties of real gases and of crystallized rubber and, once again, we do this in juxtaposition: The figures on the left-hand side of the vertical bars refer to gases and those on the right to rubber.

The observed deviations of gases and of rubber from the ideal or amorphous behavior are best represented by the shape of the isotherms in a (p,V)- or a (P,L)-diagram, *cf.* Fig. 5.9. For large values of V and T the isotherms of the gas fall below the ideal hyperbolic graphs, and for low temperatures there is even a horizontal branch. Observation shows that the real gas condenses to a liquid as it passes through the horizontal branch. As soon as only liquid is left – for small values of V – the isotherms become extremely steep, because it takes large pressures to compress a liquid to any great extent. All of this was already described in Chapter 2.

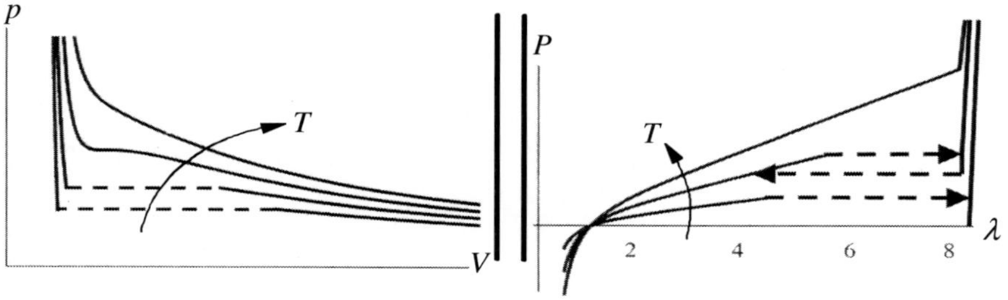

Fig. 5.9 Isotherms of real gases and crystallizing rubber

In rubber the isotherms start out for $L > L_0$ just as they do in the amorphous rubber, *cf.* Fig. 5.8. As soon, however, as the rubber begins to crystallize – for $L \geq 4L_0$ – the isotherms fall below the linear asymptotic lines of amorphous rubber. For low temperatures they even develop horizontal branches, along which the crystallization proceeds. Eventually, for $L \approx 8L_0$, the isotherms become very steep. Upon unloading a new phenomenon occurs: The horizontal branch of the unloading curve lies below the one of the loading curve. Therefore the loading-unloading behavior of rubber is hysteretic.*

At still lower temperatures this hysteretic behavior leads even to a large residual deformation after unloading, *cf.* Fig. 5.9. The curves of the figure imply, however, that the residual deformation may be recovered after a temperature rise; at high temperature rubber returns to its natural state with length $\lambda_0 = 1$.

As indicated, the deviations of a gas and of crystallizing rubber from the ideal or amorphous behavior are due to a phase transition: Liquid↔vapor in the gas and amorphous↔crystalline in rubber. In both cases the phase transition is caused by interatomic attractive forces. In other words, there is a variable potential energy of interaction and we may assume – or must assume – that the deviations from the ideal behavior are caused by a V- or an L-dependence of the internal energy U.

This assumption may be made more concrete by simple molecular considerations: For a large V the molecules of a gas are so far apart that they do not feel any mutual attractions; the interaction energy may then be put equal to zero. However, for intermediate or small volumes the molecules are so close that the atoms attract each other and the interaction energy – part of the internal energy – becomes smaller. Eventually, when the molecules are close enough to touch each other, the energy rises sharply, because the atoms cannot penetrate each other. The right hand side of Fig. 5.11 shows the resulting graph $U = U(V)$, albeit only schematically and qualitatively.

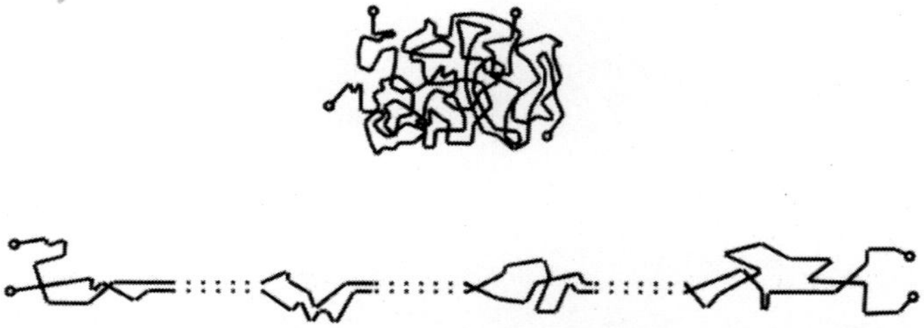

Fig. 5.10 Molecular models for two rubber molecules crystallizing under tension

For rubber the molecular model is radically different. Rubber consists of long chain molecules and in the natural state these molecules are strongly entangled, *cf.* Fig. 5.10, which shows two rubber molecules in the entangled and the stretched-out state, respectively. In the entangled state there are only a few places where the molecules come close to each other and interact energetically, and this remains so for small deformations. It follows that U is independent of deformation and we may set it equal to zero without loss of generality. For intermediate and large deformations, however, the chains are partially disentangled by the tensile force and

*Hysteresis (from the Greek) “hysteros” meaning “later.”

the lateral contraction forces them into close contact in certain ranges, *cf.* Fig. 5.10. Thus the interaction energy becomes negative for increasing L. In the ranges where the adjacent molecules are close, the atoms of the chains lie parallel to each other and we say that the rubber is crystallized there.

Eventually, when the chains are nearly fully extended by the tensile force the energy increases sharply with increasing L, because the molecules resist rupture. The corresponding graph $U = U(L)$ is shown on the right hand side of Fig. 5.12.

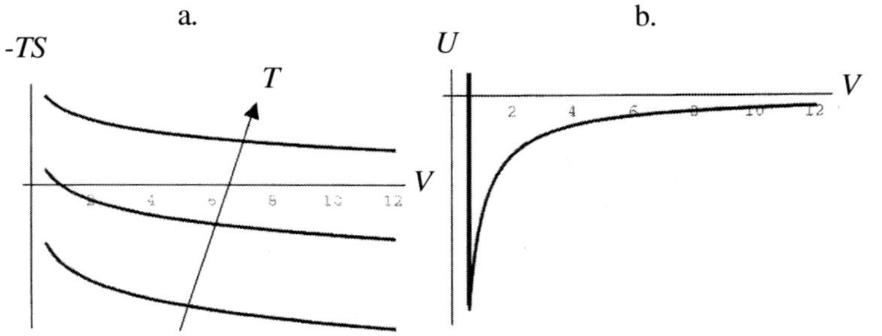

Fig. 5.11 a. $-TS$ for a gas as function of V
b. Internal energy as function of V (schematic and quantitative).

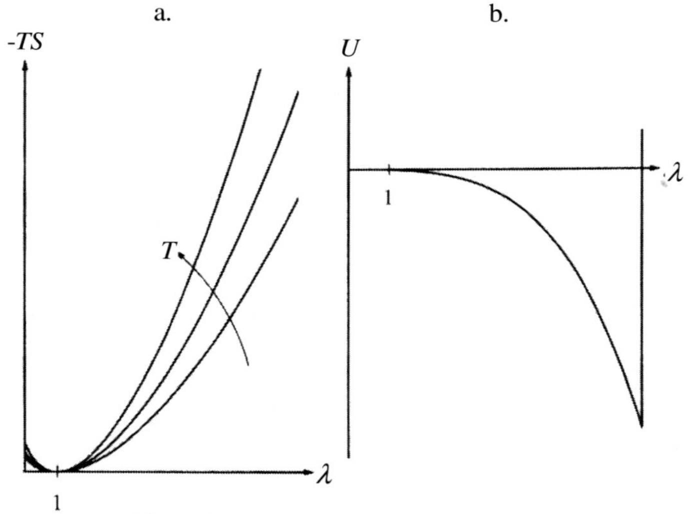

Fig. 5.12 a. $-TS$ for rubber as function
b. Internal energy as a function of λ

5.6.4 *Free energy of gases and rubber.* (p,V)*- and* (P,L)*-curves.*

We assume that the dependence of entropy on V of a gas, or on L in rubber is not changed by the molecular attraction – at least not qualitatively – so that the equations $(5.14)_1$ and $(5.23)_2$ remain valid. In this case the contribution $-TS$ of the entropy to the free energy $F = U - TS$ of a gas has the form shown in Fig. 5.11 a. for different values of T.

In order to obtain the free energy as a function of V we must add the graphs of Fig. 5.11a to the one of Fig. 5.11b. Thus we obtain the graphs of Fig. 5.13 a. We see – just barely (!) – that the free energies are non-convex for small temperatures.

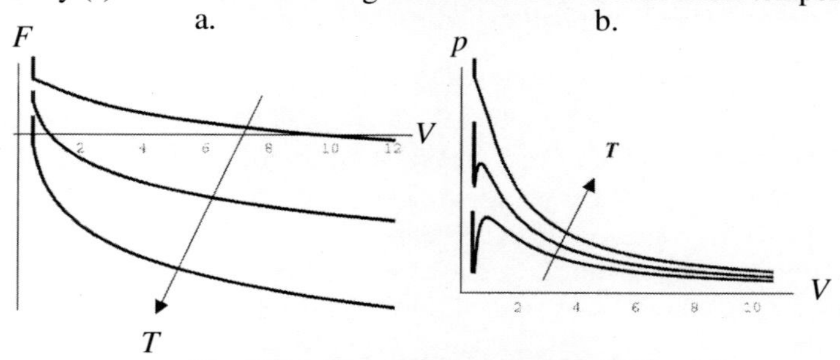

Fig. 5.13 a. Free energy as a function of V and T
b. Pressure as a function of V and T

Because of (5.30) the pressure p as a function of V is obtained from the free energy by differentiation. Fig. 5.13 b shows the result: Non-monotonic (p,V)-curves at least for small temperatures.

Similar for rubber: In order to obtain the free energy as a function of λ, we must add the graphs of Fig. 5.12 a and 5.12 b. In this manner we obtain non-convex free energies for small temperature as shown in Fig. 5.14 a. Differentiation provides the (P, λ)-curves of Fig. 5.14 b.

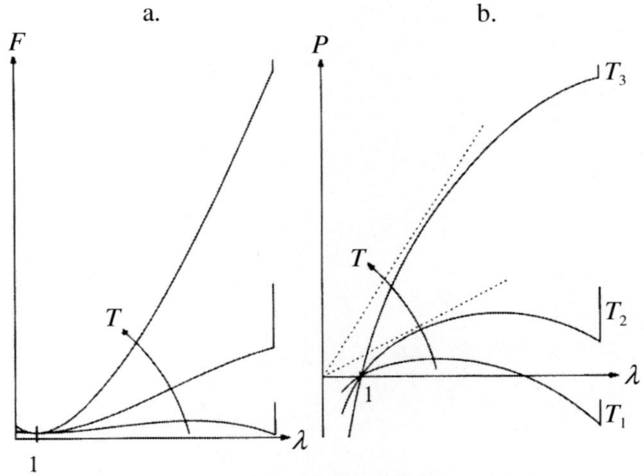

Fig. 5.14 a. Free energy as function of λ and T
b. Load as function of λ and T

A comparison of the graphs of Figs. 5.13 b and 5.14 b with the observed graphs of Fig. 5.9 shows a satisfactory agreement for high temperature, although now we have vertical isotherms for small volumes and for large values of λ; this occurs because of the cut-off of the energy curves in Figs. 5.11 b and 5.12 b. For intermediate and low temperatures, however, we do not have reversible horizontal branches in the (p,V)-diagram, nor do we have hystereses in the (P,L)-diagram.

What we do have are non-monotone (p, V)-curves and (P, λ)-curves.*

5.6.5 *Reversible and hysteretic phase transitions*

A possible, albeit very rough interpretation of the *hysteretic* (P,λ)-curves of Fig. 5.9 is as follows: One may assume that the loading of the rubber rod proceeds on the left ascending part of the (P,λ)-curves starting in the natural state $(0,1)$ and ending in the maximum of that curve. At this maximal load the transition amorphous $\rightarrow$ crystalline could occur as indicated in Fig. 5.15. Afterwards the load could increase very steeply on the right crystalline branch. Unloading along this branch would leave the rod crystallized until the sharp minimum is reached. For slightly smaller loads the reverse transition crystalline $\rightarrow$ amorphous is then performed. In this manner the rod runs through a hysteresis loop, *cf.* 5.15.

The *reversible* transition liquid$\leftrightarrow$vapor of Fig. 5.9 may be obtained from the (p,V)-graphs of Fig. 5.13 by an argument described in Paragraph 4.2.5 for a van der Waals gas. According to that argument phase equilibrium occurs on *the* horizontal line – the Maxwell line – which encloses equal areas with the isotherm above and below, *cf.* Fig. 5.15 a.

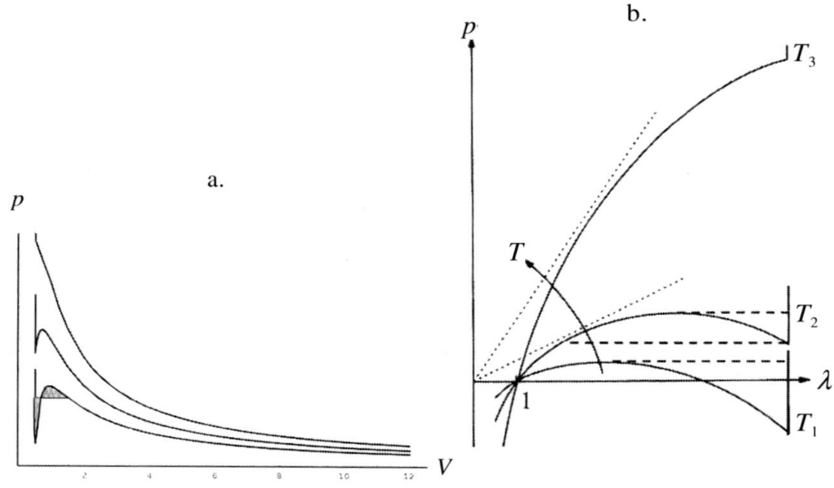

Fig. 5.15 a. Maxwell line for reversible liquid $\leftrightarrow$ vapor transition
b. Hysteresis for amorphous $\rightarrow$ crystalline phase transition and the reverse transition crystalline $\rightarrow$ amorphous

---

* Of course, the non-monotone isotherms remind us of the isotherms in a van der Waals gas, *cf.* Fig. 2.12.

With this in mind we leave the subject of hysteresis for now. Later, in Chapter 12, we shall come back to it and discuss the question why some phases transitions are hysteretic and others are reversible. Here we just mention that the answer lies in the size of the interfacial energy between different phases: A transition is reversible, if the interfacial energy is negligible, otherwise it is hysteretic.

5.7 History of the molecular interpretation of entropy

BOLTZMANN's interpretation of entropy as $S = k \ln W$ rested on the principles of classical mechanics. However, the probabilistic flavor of the equation disturbed mechanical traditionalists and they objected vigorously. BOLTZMANN had to fight on two fronts: Against the

reversibility objection and the *recurrence objection.*

Molecular processes are reversible. They may just as well run forward or backwards. And, if the entropy increases in one direction, it should decrease in the other direction. Thus LOSCHMIDT concluded that the entropy should *equally frequently* increase and decrease. This was the reversibility objection.

BOLTZMANN responded that there are near infinitely more disordered microstates than ordered ones, and that each microstate occurs just as frequently as any other one. Thus when we start in an ordered state, the entropy-increasing transition order → disorder will *nearly always* occur. Therefore the entropy should increase *more often* than it decreases.

Nota bene: "nearly" and "almost!" Indeed, the Second Law is not a deterministic law for individual particles. Rather it is a probabilistic law for many particles, which is obeyed with great probability. The American physicist Josiah Willard GIBBS was convinced by BOLTZMANN's reply. He says:

... the impossibility of an uncompensated decrease of entropy seems to be reduced to an improbability.

Josiah Willard GIBBS (1839-1903) completed the work of CLAUSIUS by extending it to mixtures. We shall present that subject in Chap. 8. GIBBS's great work on the subject carried a quote from CLAUSIUS as a motto:

Die Energie der Welt ist constant.
Die Entropie der Welt strebt einem Maximum zu.

The recurrence objection made use of a result of mechanics, which had been proved by the mathematician Jules Henri POINCARÉ (1854-1912): Once a closed mechanical system is set in motion it will – in the course of time – return arbitrarily often into the immediate neighborhood of its initial state. This result rather obviously contradicts the monotonic approach of the system to an equilibrium state of maximal entropy.

Once again the solution lies in the probabilistic character of the approach to equilibrium. POINCARÉ's theorem is correct and BOLTZMANN's entropy principle is *almost* correct, *i.e.* the initial state may be recovered, even if it has a small entropy, albeit with a very small probability indeed. We have discussed this in the context of the rubber molecule.

L.E. BOLTZMANN (1844-1903). His tombstone on Vienna's central cemetery.

Even today a satisfactory interpretation of the Second Law and of the entropy is a fascinating challenge. People speculate that the direction of time is determined by the increasing entropy of the universe. BOLTZMANN did originally suggest that mind-boggling idea.
He says: ... *in the universe which is nearly everywhere in equilibrium, and therefore dead, there must be small regions of the size of our stellar space which, during the relatively short periods of eons, deviate from equilibrium and among these equally many in which the probability of states increases and decreases. ... A creature that lives in such a period and in such a world will denote the direction of time toward less probable states as the past, the opposite direction as the future.*

BOLTZMANN and Ernst ZERMELO (1871-1953) carried out an acrimonious public debate on the probabilistic interpretation of the entropy growth.

ZERMELO: "··· Boltzmann wishes to conserve the mechanical point-of-view by interpreting the second law as a merely probabilistic law which need not always to hold."

BOLTZMANN: "For me it is impossible to understand how someone can refute the applicability of probability calculus [in thermodynamics], if some other considerations show that, in the course of eons, occasionally freak occurrences do occur. Because this is exactly what probability calculus also predicts."

Nobody admitted defeat in this discussion, but eventually Boltzmann prevailed, when physicists became used to probabilistic arguments. Now then, if the Second Law is only obeyed in a probabilistic sense, it is pure chance that reigns and then maybe it can be outwitted.

James Clerk MAXWELL (1831-1879) conceived the idea of a demon for the purpose: *"··· a being with extraordinary talents so that it can follow the orbit of all molecules. The demon guards a valve between two partial volumes of a gas with different temperatures. He permits passage only to the fast molecules that move from the cold to the warm side and only to the slow molecules in the other direction."* Obviously the demon can thus let the warm side become warmer and the cold side become colder – in contradiction to the Second Law. In the same spirit W.THOMSON (Lord KELVIN) conceived of an army of Maxwell demons, *equipped with molecular cricket bats*, and they may make temperature rise and fall contrary to what the Second Law requires.

The universal character of the formula $S = k \ln W$ often tempts non-thermodynamicists to consider entropy in *their* fields of knowledge: • Biologists calculate the growth of entropy in the distribution of species over a habitable area; • Economists consider the spread of goods and money in terms of entropy; • Ecologists consider the waste of resources as a positive entropy production; • and sociologists consider the entropy growth involved in the segregation of constituent groups of populations.

6 Steam engines and refrigerators

6.1 The history of the steam engine

It was Denis PAPIN – the inventor of the pressure cooker, *cf.* Paragraph 2.4.8 – who first condensed vapor and lifted a weight by doing so. He used a brass tube of diameter 5 cm; some water at the bottom was evaporated and the vapor pushed a piston upward which was then fixed by a latch. Afterwards the tube was taken away from the fire, the vapor condensed and a Torricelli vacuum formed in the tube. When the latch was unlocked, the air pressure pushed the piston downward and lifted a weight of 60 pounds.

The first proper heat engine was constructed by Thomas NEWCOMEN (1663-1729), a blacksmith from Dartmouth in England. The Newcomen engines were used for pumping water out of coal mines. Fig. 6.1 shows a schematic picture of the engine: A movable piston closed off a cylinder and was pulled upward by a pump rod that turned a beam with two lateral arch heads. During the upward motion the cylinder was filled with steam at a pressure of 1bar. The entering steam pushed air and some water – left over from the previous stroke – out of the cylinder through an eduction pipe and a snifting valve. The piston was sealed with leather bands and water on top. After closing the steam valve a water cock was opened inside the steam-filled cylinder, whereupon the steam condensed and the piston came down, driven by the outside air pressure. The steam valve and the injection valve were controlled by a plug rod hanging down from the rotating beam.

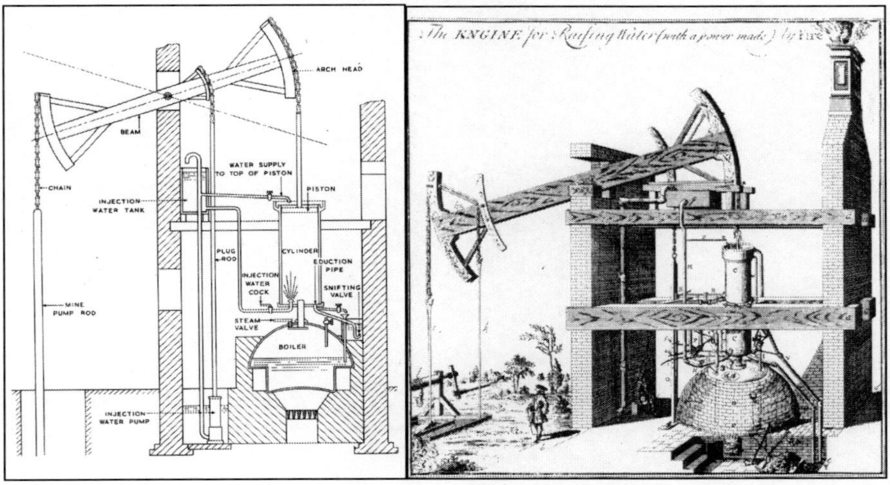

Fig. 6.1 Newcomen's steam engine (1712).

In the first Newcomen engine the cylinder had a diameter of 48 cm and the stroke of the piston had the length 1.80 m. In each downward stroke the piston lifted 40 l of water by 56 m, and the engine needed 5 s between successive strokes. It is easy to calculate that the power thus amounted to 4.5 kW. The engine was very reliable and by 1775 hundreds of them were installed, most of them in England and Scotland. The efficiency, however, was less than 1%:

*About 6 million of foot-pounds of useful work were done for each bushel of coal burnt.**

Because of the low efficiency James WATT (1763-1811) set himself the goal to improve the Newcomen engine. He was able to identify the reason for the low efficiency: Each batch of new steam had to reheat the cylinder walls which had just been cooled by the injected water. Thus a good part of the new steam condensed immediately. Therefore WATT had two objectives:

... first that the cylinder should be maintained as hot as the steam which entered it; and secondly, that when the steam was condensed, the water of which it was composed and the injection itself should be cooled as low as possible.

In pursuit of this program WATT invented the double-walled cylinder between whose walls the hot steam was conducted; and he invented the separate condenser. Fig. 6.2 shows WATT's steam engine of 1788.

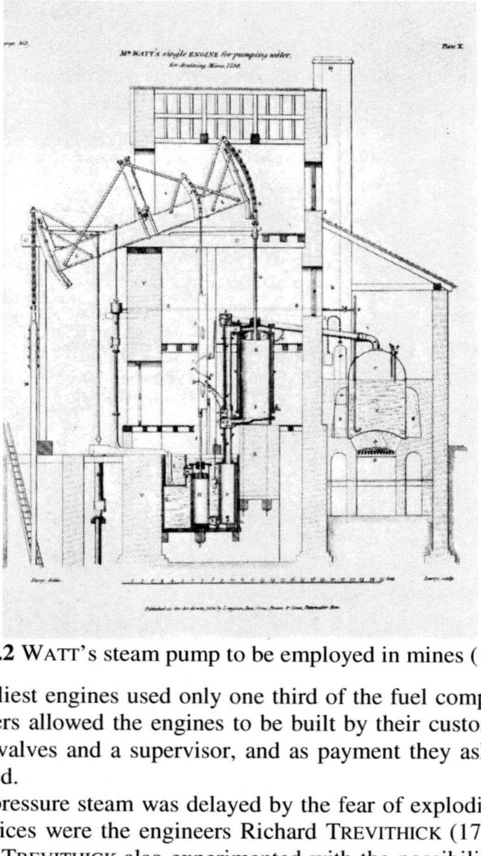

Fig. 6.2 WATT's steam pump to be employed in mines (1788).

Even WATT's earliest engines used only one third of the fuel compared to NECOMEN's. WATT and his partners allowed the engines to be built by their customers. They provided only the blueprints, valves and a supervisor, and as payment they asked for a part of the value of the fuel saved.

The use of high-pressure steam was delayed by the fear of exploding boilers. Pioneers of high pressure devices were the engineers Richard TREVITHICK (1771-1833) and Oliver EVANS (1755-1819). TREVITHICK also experimented with the possibility to close the steam

---

* R.J. Law. The steam engine. A Science Museum booklet. Her Majesty's Stationary Office, London (1965). Figs. 6.1 and 6.2 were taken from this brochure.

valve early during the working stroke and to let the steam expand afterwards. Thus less work was gained per stroke but still less steam was consumed so that the efficiency grew.

It is clear that when high pressure steam is employed, one does not need to have a low pressure in the condenser. Indeed, in this case one does not need a condenser at all: the expanded steam may simply be released from the cylinder into the atmosphere. Thus most steam locomotives do not have a condenser.

WATT has not only improved NEWCOMEN's engine, rather he has employed steam engines for many purposes other than pumping water out of mines. Thus in his "rotative engine" he has converted the up-and-down motion of the piston into the rotation of a wheel and a shaft. This engine has many uses: It could drive hammers, spinning wheels and looms, then ships and locomotives; in this manner WATT's inventions became the motor of the industrial revolution.

In 1783 WATT tested a strong horse and decided that it could raise a 150-pound weight nearly four feet in one second. He therefore defined a "horsepower" as 550 foot-pound per second. This unit is still in use. However, the unit of power in the international metric system is called one WATT. One horsepower equals 746 WATT.

6.2 Steam engines

6.2.1 *The (T,S)-diagram*

We recall Paragraph 4.1.1 where we have listed the advantages of the (T,S)-diagram for the evaluation of heats – added and withdrawn – and of the work of a cycle. Fig. 6.3 shows (T,S)-diagrams, on the left hand side a schematic one and on the right-hand side one for water and water vapor. In the latter case the entropy constant is arbitrarily chosen so that $S=0$ holds in the triple point. This choice does not restrict the applicability of the diagram for the present purposes.

Inspection shows that the isobars in the vapor domain for high temperatures grow exponentially. This is the behavior expected for an ideal gas. In the liquid range all isobars up to 10^3bar are very close to the evaporation line.

6.2.2 *Clausius-Rankine process. The essential role of enthalpy*

Thermodynamically the cyclic process in the steam engine is identical to the Joule process that occurs in the hot air engine, *cf.* Paragraph 3.4.2, and which we have called the Joule cycle in that paragraph: Both cycles consist of two isobars and two adiabates. The difference is, of course, the working agent, steam instead of air. For the steam engine the cycle is called the Clausius-Rankine process. The individual branches of the cycle are as follows, *cf.* Fig. 6.4.

$2'$ - $3'$: The feed water pump compresses water adiabatically and feeds it into the boiler.

3′-3: In the boiler the water is isobarically heated to boiling and then evaporated; subsequently it may be superheated – still at a constant pressure – in the superheater.

3-2: In the steam cylinder the steam expands adiabatically.

2-2′: The steam is condensed in the cooler.

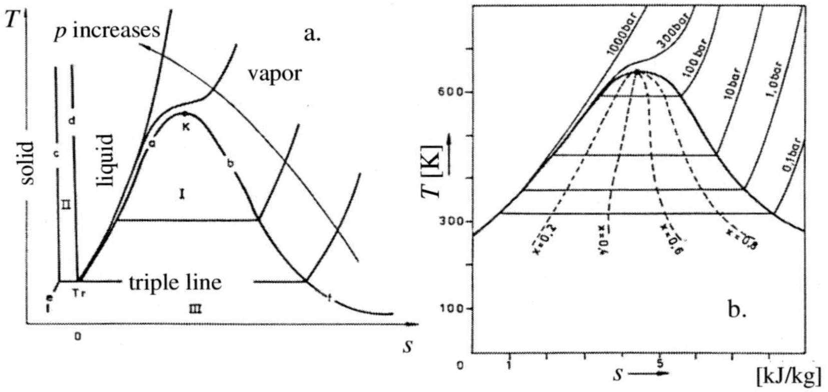

Fig. 6.3 a. (T,S)-diagram (schematic) Notation as in Fig. 2.7
b. (T,S)-diagram for water.

The elements of the hot air engine and of the steam engine correspond to each other as follows:

compressor	–	feed water pump
heat exchanger for heating	–	boiler and superheater
pneumatic engine	–	steam cylinder with piston
heat exchanger for cooling	–	condenser

Fig. 6.4 shows a schematic picture of the engine and also a standardized schematic picture in which the expansion occurs in a turbine rather than in the steam cylinder.

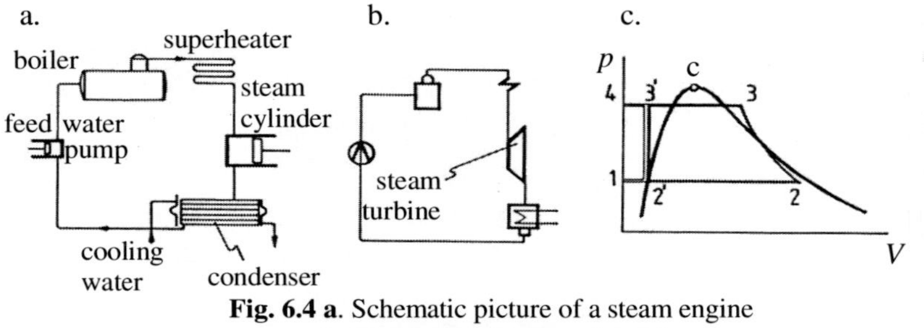

Fig. 6.4 a. Schematic picture of a steam engine
b. Standardized schematic picture
c. (p,V)-diagram with wet steam region.

The work of the feed-water pump and of the steam cylinder is represented in Fig. 6.4 by

$$W_{FP} = -\int\limits_{12'3'4} p\,dV \qquad \text{or} \qquad W_{FP} = \int\limits_{12'3'4} V\,dp = \int\limits_{2'}^{3'} V\,dp$$

$$W_{SC} = -\int\limits_{4321} p\,dV \qquad\qquad W_{SC} = \int\limits_{4321} V\,dp = \int\limits_{3}^{2} V\,dp. \tag{6.1}$$

The last step in each line reflects the observation that $dp = 0$ holds on the horizontal branches and that $V = 0$ on the branch 4-1. The branches $2' - 3'$ and 2-3 are adiabatic. Therefore the First Law for reversible processes (1.57) reads

$$\dot{Q}\,dt = dU + p\,dV = 0 \;, \text{ or with } H = U + pV: \; \dot{Q}\,dt = dH - V\,dp = 0 \tag{6.2}$$

$V\,dp$ may therefore be replaced in (6.1) by dH and we obtain

$$W_{FP} = H_{3'} - H_{2'} \text{ and } W_{SC} = H_2 - H_3. \tag{6.3}$$

The total work of the cycle is then given by the difference of enthalpies

$$W = H_2 - H_3 - \left(H_{2'} - H_{3'}\right). \tag{6.4}$$

Equation $(6.3)_2$ states that the work of the steam cylinder is equal to the enthalpy drop of the steam. In Paragraph 1.5.9 we have seen that the work of a turbine is also determined by the enthalpy difference of the flow, *cf.* (1.80). This means that the thermodynamic treatment of the steam engine is unaffected, if the steam cylinder is replaced by a turbine: The amounts of work are equal.

For the work of the feed water pump we need not necessarily know the enthalpies of liquid water, as $(6.3)_1$ would suggest. Indeed, since water is to a good approximation incompressible, we have $V_{2'} \approx V_{3'}$. The area $(1, 2', 3', 4)$, which represents the work, may therefore be written as

$$W_{FP} = m\upsilon'(p_1)(p_4 - p_1), \tag{6.5}$$

where $\upsilon'(p_1)$ is the specific volume of liquid water at the lower pressure p_1, essentially $\upsilon'(p_1) \approx 1$ l/kg.

Not only can the works involved in the operation of a steam engine be calculated from enthalpy differences, the same is true for the heats. Indeed, since the heating occurs on isobaric branches of the cycle we have

$$Q_{boiler} = H_3 - H_{3'} \text{ and } Q_{cooler} = H_{2'} - H_2.$$

6.2.3 *Clausius-Rankine process in a (T, S)-diagram*

Fig. 6.5 shows the Clausius-Rankine cycle in a (T, S)-diagram, with and without superheating, respectively. The individual branches represent

- 1-2: adiabatic compression in the feed-water pump.
- 2-3 (3′): isobaric heating and evaporation in the boiler (and superheating in a heat exchanger).
- 3-4 (3′-4′): adiabatic expansion, assumed reversible, *i.e.* isentropic.
- 4(4′)-1: condensation.

The distance of the isobar in the liquid region from the boiling line is exaggerated in Fig. 6.5; in reality, and on the scale of the figure, all isobars up to several hundred bar virtually coincide with the boiling line, *cf.* Fig. 6.3. Therefore the graphs of Fig. 6.6 represent the cycle more realistically.

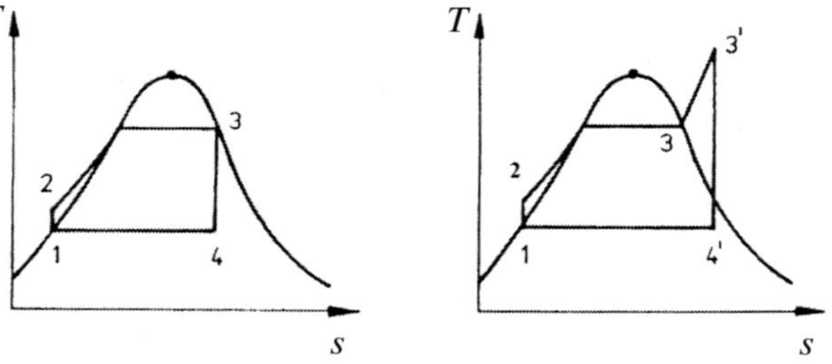

Fig. 6.5 (T,S)-diagram of the Clausius-Rankine process

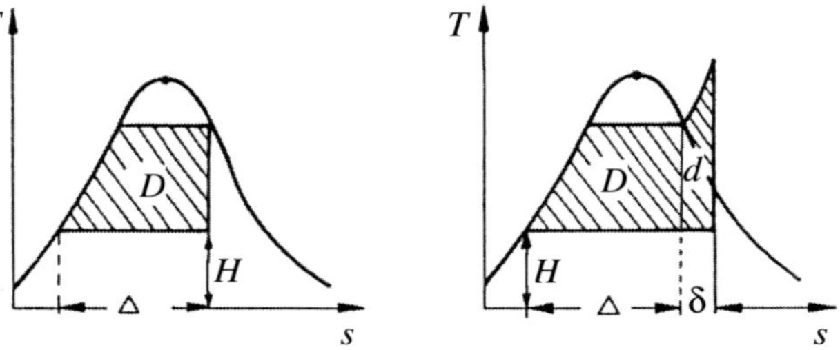

Fig. 6.6 The effect of superheating

As we already know, the (T,S)-diagram recommends itself by the fact that the area inside the process curve represents the work, while the areas below the upper and lower branches of the curve represent added and released heats, respectively. Therefore the efficiency of the process is easily visualized by a quotient of areas. For the cycles in Fig. 6.6 we have

$$e_{\text{left}} = \frac{D}{D + \Delta H} \quad \text{and} \quad e_{\text{right}} = \frac{D+d}{D+d+(\Delta+\delta)H}.$$

We may use this observation to prove that the efficiency of the process can be improved by superheating. We ask for the condition that $e_{\text{right}} > e_{\text{left}}$ holds and obtain after a short calculation

$$\frac{D}{\Delta} < \frac{d}{\delta}.$$

This means that the mean height of the quadrangle D must be smaller than that of the quadrangle d and this is indeed the case, since the isobars turn upwards after leaving the wet region.

Another practical reason for superheating – apart from increasing the efficiency – is that a full expansion downwards from point 3 to point 4, *cf.* Fig. 6.5 increases the moisture content at the end of the turbine, because point 4 lies deep in the wet steam region; this must be avoided, since water droplets may damage the turbine when hitting the blades.

On the other hand, superheating must not exceed 600°C, because this is a metallurgically safe temperature for the strength of the material of the superheater. Therefore it may be necessary to limit the superheating. In that case reheating after partial expansion may be employed in order to further prevent too much moisture in the turbine. The expansion is then split into two separate expansions that occur in a high pressure turbine and in a low pressure turbine. Between the turbines the partially expanded steam is lead through the boiler again where it is reheated to the boiler temperature, *cf.* Fig. 6.7.

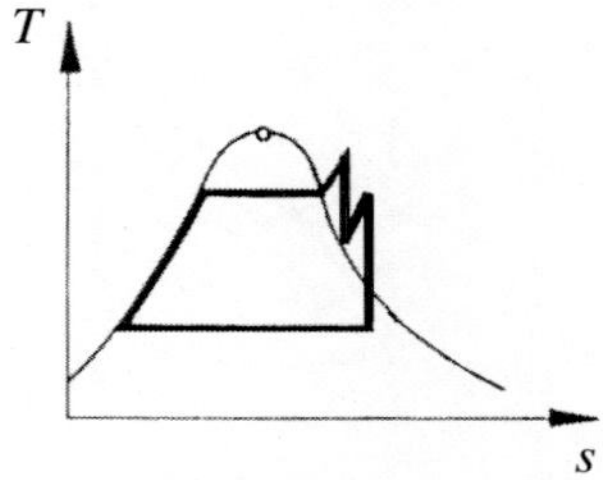

Fig. 6.7 Clausius-Rankine process with superheating and reheating to the boiler temperature

It is also clear that the expansion to a lower pressure will lead to more work, *cf.* Fig. 6.8. Of course, this measure is limited, because the pressure in the condenser cannot fall below the vapor pressure appropriate to the temperature of the cooling water. Also when the condenser pressure is too low, it becomes difficult to isolate the chamber and prevent the invasion of air.

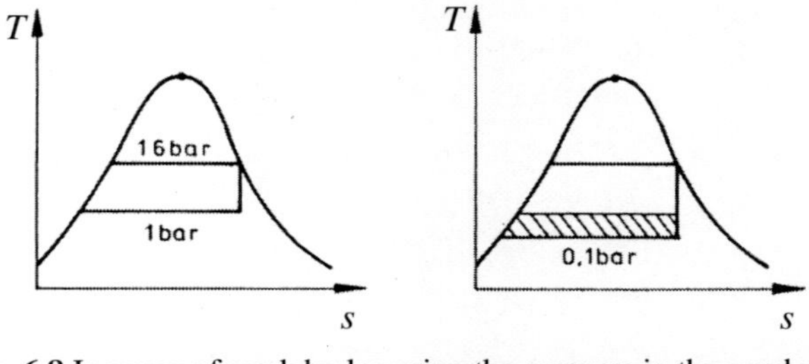

Fig. 6.8 Increase of work by lowering the pressure in the condenser

6.2.4 *The* (h,s)*-diagram*

The (h,s)-diagram is an important auxiliary tool in technical thermodynamics. It is often drawn so that the boiling line starts in the origin, which means that the additive constants in the entropy and in the enthalpy are both chosen as zero in the triple point. It is noteworthy that the isobars pass through the boiling line and through the dew line without a kink and that the slope of both lines is equal to T_C in the critical point. Fig. 6.9 a shows the (h,s)-diagram in schematic form, and with a cycle $1 2 3\,3' 4'$ which represents the Clausius-Rankine process; once again the distance of the isobar from the boiling point is exaggerated in order to be able to identify the adiabatic compression in the feed water pump.

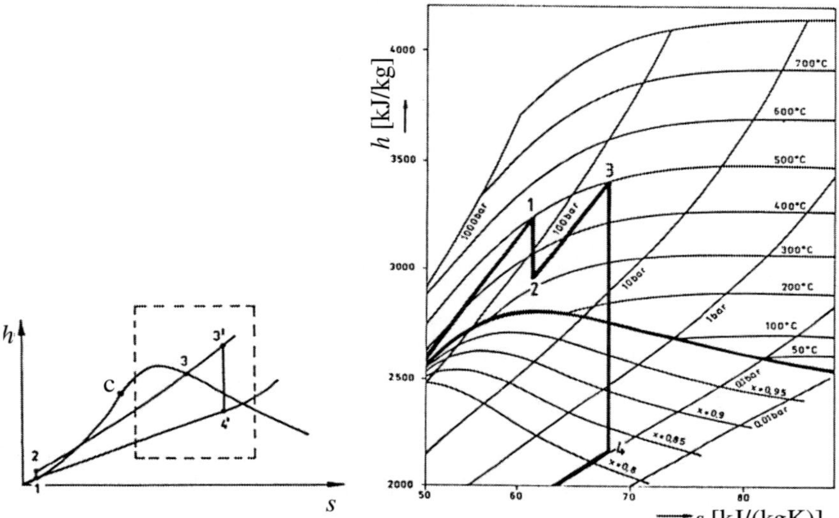

Fig. 6.9 a. Clausius-Rankine cycle in a (h,s)-diagram. Numbering as in Fig. 6.5
b. Relevant part of the (h,s)-diagram

In the (h,s) -diagram the works and heats needed for the calculation of efficiencies are easy to identify. Indeed, in Paragraph 6.2.2 we have shown that the works of the feed water pump and of the steam cylinder – or turbine -- are represented by enthalpy differences and so are the heats added or released in the boiler, superheater, and condenser. With the numbering of Fig. 6.9 a this means

$$\begin{aligned} w_{FP} &= h_2 - h_1 \quad & w_{SC} &= h_{4'} - h_{3'} \\ q_+ &= h_{3'} - h_2 \quad & q_- &= h_1 - h_2 \end{aligned} , \tag{6.6}$$

and the efficiency is given by

$$e = \frac{\left|h_{4'} - h_{3'} + h_2 - h_1\right|}{h_{3'} - h_2} . \tag{6.7}$$

However, the diagrams used by engineers show only part of the full (h,s)-diagram. This part is indicated in Fig. 6.9 a by the dashed window. The information seen in this window suffices for efficiency calculations, although the points 1 and 2 are not shown. Indeed, h_1 may be calculated from Table 2.4 – with properties of boiling water and saturated steam – as $h'(p_4)$ and h_2 follows from

$$h_2 - h_1 = v'(p_{4'})(p_{3'} - p_{4'}) \tag{6.8}$$

for the work of the feed water pump, *cf.* $(6.3)_1$ and (6.5). $v'(p_{4'})$ may again be read off from Table 2.4.

6.2.5 *Steam flow rate and efficiency of a power station*

In a small power station, with a power output of 200 MW, steam of 500°C and 200 bar is fed into the turbine. The steam leaves the high pressure turbine at a pressure of 50 bar and is then reheated to 500°C by conducting it through a heat exchanger in the boiler. After that the steam is expanded in the low pressure turbine to the condenser pressure of 0.1 bar. The temperature rise of the cooling water for the condenser must be limited to ΔT_c=30 K.

The part of the process that is visible in a conventional (h,s) -diagram is represented in Fig. 6.9 b by the solid line. From this diagram we read off the specific enthalpies of the corner points*

$$h_1 = 3250\tfrac{\text{kJ}}{\text{kg}},\ h_2 = 2900\tfrac{\text{kJ}}{\text{kg}},\ h_3 = 3450\tfrac{\text{kJ}}{\text{kg}},\ h_4 = 2150\tfrac{\text{kJ}}{\text{kg}}.$$

Hence follows the work of the two turbines

$$w_{\text{Tu}} = h_2 - h_1 + h_4 - h_3 = -1650\tfrac{\text{kJ}}{\text{kg}}.$$

The work of the feed water pump results from (6.8)

$$w_{\text{FP}} = v'(p_4)\,(p_1 - p_4) = 10^{-3}\tfrac{\text{m}^3}{\text{kg}}(200 - 0.1)\cdot 10^5\tfrac{\text{N}}{\text{m}^2} = 20\tfrac{\text{kJ}}{\text{kg}}.$$

$v'(p_4)$ must be taken from Table 2.4; its value is approximately $10^{-3}\tfrac{\text{m}^3}{\text{kg}}$, of course, irrespective of pressure. We see that w_{FP} is very small compared to the work of the turbines; it is often neglected in the calculation of the efficiency.

The heat supplied to the water and the steam in the boiler and in the heat exchangers for superheating and reheating is given by

$$q_+ = h_1 - [h'(p_4) + v'(p_4)(p_1 - p_4)] + h_3 - h_2 = 3590\tfrac{\text{kJ}}{\text{kg}},$$

where $h'(p_4) = 190\tfrac{\text{kJ}}{\text{kg}}$ was read off from Table 2.4.

Thus the efficiency is given by

$$e = \frac{-w_{\text{Tu}} - w_{\text{FP}}}{q_+} = 0.46.$$

The mass flow of steam is

$$\dot{m}_{\text{S}} = \frac{200\,\text{MW}}{w_{\text{Tu}}} = 121\tfrac{\text{kg}}{\text{s}} = 436\tfrac{\text{t}}{\text{h}}.$$

The mass flow rate $\dot{m}_{\text{C}}$ of cooling water follows from the heat q_- to be absorbed in the condenser. We have

$$q_- = h_4 - h'(p_4) = 1960\tfrac{\text{kJ}}{\text{kg}}.$$

We set $\dot{m}_{\text{C}} c_{\text{W}} \Delta T_{\text{C}} = \dot{m}_{\text{S}} q_-$ where c_{W} is the specific heat of water and $\Delta T_{\text{C}} = 30\,\text{K}$ the temperature increase of the cooling water. Thus we obtain

$$\dot{m}_{\text{C}} = 1.9\tfrac{\text{t}}{\text{s}} = 6800\tfrac{\text{t}}{\text{h}}.$$

---

* Students of technical thermodynamics are usually given a large-scale (h,s)-diagram which can be bought in university book stores, or they use electronic versions of the chart.

If the station were to work without reheating, the enthalpy of the steam after expansion from 200 bar to 0.1 bar would be equal to 1950 kJ/kg so that $w_{\text{Tu}} = 1300\,\text{kJ/kg}$. q_+ would be $3040\,\text{kJ/kg}$ in this case and therefore e follows as 43% – instead of 46%. Therefore the reheating has improved the efficiency.

It is also interesting to compare these efficiencies with the efficiency of the Carnot engine working in the same temperature range between 500°C and $T(p_4) = 47°\text{C}$. Such an engine would have the efficiency 59% according to (4.5) which would be the maximum that can be obtained for that temperature range.

6.2.6 *Carnotization*

We recall that the Carnot cycle has the largest efficiency among all cycles working in the same range of temperature. Also a reversible Carnot cycle – consisting of two isotherms and two isentropes – may be represented as a rectangle in a (T,s)-diagram, *cf.* Fig. 4.3. Having this in mind we consider again the Clausius-Rankine cycle – without superheating, *cf.* Fig. 6.10 a. This cycle is "nearly" a rectangle and one may ask why we do not make it into a Carnot cycle by stopping the condensation in the wet region at point 1 and pumping the wet steam into the boiler along the dashed isentrope. This, unfortunately, is impossible for practical reasons, because the wet steam – with its extreme differences of the density between boiling water and saturated steam – cannot be compressed in a pump without damaging the surface of a moving piston or of the turbine blades.

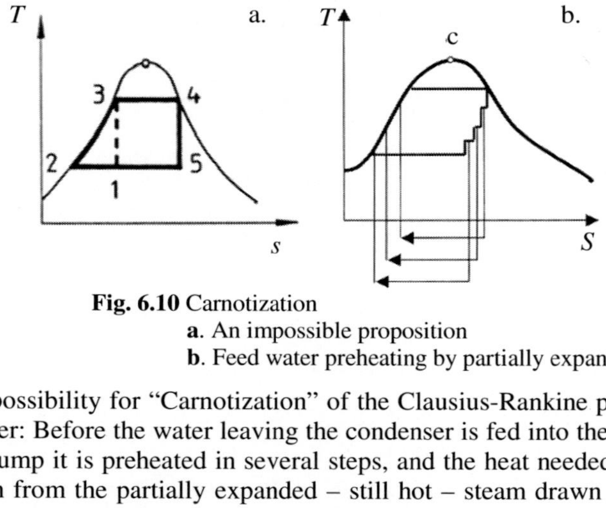

Fig. 6.10 Carnotization
a. An impossible proposition
b. Feed water preheating by partially expanded steam.

Another possibility for "Carnotization" of the Clausius-Rankine process is feasible, however: Before the water leaving the condenser is fed into the boiler by the feed water pump it is preheated in several steps, and the heat needed for this purpose is taken from the partially expanded – still hot – steam drawn from the turbine. Fig. 6.10 b shows, how the "heat packages" are transferred from the turbine to the feed water. The diagram in the figure is a (*T*,*S*) -diagram – not a (*T*,*s*) – so that with each withdrawal of steam the expansion isentrope moves to the left. If the steps are smoothed out, the cycle is represented by a parallelogram which has approximately the same area as the rectangle of Fig. 6.10 a. Thus the efficiency of the engine is that of a Carnot engine. Fig. 6.11 a shows how preheating the feed water is realized in practice. After heat has been withdrawn from the steam it enters the condenser through a throttle valve. Up to six steps of preheating have been used.

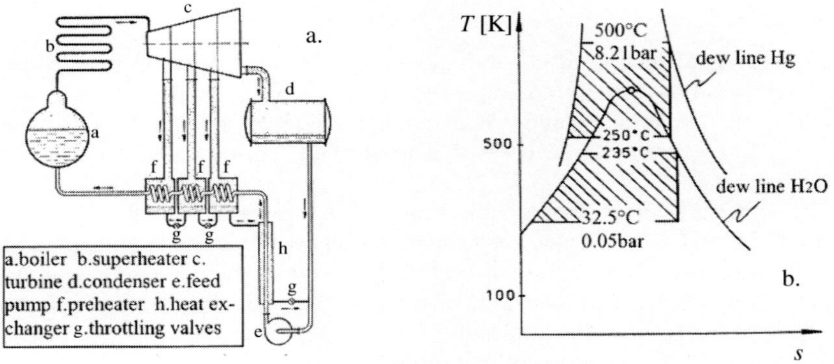

Fig. 6.11 a. Preheating feed water by partially expanded steam
b. Mercury-water cycle. The heat withdrawn from the mercury condenser is used to heat and evaporate water.

6.2.7 *Mercury-water binary vapor cycle*

Although the steam engine does not perform a Carnot cycle, its process curve is similar enough to that of a Carnot process that we may assume that a high temperature of the boiler is favorable for a large efficiency. Note that little, if anything, can be done about the temperature of the cooler; it is determined by the available cooling water which will generally be at environmental temperature.

In the case of water, however, high temperatures require high pressures. Therefore it has been proposed to use mercury as a working agent. Mercury has a critical temperature of 1460°C and a critical pressure of 1056 bar. However, at 500°C the vapor pressure is only 8.21 bar. On the other hand, at room temperature the vapor pressure is a minimal $3.6 \cdot 10^{-6}\,\text{bar}$; accordingly the density is extremely small and a condenser at room temperature would thus be very large indeed in order to accommodate a reasonable mass flux. Also it cannot be effectively sealed without involving high costs.

In order to take advantage of the low vapor pressure at high temperature and still not be hampered by a large cooler volume, a hybrid mercury-water engine was constructed with a (T,s) -diagram as shown in Fig. 6.11 b: Mercury is evaporated at 500°C and 8.21 bar and expanded to 0.1 bar where the vapor temperature equals 250°C. The mercury condenser is cooled by water of $30.6\,\text{bar}$ which evaporates at 235°C. Thus the mercury cooler acts as a boiler for the water cycle. The water steam is expanded to 0.05 bar, corresponding to 32.5°C.

The works of the mercury and the water cycles are shaded in Fig. 6.11 b. In an approximate manner the heat added is represented by the curve under the upper line of the mercury cycle so that even a cursory glance reveals that the efficiency of the combined engine is quite large.

Despite this the method is not practical, because in a pilot plant mercury was leaking everywhere; employees lost their hair and teeth and so the project was cancelled, although efficiencies higher than 50% had been reached.

6.2.8 *Combined gas-vapor cycle*

Progress in the construction of gas turbines involves water-cooled turbine blades and high-temperature-resistant coating of the blades with ceramic materials. Because of this, modern power plants – in the quest for high temperatures – make use of a combination of a Brayton cycle with air and of a Clausius-Rankine cycle in a steam engine.

Air may enter the compressor at 300K and – after compression, typically with a pressure ratio of 8:1 – it is heated isobarically by fuel combustion to temperatures of about 1300K. It is then expanded in a gas turbine and the exhaust gas of approximately 720K is used in the boiler of a steam engine to evaporate water and superheat it to 690K. The combined gas-vapor cycle is schematically shown in Fig. 6.12 and in the (T,s)-diagram of that figure.

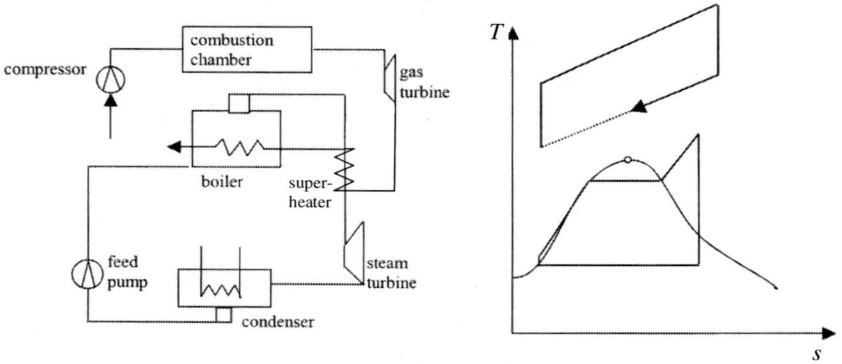

Fig 6.12 Combined gas-vapor cycle

6.3 Refrigerator and heat pump

6.3.1 *Compression refrigerator*

The cycle of a coolant in a compression refrigerator is in principle the inverse of the cycle of water in a steam engine. The purpose, however, is not the conversion of heat into work, rather it is the creation of cold by work. The most efficient way of producing cold is by evaporation; thus, if diethyl ether evaporates on the skin of the hand, the hand becomes cold, because the heat of evaporation is drawn from the skin, at least partly. This process requires no work; work is needed, however, when the cooling process is to be repeated over and over again.

In a refrigerator the refrigerant in the cooling coil boils and evaporates under small pressure and the necessary heat of evaporation is drawn from the food in storage (say). The saturated vapor is compressed – and heats up – isentropically in a compressor. Thus superheated vapor is created which is subsequently fed into a cooler – usually at the backside of the refrigerator – which may exchange heat with the surrounding air. The coolant condenses there and assumes room temperature, still under a high pressure. Afterwards it expands through a throttling valve back into the low temperature boiler. The throttling is accompanied by partial evaporation and cooling. Fig. 6.13a shows the corresponding cycle in a (T, s)-diagram. The throttling step – indicated by the dashed line – is irreversible and increases the entropy.

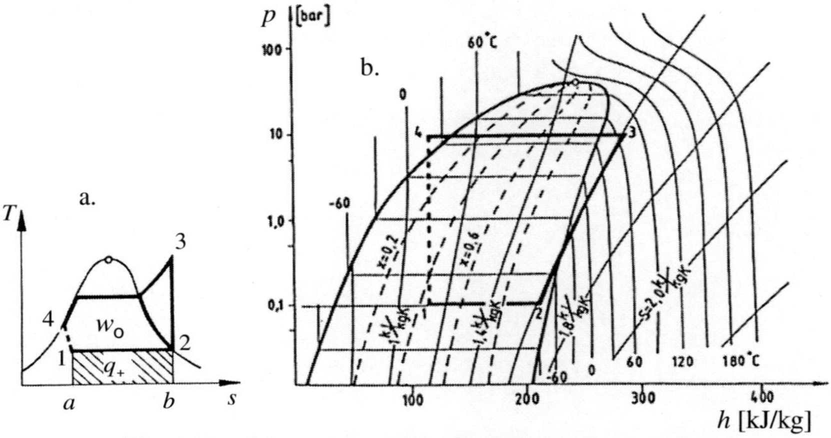

Fig. 6.13 a. Schematic cycle of refrigeration in the (T,s) -diagram
b. $(\log p, h)$ -diagram of Freon12.

For a quantitative description of refrigerators the $(\log p, h)$ -diagram is useful. Fig. 6.13 b shows such a diagram for Freon12, a popular coolant. In the diagram the additive constants in entropy and enthalpy were arbitrarily chosen so that $s = 1\,\text{kJ/(kgK)}$ and $h = 100\,\text{kJ/kg}$ hold in the state where the coolant boils at 0°C. Detailed diagrams of this type are available in bookstores or in digitized version.

The efficiency of a thermodynamic machine is always the ratio of the desired output and the required input. However, output and input are radically different for the refrigerator and the steam engine. Indeed, the desired output in the present case is the heat q_+ withdrawn from the storage room and transferred to the coolant for evaporation. And the required input is the work w_0 needed for the operation of the compressor. In the (T,s) -diagram of Fig. 6.13 a both quantities are represented as areas and we have

$$e = \frac{q_+}{w_0} = \frac{\text{area}(a12b)}{\text{area}(1234)}.$$

Clearly the value of e may be larger than 1.

6.3.2 *Calculation for a cold storage room*

We investigate a refrigerator working with Freon12 that cools a storage room by withdrawing the heating $\dot{Q}_+ = 75\,\text{kW}$. The coolant with the vapor content $x = 0.95$ and a pressure of $0.1\,\text{bar}$ enters the compressor where the pressure is raised to 10 bar. The now superheated vapor is condensed and cooled to 20°C.

Finally by adiabatic throttling – at constant enthalpy – the coolant is expanded back into the boiler room at 0.1 bar.

The corresponding graphical cycle is shown in 6.13 b by a thick line and we read off:

boiler temperature: –73°C , condenser temperature: 40°C .

The specific work of the compressor is equal to the difference of enthalpies before and after compression, see Paragraph 6.2.2. Thus we have, *cf.* Fig. 6.13 b

$$w_o = h_3 - h_2 = (280-210)\tfrac{\text{kJ}}{\text{kg}} = 70\tfrac{\text{kJ}}{\text{kg}}\,.$$

The specific cooling is equal to the difference of enthalpies before and after evaporation

$$q_+ = h_2 - h_1 = (210-110)\tfrac{\text{kJ}}{\text{kg}} = 100\tfrac{\text{kJ}}{\text{kg}}\,.$$

Therefore we obtain the efficiency

$$e = \frac{q_+}{w_o} = 1.43\,.$$

The rate of mass transfer is

$$\dot{m} = \frac{\dot{Q}_+}{q_+} = 0.75\frac{\text{kJ}}{\text{kg}} = 2.7\frac{\text{t}}{\text{h}}$$

and the power required for the compressor is

$$\dot{W} = \dot{m} w_o = 52\,\text{kW}\,.$$

6.3.3 *Absorption refrigerator*

Most of the parts of an absorption refrigerator are the same ones as those of a compression refrigerator. Thus in both cases there is an evaporator, a condenser and a throttling valve. However, the compressor is absent; the role of the compressor is played by the complex arrangement shown in Fig. 6.14 consisting of a cooled absorber, a pump, a heated generator, a rectifier and another throttling valve. Let us consider their functions, starting with the absorber which absorbs ammonia (say) coming from the evaporator of the plant – the place where the cold is generated by evaporation of ammonia under low pressure.

The ammonia vapor is absorbed by water, a reaction that is exothermic so that it releases heat. The mixture is cooled so that is may absorb more ammonia. It is then pumped to the generator where it is heated so as to produce an ammonia rich vapor. The pump serves to maintain the circulation. The pressure rises in the generator because of the evaporation of the NH_3-H_2O mixture. Thus the whole arrangement effectively works as a compressor. Water is split off the ammonia rich vapor in a rectifier so that essentially pure ammonia goes to the condenser. The hot and pressurized water runs back from the rectifier to the absorber through a throttling valve.

Absorption refrigerators were the first viable refrigerators in the 19th century before cheap electric motors were available for the compression of the evaporated gas.

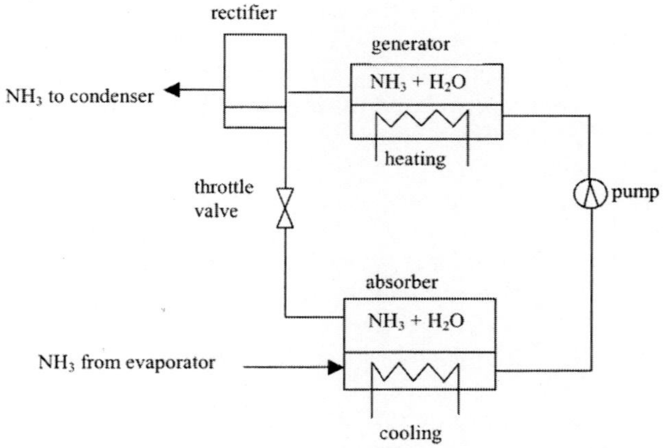

Fig. 6.14 Compression by desorption under heating in generator

6.3.4 *Refrigerants*

The coolant in a refrigerator cannot be water, since water freezes at small pressures near 0°C. The original and effective agent was ammonia NH_3, whose boiling point at 1bar is –34°C and, of course, for lower pressures it boils at even lower temperature. However, ammonia is not ideally suited, since – in the long run – it damages the pipes and the containers so that leaks will develop. In this respect fluorized hydrocarbons $C_kH_lCl_mF_n$ are better; these are usually known under commercial names, such as Freon, Frigen, Kaltron, *etc.*; they have recently acquired a bad reputation as "ozone killers."

These gases are basically hydrocarbons of the form C_kH_{2k+2} – like methane CH_4, ethane C_2H_6, propane C_3H_8 – in which one or more hydrogen atoms are replaced by chlorine or fluorine, or both. In particular, the compounds with fluorine are very stable and, if fluorine atoms are involved, the chlorine atoms are also very strongly attached to the molecule. Therefore the compounds are chemically quite inert, *i.e.* they do not react, not even when heated.

Teflon is a long chain-molecule of that type and it is so stable that the coating of a frying pan survives the heat; nor does it combine chemically with the food broiled in the pan. Fluorized hydrocarbons were invented in the 1920s in the Dupont-Nemours laboratories, and the coolant Freon-11 (CCl_3F) and Freon-12 (CCl_2F_2) of this company are often used in refrigerators. It is said that, in the early days of marketing the chemicals, a Dupont agent took a deep breath of them and then, exhaling, he blew out a candle. Thus he demonstrated the chemical stability: The gas did not react, neither inside his lungs nor in the flame of the candle.

The scientific name of coolants is $R(k-1, l+1, n)$, where the "R" stands for "refrigerant." Mono-chlorine-difluorine-methane $CHClF_2$ for instance is $R(22)$; it should be $R(022)$ but the initial zero is usually dropped. The three numbers k, l, n characterize the substance uniquely, since m is always equal to $2k+2-l-n$, because all C-bonds must be occupied.

Fluorized hydrocarbons – often mixtures of Freon-11 and Freon-12 – are frequently used as propellants in spray cans; at room temperature their vapor pressure is relatively small so that a thin and light aluminum can may serve as a safe container. In the end, of course all the industrially used freon ends up in the atmosphere and because of its stability it is not washed out by rain, nor does it combine with other substances that might form a sediment. So it diffuses throughout the air where it would seem to be innocuous, again because of its stability. However, in the upper atmosphere there *is* a process that can indeed shatter the molecules of the gas and will set the chlorine atoms free. That process is the bombardment of the molecules with high-energy ultraviolet radiation.

The lower atmosphere is shielded from the ultraviolet radiation of the sun by the stratospheric ozone layer which protects our skin from exposure to the high energy radiation. However, the ozone layer is slowly destroyed through reactions of the chlorine atom set free from a fluorized hydrocarbon. This atom reacts with ozone according to the chemical equation

$$\mathrm{Cl} + \mathrm{O}_3 \to \mathrm{ClO} + \mathrm{O}_2$$

and the chlorine oxide interacts with a rare O-atom to form O_2 and Cl again. The latter atom can again destroy an ozone molecule, *etc.* Thus even a small amount of freon in the atmosphere may be able to destroy a large amount of ozone and create the "ozone hole." As a result the damaging radiation can reach the organic molecules of plants and animals on the surface of the earth.

Therefore fluorized hydrocarbons were banned and it seems that in recent years the ozone hole is indeed becoming smaller.

6.3.5 *Heat pump*

Typically a heat pump is the same as a refrigerator. However, the purpose is different. While the required input is still the work of the compressor, the desired output is the cooling $|q_-|$ during condensation, *i.e.* the 3-4 branch in Fig. 6.13 a. In this case water may be used as the circulating agent, since usually all partial processes have temperatures above its freezing point.

The isobaric evaporation of water under low pressure – 0.1 bar (say) corresponding to a temperature of 7°C – may occur in pipes that are led through the groundwater. The condensation occurs in the living room which is to be heated. Recently an interesting variant was reported, where the evaporation took place in the cow-shed of a farm, while the condensation heated the living quarters.

The efficiency is given by

$$e = \frac{|q_-|}{w_0} = \frac{w_0 + q_+}{w_0},$$

where q_+ is the heat added to the process for the evaporation of the water. It is thus obvious that the efficiency of a heat pump is always greater than one.

We consider a heat pump working with Freon-12 so that we may use the diagram of Fig. 6.13b. Wet vapor is evaporated and superheated at 1.5 bar until it reaches a temperature of 45°C. Then follows the compression. The subsequent cooling and condensation proceeds until the coolant is fully liquid at 30°C. The process is closed by throttling. The required heating $|\dot{Q}_-|$ is 10^3 kW.

The cycle is represented in Fig. 6.15 by the thick lines. We read off all relevant data of the process as follows.

- temperature after compression: 105°C
- pressure ratio of compressor: $\frac{8}{1.5} = 5.3$
- specific work of the compressor: $w_o = h_3 - h_2 = 40\,\frac{\text{kJ}}{\text{kg}}$
- specific heat gained: $|q_-| = h_3 - h_4 = 185\,\frac{\text{kJ}}{\text{kg}}$
- efficiency: $e = 4.6$
- mass transfer: $\dot{m} = \frac{|\dot{Q}_-|}{|q_-|} = 5.5\,\frac{\text{kg}}{\text{s}} = 19.5\,\frac{\text{t}}{\text{h}}$
- power of compressor: $\dot{W} = \dot{m}\, w_o = 216\,\text{kW}$.

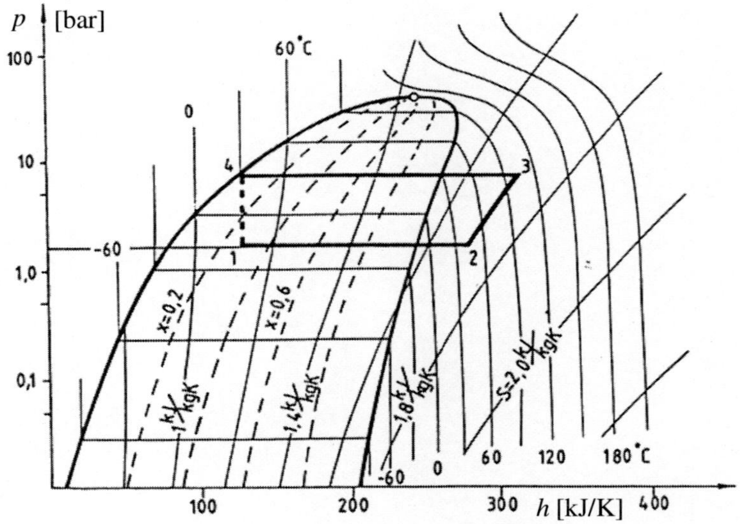

Fig. 6.15 On the layout of a heat pump

7 Heat Transfer

7.1 Non-Stationary Heat Conduction

7.1.1 *The heat conduction equation*

In a body at rest with constant mass density and momentum the equations of balance of mass and momentum are identically satisfied, if we neglect thermal expansion, and the energy balance (1.48) – without radiation – is reduced to the form

$$\rho\frac{\partial u}{\partial t}+\frac{\partial q_i}{\partial x_i}=0\,. \tag{7.1}$$

Under these circumstances the specific internal energy u is only a function of T. The heat flux q_i is given by the Fourier law $(2.3)_3$. We may therefore write

$$u=cT+\alpha \quad \text{and} \quad q_i=-\kappa\frac{\partial T}{\partial x_i}\,, \tag{7.2}$$

and we assume that the specific heat c and the thermal conductivity κ are constants. In this case we have

$$\frac{\partial T}{\partial t}=\lambda\frac{\partial^2 T}{\partial x_i \partial x_i} \quad \text{where} \quad \lambda=\frac{\kappa}{\rho c}\,. \tag{7.3}$$

λ is often called the thermal diffusivity; it is positive.*

Equation (7.3) is a partial differential equation for the determination of the temperature when initial and boundary values are given; it is called the heat conduction equation. We shall investigate the special case that T depends only on *one* spatial variable, namely $x_1=x$. In that case (7.3) reduces to

$$\frac{\partial T}{\partial t}=\lambda\frac{\partial^2 T}{\partial x^2}\,. \tag{7.4}$$

This is the one-dimensional heat conduction equation. It is one of the simplest partial differential equations and, in particular, it is linear in T, which implies that, if $T_1(x,t)$ and $T_2(x,t)$ both satisfy the equation, so does the sum $a_1T_1+a_2T_2$, when a_1, a_2 are constants.

7.1.2 *Separation of variables*

For linear differential equations a possible method of solution is the so-called *separation of variables*. The function $T(x,t)$ is written as $\vartheta(t)\chi(x)$, so that (7.4) assumes the form

$$\dot{\vartheta}(t)\chi(x)=\lambda\vartheta(t)\chi''(x)\,. \tag{7.5}$$

---

* In a gas it is c_p that enters the equation. In a liquid or solid the specific heats c_p and c_υ are equal to the extent that those materials are incompressible, *cf.* (4.36).

Dot and prime denote temporal and spatial derivatives, respectively. After dividing by $\lambda\vartheta(t)\chi(x)$ one separates the variables, *i.e.* one writes the time- and space-dependent functions on different sides of the equation, *viz.*

$$\frac{1}{\lambda}\frac{\dot{\vartheta}}{\vartheta}=\frac{\chi''}{\chi}. \tag{7.6}$$

Both sides must be equal to a constant, since, by (7.6), they can neither depend on x nor t. We call the constant $-a^2$ so as to emphasize that it should not be positive.* Thus we have two ordinary differential equations

$$\dot{\vartheta}=-\lambda a^2\vartheta \quad \text{and} \quad \chi''=-a^2\chi . \tag{7.7}$$

The solutions read

$$\vartheta(t)=C\exp\left(-\lambda a^2 t\right) \quad \text{and} \quad \chi(x)=A\cos ax+B\sin ax , \tag{7.8}$$

where the multiplicative constant C may be set equal to one without loss of generality, because this merely amounts to a redefinition of the arbitrary constants A and B.

An index n is introduced to number the possible values a_n of a and the corresponding values A_n, B_n of A and B. The general solution is then given by the sum

$$T(x,t)=\sum_n \exp\left(-\lambda a_n^2 t\right)\left[A_n\cos a_n x+B_n\sin a_n x\right]. \tag{7.9}$$

7.1.3 *Examples of heat conduction*

- *Heat conduction of an adiabatic rod of length* L

We solve a particularly simple initial and boundary value problem which – despite its simplicity – sheds some light on the procedures that must be followed for the solution of heat conduction problems. The ends of the heat conducting rod lie at the positions $x=\pm L/2$. The rod is adiabatically isolated. Along the rod we assign an initial temperature $T_0(x)$. Thus we have the initial and boundary value problem

$$\left.\frac{\partial T}{\partial x}\right|_{x=\pm l/2}=0 \quad \text{and} \quad T(x,0)=T_0(x). \tag{7.10}$$

The boundary values $(7.10)_1$ guarantee that the heat fluxes at the ends vanish so that the condition of adiabatic isolation is satisfied.

The boundary values can be satisfied, if we require that each term $\vartheta_n(t)\chi_n(x)$ in (7.9) vanishes at the ends:

---

* If the constant were positive, the temperature would exponentially increase at every point x, *cf.* (7.8). However, note Paragraph 7.1.3, where a^2 is imaginary for one of the examples.

$$\begin{aligned} &\text{for } x=-\frac{L}{2}: \quad \left[A_n \sin a_n \frac{L}{2} + B_n \cos a_n \frac{L}{2}\right] a_n = 0 \\ &\text{for } x=+\frac{L}{2}: \quad \left[-A_n \sin a_n \frac{L}{2} + B_n \cos a_n \frac{L}{2}\right] a_n = 0. \end{aligned} \tag{7.11}$$

These equations imply restrictions on the values of the constants a_n and on the coefficients A_n and B_n. Obviously one of the values a_n may be zero and we denote that one by a_0 so that we have

$$a_0 = 0\,. \tag{7.12}$$

This result satisfies the boundary conditions (7.11) for arbitrary values of A_0 and B_0. However, (7.11) also has solutions for $a_0 \neq 0$. In this case (7.11) is a linear homogeneous system of equations for A_n and B_n. A non-trivial solution requires that the determinant of the system vanishes and this happens for

$$\sin a_n L = 0 \;,\quad \text{hence} \quad a_n = \frac{n\pi}{L} \;,\quad n = \cdots, -2, -1, 1, 2, \cdots. \tag{7.13}$$

The corresponding values of A_n and B_n follow from (7.11) as

$$\begin{aligned} &A_n = 0 \qquad\qquad B_n \sim \text{arbitrary} \quad \text{for } n \sim \text{odd} \\ &A_n \sim \text{arbitrary} \quad A_n = 0 \qquad\qquad \text{for } n \sim \text{even}. \end{aligned} \tag{7.14}$$

Insertion of (7.12), (7.13), and (7.14) into the general solution (7.9) provides

$$\begin{aligned} T(x,t) = A_0 + &\sum_{\substack{n>0 \\ n\sim\text{odd}}} \exp\!\left(-\lambda\left(\tfrac{n\pi}{L}\right)^2 t\right)\left(B_n - B_{-n}\right)\sin\frac{n\pi}{L}x + \\ &\sum_{\substack{n>0 \\ n\sim\text{even}}} \exp\!\left(-\lambda\left(\tfrac{n\pi}{L}\right)^2 t\right)\left(A_n + A_{-n}\right)\cos\frac{n\pi}{L}x \end{aligned} \;. \tag{7.15}$$

Thus the boundary conditions $(7.10)_1$ are satisfied and it remains to satisfy the initial condition

$$T_0(x) = A_0 + \sum_{\substack{n>0 \\ n\sim\text{odd}}} \left(B_n - B_{-n}\right)\sin\frac{n\pi}{L}x + \sum_{\substack{n>0 \\ n\sim\text{even}}} \left(A_n + A_{-n}\right)\cos\frac{n\pi}{L}x\,. \tag{7.16}$$

From this relation the coefficients A_0, $B_n - B_{-n}$, and $A_n + A_{-n}$ may be calculated. A_0 follows by integration of (7.16) over the length of the rod

$$A_0 = \frac{1}{L}\int_{-L/2}^{L/2} T_0(x')\,\mathrm{d}x'\,. \tag{7.17}$$

The coefficients $A_n + A_{-n}$ and $B_n - B_{-n}$ also follow by integration from $-L/2$ to $L/2$, but before integration the equation (7.16) must be multiplied by $\sin\frac{n'\pi}{L}x$ and $\cos\frac{n'\pi}{L}x$, respectively. One makes use of the identities

$$\int_{-L/2}^{L/2} \sin\frac{n\pi}{L}x\,\sin\frac{n'\pi}{L}x\,\mathrm{d}x = \frac{L}{2}\delta_{nn'} \quad \text{for } n+n' \sim \text{even}$$

$$\int_{-L/2}^{L/2} \sin\frac{n\pi}{L}x\,\cos\frac{n'\pi}{L}x\,\mathrm{d}x = 0$$

$$\int_{-L/2}^{L/2} \cos\frac{n\pi}{L}x\,\cos\frac{n'\pi}{L}x\,\mathrm{d}x = \frac{L}{2}\delta_{nn'} \quad \text{for } n+n' \sim \text{even}$$

and obtains

$$A_n + A_{-n} = \frac{2}{L}\int_{-L/2}^{L/2} T_0(x')\cos\frac{n\pi}{L}x'\,\mathrm{d}x' \;,\; B_n - B_{-n} = \frac{2}{L}\int_{-L/2}^{L/2} T_0(x')\sin\frac{n\pi}{L}x'\,\mathrm{d}x' \,. \tag{7.18}$$

Insertion into (7.15) provides

$$\begin{aligned} T(x,t) = {} & \frac{1}{L}\int_{-L/2}^{L/2} T_0(x')\,\mathrm{d}x' + \\ & + 2\sum_{\substack{n>0\\ n\sim\text{odd}}} \exp\left(-\lambda\left(\tfrac{n\pi}{L}\right)^2 t\right)\frac{1}{L}\int_{-L/2}^{L/2} T_0(x')\sin\frac{n\pi}{L}x'\sin\frac{n\pi}{L}x + \\ & + 2\sum_{\substack{n>0\\ n\sim\text{even}}} \exp\left(-\lambda\left(\tfrac{n\pi}{L}\right)^2 t\right)\frac{1}{L}\int_{-L/2}^{L/2} T_0(x')\cos\frac{n\pi}{L}x'\cos\frac{n\pi}{L}x\,\mathrm{d}x'. \end{aligned} \tag{7.19}$$

The exponential factor makes terms with a large n unimportant for $t > 0$. Therefore good results may be obtained when the summation is broken off after a few terms.

Fig. 7.1 a illustrates the solution (7.19) for a particular initial condition. We choose T_0 homogeneous over both halves, but unequal such that we have

$$T_0(x) = \begin{cases} T_M = \text{const} & -L/2 < x < 0 \\ T_m = \text{const} & 0 < x < L/2 \,. \end{cases} \tag{7.20}$$

In this case the integrals in (7.19) are easily evaluated and we obtain

$$T(x,t) = \frac{T_M + T_m}{L} - 2(T_M - T_m)\sum_{\substack{n>0\\ n\sim\text{odd}}} \exp\left(-\lambda\left(\tfrac{n\pi}{L}\right)^2 t\right)\frac{1}{n\pi}\sin\frac{n\pi}{L}x \,. \tag{7.21}$$

Fig. 7.1 a shows this solution for different times t. Only few terms of the sum needed to be calculated, just enough so that the graphs were not visibly affected by the omitted ones; in the present case three terms were enough. Inspection shows that the initial temperature is quickly smoothed out and that the temperature field tends to the average value between T_M and T_m.

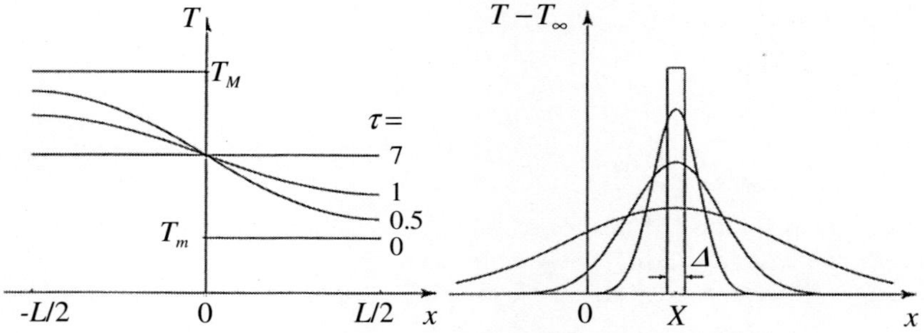

Fig. 7.1 Temperature adjustment by heat conduction
a. In an adiabatically isolated rod of length L
[$\tau = \lambda \pi^2 / L^2\, t$ is a dimensionless time]
b. Decay of a heat pole

- *Heat conduction in an infinitely long rod*

The solution for an infinitely long rod is obtained from (7.19) by letting L tend to infinity. The initial condition is chosen so as to deviate from a homogeneous temperature T_∞ for large values of $|L|$ only in a finite range of x. In this case it is possible and appropriate to replace T in the general solution (7.19) by $T - T_\infty$. Also T_0 in (7.17) ought to be replaced by $T_0 - T_\infty$ so that A_0 vanishes for $L \to \infty$ provided the initial temperature $T_0(x) - T_\infty$ is integrable.

In the limiting case $L \to \infty$ the expression $n\pi/L$ becomes a continuous variable which we denote by α. When n proceeds by $\Delta n = 2$ in the sums of (7.19) the variable α experiences an increase of $\mathrm{d}\alpha = \frac{\pi}{L}\Delta n = \frac{\pi}{L}2$. We write

$$n = \frac{L}{\pi}\alpha \text{, hence } \Delta n = 2 = \frac{L}{\pi}\mathrm{d}\alpha$$

and, accordingly, we replace the sums over n in (7.19) by integrals over α from 0 to ∞

$$\begin{aligned} T(x,t) - T_\infty &= \frac{1}{\pi}\int_0^\infty \left[\exp\left(-\lambda\alpha^2 t\right) \left(\int_{-\infty}^{\infty} [T_0(x') - T_\infty] \sin\alpha x' \,\mathrm{d}x' \right) \sin\alpha x \right] \mathrm{d}\alpha + \\ &\quad \frac{1}{\pi}\int_0^\infty \left[\exp\left(-\lambda\alpha^2 t\right) \left(\int_{-\infty}^{\infty} [T_0(x') - T_\infty] \cos\alpha x' \,\mathrm{d}x' \right) \cos\alpha x \right] \mathrm{d}\alpha \qquad (7.22) \\ &= \frac{1}{\pi}\int_{-\infty}^{\infty} [T_0(x') - T_\infty] \left[\int_0^\infty \exp\left(-\lambda\alpha^2 t\right) \cos\alpha(x' - x)\,\mathrm{d}\alpha \right] \mathrm{d}x'. \end{aligned}$$

In the last step the sequence of integrations was switched. The integral within the brackets can easily be evaluated – or looked up in an integration table – and we obtain

$$\int_0^\infty \exp\left(-\lambda\alpha^2 t\right)\cos\alpha(x'-x)\,\mathrm{d}\alpha = \frac{\sqrt{\pi}}{2\sqrt{\lambda t}}\exp\left(-\frac{(x'-x)^2}{4\lambda t}\right).$$

Hence follows

$$T(x,t)-T_\infty = \frac{1}{2\sqrt{\pi\lambda t}}\int_{-\infty}^{\infty}\left[T_0(x')-T_\infty\right]\exp\left(-\frac{(x'-x)^2}{4\lambda t}\right)\mathrm{d}x'. \tag{7.23}$$

This is the general solution of the heat conduction equation (7.4) for an infinitely long rod. Special cases result for special initial conditions.

For instance, if we choose a narrow "heat pole" at the point X , *i.e.*

$$T_0(x') = \begin{cases} T_0 & \text{for } |x'-X| < \dfrac{\Delta}{2} \\ T_\infty & \text{else}, \end{cases}$$

we obtain from (7.23)

$$T(x,t)-T_\infty = \frac{T_0-T_\infty}{2\sqrt{\pi\lambda t}}\exp\left(-\frac{(X-x)^2}{4\lambda t}\right)\Delta. \tag{7.24}$$

For each $t>0$ this is a bell-shaped curve, which becomes flatter and broader as time proceeds. Fig. 7.1 b shows how the heat pole is dispersed in time.

- *Maximum of temperature of the heat-pole-solution*

The graphs of Fig. 7.1 b show that at all positions $|x-X|>\Delta$ the temperature first increases and then decreases. For a given position the maximum is reached at time t_{max} which may be calculated from

$$\frac{\partial T(x,t)}{\partial t} = 0 \quad \text{by (7.24) as} \quad t_{max} = \frac{(x-X)^2}{2\lambda}. \tag{7.25}$$

The maximum value of T_{max} of T results by insertion of t_{max} into (7.24) and we obtain

$$T_{max} = T_\infty\left(1-\frac{1}{\sqrt{2\pi\,\mathrm{e}}}\frac{T_0-T_\infty}{T_\infty}\frac{\Delta}{x-X}\right). \tag{7.26}$$

It is noteworthy, perhaps, that the temperature T_{max} is universal, *i.e.* independent of the thermal diffusivity λ ; However, the time t_{max} does depend on λ; it is inversely proportional to the thermal conductivity.

Finally, we consider the following situation: A long wire of temperature 70°C is suddenly heated up to the temperature 1100°C over a length of 1 mm so that it melts. On the wire a rubber ring is attached, which must not become hotter than 150°C. We ask for the smallest distance $x-X$ at which the ring may be placed from the hot spot. An easy rearrangement of (7.26) gives the answer: $x - X = 3.06\,\text{mm}$. This distance is independent of the material of the wire.

A peculiarity of the solution (7.24) is revealed by the observation that for all times $t > 0$ the temperature difference $T(x,t)\text{-}T_\infty$ is non-zero everywhere. Another way of saying this is that the information about the initial heat pole has spread with infinite speed. This phenomenon is not real; it is an artifact of the Fourier law (2.3). That law is perfect for all practical heat conduction problems, but it is not strictly valid.*

- *Heat waves in the Earth*

The daily or annual temperature changes on the earth's surface may be represented by a periodic function of the form

$$T(0,t) = T_0 + \Delta T \cos\left(\frac{2\pi}{\tau} t\right). \tag{7.27}$$

We wish to determine the temperature $T(x,t)$ in the ground at depth x and at time t as predicted by the equation (7.4) of heat conduction. The generic solution is still given by (7.9). However, if this solution is required to be compatible with the boundary value (7.27) for $T(0,t)$, we can only have three non-vanishing values of a_n^2, namely

$$a_n^2 = 0 \quad \text{and} \quad a_{\pm 1}^2 = \pm \mathrm{i} \frac{2\pi}{\lambda \tau}. \tag{7.28}$$

Note that $a_{\pm 1}^2$ must be imaginary so that the time-dependent exponential factor in (7.9) can represent an oscillation with period τ. Thus the solution (7.9) assumes the form

$$\begin{aligned} T(x,t) = A_0 &+ \exp\left(-\mathrm{i} \tfrac{2\pi}{\tau} t\right)\left[A_{+1} \cos\left(\sqrt{\mathrm{i}}\, \tfrac{2\pi}{\tau} x\right) + B_{+1} \sin\left(\sqrt{\mathrm{i}}\, \tfrac{2\pi}{\tau} x\right)\right] + \\ &\exp\left(\mathrm{i} \tfrac{2\pi}{\tau} t\right)\left[A_{-1} \cos\left(\sqrt{\mathrm{i}}\, \tfrac{2\pi}{\tau} x\right) + B_{-1} \sin\left(\sqrt{\mathrm{i}}\, \tfrac{2\pi}{\tau} x\right)\right], \end{aligned}$$

and, if this solution is to be a cosine-type oscillation at $x = 0$ about the value T_0 and with the amplitude ΔT, we obtain the following values for A_n

$$A_0 = T_0 \quad \text{and} \quad A_{\pm 1} = \frac{\Delta T}{2},$$

while the values $B_{\pm 1}$ are not restricted by the boundary conditions. Therefore the solution has the form

* The phenomenon has motivated the formulation of Extended Thermodynamics in which all speeds are finite. *Cf.* I. Müller, T. Ruggeri. Springer Tracts of Natural Philosophy 37. 2nd edition 1998.

$$T(x,t)=T_0+\exp\left(\mathrm{i}\tfrac{2\pi}{\tau}t\right)\left[\tfrac{1}{2}\Delta T\cos\left(\sqrt{-\mathrm{i}}\sqrt{\tfrac{2\pi}{\tau\lambda}}x\right)+B_{-1}\sin\left(\sqrt{-\mathrm{i}}\sqrt{\tfrac{2\pi}{\lambda\tau}}x\right)\right]+ \\ \exp\left(-\mathrm{i}\tfrac{2\pi}{\tau}t\right)\left[\tfrac{1}{2}\Delta T\cos\left(\sqrt{\mathrm{i}}\sqrt{\tfrac{2\pi}{\tau\lambda}}x\right)+B_{+1}\sin\left(\sqrt{\mathrm{i}}\sqrt{\tfrac{2\pi}{\lambda\tau}}x\right)\right]. \tag{7.29}$$

We recall that $\sqrt{\pm\mathrm{i}}=\frac{1}{\sqrt{2}}(1\pm\mathrm{i})$ holds and that we have

$$\cos\alpha=\tfrac{1}{2}[\exp(\mathrm{i}\alpha)+\exp(-\mathrm{i}\alpha)]\ ,\quad \sin\alpha=\tfrac{1}{2}[\exp(\mathrm{i}\alpha)-\exp(-\mathrm{i}\alpha)]$$

and we use these relations to replace the sine and cosine expressions in (7.29). In addition we plausibly require that $T(x,t)$ for $x\to\infty$ – *i.e.* deep in the ground – remains finite. This implies

$$B_{\pm1}=\pm\mathrm{i}\frac{\Delta T}{2}$$

so that the final solution may be written in the form

$$T(x,t)=T_0+\Delta T\exp\left(-\sqrt{\tfrac{\pi}{\lambda\tau}}x\right)\cos\left(\tfrac{2\pi}{\tau}t-\sqrt{\tfrac{\pi}{\lambda\tau}}x\right). \tag{7.30}$$

This solution represents a wave which penetrates the ground with a phase velocity V and experiences a decay of its amplitude by the amount e^{-1} on the length L. We have

$$V=2\sqrt{\pi\frac{\lambda}{\tau}}\quad\text{and}\quad L=\sqrt{\frac{1}{\pi}\lambda\tau}\ . \tag{7.31}$$

For champagne storage it is important that the annual temperature change is less than $1\,\mathrm{K}$. Therefore the storage room must be at a depth x_{C} that follows from the equation

$$\Delta T\exp\left(-\sqrt{\tfrac{\pi}{\lambda\tau}}x_{\mathrm{C}}\right)\le 1\,\mathrm{K}\ .$$

If the annual temperature amplitude equals $\Delta T=20\,\mathrm{K}$ and if the thermal diffusivity of the ground has the value $2\cdot10^{-7}\,\frac{\mathrm{m}^2}{\mathrm{s}}$, we obtain $x_{\mathrm{C}}=4.2\,\mathrm{m}$ as the minimal depth of the wine cellar. The phase speed of the wave is $V\approx1\,\frac{\mathrm{mm}}{\mathrm{s}}$ and the e^{-1}-depth is $L=1.42\,\mathrm{m}$.

The solution (7.30) exhibits a depth-dependent phase shift of size $\sqrt{\frac{\pi}{\lambda\tau}}x$. At the depth $x_\pi=\sqrt{\pi\lambda\tau}$ that shift equals π and for the data given above is $x_\pi=4.4\,\mathrm{m}$. Hence it follows that the stored champagne is coolest in summer and warmest in wintertime.

It is also possible to calculate the *daily* temperature oscillations in the ground. Because of the smaller amplitude of $\Delta T=5\,\mathrm{K}$ and because of the smaller period of $\tau=1\,\mathrm{day}$, we obtain different values in this case. The depth x_{C} would be a mere $12\,\mathrm{cm}$.

7.1.4 *On the history of non-stationary heat conduction*

The pioneer of heat conduction was Jean Baptiste Joseph Baron de FOURIER (1768-1830), who published his results in 1822 in his book "Théorie analytique de la chaleur." FOURIER had a strange conception of the nature of heat. He writes:

Heat is the principle behind elasticity. The repulsive force maintains the shape of solids and the volume of liquids. Indeed, the molecules in solids would fall toward each other – following their attraction – unless the heat kept them apart.

With some good will one may give that statement a valid modern meaning, but this might mean to ascribe too much understanding to FOURIER. And then, he does not use such concepts for his derivation of the law of heat conduction $(7.2)_2$ and the heat equation (7.3). He merely assumes that heat follows a temperature gradient and he balances the in- and efflux of heat in a "corpuscle," *i.e.* an infinitesimal mass element.

His life-long preoccupation with heat had given FOURIER a fixed idea. ASIMOV writes: FOURIER believed heat to be essential to health and so he always kept his dwelling place overheated and swathed himself in layer upon layer of clothes. He died from a fall down the stairs.

In the effort to solve the heat equation FOURIER developed the theory of approximation of arbitrary functions by harmonic functions. This was an unparalleled mathematical breakthrough, whose significance is still felt today after 200 years. It is true that the method of separation of variables was not invented by FOURIER. Daniel BERNOULLI had used it to describe the oscillating string. However, FOURIER noticed that BERNOULLI's results imply

> *... that every arbitrary function may be expanded in sines and cosines.*

In our analysis we see that observation confirmed in equation (7.16) from which we could explicitly calculate the coefficients by (7.17) and (7.18). FOURIER writes

> *...that BERNOULLI's assumption is correct cannot be confirmed more convincingly than by writing down the expansion and then determining the coefficients.*

Despite this FOURIER is amazed:

> *... it is most noteworthy that convergent series for arbitrary functions can be formed Thus there are functions, which coincide in certain intervals, and which outside these intervals are quite different They are represented by curves which osculate a given curve on a finite interval and deviate in other intervals.*

7.2 Heat Exchangers

7.2.1 *Heat transport coefficients and heat transfer coefficient*

In heat exchangers two fluids – one warm and one cold – flow past each other in separate channels on different sides of a heat conducting wall. Fig. 7.2 shows the channels and the wall – shaded – perpendicular to the x-direction. The flow occurs parallel to the y-direction. The fluids have different temperatures so that there is a normal heat flux $q_x(y)$ through the wall. If $T_L(y)$ and $T_R(y)$ are the temperatures of the fluids on either side of the wall, the heat flux in the wall reads, *cf.* (2.10)

$$q_x(y) = \frac{\kappa}{D}\left[T_L(y) - T_R(y)\right]. \tag{7.32}$$

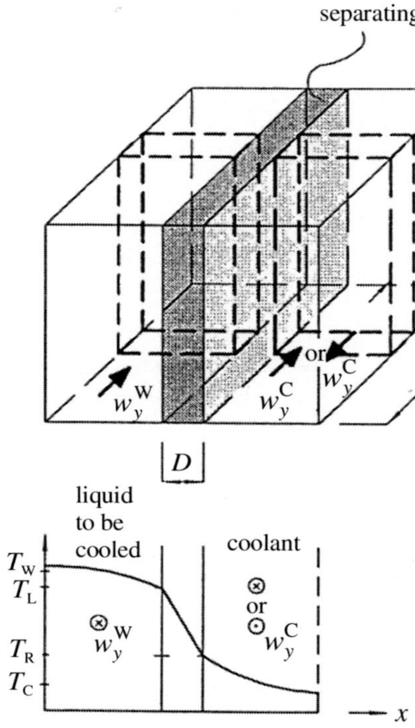

Fig. 7.2 Exchange of heat between two flow channels
a. Control volumes in the flow channels (dashed)
b. Temperature profile in a heat exchanger for one *y*

In the two channels the heat fluxes are not so easily determined. Of course, they depend on the temperature gradients and on the heat conductivities of the fluids. But the temperature gradients inside the channels are generally unknown. They are usually strongly non-homogeneous and depend on the type of flow – laminar or turbulent – in a complicated manner. A plausible approximate assumption, basic to the theory of heat exchangers, states that the heat fluxes may be written as

$$q_x(y) = \alpha_W\left[T_W(y) - T_L(y)\right] \text{ and } q_x(y) = \alpha_C\left[T_R(y) - T_C(y)\right], \tag{7.33}$$

where $T_W(y)$ and $T_C(y)$ are the mean temperatures of the fluids in the warm and in the cold channel at a given y, respectively. α_W and α_C are empirical coefficients, called *heat transport coefficients*.

Since q_x has the same value in all three equations, (7.32), (7.33), we conclude that

$$q_x(y) = \frac{1}{\frac{1}{\alpha_\mathrm{W}} + \frac{\kappa}{D} + \frac{1}{\alpha_\mathrm{C}}}\left[T_\mathrm{W}(y) - T_\mathrm{C}(y)\right] \equiv k\left[T_\mathrm{W}(y) - T_\mathrm{C}(y)\right] \tag{7.34}$$

holds, where k – defined by $(7.34)_2$ – is called *the heat transfer coefficient.* For actual heat exchangers that coefficient must be measured and its values are tabulated for the most common types.

7.2.2 *Temperature gradients in the flow direction*

In order to determine how $T_\mathrm{W}(y)$ and $T_\mathrm{C}(y)$ depend on the flow direction y, we write the equation of the balance of energy for the thin cross-sectional slices of thickness dy of the channels. For a stationary flow, and ignoring gravitation, radiation, and friction, we obtain

$$\int_{\partial V} (\rho h w_i + q_i) n_i \mathrm{d}A = 0 \tag{7.35}$$

where ∂V is the surface which is represented by one or the other of the dashed slices in Fig. 7.2. We assume that

- the heat flux in the flow direction is negligible, and that
- among the narrow lateral surfaces of the slices an appreciable heat flux occurs only through the strip of the separating wall.

In that case we obtain on the warm and cold sides respectively

$$\int_{\partial V_\mathrm{W}} q_i n_i \mathrm{d}A = k(T_\mathrm{W} - T_\mathrm{C})L_z \mathrm{d}y \qquad \Bigg\| \qquad \int_{\partial V_\mathrm{C}} q_i n_i \mathrm{d}A = -k(T_\mathrm{W} - T_\mathrm{C})L_z \mathrm{d}y\,. \tag{7.36}$$

It is because of these lateral heat fluxes that the temperatures, hence the enthalpies h, depend on y. In an approximate manner we may write

$$\begin{aligned} &\int_{\partial V_\mathrm{W}} \rho h w_i n_i \mathrm{d}A \\ &= \dot m_\mathrm{W}\left[h_\mathrm{W}(y+\mathrm{d}y) - h_\mathrm{W}(y)\right] \\ &= \dot m_\mathrm{W} c_\mathrm{W} \frac{\mathrm{d}T_\mathrm{W}}{\mathrm{d}y}\mathrm{d}y \end{aligned} \qquad \Bigg\| \qquad \begin{aligned} &\int_{\partial V_\mathrm{C}} \rho h w_i n_i \mathrm{d}A = \\ &= \pm \dot m_\mathrm{C}\left[h_\mathrm{C}(y+\mathrm{d}y) - h_\mathrm{C}(y)\right]. \\ &= \pm \dot m_\mathrm{C} c_\mathrm{C} \frac{\mathrm{d}T_\mathrm{C}}{\mathrm{d}y}\mathrm{d}y \end{aligned} \tag{7.37}$$

In (7.37) we have introduced the specific heats $c = (\partial h/\partial T)_p$. The $\pm$-sign refers to the equal and opposite flow directions in the two channels, respectively: Accordingly we speak of parallel and anti-parallel heat exchangers. In order to fix the ideas we consider the flow on the warm side to be in the y-direction.

Insertion of (7.36) and (7.37) into (7.35) leads to

$$\frac{\mathrm{d}T_\mathrm{W}}{\mathrm{d}y} = -\frac{kL_z}{\dot m_\mathrm{W} c_\mathrm{W}}(T_\mathrm{W} - T_\mathrm{C}) \qquad \Bigg\| \qquad \frac{\mathrm{d}T_\mathrm{C}}{\mathrm{d}y} = \pm\frac{kL_z}{\dot m_\mathrm{C} c_\mathrm{C}}(T_\mathrm{W} - T_\mathrm{C})\,. \tag{7.38}$$

Here and below the upper sign refers to parallel flow, the lower one to anti-parallel flow.

It is obvious from (7.38) that the temperature difference $T_{\mathrm{W}} - T_{\mathrm{C}}$ obeys the single differential equation

$$\frac{\mathrm{d}(T_{\mathrm{W}} - T_{\mathrm{C}})}{\mathrm{d}y} = -\left(\frac{kL_z}{\dot{m}_{\mathrm{W}} c_{\mathrm{W}}} \pm \frac{kL_z}{\dot{m}_{\mathrm{C}} c_{\mathrm{C}}}\right)(T_{\mathrm{W}} - T_{\mathrm{C}}). \tag{7.39}$$

It follows that – in both cases – the temperature difference is exponential. For anti-parallel flow directions there is the possibility that $T_{\mathrm{W}}(y) - T_{\mathrm{C}}(y)$ is constant along the heat exchanger. This happens when $\dot{m}_{\mathrm{W}} c_{\mathrm{W}} = \dot{m}_{\mathrm{C}} c_{\mathrm{C}}$ holds.

7.2.3 *Temperatures along the heat exchanger*

For simplicity of notation we put

$$\frac{1}{\lambda_{\mathrm{W}}} = \frac{kL_z}{\dot{m}_{\mathrm{W}} c_{\mathrm{W}}} \quad \text{and} \quad \frac{1}{\lambda_{\mathrm{C}}} = \frac{kL_z}{\dot{m}_{\mathrm{C}} c_{\mathrm{C}}} \tag{7.40}$$

and obtain from (7.39)

$$T_{\mathrm{W}} - T_{\mathrm{C}} = [T_{\mathrm{W}}(0) - T_{\mathrm{C}}(0)]\exp\left[-\left(\tfrac{1}{\lambda_W} \pm \tfrac{1}{\lambda_C}\right)y\right]. \tag{7.41}$$

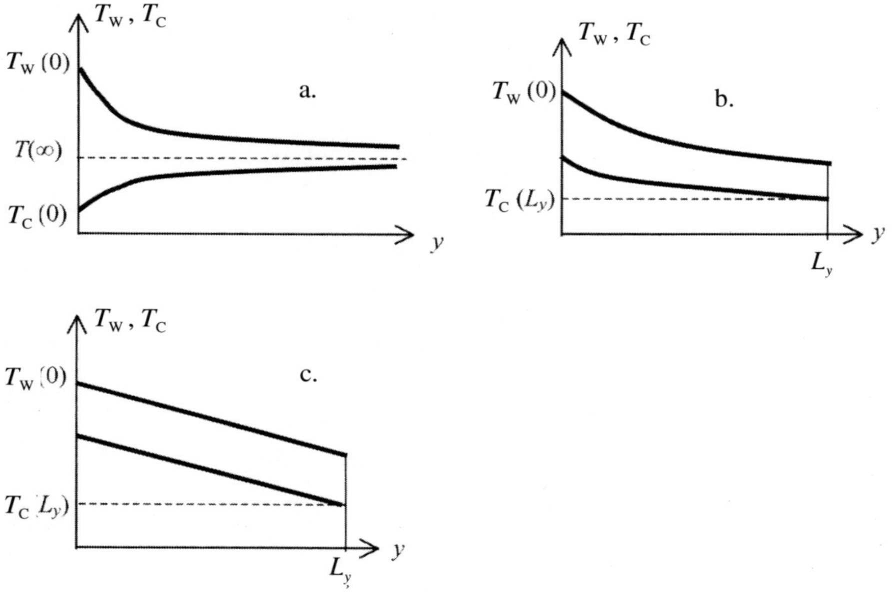

Fig 7.3 Temperature of heat exchangers
a. Parallel flow
b. Anti-parallel flow
c. Linear graphs for anti-parallel flow

Insertion into (7.38) and integration gives

$$\begin{aligned} T_{\mathrm{W}}(y) &= T_{\mathrm{W}}(0) - \frac{\lambda_{\mathrm{C}}}{\lambda_{\mathrm{C}} \pm \lambda_{\mathrm{W}}}[T_{\mathrm{W}}(0) - T_{\mathrm{C}}(0)]\left(1 - \exp\left[-\left(\tfrac{1}{\lambda_{\mathrm{W}}} \pm \tfrac{1}{\lambda_{\mathrm{C}}}\right)y\right]\right) \\ T_{\mathrm{C}}(y) &= T_{\mathrm{C}}(0) \pm \frac{\lambda_{\mathrm{W}}}{\lambda_{\mathrm{C}} \pm \lambda_{\mathrm{W}}}[T_{\mathrm{W}}(0) - T_{\mathrm{C}}(0)]\left(1 - \exp\left[-\left(\tfrac{1}{\lambda_{\mathrm{W}}} \pm \tfrac{1}{\lambda_{\mathrm{C}}}\right)y\right]\right) \end{aligned} . \quad (7.42)$$

Fig. 7.3 a shows the graphs of these functions for parallel flow of the fluids in the channels. In this case there is an asymptotic common value for the two temperatures for $y \to \infty$, namely

$$T_\infty = \frac{\lambda_{\mathrm{W}}}{\lambda_{\mathrm{C}} + \lambda_{\mathrm{W}}} T_{\mathrm{W}}(0) + \frac{\lambda_{\mathrm{C}}}{\lambda_{\mathrm{C}} + \lambda_{\mathrm{W}}} T_{\mathrm{C}}(0) \quad \text{hence} \quad \frac{T_{\mathrm{W}}(0) - T_\infty}{T_{\mathrm{C}}(0) - T_\infty} = -\frac{\lambda_{\mathrm{C}}}{\lambda_{\mathrm{W}}}. \quad (7.43)$$

The last equation means that in the end the warm fluid has lost as much heat as the cold one has absorbed. The cooling effect $T_{\mathrm{W}}(0) - T_\infty$ depends on the mass transfers and specific heats of the fluids, and on the area of the heat exchanging wall.

For anti-parallel flow the equations (7.42) are perhaps not the most useful expressions for the temperature, because for the cold channel $T_{\mathrm{C}}(0)$ is the temperature *at the end* of the heat exchanger and that temperature may be unknown. It seems more natural in this case to represent the temperatures in terms of $T_{\mathrm{W}}(0)$ and $T_{\mathrm{C}}(L_y)$, the temperatures of both fluids at their point of entry into the heat exchanger. If this is done, an easy calculation results in the formulae

$$\begin{aligned} T_{\mathrm{W}}(y) &= \frac{\left(\frac{\lambda_{\mathrm{C}}}{\lambda_{\mathrm{W}}} \exp[-(\text{-})y] - \exp[-(\text{-})L_y]\right) T_{\mathrm{W}}(0) + \frac{\lambda_{\mathrm{C}}}{\lambda_{\mathrm{W}}}(1 - \exp[-(\text{-})y]) T_{\mathrm{C}}(L_y)}{\frac{\lambda_{\mathrm{C}}}{\lambda_{\mathrm{W}}} - \exp[-(\text{-})L_y]} \\ T_{\mathrm{C}}(y) &= \frac{(\exp[-(\text{-})y] - \exp[-(\text{-})L_y]) T_{\mathrm{W}}(0) + \left(\frac{\lambda_{\mathrm{C}}}{\lambda_{\mathrm{W}}} - \exp[-(\text{-})y]\right) T_{\mathrm{C}}(L_y)}{\frac{\lambda_{\mathrm{C}}}{\lambda_{\mathrm{W}}} - \exp[-(\text{-})L_y]} . \end{aligned} \quad (7.44)$$

For conciseness the exponents in these equations are abbreviated; they are equal to those of (7.42) for anti-parallel flows. Fig. 7.3 b shows typical graphs for this case.

An interesting special case is given by $\lambda_{\mathrm{C}} = \lambda_{\mathrm{W}}$: Numerators and denominators in the equations (7.44) vanish. Therefore we treat the case in the limit

$$\frac{\lambda_{\mathrm{C}}}{\lambda_{\mathrm{W}}} = 1 + \alpha \quad \text{and} \quad \frac{1}{\lambda_{\mathrm{C}}} - \frac{1}{\lambda_{\mathrm{W}}} = \frac{1}{\lambda}\alpha$$

for a small α. λ is the limiting value of both λ_{C} and λ_{W}. We obtain

$$\begin{aligned} T_{\mathrm{W}}(y) &= T_{\mathrm{W}}(0) - [T_{\mathrm{W}}(0) - T_{\mathrm{C}}(L_y)] \frac{1}{1 + \frac{\lambda}{L_y}} \frac{y}{L_y} \\ T_{\mathrm{C}}(y) &= T_{\mathrm{C}}(L_y) + [T_{\mathrm{W}}(0) - T_{\mathrm{C}}(L_y)] \frac{1}{1 + \frac{\lambda}{L_y}} \left(1 - \frac{y}{L_y}\right). \end{aligned} \quad (7.45)$$

Both functions are linear and their slopes are equal, *cf.* Fig. 7.3 c. Therefore one might say that the heat exchanger is equally effective for all values of y. And obviously the temperature $T_C(0)$ at the exit of the *cold* flow may be higher than the temperature $T_W(L_y)$ at the exit of the *warm* flow.

Of particular interest is the hypothetical limit $\lambda/L_y \to 0$, where both temperatures are equal for all values of y. We have

$$T_W(y) = T_C(y) = T_W(0) - [T_W(0) - T_C(L_y)]\frac{y}{L_y}. \tag{7.46}$$

In this the case we speak of *reversible heat exchange*, because an exchange of heat at no temperature difference, or a minimal one, is reversible. What we need in this case is

either a large area L_yL_z of the separating wall
or a large heat transfer coefficient
or a small mass transfer rate $\dot{m}$
or a small specific heat c

and, of course, reversible heat exchange requires the employment of anti-parallel flows in the channels.

Regeneration in cycles – like those of ERICSON, STIRLING and BRAYTON discussed in Chap. 3 – usually employ heat exchangers with anti-parallel flow and attempt to realize reversible heat exchange as closely as possible, *e.g.* Fig. 3.11.

7.3 Radiation

7.3.1 *Coefficients of spectral emission and absorption*

A body cools when it emits heat by radiation and it heats up when absorbing heat. Reflection and transmission of radiation have no heating effect.

Heat radiation consists of electro-magnetic waves with a larger wave length than that of the red color of visible light. Therefore one speaks of *infrared radiation*; it covers the range of wave lengths

$$10^{-6}\,\mathrm{m} < \lambda < 10^{-4}\,\mathrm{m}. \tag{7.47}$$

What happens is this: During absorption the electric field of the radiation induces motion in the electric charges of the atoms or molecules. Thus they acquire kinetic energy so that the body heats up. The emission is due to the fact that accelerated electric charges – in particular oscillating molecular charges or electrons in a metal – slow down by emitting an electro-magnetic field, *i.e.* they lose their kinetic energy and the body cools.

Radiation may also lead to heating outside the range (7.47) of wave lengths by special effects. This is put in evidence by the microwave oven. The microwave has a wavelength of about 12cm and that wave length is capable of setting water molecules into rotation thus increasing their kinetic energy and eventually heating up water-containing food nearly homogeneously in a short time. Ceramic or metallic dishes remain unaffected, since they do not contain water.* However, metal reflects the microwave so that it is counter-productive to keep the food in a metal container when we wish to heat it.

In radiation thermodynamics we call a body

- black, if it absorbs all incident radiation,
- white, if it reflects all incident radiation,
- gray, if it absorbs equal proportions of all incident wave lengths.

These assignments of color are *by analogy* with visible light only. Thus we must not think that a black body for heat radiation should also *look* black. Indeed, the sun, or a white hot furnace are not bad examples for black bodies *in heat radiation* and yet we have to avert our eyes from their brightness *in the range of visible light.*

The most important quantities for the thermodynamic effects of radiation are the densities J of the energy fluxes emitted or absorbed by the body. Both are composed from spectral contributions $J_\lambda \mathrm{d}\lambda$ of all occurring wave lengths so that we have

$$J = \int_0^\infty J_\lambda \mathrm{d}\lambda \,.$$

We call J_λ the density of the spectral energy flux.

For the radiation emitted by a black body J_λ depends only of the temperature of the body; it is *independent of the material* of the body. Therefore we say that J_λ is a universal function of λ and T. Measurements show the graphs of Fig. 7.4 for this function and its analytic form is given by the Planck formula

$$J_\lambda^{\mathrm{B}}(T) = C\frac{1}{\lambda^5}\frac{1}{\exp\left(\frac{hc}{\lambda kT}\right)-1}, \text{ where}$$

$$C = 5.66\cdot 10^{-8}\frac{\mathrm{W}}{\mathrm{m}^2\mathrm{K}^4}\left(\frac{hc}{k}\right)^4 \frac{1}{\int\limits_0^\infty \frac{x^3\mathrm{d}x}{\exp(x)-1}}. \tag{7.48}$$

h is the Planck constant; h and the Boltzmann constant k are universal constants and h has the value $h = 6.6252\cdot 10^{-34}\,\mathrm{Js}$. The index B stands for black body radiation.

* Of course a ceramic plate carrying the heated substance will become warm itself or even hot by conduction.

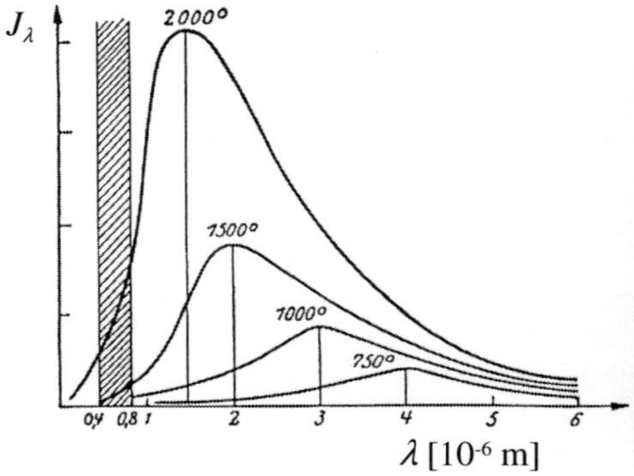

Fig. 7.4 Planck distribution. The shaded part represents the visible spectrum

If the body which emits the radiation is not black, J_λ is different from (7.48), both in the λ- and the T-dependence, and it depends on the material. We characterize the deviation from black body radiation by the spectral emission coefficient $\varepsilon(\lambda,T)$ and write

$$J_\lambda(T) = \varepsilon(\lambda,T)\, J_\lambda^{\mathrm{B}}(T). \tag{7.49}$$

In words: The spectral emission coefficient $\varepsilon(\lambda,T)$ is defined as the ratio of the emitted spectral energy flux and the spectral energy flux of a black body at the same temperature. $\varepsilon(\lambda,T)$ is a function whose form depends on the material of the body, and on the nature of its surface, *e.g.* metal, polished or rough, or wood.

The spectral energy flux density J_λ absorbed by a body is the fraction $A(\lambda,T)$ of the incoming spectral energy flux H_λ. This fraction is called the absorption number; it depends on the material of the body, and generally on λ and T. We write

$$J_\lambda = A(\lambda,T) H_\lambda; \tag{7.50}$$

the rest of H_λ is reflected or transmitted.

7.3.2 *Kirchhoff's law*

$\varepsilon(\lambda,T)$ and $A(\lambda,T)$ are two constitutive functions, *i.e.* they depend on the material of the bodies involved. One determines emission, the other absorption, and *a priori* it is not clear that there should be any relation between the two functions. And yet, it turns out that they are equal. This is the contents of Kirchhoff's law:

$$\varepsilon(\lambda,T) = A(\lambda,T). \tag{7.51}$$

Gustaf Robert KIRCHHOFF (1824-1887) discovered that each element, when heated to incandescence, sends out light of frequencies that are characteristic for the element. He also observed that when light passes through a thin layer of an element – or through its vapor – it loses exactly those frequencies which the hot element emits. So, since the sunlight lacks the frequencies that hot sodium emits, KIRCHHOFF concluded that the solar atmosphere must contain sodium vapor.

For simplicity we prove Kirchhoff's law for the particularly simple case of two parallel plates 1 and 2 with temperatures $\varepsilon_1(\lambda,T_1)$, $\varepsilon_2(\lambda,T_2)$ and absorption numbers $A_1(\lambda,T_1)$, $A_2(\lambda,T_2)$. Between the plates radiation of frequency λ is exchanged: From 1 to 2 we have an energy flux H_λ^1 and in the opposite direction we have H_λ^2. These fluxes consist of emitted and reflected parts and we may write

$$H_\lambda^1 = \varepsilon_1 J_\lambda^{\mathrm{B}}(T_1) + (1-A_1) H_\lambda^2\,,\quad H_\lambda^2 = \varepsilon_2 J_\lambda^{\mathrm{B}}(T_2) + (1-A_2) H_\lambda^1\,, \tag{7.52}$$

when we drop arguments for this intermediate calculation. Hence follows

$$H_\lambda^1 = \frac{\varepsilon_1 J_\lambda^{\mathrm{B}}(T_1) + (1-A_1)\varepsilon_2 J_\lambda^{\mathrm{B}}(T_2)}{1-(1-A_1)(1-A_2)}\,,\quad H_\lambda^2 = \frac{\varepsilon_2 J_\lambda^{\mathrm{B}}(T_2) + (1-A_2)\varepsilon_1 J_\lambda^{\mathrm{B}}(T_1)}{1-(1-A_1)(1-A_2)}\,. \tag{7.53}$$

We note that H_λ^1 and H_λ^2 depend on T_1, T_2 in a complicated manner. The net flux $Q_\lambda^{1\to 2}$ from plate 1 to plate 2 results as

$$Q_\lambda^{1\to 2} = H_\lambda^1 - H_\lambda^2 = \frac{\varepsilon_1 A_2 J_\lambda^{\mathrm{B}}(T_1) - \varepsilon_2 A_1 J_\lambda^{\mathrm{B}}(T_2)}{A_1 + A_2 - A_1 A_2}\,.$$

In equilibrium we must have $Q_\lambda^{1\to 2} = 0$ and the temperatures of the plates must be equal. Thus we obtain

$$\frac{\varepsilon_1(\lambda,T)}{A_1(\lambda,T)} = \frac{\varepsilon_2(\lambda,T)}{A_2(\lambda,T)}\,.$$

In case that plate 2 is black, so that numerator and denominator on the right hand side are both equal to one, we obtain $\varepsilon_1(\lambda,T) = A_1(\lambda,T)$ which proves Kirchhoff's law.

We note that (7.51) is not an equilibrium property. It represents a relationship between two constitutive functions and has nothing to do with equilibrium, even though we have proved the equality by making use of the argument that $Q_\lambda^{1\to 2} = 0$ vanishes in equilibrium and that in equilibrium $T_1 = T_2$ holds.

7.3.3 *Averaged emission coefficient and averaged absorption number*

For technical purposes the above spectral consideration are often not very important, because as a rule the radiation is a superposition of many frequencies. Consequently Kirchhoff's law loses much of its stringency. Let us consider this:

The emitted energy flux density J^{B} of a black body results from (7.48) by integration over all wavelengths which yields

$$J^{\mathrm{B}}(T)=\int_0^\infty J_\lambda^{\mathrm{B}}\mathrm{d}\lambda=\sigma T^4 \ , \ \text{where} \ \ \sigma=5.67\cdot 10^{-8}\frac{\mathrm{N}}{\mathrm{m}^2\mathrm{K}^4}. \tag{7.54}$$

This law, which had been found experimentally – long before PLANCK – by the physicist Josef STEFAN (1835-1893), is known as the Stefan-Boltzmann law, or T^4-law. Note that doubling the temperature increases the radiative energy flux by the factor 16.

Integration of (7.49) over λ provides the energy flux density $J_{\mathrm{em}}(T)$ emitted by a non-black body

$$J_{\mathrm{em}}(T)=\int_0^\infty \varepsilon(\lambda,T)J_\lambda^{\mathrm{B}}(T)\mathrm{d}\lambda=\frac{\int_0^\infty \varepsilon(\lambda,T)J_\lambda^{\mathrm{B}}(T)\mathrm{d}\lambda}{\int_0^\infty J_\lambda^{\mathrm{B}}(T)\mathrm{d}\lambda}\sigma T^4=\varepsilon(T)\sigma T^4 . \tag{7.55}$$

Equation $(7.55)_3$ defines $\varepsilon(T)$ as an average value of $\varepsilon(\lambda,T)$. It is called the averaged emission coefficient and is, again, a constitutive property.

Integration of (7.50) over λ provides the energy flux density absorbed by a body of temperature T

$$J_{\mathrm{abs}}(T)=\int_0^\infty A(\lambda,T)H_\lambda\mathrm{d}\lambda=\frac{\int_0^\infty A(\lambda,T)H_\lambda\mathrm{d}\lambda}{\int_0^\infty H_\lambda\mathrm{d}\lambda}H=A(T)H . \tag{7.56}$$

$H=\int_0^\infty H_\lambda\mathrm{d}\lambda$ is the incident energy flux and $A(T)$, defined by $(7.55)_3$ is the average absorption number.

In order to see how much of the equality $\varepsilon(\lambda,T)=A(\lambda,T)$ survives in the average values $\varepsilon(T)$ and $A(T)$ we repeat the argument of Paragraph 7.3.2 concerning two plates at temperatures T_1, T_2. We consider radiation composed of many wave lengths and therefore the energy fluxes may be written as

$$H^1=\int_0^\infty H_\lambda^1\mathrm{d}\lambda \quad \text{and} \quad H^2=\int_0^\infty H_\lambda^2\mathrm{d}\lambda .$$

Thus, by integration over (7.52) and use of (7.55), (7.56) it follows that

$$H^1=\varepsilon_1(T)\sigma T_1^4+(1-A_1)H^1 \ , \ \ H^2=\varepsilon_2(T)\sigma T_2^4+(1-A_2)H^2 \ , \tag{7.57}$$

where we have set

$$A^1 = \frac{\int_0^\infty A_1(\lambda, T_1) H_\lambda^2 \mathrm{d}\lambda}{H^2}, \quad A^2 = \frac{\int_0^\infty A_2(\lambda, T_2) H_\lambda^1 \mathrm{d}\lambda}{H^1} \quad \text{or, by (7.51)}$$

$$A^1 = \frac{\int_0^\infty \varepsilon_1(\lambda, T_1) H_\lambda^2 \mathrm{d}\lambda}{H^2}, \quad A^2 = \frac{\int_0^\infty \varepsilon_2(\lambda, T_2) H_\lambda^1 \mathrm{d}\lambda}{H^1}. \tag{7.58}$$

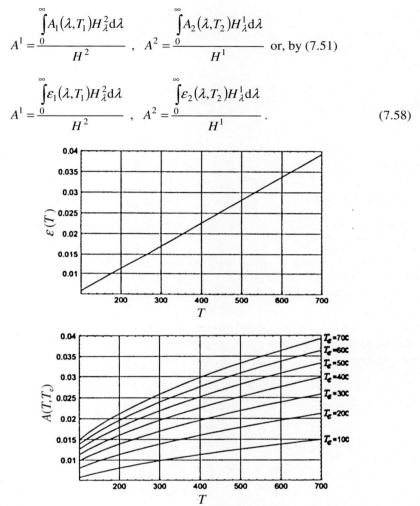

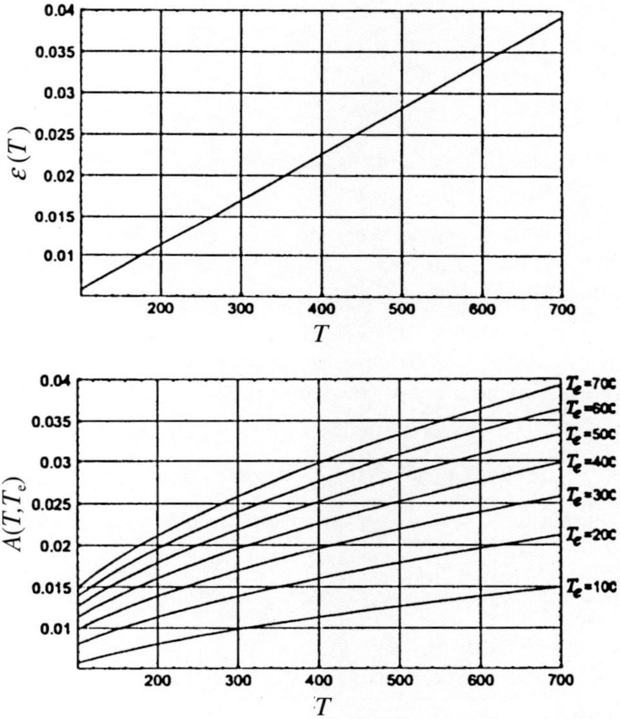

Fig. 7.5 Emission coefficient $\varepsilon(T)$ and absorption number $A(T, T_e)$ for polished aluminum of temperature T receiving radiation from a black body of temperature T_e. Note that for $T = T_e$ we have $\varepsilon(T) = A(T, T_e)$ as required by (7.60) [graphs copied from Rosenov, W.H., Hartnett, J.P., Handbook of Heat Transfer, McGraw Hill (1973)].

It follows that A_1 and A_2 depend on both T_1 and T_2; indeed, by (7.53), H_λ^1 and H_λ^2 both depend on these temperatures.

To be sure if the bodies are gray, so that ε is independent of λ, A_1 and A_2 are only functions of T_1 or T_2, respectively, and (7.58) shows that Kirchhoff's law also holds for the average quantities

$$A(T) = \varepsilon(T).$$

This is not generally true, however. Generally we obtain from (7.57) for H^1 and H^2 and by calculation of $Q^{1\to 2} = H^1 - H^2$

$$Q^{1\to 2} = \frac{\varepsilon_1(T_1)A_2(T_2,T_1)\sigma T_1^4 - \varepsilon_2(T_2)A_1(T_1,T_2)\sigma T_2^4}{A_1(T_1,T_2) + A_2(T_2,T_1) - A_1(T_1,T_2)\,A_2(T_2,T_1)}. \tag{7.59}$$

For equilibrium with $Q^{1\to 2} = 0$ and $T_1 = T_2 = T$ it follows

$$\frac{\varepsilon_1(T)}{A_1(T,T)} = \frac{\varepsilon_2(T)}{A_2(T,T)},$$

and, if plate 2 is black, so that ε_2 and A_2 are both equal to one

$$\varepsilon_1(T) = A_1(T,T). \tag{7.60}$$

This is Kirchhoff's law between the averaged emission coefficient and the average absorption number. In contrast to (7.51) *it is only valid in equilibrium.*

For non-equilibrium the values of $\varepsilon(T_1)$ and $A(T_1,T_2)$ have been measured for non-black and non-gray bodies and they may be looked up in technical handbooks. Fig. 7.5 shows the graphs for polished aluminum of temperature T that receives radiation from a black body of temperature T_e.

Some values of ε for room temperature are listed in Table 7.1.

Table 7.1 Emission coefficients ε (300K)

aluminium polished 0.05 oxydized 0.15	gypsum 0.85 glass 0.94
copper polished 0.1 oxydized 0.8	graphite 0.42 rubber 0.92
nickel oxydized 0.31	wood 0.85 soot 0.95
iron rusted 0.7	paper 0.85

7.3.4 *Examples of thermodynamics of radiation*

- *Temperature of the sun and its planets*

The most conspicuous radiation phenomenon is the solar radiation. The sun emits energy at a rate of $3.7 \cdot 10^{26}\,\mathrm{W}$. With the solar radius being $R_{\text{sun}} = 0.7 \cdot 10^9\,\mathrm{m}$ this amounts to an energy flux density

$$J_{\text{sun}} = 60.08 \cdot 10^6 \, \frac{\text{W}}{\text{m}^2} \,, \tag{7.61}$$

and if the sun is considered as a black body so that (7.54) holds, we obtain for the temperature of the solar surface

$$T_{\text{sun}} = 5700\,\text{K} \,. \tag{7.62}$$

Table 7.2 Temperature of planets

planet	Earth	Mars	Jupiter
$R\,[\text{m}]$	$6.3 \cdot 10^6$	$3.4 \cdot 10^6$	$71.8 \cdot 10^6$
$d\,[\text{m}]$	$1.5 \cdot 10^{11}$	$2.3 \cdot 10^{11}$	$7.7 \cdot 10^{11}$
$J\left[\frac{\text{W}}{\text{m}^2}\right]$	1330	557	50
$T\,[\text{K}]$	276	222	121

A planet at the distance d from the sun is exposed to the fraction $J = J_{\text{sun}}\left(R_{\text{sun}}/d\right)^2$ of the energy flux J_{sun}. For three planets these values are listed in Table 7.2. The data of this table permit a rough estimate of planetary temperatures. As an example we consider the Earth, which on its day side with the cross-section πR^2, *cf.* Table 7.2, receives an energy of $1330\,\text{W}/\text{m}^2$ – the so-called solar constant. Thus in 24 hours the Earth receives

$$J\,\pi R^2\,24\text{h} = 1.43 \cdot 10^{22}\,\text{J} \,. \tag{7.63}$$

In the same time the Earth radiates heat with its total surface $4\pi R^2$ and, if it were a black body with the mean temperature T, it would lose energy during a day by the amount

$$5.66 \cdot 10^{-8} \frac{\text{W}}{\text{m}^2\text{K}^4} T^4 4\pi R^2 \cdot 24\text{h} = 2.44 \cdot 10^{12}\,\text{J}\left(\frac{T}{\text{K}}\right)^4 . \tag{7.64}$$

Under stationary conditions the values (7.63) and (7.64) must be equal. Hence follows a mean temperature

$$T = 276\,\text{K} \,. \tag{7.65}$$

This may seem too low, but for a rough estimate it is not bad. We have neglected that the Earth has a smaller emission coefficient than one, and that part of the solar radiation is reflected from the Earth, primarily by the clouds.

The value (7.65) has been inserted into Table 7.2 and the corresponding values for Mars and Jupiter are also listed there. These temperatures are sometimes called *astronomical temperatures*; their values drop as the distance d increases, of course.

- *A comparison of radiation and conduction*

We recall from Paragraphs 7.3.2 and 7.3.3 the radiative flux $Q^{1\to 2}$ between two parallel plates. The value is given by (7.59). If the temperature difference between plate 1 and plate 2 is small, the difference between $A_1(T_1,T_2)$ and $\varepsilon_1(T_1)$ and between $A_2(T_2,T_1)$ and $\varepsilon_2(T_2)$ is also small and we may ignore it. In this case (7.59) is simplified to give

$$Q^{1\to 2} = \frac{1}{\frac{1}{\varepsilon_2}+\frac{1}{\varepsilon_1}-1}\sigma\left(T_1^4 - T_2^4\right). \tag{7.66}$$

If, in particular, one of the plates – plate 2 (say) – is black, or if both are black, we obtain

$$Q^{1\to 2} = \varepsilon_1\sigma\left(T_1^4 - T_2^4\right) \quad \text{or} \quad Q^{1\to 2} = \sigma\left(T_1^4 - T_2^4\right),$$

respectively.

Again, if the temperatures are close, the energy flux (7.66) may be linearized in $T_1 - T_2$ and we have

$$Q_{\text{Rad}}^{1\to 2} = \frac{T_1^3}{\frac{1}{\varepsilon_2}+\frac{1}{\varepsilon_1}-1}\sigma\left(T_1 - T_2\right). \tag{7.67}$$

This equation is easily compared with the corresponding formula for heat conduction between two plates of distance D which, by (2.10), reads

$$Q_{\text{Cond}}^{1\to 2} = \frac{\kappa}{D}\left(T_1 - T_2\right). \tag{7.68}$$

For a comparison of the relative significance of the heat transfers by radiation and conduction we consider two plates of distance $D = x\,\text{m}$, first both of wood with $\varepsilon = 0.9$, and then both of polished metal with $\varepsilon = 0.05$, and air in-between. Let the corresponding temperatures be $T_1 = 303\,\text{K}$ and $T_2 = 283\,\text{K}$. With $\kappa_{\text{air}} = 0.025\,\text{W}/(\text{mK})$ we thus obtain

$$Q_{\text{Rad}}^{\text{WW}} = \frac{4\cdot 303^3}{1.2}\cdot 5.66\cdot 10^{-8}\cdot 20\frac{\text{W}}{\text{m}^2} = 106.8\frac{\text{W}}{\text{m}^2}$$

$$Q_{\text{Rad}}^{\text{MM}} = \frac{4\cdot 303^3}{39}\cdot 5.66\cdot 10^{-8}\cdot 20\frac{\text{W}}{\text{m}^2} = 3.3\frac{\text{W}}{\text{m}^2}$$

$$Q_{\text{Cond}}^{\text{WW}} = \frac{0.025}{x}\cdot 20\frac{\text{W}}{\text{m}^2} = \begin{cases} 464\frac{\text{W}}{\text{m}^2} & \text{for } x = 0.001 \\ 4.64\frac{\text{W}}{\text{m}^2} & \text{for } x = 0.1. \end{cases}$$

We conclude that the heat exchange by radiation may be of the same order of magnitude as that by conduction. This is true in particular for a transfer between more distant plates, because the efficiency of conduction drops in that case, while radiation is unaffected. Of course, radiation is the only "mechanism" for heat transfer through a vacuum. We also see that radiative transfer between polished metals is two orders of magnitude smaller than between plates with large emission coefficients. This is the reason why the internal surfaces of thermo-flasks are polished. For good measure, in order to prevent conduction, the space between these plates is evacuated.

7.3.5 *On the history of heat radiation*

The understanding and correct description of heat radiation was one of the major problems of 19th century physics. Many outstanding physicists contributed to its solution, such as KIRCHHOFF, STEFAN, BOLTZMANN, WIEN, PLANCK, and EINSTEIN.

BOLTZMANN derived the T^4-law for the energy flux density of a black body by identifying such a body with a radiation-filled cavity. This was KIRCHHOFF's idea and, indeed, a small hole in the wall of the cavity with absorbing walls simulates a black body, because none of the incoming radiation is reflected. The radiation emitted by the cavity through the hole is isotropic and we have, *cf.* Fig. 7.6

$$J^{\mathrm{B}} = \int_0^{\pi/2}\int_0^{2\pi} e\, c\cos(\vartheta)\frac{\sin(\vartheta)\mathrm{d}\vartheta\mathrm{d}\varphi}{4\pi} = \frac{c}{4}e\,. \tag{7.69}$$

e is the constant energy density of the radiation and c is its speed of propagation, the speed of light. For the determination of e BOLTZMANN argued as follows: He knew from STEFAN that e depends only on T, the wall temperature of the cavity. Also he knew from electrodynamics that the pressure p of radiation is equal to $\frac{1}{3}e$. With considerable courage – or deep insight – he wrote down a Gibbs equation for the radiation

$$\mathrm{d}S = \frac{1}{T}\left[\mathrm{d}(eV) + p\,\mathrm{d}V\right] \quad \text{or with} \quad e = e(T)$$

$$\mathrm{d}S = \frac{1}{T}\frac{\mathrm{d}e}{\mathrm{d}T}V\,\mathrm{d}T + \frac{1}{T}(e+p)\mathrm{d}V \quad \text{or with} \quad e = 3p$$

$$\mathrm{d}S = \frac{1}{T}\frac{\mathrm{d}e}{\mathrm{d}T}V\,\mathrm{d}T + \frac{1}{T}\frac{4}{3}e\,\mathrm{d}V\,.$$

Hence follows the integrability condition, *cf.* Paragraph 4.2.1

$$\frac{\partial}{\partial V}\left(\frac{1}{T}\frac{\mathrm{d}e}{\mathrm{d}T}V\right) = \frac{\mathrm{d}}{\mathrm{d}T}\left(\frac{1}{T}\frac{4}{3}e\right) \quad \Rightarrow \quad \frac{1}{e}\mathrm{d}e = 4\frac{1}{T}\mathrm{d}T\,,$$

and therefore by integration $e = CT^4$. The constant C could not be calculated by BOLTZMANN, it was measured by Josef STEFAN (1835-1893).

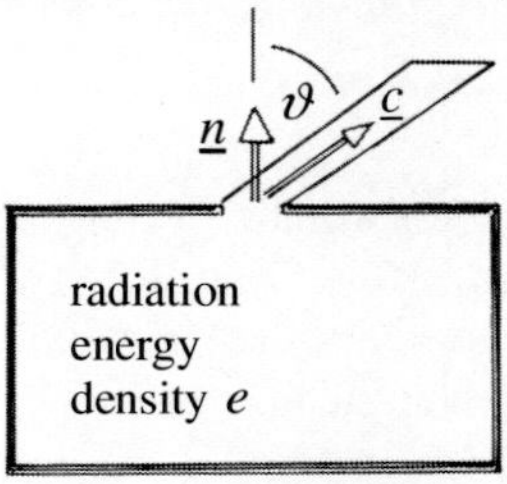

Fig. 7.6 Energy density and energy flux density of cavity radiation.

There remained the problem of understanding the distribution of e – or J – over the wave lengths which had been measured, *cf.* Fig. 7.4. This problem was solved by PLANCK and the ingredients of his solution provided the stimulus for a completely new way of thinking in physics: quantum physics.

The graphs of Fig. 7.4 represent the spectral energy flux densities $J_{\lambda}^{B}(T)$ as functions of the wavelength λ and of temperature. From measurements PLANCK knew that the graphs

- fall off for large λ as $1/\lambda^4$
- increase for small λ as $\exp\left(-\frac{\alpha}{\lambda}\right)/\lambda^5$
- have maxima at values that shift with $1/T$.

PLANCK succeeded to interpolate between the two limiting cases for large and small λ and arrived at the Planck formula, *cf.* (7.48).

More important was that this formula could be related to the properties of the "oscillators" in the walls of the cavity (say) that absorbed and emitted the radiation. PLANCK could do this by assuming that the possible energy values of the oscillators are not continuous, but discrete, so that absorption and emission of energy could only occur by "quanta" of the amount hc/λ, where h was a universal constant, now called Planck's constant.

Max Ernst Ludwig PLANCK (1858-1947) was an eminent thermodynamicist even before he found the correct radiation formula. He was honored in early post-war Germany when his head was used on early 2 deutschmark coins. This did not last long, however, because soon a more worthy politician was found to replace him. PLANCK was also a little bit of a philosopher. After having been involved in a bitter scientific struggle he summarized his experience thus:

> *The only way to get revolutionary advances in science accepted is to wait for the old scientists to die.*

7.4 Utilization of Solar Energy

7.4.1 *Availability*

According to the values of Table 7.2 the total power of solar radiation to the Earth is given by, *cf.* (7.63)

$$J\,\pi R^2 = 1.66 \cdot 10^{11}\,\text{MW}\,.$$

It is true that 28% of this amount is reflected, primarily the short wave part which is why the Earth appears blue, when viewed from the moon. A further 25% of solar radiation is absorbed: 3% in the stratosphere by the "good ozone",* and 22% in the troposphere by the clouds. Therefore 47% of the total solar power reach the surface of the Earth, 22% directly and 25% after scattering in the air and in the clouds.

The human demand of energy is estimated to be equal to $21 \cdot 10^6\,\text{MW}$, about 0.013% of the total solar power available. Therefore clever thermodynamicists try to implement the so far unexploited bulk of solar radiation for heating purposes, or for conversion into mechanical or electric energy.

---

* The "good ozone" in the stratosphere shields us from the ultraviolet radiation; it is being threatened by fluorized hydrocarbons, *cf.* Paragraph 6.3.4. The "bad ozone" occurs at the Earth's surface, it irritates our mucous membranes.

7.4.2 *Thermosiphon*

A fairly modest appliance – from the thermodynamic point-of-view – is the thermo-siphon, shown schematically in Fig. 7.7. It serves for solar heating. A collector with water-filled heating coils absorbs the solar heating $\dot{Q}$. Heated water is lighter than cold water and therefore it rises and creates a mass flux $\dot{m}$ which is related to the heating $\dot{Q}$ and to the temperature rise $T_T - T_B$ by

$$\dot{Q} = \dot{m} c_W \left(T_T - T_B\right). \tag{7.70}$$

If $\dot{Q}$ is known, this relation contains two unknowns, *viz.* $\dot{m}$ and $T_T - T_B$. We proceed to determine $T_T - T_B$ as a function of $\dot{Q}$. This project can succeed, because $\dot{m}$ itself is determined by $\dot{Q}$ and by the buoyancy force on the heated water.

The density difference between top and bottom, *cf.* Fig. 7.7, is given by

$$\frac{\rho_B - \rho_T}{\rho_B} = \alpha\left(T_T - T_B\right) \quad \text{where } \alpha \text{ is the thermal expansion coefficient.} \tag{7.71}$$

If we assume that the temperature and density difference depend linearly on the height y, we obtain the pressures in the cold water pipe and the warm water pipe behind the heater and in front of it, respectively

$$p_{CW} = p_B + \rho_B g \left(h_1 + h_2\right)$$

$$p_{WW} = p_B + \int_0^{h_1} \left(\rho_B + \frac{\rho_T - \rho_B}{h_1} y \right) g \, dy + \rho_T g \, h_2 .$$

After integration we obtain the pressure difference

$$\Delta p = p_{CW} - p_{WW} = \left(\rho_B - \rho_T\right) g \left(\tfrac{1}{2} h_1 + h_2\right),$$

or by (7.71)

$$\Delta p = \alpha \, \rho_B g \left(T_T - T_B\right)\left(\tfrac{1}{2} h_1 + h_2\right),$$

so that the pressure in the cold pipe is greater than the one in the warm pipe.

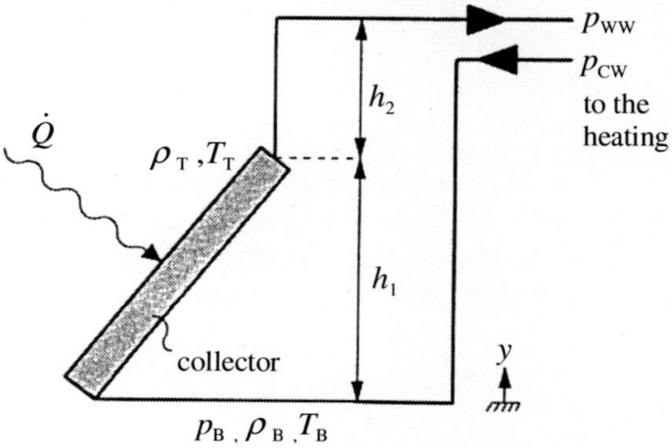

Fig. 7.7 Thermo-siphon, schematic

By the law of Hagen-Poiseuille* a pressure difference Δp of a fluid in a circular pipe of radius R and length L creates a volume transfer $\dot{V}$

$$\dot{V} = \frac{\pi R^4}{8\eta L}\Delta p\,.$$

η is the viscosity of the fluid. To a good approximation the mass transfer rate is equal to $\dot{m} = \rho_B \dot{V}$ and therefore we may write $\dot{m}$ in terms of $T_T - T_B$ as

$$\dot{m} = \frac{\pi R^4}{8\eta L}\alpha\,\rho_B^2 g\left(T_T - T_B\right)\left(\tfrac{1}{2}h_1 + h_2\right). \tag{7.72}$$

Elimination of $\dot{m}$ between (7.70) and (7.72) gives the desired relation between $T_T - T_B$ and $\dot{Q}$

$$T_T - T_B = \sqrt{\frac{\dot{Q}}{c_W}\frac{8\eta L}{\pi R^4}\frac{1}{\alpha\rho_B^2 g}\frac{1}{\frac{1}{2}h_1 + h_2}}\,. \tag{7.73}$$

We suppose that the collector has the area $1\,\text{m}^2$ and that the amount of $0.47\cdot 1330\,\text{W}/\text{m}^2$ is absorbed. For this case the temperature rise may be calculated from the known constitutive properties of water and from given dimensions of the appliance. We set

$$c_W = 4.18\tfrac{\text{kJ}}{\text{kgK}}\ ,\quad \eta = 10^{-3}\tfrac{\text{Ns}}{\text{m}^2}\ ,\quad R = 1\text{cm}\ ,\quad L = 10\text{m}\ ,\quad \alpha = 2\cdot 10^{-4}\tfrac{1}{\text{K}}\ ,$$

$$\rho_B = 10^3\tfrac{\text{kg}}{\text{m}^3}\ ,\quad g = 9.81\tfrac{\text{m}}{\text{s}^3}\ ,\quad h_1 = \frac{1}{\sqrt{2}}\text{m}\ ,\quad h_2 = 0.5\,\text{m}$$

and obtain

$$T_T - T_B = 12\,\text{K}\,.$$

This is not much. For a bigger value one may think of making $\dot{Q}$ larger by increasing the surface of the collector. Note, however, that $T_T - T_B$ is not proportional to $\dot{Q}$, so that doubling the area leads to an increase of $T_T - T_B$ by a factor of only $\sqrt{2}$.

7.4.3 *Green house*

Radiation has its main energy concentrated in the short-wave or long-wave range depending on whether it is emitted by a hot or by a cold body, *cf.* Fig. 7.4. Solar radiation passes more or less unabsorbed through the glass panels of a green house; it is absorbed by the soil on the bottom, and afterwards the heated soil heats up the air in the green house. Roof and walls ensure that the heat remains inside.

---

* We learn about this law in fluid mechanics. Strictly speaking the law holds only for incompressible fluids with $\rho = \text{const}$. Its use in the present case leads to a negligibly small mistake.

But this is not all. The heated soil itself sends out a long wave radiation appropriate to its relatively small temperature – small compared to the solar temperature. The glass of the roof, however, does not allow the long-wave radiation to pass, it reflects it to the soil and contributes to the heating of the soil.

The different transparency of glass for different wave lengths is illustrated in Fig. 7.8. The energy flux maximum of solar radiation lies around $0.5\,\mu\mathrm{m}$ – corresponding to a solar temperature of 6000 K – in the range where glass is transparent to more than 95%. In contrast to that the maximum of the Earth's radiation lies in the range between $300\,\mathrm{K}$ and $400\,\mathrm{K}$ and it occurs at $8\,\mu\mathrm{m}$ where, by Fig. 7.8, glass is virtually opaque.

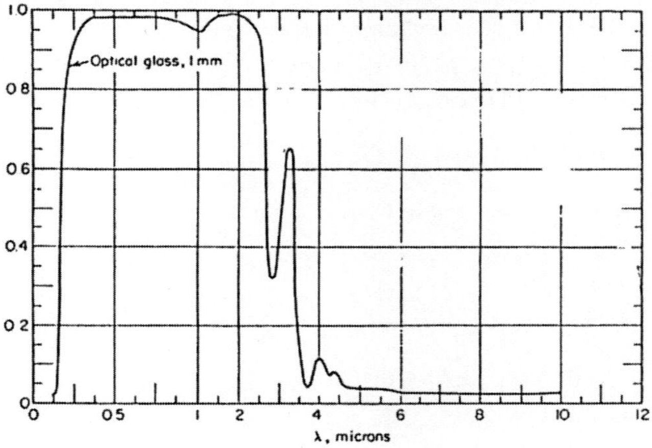

Fig. 7.8 Transparency of glass for radiation of different wavelengths [graph adapted from Rohsenow and Hartlett. Handbook of Heat Transfer, McGrawHill (1973)].

If we assume that 90% of the solar radiation of $0.47 \cdot 1330\,\mathrm{W/m^2}$ arrives at the bottom of the green house and that a mere 10% of the soil radiation passes the roof in the other direction, we obtain for stationary conditions

$$0.9 \cdot 0.47 \cdot 1330 \frac{\mathrm{W}}{\mathrm{m}^2} = 0.1 \sigma T_\mathrm{B}^4 .$$

This is an equation for T_B, the soil temperature of the green house temperature. We insert the value $\sigma = 5.67 \cdot 10^8 \frac{\mathrm{W}}{\mathrm{m}^2\mathrm{K}^2}$ of the Stefan-Boltzmann constant and obtain $T_\mathrm{B} = 563\,\mathrm{K}$ or nearly 300°C. Of course, a green house does not really become that hot; the reason for a lower temperature is the considerable heat efflux by conduction through the walls and through the roof, which we have ignored in the above calculation.

7.4.4 *Focusing collectors. The burning glass*

One may increase the temperature of a body that absorbs solar radiation by focusing the radiation. For this purpose the receiver of the radiation may be constructed as a parabolic mirror or as a burning glass, *cf.* Fig. 7.9. In this manner one focuses the radiation onto a small absorber. In order to characterize the efficiency of the system receiver-absorber one defines the ratio of concentration

$$e = \frac{\text{area of receiver } A_\mathrm{R}}{\text{area of absorber } A_\mathrm{A}} .$$

We proceed to calculate the temperature of the absorber under stationary conditions for a given concentration ratio. For simplicity we assume that the absorber is a black body, as is the sun, at least approximately.
First of all we recall that the energy flux $P_{\mathrm{S}\to\mathrm{A}}$ from the sun to the receiver – and then to the absorber – is given by, *cf.* Table 7.2 for the notation

$$P_{\mathrm{S}\to\mathrm{A}} = \frac{R_\mathrm{sun}^2}{d^2}\sigma T_\mathrm{sun}^4 A_\mathrm{R} = \frac{1}{46000}\sigma T_\mathrm{sun}^4 A_\mathrm{R} .$$

The absorber also radiates according to its temperature T_A . Its energy flux is

$$P_{\mathrm{A}\to} = \sigma T_\mathrm{A}^4 A_\mathrm{R} .$$

Under stationary conditions we must have $P_{\mathrm{S}\to\mathrm{A}} = P_{\mathrm{A}\to}$, and therefore we obtain

$$T_\mathrm{A} = T_\mathrm{sun}\sqrt[4]{\frac{A_\mathrm{R}/A_\mathrm{A}}{46000}} = T_\mathrm{sun}\sqrt[4]{\frac{e}{46000}} . \tag{7.74}$$

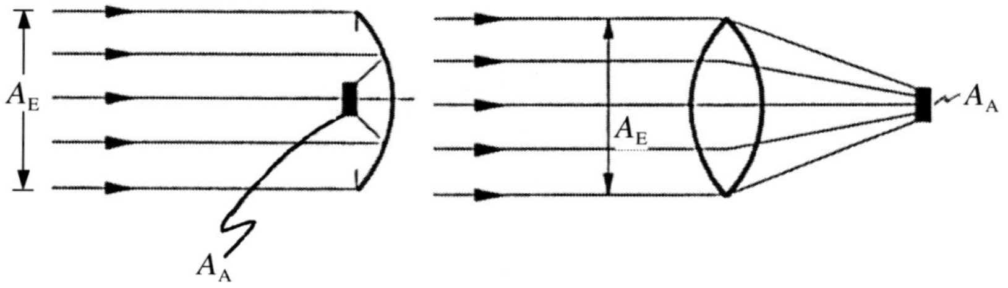

Fig. 7.9 Concentrating collectors
a. Parabolic channel or parabolic dish
b. Burning glass

Achievable values for the concentration ratios are

$e = 80$ for the parabolic channel

$e = 200$ for the parabolic dish or the burning glass.

Therefore we may reach temperatures $T_\mathrm{A} = 1150\mathrm{K}$ or $T_\mathrm{A} = 1450\mathrm{K}$, respectively.

8 Mixtures, solutions, and alloys

8.1 Chemical potentials

8.1.1 *Characterization of mixtures*

We consider mixtures, or solutions, or alloys with ν constituents, which are characterized by a Greek index. Thus p_α is the partial pressure of constituent α, ρ_α its density, and u_α or s_α are the specific values of the internal energy and entropy, respectively. Pressure, density, internal energy density, and entropy density *of the mixture* are additively composed of the corresponding partial quantities

$$p = \sum_{\alpha=1}^{\nu} p_\alpha \,, \quad \rho = \sum_{\alpha=1}^{\nu} \rho_\alpha \,, \quad \rho u = \sum_{\alpha=1}^{\nu} \rho_\alpha u_\alpha \,, \quad \rho s = \sum_{\alpha=1}^{\nu} \rho_\alpha s_\alpha \,. \tag{8.1}$$

This does not mean, however, that ρ_α, u_α, and s_α are given by the constitutive functions of the pure constituent α. Indeed, ρ_α, u_α, and s_α will generally depend on p_α and T, *and on all the other* partial pressures p_β.

Depending on the field of application of mixture theory – chemistry, physicochemistry, chemical engineering, metallurgy – the presence and contribution of the constituent α to the mixture is characterized differently:

m_α	mass
$\rho_\alpha = \frac{m_\alpha}{V}$	mass density (V – volume of mixture for p and T)
$\upsilon_\alpha = \frac{1}{\rho_\alpha}$	specific volume
$c_\alpha = \frac{m_\alpha}{m}$	concentration
N_α	particle number
$\nu_\alpha = \frac{N_\alpha}{A}$	mol number (A – Avogadro number)
$n_\alpha = \frac{N_\alpha}{V}$	particle number density
$X_\alpha = \frac{\nu_\alpha}{\nu}$	mol fraction ($\nu = \sum_{\alpha=1}^{\nu} \nu_\alpha$ – mol number of mixture)
$\frac{V_\alpha}{V}$	volume fraction (V_α – volume of pure constituent α for p and T)
$\frac{p_\alpha}{p}$	pressure ratio.

Despite this variety the list *does not* contain the most important partial quantity. This most important quantity is the chemical potential μ_α; it measures the presence of constituent α in the mixture in a similar manner as temperature measures the "presence of heat" in a body. We proceed to explain this quantity.

8.1.2 *Chemical potentials. Definition and relation to Gibbs free energy*

The chemical potential of constituent γ in a mixture is defined as *the* quantity which is continuous at a wall that is permeable for that constituent. We denote it by μ_γ. Because of its continuity the chemical potential μ_γ is measurable – at least in principle – namely under the condition that for all mixtures we can find a γ-permeable membrane, which holds back all other constituents. We postpone the discussion of an actual measurement of μ_γ until Paragraph 8.1.4.

We may relate the chemical potential μ_γ to the Gibbs free energy G of the mixture as follows: We recall the stability considerations of Paragraph 4.2.7 and apply them to the situation shown in Fig. 8.1, where two mixtures – both containing constituent γ – are kept apart by a semi-permeable membrane, permeable for constituent γ. The pressures p^{I} and p^{II} are kept constant and homogeneous within the compartments I and II, and T is homogeneous everywhere, and there is no kinetic energy. We refer to (4.59) and neglect the potential energy. In this case we may write (4.59) in the form

$$\frac{\mathrm{d}\left(U - TS + p^{\mathrm{I}}V^{\mathrm{I}} + p^{\mathrm{II}}V^{\mathrm{II}}\right)}{\mathrm{d}t} \leq -S\frac{\mathrm{d}T}{\mathrm{d}t} + V^{\mathrm{I}}\frac{\mathrm{d}p^{\mathrm{I}}}{\mathrm{d}t} + V^{\mathrm{II}}\frac{\mathrm{d}p^{\mathrm{II}}}{\mathrm{d}t}.$$

For fixed values of T, p^{I}, and p^{II} this means that the quantity $U - TS + p^{\mathrm{I}}V^{\mathrm{I}} + p^{\mathrm{II}}V^{\mathrm{II}}$ tends to a minimum as equilibrium is approached. With $U = U^{\mathrm{I}} + U^{\mathrm{II}}$ and $S = S^{\mathrm{I}} + S^{\mathrm{II}}$ this quantity is recognized as the sum of the Gibbs free energies G^{I} and G^{II}; respectively these depend on

$$\left(T, p^{\mathrm{I}}, m_1^{\mathrm{I}}, \cdots, m_\gamma^{\mathrm{I}}, \cdots, m_{\nu^{\mathrm{I}}}^{\mathrm{I}}\right) \text{ and } \left(T, p^{\mathrm{II}}, m_1^{\mathrm{II}}, \cdots, m_\gamma^{\mathrm{II}}, \cdots, m_{\nu^{\mathrm{II}}}^{\mathrm{II}}\right).$$

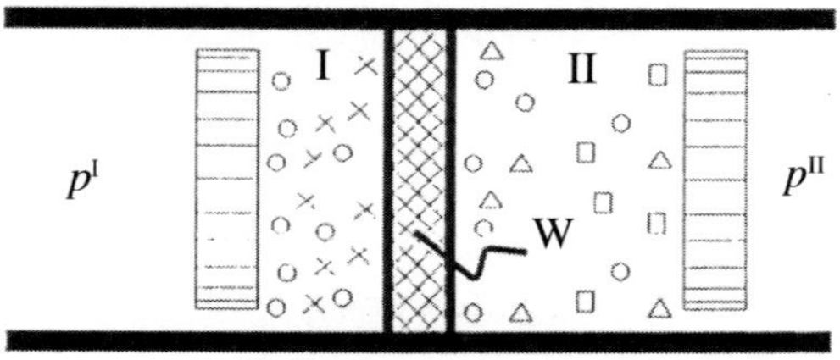

Fig. 8.1 A semi-permeable wall W between two mixtures, both containing constituent γ

The total Gibbs free energy thus reads

$$G = G^{\mathrm{I}}\left(T, p^{\mathrm{I}}, m_1^{\mathrm{I}}, \cdots, m_\gamma^{\mathrm{I}}, \cdots, m_{\nu^{\mathrm{I}}}^{\mathrm{I}}\right) + G^{\mathrm{II}}\left(T, p^{\mathrm{II}}, m_1^{\mathrm{II}}, \cdots, m_\gamma^{\mathrm{II}}, \cdots, m_{\nu^{\mathrm{II}}}^{\mathrm{II}}\right).$$

The only variables in this expression are the masses m_γ^{I} and m_γ^{II}, and there is only one *independent* variable, namely m_γ^{I} (say), because we have $m_\gamma^{\mathrm{I}} + m_\gamma^{\mathrm{II}} = m_\gamma$ and m_γ is constant. A necessary condition for equilibrium, *i.e.* for a minimum of G is therefore

$$\frac{\partial G}{\partial m_\gamma^{\mathrm{I}}}=0 \;\Rightarrow\; \frac{\partial G^{\mathrm{I}}}{\partial m_\gamma^{\mathrm{I}}}=\frac{\partial G^{\mathrm{II}}}{\partial m_\gamma^{\mathrm{II}}}.$$

This means that $\partial G/\partial m_\gamma$ is continuous at the semi-permeable wall in equilibrium. Because of the definition of the chemical potential we may therefore conclude that we have*

$$\mu_\gamma=\frac{\partial G}{\partial m_\gamma}. \tag{8.2}$$

Therefore the equilibrium condition may be written in the form

$$\mu_\gamma^{\mathrm{I}}\left(T,p^{\mathrm{I}},m_\beta^{\mathrm{I}}\right)=\mu_\gamma^{\mathrm{II}}\left(T,p^{\mathrm{II}},m_\beta^{\mathrm{II}}\right). \tag{8.3}$$

This condition will enable us to calculate the masses in the compartments when all the other variables are known. However, of course, this is feasible only when the form of the functions μ^{I} and μ^{II} are known. In Paragraph 8.1.4 we shall show how this knowledge can be obtained.

The chemical potential is not a very suggestive, intuitively accessible quantity. The equilibrium conditions in mixtures would be more plausible, if the partial pressure p_γ, or the partial density ρ_γ were continuous at a semi-permeable wall. This, however, is not in general the case – or at most approximately – and therefore we have to acquaint ourselves with the chemical potential, whether we like it or not.

8.1.3 *Chemical potentials; eight useful properties*

We repeat (8.2) in order to start the list of useful properties of the chemical potential which covers the equation (8.4) through (8.11)

$$\mu_\gamma=\left(\frac{\partial G}{\partial m_\gamma}\right)_{T,p,m_{\beta\neq\gamma}}. \tag{8.4}$$

Because of this relation the chemical potentials must obey integrability conditions, namely

$$\frac{\partial \mu_\gamma}{\partial m_\beta}=\frac{\partial \mu_\beta}{\partial m_\gamma}. \tag{8.5}$$

Some properties of the chemical potentials are consequences of the additivity of the Gibbs free energy. Additivity means that, if the body is increased by increasing its masses m_β by a factor z, the Gibbs free energy is also increased by the same factor so that we may write

$$G\left(T,p,z\,m_\beta\right)=z\,G\left(T,p,m_\beta\right).$$

* Strictly speaking, of course, μ_γ could be a universal function of $\partial G/\partial m_\gamma^{\mathrm{I}}$, but we simplify the argument, thus avoiding an non-essential cumbersome argument.

Mathematically this means that G is a homogeneous function of the m_β's of degree one. This is a simple special case of homogeneous functions of degree n which have the property

$$F(z\,x_1, z\,x_2, \cdots, z\,x_\nu) = z^n F(x_1, x_2, \cdots, x_\nu).$$

By a theorem due to Leonard EULER homogeneous functions satisfy the relation

$$\sum_{\beta=1}^{\nu} \frac{\partial F}{\partial x_\beta} x_\beta = nF(x_1, x_2, \cdots, x_\nu).$$

It follows that G may be written as $G = \sum_{\beta=1}^{\nu} \frac{\partial G}{\partial m_\beta} m_\beta$ or, by (8.4)

$$G = \sum_{\gamma=1}^{\nu} \mu_\gamma m_\gamma\,. \tag{8.6}$$

It is obvious from (8.4) that the μ_γ's themselves are homogeneous functions of degree zero, *i.e.* they do not change when m_γ – and therefore G – are multiplied by z. Consequently EULER's theorem implies

$$\sum_{\beta=1}^{\nu} \frac{\partial \mu_\gamma}{\partial m_\beta} m_\beta = 0 \quad \text{or, by (8.5)} \quad \sum_{\beta=1}^{\nu} \frac{\partial \mu_\beta}{\partial m_\gamma} m_\beta = 0\,. \tag{8.7}$$

The equation $(8.7)_2$ is called the *Gibbs-Duhem relation.*

Homogeneity of degree zero of the chemical potentials, *i.e.* the relation

$$\mu_\beta(T, p, z\,m_1, \cdots, z\,m_\nu) = \mu_\beta(T, p, m_1, \cdots, m_\nu)$$

obviously means that μ_β can only depend on such combinations of m_β which do not change when the body is increased, such as the concentrations c_β or the mol fractions X_β. Thus μ_β depends only on $\nu+1$ rather than $\nu+2$ variables. And *the Gibbs-Duhem relation is the mathematical representation of this observation*: If μ_β is written as a function of m_γ, the derivatives $\partial\mu_\beta/\partial m_\gamma$ are linearly dependent through the Gibbs-Duhem equation.

As a corollary of (8.6) it follows that in a pure constituent – constituent γ (say) – where $\nu = 1$ holds, we must have

$$\mu_\gamma(T, p) = g_\gamma(T, p), \tag{8.8}$$

so that the chemical potential of a pure constituent is equal to its specific Gibbs free energy.

The Gibbs equation for a mixture reads by (4.32), (4.33), and (8.2)

$$\mathrm{d}G = -S\,\mathrm{d}T + V\,\mathrm{d}p + \sum_{\gamma=1}^{\nu} \mu_\gamma \mathrm{d}m_\gamma\,. \tag{8.9}$$

It implies integrability conditions of the form

$$\frac{\partial S}{\partial m_\delta} = -\frac{\partial \mu_\delta}{\partial T} \quad \text{and} \quad \frac{\partial V}{\partial m_\delta} = \frac{\partial \mu_\delta}{\partial p}\,. \tag{8.10}$$

Note that (8.9) contains a non-trivial extension of the thermodynamics of simple fluids. Indeed, by

$$G = U - TS + pV \quad \text{and} \quad \dot{Q}\,\mathrm{d}t = \mathrm{d}U + p\,\mathrm{d}V\ ,$$

we may write (8.9) in the form

$$\frac{\mathrm{d}S}{\mathrm{d}t} = \frac{\dot{Q}}{T} - \sum_{\gamma=1}^{\nu} \frac{\mu_\gamma}{T}\frac{\mathrm{d}m_\gamma}{\mathrm{d}t}\,. \tag{8.11}$$

We conclude that a reversible change of entropy needs not necessarily be connected with heating, it may be achieved by mass transfers weighed by chemical potentials.

8.1.4 *Measuring chemical potentials*

The measurement – at least in principle – of the chemical potentials rests on their continuity at semi-permeable walls. A mixture of two or more constituents – including constituent γ – is put in contact with the pure constituent γ at a γ-permeable wall. By (8.3) and (8.8) we have

$$\mu_\gamma\left(T, p^{\mathrm{I}}, m_\beta^{\mathrm{I}}\right) = g_\gamma\left(T, p^{\mathrm{II}}\right),$$

Where p^{I} and p^{II} are the pressures in the mixture and in the pure constituent γ, respectively. $g_\gamma(T, p)$ is a known function – known from (p,V,T)-measurements and from c_v-, or c_p-measurements. We prove this statement below, in "small print" and we shall see that the statement must be qualified, albeit in an inessential manner. First, however, we proceed by describing the measurement of the chemical potential μ_γ.

We assume that $T, p^{\mathrm{I}}, p^{\mathrm{II}}$, and m_β, $\beta = 1, 2, \cdots, \nu$ are all given. Hence follows the value $g_\gamma\left(T, p^{\mathrm{II}}\right)$ and therefore the value of μ_γ. The latter belongs to the variables T, p^{I}, $m_1^{\mathrm{I}}, \cdots, m_\gamma^{\mathrm{I}}, \cdots, m_\nu^{\mathrm{I}}$ which are all known, except m_γ^{I}. But m_γ^{I} may be determined from $m_\gamma^{\mathrm{I}} = m_\gamma - m_\gamma^{\mathrm{II}}$ and m_γ^{II} can be measured as the mass of the pure constituent. Thus we have determined $\mu_\gamma\left(T, p^{\mathrm{I}}, m_\beta^{\mathrm{I}}\right)$ for the $(\nu+2)$-tuple $T, p^{\mathrm{I}}, m_\beta^{\mathrm{I}}$.

After that we must change the parameters and determine μ_γ for the new $(\nu+2)$-tuple, *etc.* Obviously this is a cumbersome and impractical procedure. It is only relevant in principle and confirms the above-stated basic measurability of the chemical potentials. Even if we had semi-permeable walls for all constituents in all mixtures, the large number of measurements makes the method impractical.

Actual measurements of chemical potentials proceed in a complex manner by a combination of simplifying assumptions, approximations, and extrapolations. A great help is the fact that *we can indeed calculate* these functions for mixtures of ideal gases and then – by extrapolation – for ideal mixtures. In any case, useful conclusions from the continuity of chemical potentials at semi-permeable walls cannot be drawn before we have the form of these functions. We shall come back to this later.

It remains to demonstrate – as announced – the measurability of $g_\gamma(T,p)=h_\gamma(T,p)-Ts_\gamma(T,p)$ for the pure constituent γ.

We assume that the thermal equation of state $p=p(\upsilon,T)$ – or $\upsilon=\upsilon(p,T)$ – is known as well as the specific heat $c_p=c_p(p_0,T)$ for *one* p_0 as a function of T. Hence follows

$$\mathrm{d}h=\left(\frac{\partial h}{\partial T}\right)_p \mathrm{d}T+\left(\frac{\partial h}{\partial p}\right)_T \mathrm{d}p\,, \tag{8.12}$$

because we have

$$\left(\frac{\partial h}{\partial T}\right)_p=c_p(p,T) \quad\text{and}\quad \left(\frac{\partial h}{\partial p}\right)_T=\upsilon-T\left(\frac{\partial \upsilon}{\partial T}\right)_p\,, \tag{8.13}$$

the latter from the integrability condition for $\mathrm{d}s=\frac{1}{T}(\mathrm{d}h-\upsilon\mathrm{d}p)$. Also, by $(8.13)_2$, we have

$$\left(\frac{\partial c_p}{\partial p}\right)_T=-T\left(\frac{\partial^2 \upsilon}{\partial T^2}\right)_p$$

so that (8.12) reads

$$\mathrm{d}h=\left[c_p(p_0,T)-\int_{p_0}^{p} T\left(\frac{\partial^2 \upsilon}{\partial T^2}\right)_p \mathrm{d}p\right]\mathrm{d}T+\left[\upsilon-T\left(\frac{\partial \upsilon}{\partial T}\right)_p\right]\mathrm{d}p$$

from which, by integration, we calculate $h(T,p)$ to within an additive constant.

Knowing this we may calculate $s(T,p)$ by integration of $\mathrm{d}s=\frac{1}{T}(\mathrm{d}h-\upsilon\mathrm{d}p)$ to within another constant of integration. Hence follows

$$g(T,p)=h(T,p)-Ts(T,p) \quad \text{to within a function} \quad \alpha-T\beta$$

where α and β are the additive constants in enthalpy and entropy.

This argument may be simplified further, if we choose p_0 so small – effectively zero – that ideal gas conditions prevail so that the specific heat is known as $c_p=\frac{\kappa}{\kappa-1}\frac{R}{M}$ and need not be measured. In this case we have

$$c_p(p_0,T)=\frac{\kappa}{\kappa-1}\frac{R}{M}-\int_0^p T\left(\frac{\partial^2 \upsilon}{\partial T^2}\right)_p \mathrm{d}p$$

and one may determine h, s, and g from (p,υ,T)-measurements alone.

8.2 Quantities of mixing. Chemical potentials of ideal mixtures

8.2.1 *Quantities of mixing*

We consider a mixture being formed from its pure constituents at the temperature T and pressure p. Fig. 8.2 shows the beginning and the end of the mixing process in a schematic manner. The weight of the pistons guarantees the same pressure p in each of the volumes V_α before mixing and in the eventual volume V after the opening of the valves and when homogeneous mixing has occurred. Also, the temperature T is kept constant during the mixing process.

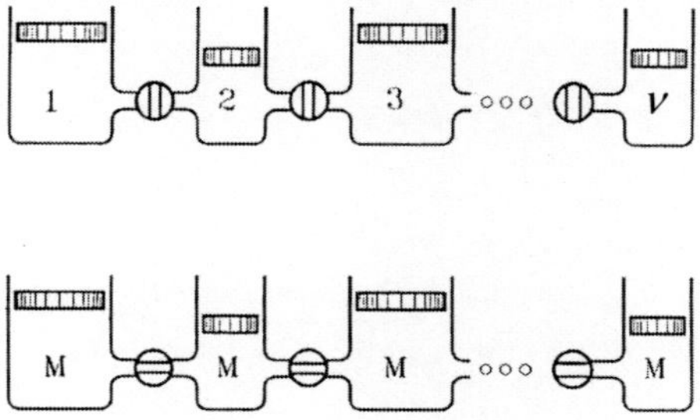

Fig. 8.2 Constituents as pure substances with $p, T, V_\alpha, U_\alpha, S_\alpha$ (top) and as a mixture (bottom). Note that the total volume may change during the mixing process.

Before and after mixing we thus have

$$V_{\mathrm{B}} = \sum_{\alpha=1}^{\nu} m_\alpha \upsilon_\alpha(T, p) \qquad V_{\mathrm{E}} = m_\gamma \upsilon_\gamma(T, p_\beta) \text{ for all } \gamma$$

$$U_{\mathrm{B}} = \sum_{\alpha=1}^{\nu} m_\alpha u_\alpha(T, p) \qquad U_{\mathrm{E}} = \sum_{\alpha=1}^{\nu} m_\alpha u_\alpha(T, p_\beta) \tag{8.14}$$

$$S_{\mathrm{B}} = \sum_{\alpha=1}^{\nu} m_\alpha s_\alpha(T, p) \qquad S_{\mathrm{E}} = \sum_{\alpha=1}^{\nu} m_\alpha s_\alpha(T, p_\beta),$$

where the specific values υ_α, u_α, s_α at the end may depend on T and on all partial pressures.

Thus we may identify quantities of mixing as follows

$$V_{\mathrm{Mix}} = m_\gamma \upsilon_\gamma(T, p_\beta) - \sum_{\alpha=1}^{\nu} m_\alpha \upsilon_\alpha(T, p) \text{ for all } \gamma$$

$$U_{\mathrm{Mix}} = \sum_{\alpha=1}^{\nu} m_\alpha \left[u_\alpha(T, p_\beta) - u_\alpha(T, p)\right] \tag{8.15}$$

$$S_{\mathrm{Mix}} = \sum_{\alpha=1}^{\nu} m_\alpha \left[s_\alpha(T, p_\beta) - s_\alpha(T, p)\right].$$

The enthalpy of mixing H_{Mix} and the Gibbs free energy of mixing G_{Mix} result from (8.15) by combination as follows

$$H_{\mathrm{Mix}} = U_{\mathrm{Mix}} + pV_{\mathrm{Mix}} \quad \text{and} \quad G_{\mathrm{Mix}} = U_{\mathrm{Mix}} + pV_{\mathrm{Mix}} - TS_{\mathrm{Mix}}. \tag{8.16}$$

H_{Mix} is also called the heat of mixing, because we conclude from the First Law in the form $\dot{Q}\,\mathrm{d}t = \mathrm{d}H - V\,\mathrm{d}p$ that $H_{\mathrm{E}} - H_{\mathrm{B}}$ is equal to the heat exchanged during the mixing process.

In general, all quantities of mixing are non-zero and, in particular, the volume of the homogeneous mixture may be different from the sum of volumes before mixing. Also the mixing process is generally accompanied by the absorption or the release of heat.

8.2.2 *Quantities of mixing of ideal gases*

For ideal gases V_{Mix} and U_{Mix} – hence H_{Mix} – are zero and S_{Mix} has an explicit form. This follows from *Dalton's law*, which states that in a mixture of ideal gases the partial quantities p_α, u_α, and s_α depend only on "their own density," and on temperature, of course. Moreover, the dependence on ρ_α and T is the same as in the pure ideal gases, *viz.*

$$p_\alpha = \rho_\alpha \frac{R}{M_\alpha} T \,,\; u_\alpha = z_\alpha \frac{R}{M_\alpha} T + \alpha_\alpha \,,\; s_\alpha(T,\rho_\alpha) = z_\alpha \frac{R}{M_\alpha} \ln T - \frac{R}{M_\alpha} \ln \rho_\alpha + \beta'_\alpha \,, \text{ or}$$
$$s_\alpha(T,p_\alpha) = (z_\alpha + 1)\frac{R}{M_\alpha} \ln T - \frac{R}{M_\alpha} \ln p_\alpha + \beta_\alpha \,. \tag{8.17}$$

Hence follows

$$m_\alpha v_\alpha(T,p_\beta) - \sum_{\delta=1}^{\nu} m_\delta v_\delta(T,p) = \frac{N_\alpha kT}{p_\alpha} - \frac{NkT}{p} = 0$$
$$u_\alpha(T,p_\beta) - u_\alpha(T,p) = 0$$
$$s_\alpha(T,p_\beta) - s_\alpha(T,p) = -\frac{R}{M_\alpha} \ln \frac{p_\alpha}{p} \,,$$

and we have

$$V_{\text{Mix}} = 0 \,,\; U_{\text{Mix}} = 0 \,,\; S_{\text{Mix}} = -\sum_{\alpha=1}^{\nu} m_\alpha \frac{R}{M_\alpha} \ln \frac{p_\alpha}{p} \,. \tag{8.18}$$

H_{Mix} vanishes and G_{Mix} is given by

$$G_{\text{Mix}} = \sum_{\alpha=1}^{\nu} m_\alpha \frac{R}{M_\alpha} T \ln \frac{p_\alpha}{p} \,. \tag{8.19}$$

Equation $(8.18)_3$ shows that S_{Mix} is composed from the entropy changes of the individual pure gases during their expansion from V_α to V. By $p/p_\alpha = V/V_\alpha$ and $m_\alpha = N_\alpha M_\alpha m_0$ with $R m_0 \frac{\text{mol}}{\text{g}} = k$, *cf.* (P.6), we may write S_{Mix} in the form

$$S_{\text{Mix}} = \sum_{\alpha=1}^{\nu} N_\alpha k \ln \frac{V}{V_\alpha} = \sum_{\alpha=1}^{\nu} N_\alpha k \ln \frac{N}{N_\alpha} \,. \tag{8.20}$$

By Avogadro´s law this is a universal expression, *i.e.* it is independent of the nature of the ideal gases that are being mixed.

This observation leads to the Gibbs paradox. Indeed, if all individual cylinders in Fig. 8.2 (top) are filled with the *same* ideal gas, there is no mixing after the valves are opened; and yet the entropy of mixing is still given by (8.20). This is a curious and somewhat disturbing feature which, however, fortunately does not influence subsequent results.

8.2.3 *Ideal mixtures*

Ideal mixtures are characterized by a particularly simple form of their quantities of mixing. We have

$$U_{\text{Mix}}^{\text{id}} = 0\,, H_{\text{Mix}}^{\text{id}} = 0\,, V_{\text{Mix}}^{\text{id}} = 0\,, S_{\text{Mix}}^{\text{id}} = -\sum_{\alpha=1}^{\nu} m_\alpha \frac{R}{M_\alpha} \ln \frac{N}{N_\alpha}\,, G_{\text{Mix}}^{\text{id}} = -TS_{\text{Mix}}^{\text{id}}\,. \quad (8.21)$$

Some solutions of liquids – particularly dilute solutions – and even some alloys really do have such quantities of mixing. This is quite surprising, in particular concerning the expression for $S_{\text{Mix}}^{\text{id}}$, which has the form that was derived for ideal gas mixtures by the use of Dalton's law for ideal gases.

Maybe the surprise is less acute when we realize that for gases, liquids, and solids the expression $S_{\text{Mix}}^{\text{id}}$ may be derived directly from BOLTZMANN´s expression $S = k \ln W$, where W is the number to realize a state, *cf.* Chap. 5. Indeed, if we discount interchanges of identical molecules, we have

$$W = 1 \qquad \Rightarrow \quad S = 0 \qquad \text{before mixing}$$

$$W = \frac{N!}{\prod_{\beta=1}^{\nu} N_\beta!} \quad \Rightarrow \quad S = k\left(N \ln N - \sum_{\beta=1}^{\nu} N_\beta \ln N_\beta \right) \quad \text{after mixing ;}$$

in the latter expression the Stirling formula was used to simplify the factorial expressions. The difference of these entropies is exactly equal to $(8.21)_4$ for the entropy of mixing of ideal gases, although no ideal gas properties were used in the argument.

8.2.4 *Chemical potentials of ideal mixtures*

Using the above arguments for ideal mixtures we may write

$$G = \sum_{\alpha=1}^{\nu} m_\alpha g_\alpha(T,p) - TS_{\text{Mix}}^{\text{id}}, \quad \text{or by } (8.21)_4$$

$$G = \sum_{\alpha=1}^{\nu} m_\alpha \left[g_\alpha(T,p) - \frac{R}{M_\alpha} T \ln \frac{\sum_{\beta=1}^{\nu} m_\beta / M_\beta}{m_\alpha / M_\alpha} \right]. \quad (8.22)$$

$g_\alpha(T,p)$ is the specific Gibbs free energy of the pure constituent α at T and p. Hence follows the chemical potential μ_γ by differentiation according to (8.2)

$$\mu_\gamma = g_\gamma(T,p) + \frac{R}{M_\gamma} T \ln \frac{m_\gamma / M_\gamma}{\sum_{\beta=1}^{\nu} m_\beta / M_\beta}. \quad (8.23)$$

This is an explicit relation for μ_γ because $g_\alpha(T,p)$ may be considered as known from (p,V,T)-measurements on the pure constituent, *cf.* Paragraph 8.1.4. Alternative forms of (8.23) may be written by using the following sequence of identities

$$\frac{m_\gamma/M_\gamma}{\sum_{\beta=1}^{\nu} m_\beta/M_\beta} = \frac{c_\gamma/M_\gamma}{\sum_{\beta=1}^{\nu} c_\beta/M_\beta} = \frac{n_\gamma}{\sum_{\beta=1}^{\nu} n_\beta} = \frac{N_\gamma}{\sum_{\beta=1}^{\nu} N_\beta} = \frac{\nu_\gamma}{\sum_{\beta=1}^{\nu} \nu_\beta} = X_\gamma . \tag{8.24}$$

The simplest form of the chemical potential of an ideal mixture thus results in terms of the mol fractions X_γ. We have

$$\mu_\gamma = g_\gamma(T,p) + \frac{R}{M_\gamma} T \ln X_\gamma . \tag{8.25}$$

8.3 Osmosis

8.3.1 *Osmotic pressure in dilute solutions. Van't Hoff's law*

The transport of fluids through semi-permeable walls is called osmosis (Greek: *osmos* = pushing). The osmotic pressure is an important characteristic of the phenomenon. Roughly speaking this is the pressure exerted on the wall by the constituents that cannot pass. At a wall permeable for the solvent Sv – and which separates a solution S from the pure solvent PSv – the osmotic pressure may be measured as $P = p^{\mathrm{S}} - p^{\mathrm{PSv}}$. We consider the solvent in the solution as constituent ν and obtain from (8.3)

$$\mu_\nu^{\mathrm{PSv}}\left(p^{\mathrm{PSv}}, T\right) = \mu_\nu^{\mathrm{S}}\left(p^{\mathrm{S}}, T, m_1, m_2, \cdots, m_{\nu^{\mathrm{S}}}\right), \tag{8.26}$$

which means equality of the chemical potentials of the solvent on the two sides of the wall. In the pure solvent, by (8.8), we have $\mu_\nu^{\mathrm{PSv}}\left(p^{\mathrm{PSv}}, T\right) = g_\nu\left(p^{\mathrm{PSv}}, T\right)$. Thus we have an implicit equation for p^S; an exploitation requires the explicit knowledge of the function μ_ν^S.

So as to be able to be specific we assume that the solution is ideal so that, by (8.23), (8.24)₃, we may write (8.26) in the form

$$g_\nu^{\mathrm{PSv}}\left(p^{\mathrm{PSv}}, T\right) = g_\nu\left(p^{\mathrm{S}}, T\right) - \frac{R}{M_\nu} T \ln \frac{\sum_{\beta=1}^{\nu-1} n_\beta + n_\nu^{\mathrm{S}}}{n_\nu^{\mathrm{S}}} . \tag{8.27}$$

We simplify this relation in two ways. First, we set

$$\ln \frac{\sum_{\beta=1}^{\nu-1} n_\beta + n_\nu^{\mathrm{S}}}{n_\nu^{\mathrm{S}}} = \ln\left(1 + \sum_{\beta=1}^{\nu-1} n_\beta / n_\nu^{\mathrm{S}}\right) \approx \sum_{\beta=1}^{\nu-1} \frac{n_\beta}{n_\nu^{\mathrm{S}}} . \tag{8.28}$$

which takes into account that the solution is dilute so that $\sum_{\beta=1}^{\nu-1} n_\beta << n_\nu^{\mathrm{S}}$ holds.

Second, we expand $g_\nu\left(p^{\mathrm{S}}, T\right)$ about p^{PSv} into a Taylor series

$$g_\nu\left(p^{\mathrm{S}},T\right)=g_\nu\left(p^{\mathrm{PSv}},T\right)+\left(\frac{\partial g_\nu}{\partial p}\right)_{p^{\mathrm{PSv}}}\left(p^{\mathrm{S}}-p^{\mathrm{PSv}}\right)+\cdots. \tag{8.29}$$

The dots indicate higher order terms. However, we recall $(4.32)_4$ which implies

$$\left(\frac{\partial g_\nu}{\partial p}\right)_{p,T}=\upsilon_\nu(p,T)=\frac{1}{\rho_\nu(p,T)}. \tag{8.30}$$

For an *incompressible* pure solvent higher order terms therefore vanish, since υ_ν, or ρ_ν, is independent of p and T.

By combining (8.27) through (8.30) we thus obtain

$$\left(p^{\mathrm{S}}-p^{\mathrm{PSv}}\right)\frac{1}{\rho_\nu}=\sum_{\beta=1}^{\nu-1}\frac{n_\beta}{n_\nu^{\mathrm{S}}}\frac{R}{M_\nu}T \quad \text{or by } n_\nu^{\mathrm{S}}m_\nu=\rho_\nu^{\mathrm{S}} \text{ and } Rm_0\tfrac{\mathrm{mol}}{\mathrm{g}}=k$$

$$p^{\mathrm{S}}-p^{\mathrm{PSv}}=\frac{\rho_\nu}{\rho_\nu^{\mathrm{S}}}\sum_{\beta=1}^{\nu-1}n_\beta kT \quad \text{or, by } \rho_\nu^{\mathrm{S}}\approx\rho_\nu \tag{8.31}$$

$$p^{\mathrm{S}}-p^{\mathrm{PSv}}=\sum_{\beta=1}^{\nu-1}n_\beta kT. \tag{8.32}$$

Jakobus Henricus VAN'T HOFF (1852-1911) was an eminent chemist whose first feat was the explanation of the optical activity of substances in solution. VAN'T HOFF argued that molecules may be arranged spatially, rather than in a plane, and thus could envisage some molecules as mirror images of others. If one type dominates, – the right handed one (say) –, the asymmetry can rotate the plane of vibration of light. His work was vigorously attacked at a time when "structural formulae" of molecules were still viewed with suspicion.

Nor did he fare much better with his work on dilute solutions which we have explained above. His idea of the ideal mixture was not accepted at first – until it proved to give good results. In the end he prevailed, however, and in 1901 he was awarded the first Nobel prize in chemistry.

This is van't Hoff's law for the osmotic pressure of dilute solutions. Its similarity to the pressure of a mixture of $\nu-1$ ideal gases is remarkable. This similarity again reflects the similarity between the entropies of mixing of ideal gases and ideal solutions which we have discussed in Paragraph 8.2.3.

If the solution were a mixture of ideal gases, the osmotic pressure would have to be interpreted as the pressure on the wall of those constituents that cannot penetrate the wall. Even for solutions of liquids this may provide a good intuitive interpretation.

Concerning the approximation $(8.31)_2$ a little calculation shows that the exact formula is given by

$$\rho_\nu^{\mathrm{S}}=\rho_\nu\frac{1}{1+\sum_{\beta=1}^{\nu-1}\frac{V_\beta}{V}},$$

where V_β/V are the volume fractions of the solutes, *cf.* Paragraph 8.1.1. The approximation is therefore consistent with the assumption of a dilute solution.

8.3.2 *Applications of osmosis*

- *Pfeffer's tube*

Pfeffer's tube is a narrow cylinder which is closed at one end by a semi-permeable wall – permeable for water. This end is dipped into a water reservoir and it will then be filled with water so that the water levels inside the tube and outside are equal, *cf.* Fig. 8.3 a. Now if some salt NaCl is dissolved in the small amount of water inside the tube, one observes that water is "pulled" into the tube from the reservoir through the permeable wall, such that in the end the solution in the tube stands considerably higher than in the reservoir, *cf.* Fig. 8.3 b.

An inexact, but intuitively appealing interpretation of the phenomenon says that the partial water pressure at the bottom of the solution is smaller than the pressure of the pure water at that point on the other side of the wall. According to this interpretation equilibrium occurs when the hydrostatic water pressures are equal on both sides of the wall. This is not a bad visualization of the phenomenon.

However, we know that it is not the partial pressures that must be equilibrated; rather the chemical potentials of the solvent must be equal on both sides. Therefore in equilibrium we must have

$$\mu_\nu^{\mathrm{PSv}}\left(p_0+\rho_\nu g\,H_1,T\right)=\mu_\nu^{\mathrm{S}}\left(p_0+\rho^{\mathrm{S}} g\,H_2,T,m_1,m_2,\cdots,m_{\nu^{\mathrm{S}}}\right).$$

This condition may be exploited in the same manner as (8.26) was exploited. However, by doing so we would merely repeat the arguments of Paragraph 8.3.1. Therefore we proceed differently, taking the result of the preceding paragraph as our starting point. We use van't Hoff's law (8.32) in order to calculate the height H_2 of the solution in a tube of cross-section A. Obviously we must set

$$p^{\mathrm{S}}=p_0+\rho^{\mathrm{S}} g\,H_2 \quad\text{and}\quad \rho^{\mathrm{PSv}}=p_0+\rho_\nu g\,H_1\,.$$

Insertion into (8.32) gives

$$\rho^{\mathrm{S}} H_2-\rho_\nu H_1=\sum_{\beta=1}^{\nu-1}\frac{N_\beta}{V}kT\frac{1}{g}\,,\quad\text{where}\quad V=AH_2\,.$$

We think of the total mass of water m_{W} as given and obtain, always with $\rho_\nu^{\mathrm{S}}\approx\rho_\nu$, and with A_{R} as the cross-section of the reservoir

$$\rho_\nu H_1\left(A_{\mathrm{R}}-A\right)=m_{\mathrm{W}}-\rho_\nu H_2 A\,.$$

ρ^{S} is the sum of $\rho_\nu^{\mathrm{S}}\approx\rho_\nu$ and $\sum\limits_{\beta=1}^{\nu-1}\rho_\beta=\sum\limits_{\beta=1}^{\nu-1}\frac{m_\beta}{AH_2}$. Thus (8.32) leads to a quadratic equation for H_2, *viz.*

$$H_2^2-\frac{1}{\rho_\nu A_{\mathrm{R}}}\left(m_{\mathrm{W}}-\frac{A_{\mathrm{R}}-A}{A}\sum_{\beta=1}^{\nu-1}m_\beta\right)H_2-\frac{1}{\rho_\nu}\frac{A_{\mathrm{R}}-A}{A_{\mathrm{R}}A}\sum_{\beta=1}^{\nu-1}N_\beta kT\frac{1}{g}=0\,.$$

The solution reads

$$H_2 = \frac{1}{2\rho_v A_R}\left(m_W - \frac{A_R - A}{A}\sum_{\beta=1}^{\nu-1} m_\beta\right) \overset{+}{(-)} \sqrt{\frac{1}{4\rho_v^2 A_R^2}(\ldots)^2 + \frac{1}{\rho_v}\frac{A_R - A}{A_R A}\sum_{\beta=1}^{\nu-1} N_\beta k T \frac{1}{g}} . \tag{8.33}$$

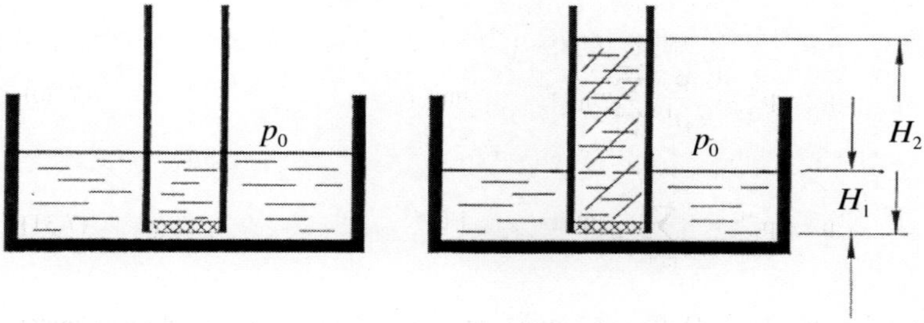

Fig. 8.3 Pfeffer's tube. Reservoir cross-section A_R , tube cross-section A

We exploit equation (8.33) for a specific example. We set

$m_W = 2\,\mathrm{kg}$, $A_R = 2 \cdot 10^{-2}\,\mathrm{m}^2$, $A = 1\,\mathrm{cm}^2$, $\rho_v = 10^3 \frac{\mathrm{kg}}{\mathrm{m}^3}$, $T = 298\,\mathrm{K}$, $g = 9.81 \frac{\mathrm{m}}{\mathrm{s}^2}$,

and we dissolve 1g NaCl. Upon solution in water the "salt molecules" NaCl dissociate into Na^+- and Cl^--ions so that we have with $M_{\mathrm{NaCl}} = 58.5 \frac{\mathrm{g}}{\mathrm{mol}}$

$$\sum_{\beta=1}^{\nu-1} N_\beta = 2 \frac{m_{\mathrm{Salt}}}{58.5\, m_0} = 2.06 \cdot 10^{22} .$$

Insertion gives

$$H_2 = 9.34\,\mathrm{m} \ ! \tag{8.34}$$

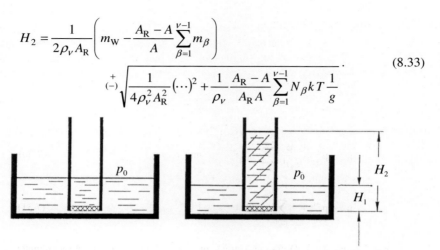

Wilhelm Pfeffer (1845-1920) was a botanist who pioneered the use of semi-permeable membranes between solutions of organic substances – like proteins – and water. He could measure the osmotic pressure and showed that it depends on the size of the dissolved molecules, or their mol mass. Thus he was first to suggest that organic substances consist of giant molecules.

H_1 turns out to be equal to 5.4 cm so that just about half of the water has been "pushed" or "pulled" into the tube by osmosis.

Of course, we must support the tube in order to maintain mechanical equilibrium. The supporting force results from

$$F = \left(\rho^S g H_2 - \rho_v g H_1\right) A + m_T g = 9.2\,\mathrm{N} + m_T g ,$$

where m_T is the mass of the tube.

It is said that plants use osmosis to pull water through the ducts of the trunk into their leaves: The roots of the plants lie in the water and inside the roots we find the watery sap of nutritious solutes. The skins of the roots are permeable for water.

- *A "perpetuum mobile" based on osmosis*

A long pipe is dipped into sea water from board of a ship. The bottom of the pipe is fixed to the ground at depth H and it is closed to the salt ions by a semi-permeable wall which is permeable for water. Let the salt density in sea water be given by $\rho_{\text{Salt}} = 35\frac{\text{g}}{l}$ and let the temperature be 5°C. We consider both as independent of depth.

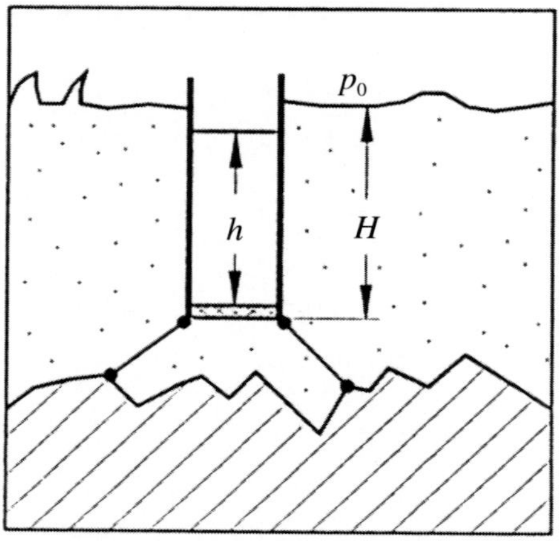

Fig. 8.4 Semi-permeable wall dipped into sea water

First of all we ask for the osmotic pressure at the semi-permeable wall. It may be calculated from van't Hoff's law (8.32)

$$p^{\text{S}} - p^{\text{PSv}} = 2n_{\text{Salt}}kT = 2\frac{\rho_{\text{Salt}}}{58.5m_0} = 27.66\,\text{bar}\,. \tag{8.35}$$

Next we determine the height h of the pure water column in the tube, *cf.* Fig.8.4. The index ν, as always, characterizes the solvent water. With

$$p^{\text{S}} = p_0 + \rho^{\text{S}} g\,H = p_0 + \left(\rho_\nu^{\text{S}} + \rho_{\text{Salt}}\right)gh \quad \text{and} \quad p^{\text{PSv}} = p_0 + \rho_\nu g\,h$$

we obtain from (8.35) and with $\rho_\nu^{\text{S}} \approx \rho_\nu$

$$h = \left(1 + \frac{\rho_{\text{Salt}}}{\rho_\nu}\right)H - 2\frac{\rho_{\text{Salt}}}{\rho_\nu}\frac{kT}{58.5m_0 g}\,. \tag{8.36}$$

It is instructive to consider some special cases:

For small values of H the height of the pure water level h is negative which means that no water can enter into the pipe unless the pipe is immersed into the salt water to a certain depth. The necessary depth follows from the given data and it turns out to be $H = 271\text{m}$!

Another instructive question is: When is $h > H$? From (8.36) we conclude that the answer is $H \geq 8000\text{m}$. This means that we must look for a deep ocean indeed. But still, if we are there, we might think of using the height difference $h - H$ for producing mechanical energy, – or having a shower in fresh water.

This sounds like a "*perpetuum mobile*" and, indeed, the process works only, if the salt concentration is independent of depth.* And it is the influence of solar radiation that keeps ρ_{Salt} at least approximately homogeneous, because the sun creates temperature differences that lead to large-scale convection in the seas. Without the solar influence the salt concentration would turn into a non-homogeneous equilibrium and h would always be smaller than H , or at most equal.

- *Physiological salt solution*

The thermodynamic layman is often impressed by the size of the osmotic pressure. This is particularly so for the blood cells which – against pure water – exhibit an osmotic pressure of 7.7 bar. The cell membranes could not possibly sustain a pressure difference of that size; they would burst.

Inside the body, however, the cells do not burst, because they are surrounded by a fluid that also contains solutes which cannot penetrate the cell wall. These exert an opposite osmotic pressure of 7.7 bar.

One must be careful, however, not to inject a patient with pure water in considerable quantity. And, indeed, the ubiquitous "drips" with which patients are treated in hospitals contain the so-called physiological salt solution, a mixture of water and salt of the density

$$\rho_{\text{Salt}} = M_{\text{NaCl}} \frac{7.7\,\text{bar}}{2RT} \underset{\text{for } T=36°\text{C}}{=} 8.8\frac{\text{g}}{l}\,.$$

Physicians say that this solution is *isotonic* to the contents of the cell; it exerts the same pressure on the cell membrane from outside as the contents of the cell from the inside.

Osmosis also plays an essential role when fruits are to be preserved by placing them in a strongly concentrated sugar solution, or when meat is preserved by salting it. The highly concentrated external solution pulls water out of the ubiquitous bacteria that let food rot. Thus the bacteria die and the food is safe, even though sometimes unrecognizable. Indeed, the food loses water, too, and the well-rounded grape, for example, turns into a rugose raisin.

- *Osmosis as a competition of energy and entropy*

Thermodynamicists understand the process of osmosis as a competition between energy that "wishes" to approach a minimum and entropy that tends to a maximum. For illustration we consider Pfeffer's tube once again and determine the

* In actual fact the salt concentration – and the temperature – in the sea are quite inhomogeneous.

available free energy $\mathcal{A}$ discussed in Paragraph 4.2.7. In the present case, if we neglect the kinetic energy, we conclude from (4.61) that*

$$\mathcal{A} = F + E_{\text{Pot}} + p_0 V \quad \text{tends to a minimum.} \tag{8.37}$$

The free energy $F = U - TS$, the potential energy E_{Pot}, and the volume are all composed of two parts, the part in the reservoir and the part in the tube. And p_0 is the approximately equal atmospheric pressure on the surface of the solution and the pure solvent.

We use the notation of the preceding paragraphs and the same assumptions, for instance the assumption that the solution is ideal and that $\rho_V^{\text{S}} \approx \rho_V$ holds. Water is incompressible so that u_V and s_V depend only on T. Thus we obtain

$$F^{\text{PSv}} + p_0 V^{\text{PSv}} = \left(m_V - m_V^{\text{S}}\right)\left(u_V(T) - Ts_V(T) + \frac{p_0}{\rho_V}\right),$$

$$F^{\text{S}} + p_0 V^{\text{S}} = m_V^{\text{S}}\left(u_V(T) - Ts_V(T) + \frac{p_0}{\rho_V}\right) + m_{\text{Salt}}\left[u_{\text{Salt}}(T) - Ts_{\text{Salt}}(T)\right] -$$

$$-T\left(m_V^{\text{S}} \frac{R}{M_V} \ln \frac{\frac{m_V^{\text{S}}}{M_V} + 2\frac{m_{\text{Salt}}}{M_{\text{Salt}}}}{\frac{m_V^{\text{S}}}{M_V}} + 2m_{\text{Salt}} \frac{R}{M_{\text{Salt}}} \ln \frac{\frac{m_V^{\text{S}}}{M_V} + 2\frac{m_{\text{Salt}}}{M_{\text{Salt}}}}{\frac{m_{\text{Salt}}}{M_{\text{Salt}}}}\right),$$

$$E_{\text{Pot}}^{\text{PSv}} = \frac{g}{2(A_{\text{R}} - A)\, u_V}\left(m_V - m_V^{\text{S}}\right)^2,$$

$$E_{\text{Pot}}^{\text{S}} = \frac{g}{2A\rho_V}\left(m_V - m_{\text{Salt}}\right) m_V^{\text{S}}. \tag{8.38}$$

The available free energy $\mathcal{A}$ is the sum of all four expressions, and inspection shows that m_V^{S}, the mass of water in the tube, is the only variable. In Fig. 8.5 we draw the parts of $\mathcal{A}$ that depend on m_V^{S}. These are the potential energy and two logarithmic terms that result from the entropy of mixing. The range of interest is

$$\frac{A}{A_{\text{R}}} \leq \frac{m_V^{\text{S}}}{m_V} < 1,$$

where the left limit represents the initial state before the salt is dissolved and the right limit corresponds to the hypothetical case that all the water was pulled into the pipe. We set

$$m_V = 2\,\text{kg}, \quad A = 1\,\text{cm}^2, \quad A_{\text{R}} = 2 \cdot 10^{-2}\,\text{m}^2, \quad m_{\text{Salt}} = 1\,\text{g}, \quad \rho_V = 1\frac{\text{g}}{\text{cm}^3},$$

* Recall p in (4.61) is the pressure on the surface of the system. Here that pressure is p_0, the atmospheric pressure.

and obtain the curves denoted by O and P in Fig. 8.5. They represent the entropy of mixing – which "drives" osmosis – and the potential energy, respectively. Both curves have been adjusted in height so that they pass through the origin.

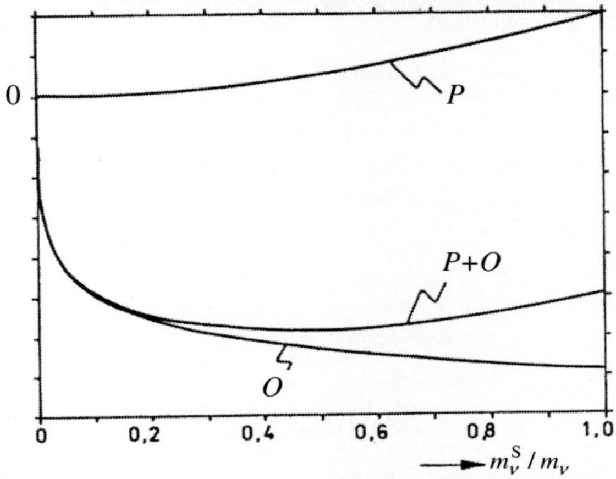

Fig. 8.5 Competition of osmosis and potential energy in Pfeffer's tube

We conclude that the osmotic curve has its minimum for $m_\nu^S = m_\nu$ when all water has been pulled into the solution: The entropy is then maximal. In contrast the potential energy has its minimum at very small values of m_ν^S, namely – roughly – when the liquid in the reservoir and the tube have the same level. The two opposing tendencies compensate one another where the curve $O+P$ has a minimum. This happens in the present case at $m_\nu^S \approx 0.47\,m_\nu$. The corresponding height of the solution in the tube is equal to 9.34m. This is the same value, of course, as the one calculated before, *cf.* (8.34).

An interesting aspect of the minimum of $\mathcal{A}$ in the present case is this: The salt profits because its entropy of mixing becomes large. The water, however, pays the cost, because it has to climb to a higher level of potential energy. We conclude that nature does not allow its constituents to be selfish. The *mixture* gains by achieving a minimal available free energy. In a manner of speaking we may also say that the trend of $\mathcal{A}$ to a minimum is so strong that it is able to "shape" the system: it is not satisfied by mixing the salt in the originally available small amount of water in the tube.

- *Desalination*

It is conceivable to use semi-permeable walls for producing fresh water from sea water by an engine as shown in Fig. 8.6 a with a water permeable piston. Let us suppose that the pressure behind the piston and the water pressure at the surface of the sea are both equal to p_0. Gravitational effects are ignored. When the piston moves upwards, seawater eventually fills the cylinder of total volume V_0 through the intake valve which is then closed. The seawater – considered incompressible – is then put under the pressure p_1=27.66 bar which, by (8.35), is its osmotic pressure at 5°C. No fresh water can penetrate the piston until that pressure is exceeded. However, for a higher pressure $p > p_1$ fresh water seeps through the piston so that the cylinder volume V decreases and the solution becomes more salty; in a manner of speaking the salt is compressed. The (p, V)-relation is obviously

$$p - p_0 = 2\frac{m_{\text{Salt}}}{V}\frac{R}{M_{\text{NaCl}}}T\,,$$

provided that van't Hoff's law still holds which we assume. The compression ends at the pressure p_2 and volume V_2 whereupon the pressure drops to p_0 by the opening of the exit valve. Subsequently the brine is pushed out as the piston completes its cycle and a new cycle begins.

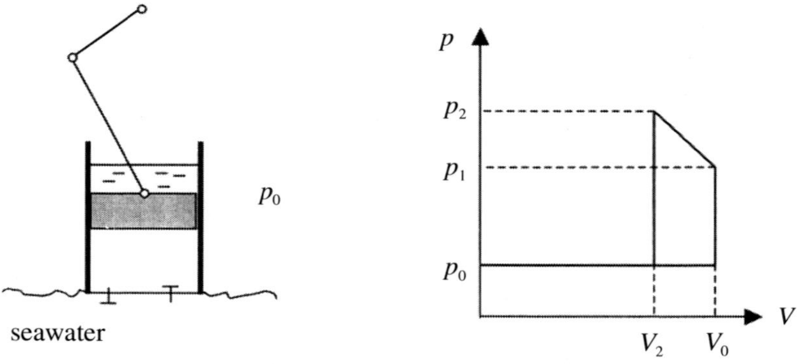

Fig 8.6 On desalination with a semi-permeable piston

The work during the intake of the seawater and the work for pushing out the brine are neglected, since they require minimal pressure differences. Therefore the work applied to the piston is

$$W_{\circ} = -\int_{V_0}^{V_2}(p - p_0)\,\mathrm{d}V = -2m_{\text{Salt}}\frac{R}{M_{\text{NaCl}}}T\ln\frac{V_2}{V_0}\,.$$

We let V_0 be 2 l in which case m_{Salt} is equal to 2·35 g, and we set $V_2 = 1\ l$ so that one liter of fresh water is produced per cycle. In that case we obtain

$$W_{\circ} = 3834\ \text{J}.$$

It is perhaps appropriate to compare this value with the heat of evaporation of 1 l of water of 5°C which, by Table 2.4, is equal to 2490 kJ. That is roughly what we need to get fresh water vapor from the evaporation of salt water; but of course, much of this energy may be regained by the subsequent condensation, if the process is conducted properly.

8.4 Mixtures in different phases

8.4.1 *Gibbs phase rule*

In a mixture of ν constituents $\alpha = 1, 2, \cdots, \nu$, which are present in f phases $h = 1, 2, \cdots, f$, the Gibbs free energy is the sum of the Gibbs free energies of the phases. By (8.6) we have in each phase $G^h = \sum_{\alpha=1}^{\nu} \mu_\alpha^h m_\alpha^h$, and therefore the total Gibbs free energy may be written in the form

$$G = \sum_{h=1}^{f} G^h = \sum_{h=1}^{f} \sum_{\alpha=1}^{\nu} \mu_\alpha^h m_\alpha^h . \tag{8.39}$$

μ_α^h depends only on the masses m_β^h of the constituents in phase h. We consider the case of fixed values T, p, and m_α ($\alpha = 1, 2, \cdots, \nu$) and ask for the distribution of these fixed masses over the phases in equilibrium. Equilibrium is characterized by a minimum of G in this case and, since we have the ν constraints $m_\alpha = \sum_{h=1}^{f} m_\alpha^h$, we need ν Lagrange multipliers λ_α and look for the minimum of the function

$$\phi = \sum_{h=1}^{f} \sum_{\alpha=1}^{\nu} \mu_\alpha^h m_\alpha^h - \sum_{\alpha=1}^{\nu} \lambda_\alpha \left(\sum_{h=1}^{f} m_\alpha^h - m_\alpha \right)$$

without constraints. Necessary conditions are that the derivatives of ϕ with respect to all m_β^j ($\beta = 1, 2, \cdots, \nu$) ($j = 1, 2, \cdots, f$) vanish. By use of the Gibbs-Duhem relation (8.7) we obtain

$$\mu_\beta^j = \lambda_\beta \quad (\beta = 1, 2, \cdots, \nu)\,(j = 1, 2, \cdots, f).$$

This is Gibbs's phase rule. Since the right hand side is independent of the phase, the equations state that in equilibrium each constituent has the same value of the chemical potential in all phases. We may thus write

$$\mu_\beta^j = \mu_\beta^f \quad (\beta = 1, 2, \cdots, \nu)\,(j = 1, 2, \cdots, f-1) \tag{8.40}$$

or, more explicitly, by listing the variables

$$\mu_\beta^j\left(T, p, X_\gamma^j\right) = \mu_\beta^f\left(T, p, X_\gamma^f\right). \tag{8.41}$$

8.4.2 *Degrees of freedom*

Gibbs's phase rule provides $\nu(f-1)$ equilibrium conditions for the $2 + (\nu - 1)f$ variables T, p, X_γ^j. Therefore we are left with

$$\begin{aligned} F &= 2 + (\nu - 1)f - \nu(f - 1) \\ F &= 2 + \nu - f \end{aligned} \tag{8.42}$$

degrees of freedom to move in a $\left(p, T, X_\gamma^j\right)$-diagram without disturbing the phase equilibrium, *i.e.* without losing a phase.

For a single body with $\nu = 1$ we know about this from earlier considerations. In this case we have $F = 3 - f$ and therefore

$$F = \begin{cases} 2 & \\ 1 \text{ , when we have } & 2 \text{ phases.} \\ 0 & 3 \end{cases}$$

(with $F = 2, 1, 0$ corresponding to 1, 2, 3 phases respectively)

Inside the vapor region, or the liquid region, where only one phase exists, we may change p and T arbitrarily – albeit locally – without leaving these regions. In an liquid-vapor equilibrium, however, each change of p requires a definite change of T, if we wish to maintain the phase equilibrium; the (p,T)-pairs must all lie on the saturated vapor curve, *cf.* Paragraph 2.4.3. If we violate this rule in a process of changing p and T, we destroy the phase equilibrium and lose one phase. Finally the triple point is characterized by a single pair of (p,T). Any change of either p or T, or both destroys the three-phase-equilibrium of solid, liquid, and vapor.

For binary and ternary mixtures with $\nu = 2$ and $\nu = 3$ we shall later encounter further examples for Gibbs's phase rule and, in particular, for the rule about degrees of freedom. Sometimes the latter itself is called Gibbs's phase rule.

8.5 Liquid-vapor equilibrium (ideal)

8.5.1 *Ideal Raoult law*

When there are only two phases – *e.g.* liquid and vapor – Gibbs's phase rule (8.41) is particularly simple. We characterize the two phases by ' for liquid and " for vapor, and write

$$\mu'_\alpha\left(T, p, X'_\beta\right) = \mu''_\alpha\left(T, p, X''_\beta\right). \tag{8.43}$$

In particular for a binary mixture, *i.e.* a mixture of only two constituents, these are two equations for the four unknowns p, T, X'_1, and X''_1. For a given pair (p,T) they may serve to calculate X'_1 and X''_1. An explicit calculation, however, obviously requires explicit knowledge of the functions μ'_α and μ''_α of their variables. In order to obtain such knowledge we make simplifying assumptions as follows

i.) The liquid solution is ideal so that

$$\mu'_\alpha\left(T, p, X'_\beta\right) = g'_\alpha(T, p) + \frac{R}{M_\alpha} T \ln X'_\alpha \,. \tag{8.44}$$

ii) The pure liquids are incompressible; in this case the functions $g'_\alpha(T, p)$ are linear in p and we may write

$$g'_\alpha(T, p) = g'_\alpha(T, p_\alpha(T)) + \upsilon'_\alpha[p - p_\alpha(T)], \tag{8.45}$$

where $p_\alpha(T)$ is the pressure of saturated steam of the pure constituent α, which we may read off from the saturation vapor-curve or from Table 2.4. υ'_α are the constant specific volumes of the liquids.

iii.) The vapor is a mixture of ideal gases so that we have

$$\mu''_\alpha\left(T, p, X''_\beta\right) = g''_\alpha(T, p) + \frac{R}{M_\alpha} T \ln X''_\alpha$$

Or, since the specific Gibbs free energy of an ideal gas is a logarithmic function of p, *cf.* (8.17)

$$\mu''_\alpha\left(T,p,X''_\beta\right)=g''_\alpha\left(T,p_\alpha(T)\right)+\frac{R}{M_\alpha}T\ln\frac{p}{p_\alpha(T)}+\frac{R}{M_\alpha}T\ln X''_\alpha . \tag{8.46}$$

The use of $p_\alpha(T)$ as reference pressures in (8.45) and (8.46) is motivated by the wish to make use of the relation $g'_\alpha(T,p_\alpha(T))=g''_\alpha(T,p_\alpha(T))$, *cf.* (4.39). In this way it is possible to eliminate the functions $g'_\alpha(T,p)$ and $g''_\alpha(T,p)$ entirely from Gibbs's phase rule.

iv.) The specific volumes v'_α of the pure liquids can be neglected in comparison with those of the vapors

$$v'_\alpha \ll \frac{1}{p_\alpha(T)}\frac{R}{M_\alpha}T . \tag{8.47}$$

From (8.43) it follows by use of the assumptions i.) through iii.) that we have

$$v'_\alpha\left[p-p_\alpha(T)\right]+\frac{R}{M_\alpha}T\ln X'_\alpha=\frac{R}{M_\alpha}T\ln\frac{p}{p_\alpha(T)}+\frac{R}{M_\alpha}T\ln X''_\alpha .$$

Division by $\frac{R}{M_\alpha}T$ and use of the assumption iv.) provides the simple equation

$$X'_\alpha p_\alpha(T)=X''_\alpha p . \tag{8.48}$$

This is the ideal Raoult law. For given p and T it gives ν relations between the $2(\nu-1)$ independent mol fractions X'_α and X''_α. In particular, for $\nu=2$ the law permits the calculation of the only two independent mol fractions X'_1 and X''_1.

Francois Marie RAOULT (1830-1901) was one of the founders of physical chemistry. From his experiments with solutions he found that the partial pressure of a solvent vapor over a solution is proportional to the mol-fraction of the solvent in the solution. This is what our results (8.49), (8.52) imply.

RAOULT suffered the same contretemps as many other thermodynamicists of the late 19th century: They had been anticipated by the great American physicist Josiah Willard GIBBS whose works remained unknown in Europe for a long time.

8.5.2 *Ideal phase diagrams for binary mixtures.*

For a binary mixture in two phases Raoult's law (8.48) is reduced to only two equations which may be written in the form

$$\begin{aligned} X'_1p_1(T)&=X''_1p \\ (1-X'_1)p_2(T)&=(1-X''_1)p \end{aligned} \quad\Rightarrow\quad \begin{aligned} p_1(T)X'_1-p\,X''_1&=0 \\ -p_2(T)X'_1+pX''_1&=p-p_2(T). \end{aligned} \tag{8.49}$$

For a given pair (p,T) these are two linear equations for the determination of X'_1 and X''_1. The solution reads

$$X'_1=\frac{p-p_2(T)}{p_1(T)-p_2(T)} \quad\text{and}\quad X''_1=-\frac{p_1(T)p_2(T)}{p_1(T)-p_2(T)}\frac{1}{p}+\frac{p_1(T)}{p_1(T)-p_2(T)} . \tag{8.50}$$

For a fixed p the first equation represents a linear function of p, the second one is a hyperbolic function of p. It is customary and convenient to draw both functions in one single diagram with p as the ordinate and X_1', or X_1'' as abscissa, *cf.* Fig. 8.7. Thus

$$p = p_2(T) + [p_1(T) - p_2(T)]X_1', \text{ and } p = \frac{p_2(T)}{1 - \left[1 - \frac{p_2(T)}{p_1(T)}\right] X_1''}. \qquad (8.51)$$

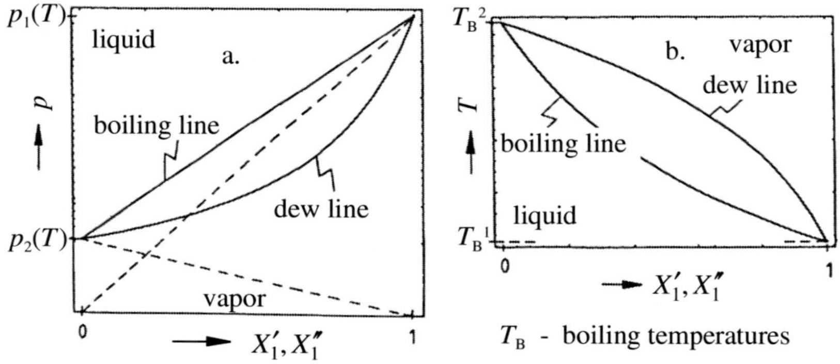

Fig. 8.7 Phase diagrams of a binary mixture
a. (p, X_1)-diagram. Dashed lines: Partial vapor pressures
b. (T, X_1)-diagram.

The graphs of Fig. 8.7 are drawn for the case $p_1(T) > p_2(T)$. The *boiling line* determines the mol fraction X_1' of the boiling mixture. The *dew line* determines the mol fraction X_1'' of the saturated vapor. Above the graphs, – for high pressure – we have the liquid region, and below the graphs, – for low pressure –, we have superheated vapor. Between the graphs we find a liquid-vapor two-phase region. For fixed p the mol fraction X_1' is smaller than X_1'', *i.e.* the constituent 1 is enriched in the vapor; this fact is the physical basis for the method of separating constituents: One siphons off the 1-rich vapor, condenses it and evaporates again so that the new vapor is even more 1-rich, *etc.* More about this later.

The dashed lines in Fig. 8.5 a. represented by the functions pX_1'' and $p(1 - X_1'')$, are the partial pressures p_1'' and p_2'' of constituent 1 and 2, respectively, in the vapor. Indeed, we have

$$pX_1'' = p\frac{n_1''}{\sum_{\beta=1}^{\nu} n_\beta''} = p\frac{p_1''/(kT)}{p\ /(kT)} = p_1'' , \quad p(1 - X_1'') = p\frac{n_2''}{\sum_{\beta=1}^{\nu} n_\beta''} = p\frac{p_2''/(kT)}{p\ /(kT)} = p_2'' . \qquad (8.52)$$

Elimination of p between the two equations (8.52) gives

$$X_1'' = \frac{p_1(T)}{p_2(T)} \frac{X_1'}{1 + \left[\frac{p_2(T)}{p_1(T)} - 1\right] X_1'} . \tag{8.53}$$

Fig. 8.8 shows X_1'' as a function of X_1' as written in (8.53). Inspection shows – always for the case $p_1(T) > p_2(T)$ – that X_1'' is larger than X_1' which could already be read off from Fig. 8.7 by the observation that the boiling line lies above the dew line in the (p,X)-diagram.

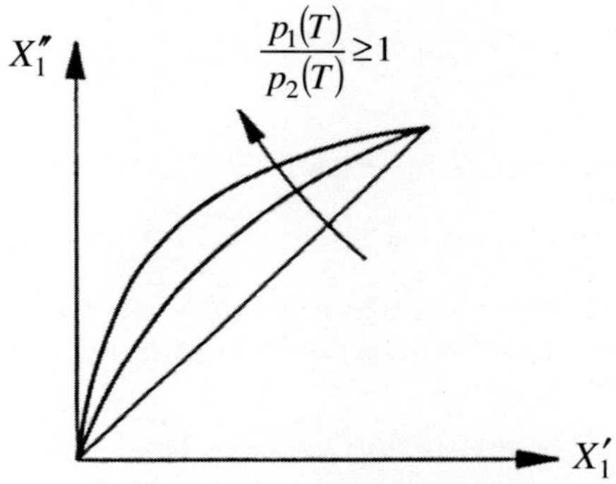

Fig. 8.8 X_1'' as a function of X_1'.

Since $p_1(T)$ and $p_2(T)$ are the saturation pressures of the constituents, and since they are known from measurements and represented by saturated vapor-curves of the type shown in Fig. 2.5, it is possible to use the equations (8.50) in order to calculate the boiling line (T, X_1') and the dew line (T, X_1'') in a (T,X_1)-diagram for a fixed pressure. In this manner one obtains the graphs of Fig. 8.7 b. Of course, the construction of these curves can only be realized numerically, or graphically, since the saturated pressure-curves are not known analytically. The boiling temperature $T_B^1(p)$ usually lies below $T_B^2(p)$ when $p_1(T) > p_2(T)$ holds.

8.5.3 *Evaporation in the* (p,T)*-diagram*

We refer to Fig. 8.9 and consider an evaporation process by lowering the pressure. Starting point is *A* in the liquid region. By lowering the pressure we arrive at the boiling line in point *B* and the evaporation starts. The initial vapor bubble has the composition of point *C, i.e.* it is 1-rich or, anyway, richer in constituent 1 than the liquid. Therefore, because of the evaporation the liquid state moves to the left, it becomes richer in constituent 2. In this new state the liquid cannot continue to evaporate before the pressure drops back to that of the boiling line. The new vapor is not quite as 1-rich as the first vapor bubble, but still richer in 1 than the initial liquid so that the liquid becomes still poorer in constituent 1. If the vapor created in these consecutive bubbles is mixed, the evaporation process proceeds – at constantly decreasing pressure – until the vapor has the mol-fraction of the original liquid and point *D* has been reached; the last liquid droplet has the composition of point E, *i.e.* it is rich in constituent 2. In summary, the state of the liquid has moved downward along the dew line from point B to E, while the vapor moved downward along the boiling line from C to D. The arrows in the figure illustrate these changes of state.

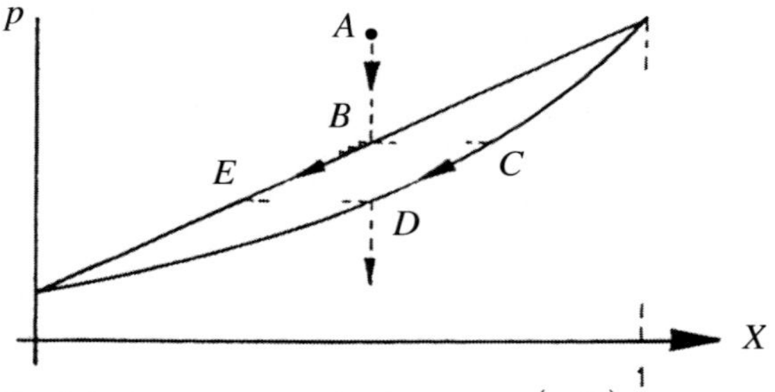

Fig. 8.9 The evaporation process in the (p,V)-diagram

If one removes the vapor so that there is no mixing between consecutive vapor bubbles, the liquid will continue to become richer in constituent 2 and finally ends up as the pure constituent 2.

8.5.4 *Saturation pressure decrease and boiling temperature increase*

In a dilute solution of constituents $\alpha = 1,2,\cdots,\nu-1$ with little volatility the mol fractions X''_α ($\alpha \neq \nu$) are very small so that $X''_\nu \approx 1$ holds. Therefore Raoult's law (8.48) for the solvent $\alpha = \nu$ reads

$$p = p_\nu(T) X'_\nu \,, \tag{8.54}$$

and it follows that the saturated vapor pressure above the dilute solution is nearly equal to the saturated pressure $p_\nu(T)$ above the pure solvent; nearly, but not exactly. We calculate the difference from (8.54) under the assumption that $\sum_{\beta=1}^{\nu-1} n'_\beta << n'_\nu$ holds, which is appropriate for a dilute solution, and we obtain

$$\frac{p - p_\nu(T)}{p_\nu(T)} = -\sum_{\beta=1}^{\nu-1} \frac{n'_\beta}{n'_\nu} \,, \tag{8.55}$$

so that the difference grows linearly with the number of dissolved molecules.

This effect is called the *decrease of saturation pressure of a solution*. If we consider seawater with 35 g of NaCl in 1 l of water at 100°C, the vapor pressure over the boiling seawater amounts to 995 hPa instead of 1013 hPa over pure water. For the calculation we must remember that NaCl dissociates into ions in a water solution.

One may interpret (8.55) by saying that the saturation vapor curve of the solution lies *below* that of the pure solvent. From the character of saturation curves as monotonically increasing convex curves it is clear that the new saturation vapor curve is also shifted *to the right-hand side*, *cf.* Fig. 8.10. This means that for a fixed pressure the boiling temperature $T_\nu(p)$ the solution is a little higher than that of the pure solvent. By use of the Clausius-Clapeyron equation (4.42) one easily calculates the *increase of the boiling temperature* to be equal to

$$\frac{T - T_\nu(p)}{T_\nu(p)} = \frac{\frac{R}{M_\nu} T_\nu(p)}{r_\nu(T)} \sum_{\beta=1}^{\nu-1} \frac{n'_\beta}{n'_\nu} . \tag{8.56}$$

For seawater under the normal pressure of 1013 hPa the increase amounts to 0.6 K.

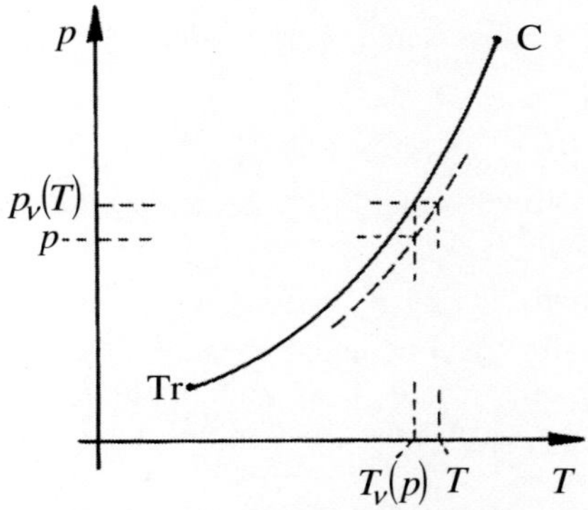

Fig. 8.10 Saturation vapor curve of a pure solvent (solid), and of a dilute solution (dashed).

8.6 Distillation, an application of Raoult's law

8.6.1 *mol as a unit*

While in most fields of physics and engineering the amount of matter is usually – and satisfactorily – measured by its mass, this is not so in inorganic chemistry and in large parts of chemical engineering. In these fields the amount of matter is characterized by the number N of molecules, or – since this is usually a prohibitively large number – by the number ν of mols. We recall Section P.2, where the mol was introduced. It has, *cf.* (P.5)

$$A = 6.023 \cdot 10^{23} \frac{1}{\text{mol}} \text{ molecules and the mass } M = \frac{m}{m_0} \frac{\text{g}}{\text{mol}} .$$

ν has the dimension mol and the mass m of a body is given by

$$m = \nu M . \tag{8.57}$$

Instead of the mass-specific quantities υ, u, h, g, and s used heretofore we shall now use the mol-specific quantities $\tilde{\upsilon}$, $\tilde{u}$, $\tilde{h}$, $\tilde{g}$, and $\tilde{s}$. They are defined by

$$V = \nu \tilde{\upsilon} , U = \nu \tilde{u} , H = \nu \tilde{h} , G = \nu \tilde{g} , \text{ and } S = \nu \tilde{s} , \tag{8.58}$$

and they have the dimensions m^3/mol, J/mol, and $\text{J}/(\text{mol}\,\text{K})$. The relation between the two types of specific quantities is thus given by

$$\tilde{a} = a\frac{m}{\nu} = a\,M\ . \tag{8.59}$$

Obviously a relation like this holds for the chemical potentials as well and we have $\tilde{\mu}_\alpha = \mu_\alpha M_\alpha$.

8.6.2 *Simple application of Raoult's law*

We consider a mixture of propane C_3H_8 and butane C_4H_{10} with mol numbers ν_1 and ν_2, respectively. The molar masses and molar volumes of the respective boiling liquids are

$$M_1 = 44\tfrac{\mathrm{g}}{\mathrm{mol}},\ \tilde{\upsilon}_1' = 7.5\cdot 10^{-5}\tfrac{\mathrm{m}^3}{\mathrm{mol}},\ M_2 = 58\tfrac{\mathrm{g}}{\mathrm{mol}},\ \tilde{\upsilon}_2' = 9.7\cdot 10^{-5}\tfrac{\mathrm{m}^3}{\mathrm{mol}}. \tag{8.60}$$

For the mixture the volume of mixing is known to be zero and the saturation pressures at $T = 293\,\mathrm{K}$ are given by

$$p_1(293\,\mathrm{K}) = 8.29\,\mathrm{bar} \quad \text{and} \quad p_2(293\,\mathrm{K}) = 2.06\,\mathrm{bar} \tag{8.61}$$

so that propane is "more volatile" than butane. Propane is the low-boiling liquid and butane is the high boiling one, where "low" and "high" refer to temperature, not pressure (!). Of course, this is due to the smaller mol mass of propane but that aspect does not need to concern us.

We consider a container with $V = 1\,\mathrm{m}^3$ filled with a propane-butane mixture with $X_1 = 1/2$, *i.e.* $\nu_1 = \nu_2$, and we want to determine the range of values of $\nu = \nu_1 + \nu_2$ such that both phases, liquid and vapor, are present in V.

The maximal value $\nu_{\max}$ is obviously the one for which

$$V'' = V - \nu_{\max}\frac{1}{2}\left(\tilde{\upsilon}_1' + \tilde{\upsilon}_2'\right) = 0$$

holds so that there is no vapor space. For the data given above this means $\nu_{\max}$ =11.6 10^3 mol.

The minimum value $\nu_{\min}$ results from the ideal gas law $\nu_{\min} = \frac{pV}{RT}$ when we insert the pressure

$$p = 2\frac{p_1(T)\,p_2(T)}{p_1(T) + p_2(T)}$$

which – by $(8.50)_2$ – corresponds to the value $X_1'' = 1/2$ so that no liquid is present. For the above data we have $p = 3.3\,\mathrm{bar}$ and therefore $\nu_{\min} = 133\,\mathrm{mol}$.

8.6.3 *Batch distillation*

An initially liquid ideal binary solution with the mol number $\nu_B(t_0)$ and the mol fraction $X_{1B}(t_0)$ is heated in a boiler B and partially evaporated. The vapor is condensed by cooling in a pipe leading into a distillate container D, *cf.* Fig. 8.11 a. The vapor entering the pipe carries the molar rates of transfer $\dot{\nu}''$ and $\dot{\nu}_1''$, the latter for the transfer rate of constituent 1. The objective of the set-up is either the enrichment of the low-boiling constituent – here constituent 1 – in the distillate, or the enrichment of the high-boiling constituent in the boiler, or both. The process occurs at constant pressure, the temperature in the boiler, however, rises. We start the calculation at time t_0, when the solution has just begun to boil. It is then characterized by the initial values $\nu'_B, \nu'_{1B} = \nu'_B X'_{1B}, T$, all at time t_0, *cf.* Fig. 8.11 b. The objective is to find the values $\nu'_B(t_E)$ at the final time t_E such that

i) either the remaining liquid is enriched in the high boiling constituent to the mol fraction $X'_{2B}(t_E) = 1 - X'_{1B}(t_E)$,

ii) or the distillate has the mol fraction $X'_{1D}(t_E)$.

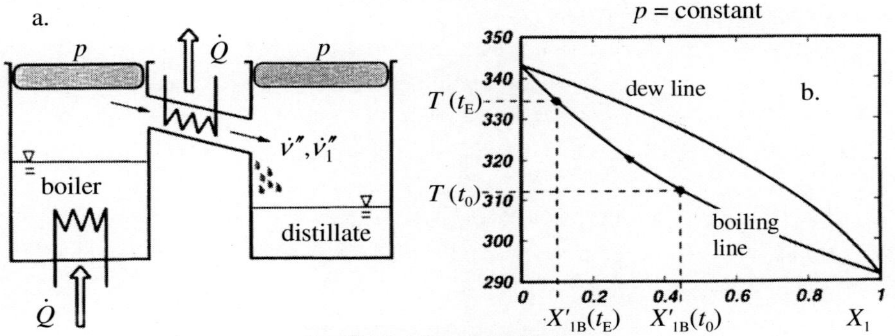

Fig. 8.11 a. Batch distillation
b. (T, X_1)-diagram of the distillation process $X'_{1B}(t_0) \to X'_{1B}(t)$.

In order to solve these problems we balance the mol numbers ν'_B and ν'_{1B} in the boiler, whose surface is passed by the vapor rates $\dot{\nu}'', \dot{\nu}''_1$. Under sustained heating the rate of change of ν'_B, the mol number of the boiling liquid, is balanced by $\dot{\nu}''$, the vapor transfer through the pipe. The same holds for ν'_{1B}, which is balanced by $\dot{\nu}''_1$, so that we have

$$\frac{d\nu'_B}{dt} = -\dot{\nu}'' \quad \text{and} \quad \frac{d\nu'_{1B}}{dt} = -\dot{\nu}''_1 .$$

We use $\dfrac{d\nu'_{1B}}{dt} = \dfrac{d\nu'_B X'_{1B}}{dt} = \dfrac{d\nu'_B}{dt} X'_{1B} + \nu'_B \dfrac{dX'_{1B}}{dt}$ and $\dot{\nu}''_1 = \dot{\nu}'' X''_{1B}$ in the second equation and eliminate $\dot{\nu}''$ between the two equations. Thus we obtain

$$\frac{d\nu'_B}{\nu'_B} = \frac{dX'_{1B}}{X''_{1B} - X'_{1B}} \quad \text{or, by integration:} \quad \ln \frac{\nu'_B(t)}{\nu'_B(t_0)} = \int_{X'_{1B}(t_0)}^{X'_{1B}(t)} \frac{dX'_{1B}}{X''_{1B} - X'_{1B}} .$$

Insertion of X''_{1B} from (8.53) gives

$$\nu'_B(t) = \nu'_B(t_0) \exp\left[\int_{X'_{1B}(t_0)}^{X'_{1B}(t)} \frac{1}{z-1} \left(\frac{1}{X'_{1B}} + \frac{z}{1 - X'_{1B}} \right) dX'_{1B} \right], \tag{8.62}$$

where $z = p_1(T)/p_2(T)$ has been introduced as a convenient abbreviation. z is obviously a function of T, and T is a function of X'_{1B} as described by the boiling line in the phase diagram, *cf.* Fig. 8.11 b.

Since we do not know an analytic form for the boiling line $T(X'_{1B})$, the exponential expression in (8.62) cannot be evaluated analytically; it must be calculated numerically as a function of its two variables $X'_{1B}(t)$ and $X'_{1B}(t_0)$. We denote the solution by $Q(X'_{1B}(t), X'_{1B}(t_0))$ and suppose that we have calculated it.

The first objective i) is then easily reached: In this case we may read off – immediately from (8.62) – at which mol number $\nu_B(t_E)$ in the boiler the distillation must be stopped. We calculate it as

$$\nu'_B(t_E) = \nu'_B(t_0) Q(X'_{1B}(t_E), X'_{1B}(t_0)), \tag{8.63}$$

For the solution of the more difficult second objective ii) we first need to calculate the mol fraction $X'_{1B}(t_E)$ in the boiler as a function of $X'_{1D}(t_E)$, the desired mol fraction of the distillate. Only afterwards can we calculate $\nu'_B(t_E)$ from the equation

$$\nu'_B(t_E) = \nu'_B(t_0) Q(X'_{1B}(X'_{1D}(t_E)), X'_{1B}(t_0)). \tag{8.64}$$

We explain how to proceed in this case: From the conservation of mol numbers $\nu'_B(t) + \nu'_D(t) = \nu'_B(t_0)$ and $\nu'_{1B}(t) + \nu'_{1D}(t) = \nu'_{1B}(t_0)$ we obtain with $\nu'_1 = \nu' X'_1$ and (8.63)

$$X'_{1D}(t_E) = \frac{1}{1-Q} X'_{1B}(t_0) - \frac{Q}{1-Q} X'_{1B}(t_E). \tag{8.65}$$

It is true that this is not the desired relation $X'_{1B}(X'_{1D}(t_E))$ but rather its inverse $X'_{1D}(X'_{1B}(t_E))$. Note that $X'_{1B}(t_E)$ also occurs as a variable in Q so that an explicit solution of (8.65) for $X'_{1B}(t_E)$ is impossible. Therefore (8.65) must be inverted numerically, or graphically, before we can use (8.64) to determine the value $\nu'_B(t_E)$ at which the distillate has the required mol fraction $X'_{1D}(t_E)$.

For technically important mixtures the necessary numerical solutions have been calculated and are graphically represented. The graphs are handed out to students of chemical engineering who wish to investigate distillation problems.

In order to show explicitly what is involved, we focus the attention on a somewhat hypothetical case which is, nevertheless, instructive. We consider a situation in which z is independent of T and equal to 4. And we start with

$$X'_{1B}(t_0) = \frac{1}{2}, \quad \nu'_B(t_0) = 1. \tag{8.66}$$

In this case we can solve the integral in (8.62) and obtain

$$\nu'_B(t) = \nu'_B(t_0) \sqrt[3]{\frac{X'_{1B}(t)}{[1 - X'_{1B}(t)]^4}}. \tag{8.67}$$

This identifies Q as the cubic root. It is then an easy matter to represent $X'_{1D}(t)$ in terms of $X'_{1B}(t)$ by evaluation of (8.65) and we obtain

$$X'_{1D}(t) = \frac{1}{1-\sqrt[3]{\frac{X'_{1B}(t)}{[1-X'_{1B}(t)]^4}}} X'_{1B}(t_0) - \frac{\sqrt[3]{\frac{X'_{1B}(t)}{[1-X'_{1B}(t)]^4}}}{1-\sqrt[3]{\frac{X'_{1B}(t)}{[1-X'_{1B}(t)]^4}}} X'_{1B}(t). \tag{8.68}$$

For the initial values (8.66) the relations (8.67) and (8.68) are represented graphically in Fig. 8.12.

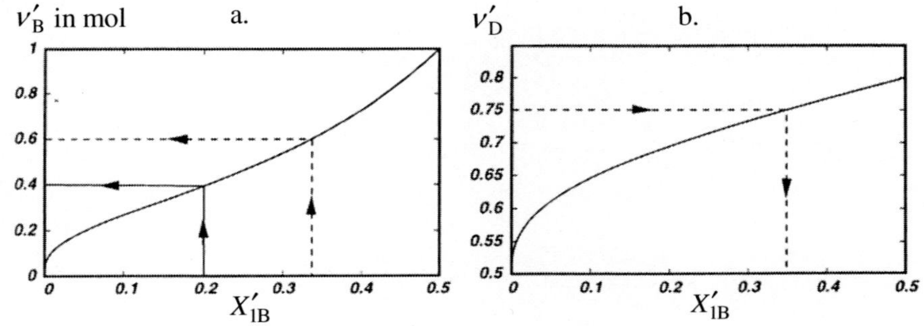

Fig. 8.12 a. The graph $\nu'_B(X'_{1B})$ for $X'_{1B}(t_0) = \frac{1}{2}$, $\nu'_B(t_0) = 1$

b. The graph $X'_{1D}(X'_{1B})$ for the same data.

The easy problem is now readily solved, it needs only the graph $\nu'_B(X'_{1B})$. For example, if we wish to finish the evaporation with $X'_{2B} = 0.8$ – or $X'_{1B} = 0.2$ – we find the appropriate value $\nu'_B(t_E)$ to be equal to $0.4\,\text{mol}$, *i.e.* we must evaporate 60% of the original mixture. The arrows on the solid lines in Fig. 8.12 a show how the graph is used.

For the other problem, when the distillate is to be enriched to a mol fraction of X'_{1D} =0.75, Fig. 8.12 b shows that the corresponding value of X'_{1B} is equal to 0.33 and Fig. 8.12 a permits the determination of $\nu'_B(t_E) = 0.6\,\text{mol}$ so that 40% of the original mixture must be evaporated. *Cf.* the dashed lines in the graphs.

8.6.4 *Continuous distillation and the separating cascade*

Continuous distillation – as opposed to batch distillation – is a stationary process in which the molar rates of change of ν'_B and ν'_{1B}, evaporated in the boiler, are replenished by the feedstock solution which needs to be separated into its constituents. Thus there is a continuous flow $\dot{\nu}, \dot{\nu}_1 = \dot{\nu} X$ * of feedstock with the initial temperature T_i, *cf.* Fig. 8.13.

The boiler temperature is $T_{(0)}$ and the 1-rich vapor with X'' is siphoned off continuously and so is the 2-rich residue with X'. The out-flowing fractions

* It is now convenient to drop the index 1 on X_1. We understand that X is the mol fraction of the low boiling constituent.

$\frac{\dot{\nu}''}{\dot{\nu}}, \frac{\dot{\nu}'}{\dot{\nu}}$ and their mol fractions X'' and X' may be read off from the (T,X)-diagram by the "center of mass rule", as shown in the figure.

Obviously a single separating unit – irrespective of the choice of temperature – cannot produce a vapor with $X'' > X^{\max}$ or a residual liquid with $X' < X^{\min}$. And when the unit does produce a vapor X'' *close* to $X^{\max}$, the vapor flow is minimal.

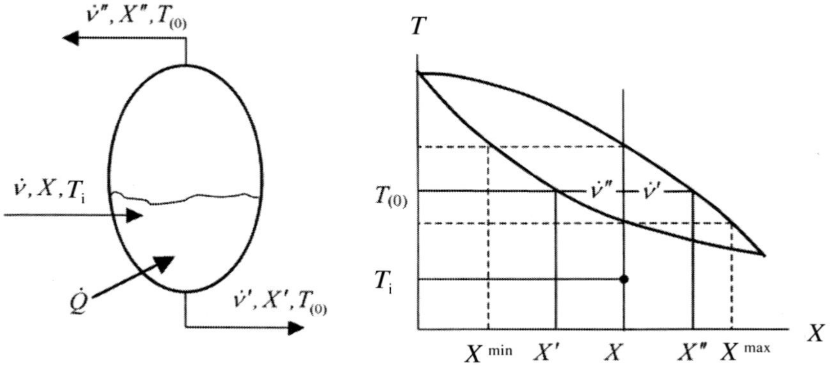

Fig. 8.13 Continuous separating unit with temperature $T_{(0)}$

It may be desirable, however, to have a more complete separation *and* sizable out-flows of both: 1-rich distillate and 2-rich residue. For that purpose we need a *separating cascade* which we proceed to describe.

The basic idea is, of course, to condense the 1-rich flow with $X''_{(0)}$, leaving the original separating unit (0), and feed the distillate with $X'_{(0)} = X''_{(0)}$ into another separating unit (1) which works at $T_{(1)}$, *i.e.* it produces a vapor with $X''_{(1)}$ and a residue with $X'_{(1)}$. The vapor is again condensed and may be led into yet another unit (2) which works at $T_{(2)}$ and produces a vapor with $X''_{(2)}$ and a residue with $X'_{(2)}$. Thus we obtain a final distillate with the mol fraction $X''_{(2)}$ which is larger than $X^{\max}_{(0)}$. Nor are the residues with $X'_{(1)}$ and $X'_{(2)}$ wasted. They are mixed with the feed of the respective previous units; and it is then appropriate to choose the temperatures $T_{(1)}$ and $T_{(2)}$ so that $X'_{(1)} = X$ holds and $X'_{(2)} = X''_{(0)}$, *cf.* Fig 8.14 which shows a schematic picture of a five-unit cascade. In this manner one avoids contamination of the feed of a unit with the residue from the next one, or else: One avoids to mix an already refined mixture with a less refined one from the previous unit. In the figure the primary unit (0) is complemented by two units (1) and (2) for the enrichment of the low-boiling constituent, and by two units – namely units (-1) and (-2) – for the enrichment of the high boiling constituent.

The roles of the units (-1) and (-2) are similar to those of the units (1) and (2). The vapor of unit (-1) is mixed – after condensation – with the feed stock X, and the vapor of the unit (-2), after condensation, is mixed with the feed of unit (0) which has $X'_{(0)}$. When the temperatures $T_{(-1)}$ and $T_{(-2)}$ are properly chosen, the mixing processes are between liquids with $X''_{(-1)} = X$ and $X''_{(-2)} = X'_{(0)}$. The final residue of mol fraction $X'_{(-2)}$ is led off to serve whatever purpose it is intended to have. $X'_{(-2)}$ is lower than the minimal mol fraction of the central unit (0).

With many units in the cascade one may achieve both: a pure low boiling constituent and a pure high boiling one, *i.e.* complete separation.

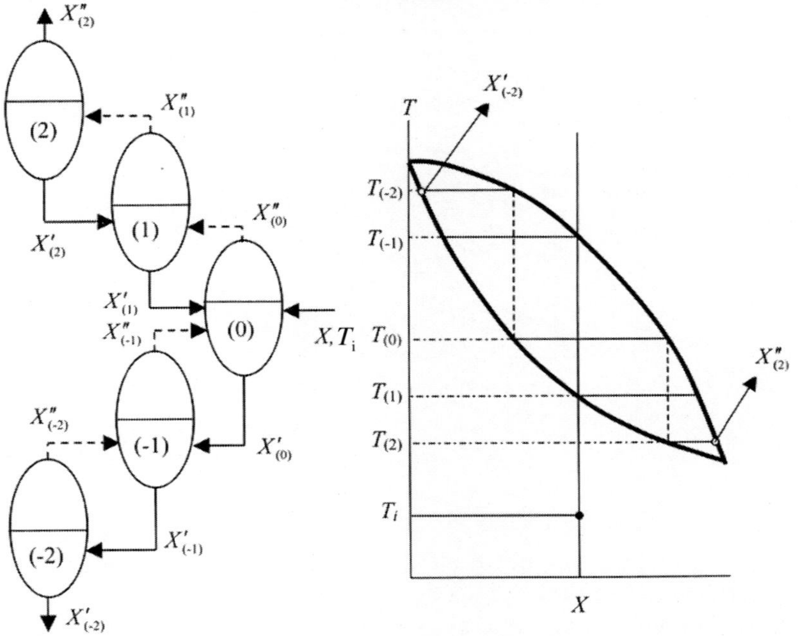

Fig. 8.14 Schematic view of a five-unit cascade and the corresponding (T,X)-diagram. All units are heated and complete condensation occurs on the dashed lines between the units

8.6.5 *Rectification column*

In the discussion of distillation in the preceding two paragraphs we have ignored the energetic aspects involved in the repeated evaporations and condensations. We leave those aspects to special books on chemical engineering, and so we continue to ignore them, except for saying that the cascade process is not efficient energetically. Thus for instance the vapor coming out of unit (0) is first fully condensed and then partially re-evaporated in unit (1).

Rather obviously it would be better to let the vapor with $X''_{(n)}$ of a generic unit (n) mix with the liquid in unit $(n+1)$ without first condensing it. If this is done in all units, ideally it is only the lowest unit – the one with the highest temperature – that needs heating for evaporation. And only the vapor from the highest unit, – the one with the lowest temperature – needs to be condensed. All intermediate units boil at their appropriate temperatures, heated by the vapor reaching them from the neighboring lower unit. That is the principle of the *rectification column.* There are various ingenious designs; one of them is sketched in Fig. 8.15.

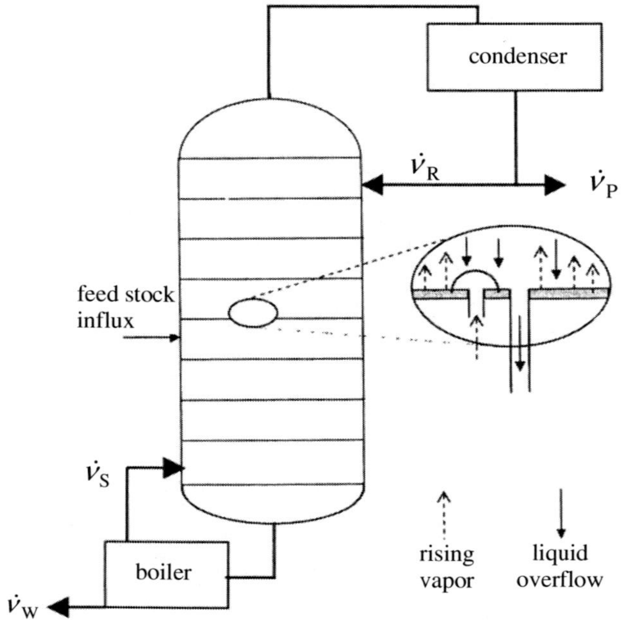

Fig. 8.15 Schematic view of a rectification column

The vapor rising from unit (n) is led through the liquid solution of unit (n+1) and there it condenses partially, primarily the high-boiling constituent, of course. After passing upwards through several – or many – such units the vapor arrives at the top where it contains essentially only the low-boiling constituent. Similarly the liquid solution of unit (n+1), enriched in the high boiling constituent by the partial vapor condensation coming from unit (n), overflows the rim of its level and drops into the solution of unit (n), enriching it in the high-boiling constituent beyond the degree of enrichment that was the result of prior evaporation. After several such steps the liquid at the bottom becomes nearly pure in the high-boiling constituent and is led out.

At the top the nearly pure vapor is led out and condensed. The resulting nearly pure liquid is partly led out from the plant altogether as the *product flow* $\dot{V}_P$. The rest is led back to the uppermost level and falls down, losing more and more of the low-boiling constituent on every step. At the sump the nearly pure liquid – rich in the high-boiling constituent – is partially evaporated and the vapor flow $\dot{V}_S$ is led back to the lowest unit, while the liquid *"waste" flow* $\dot{V}_W$ is led out of the plant.

Rectification columns of this type have been built which contain 30 levels and are 30 m high at a diameter of 5 m.

8.7 Liquid-vapor equilibrium (real)

8.7.1 *Activity and fugacity*

We recall that the chemical potentials of an ideal mixture are given by the equation

$$\mu_\alpha(T,p,X_\beta)=g_\alpha(T,p)+\frac{RT}{M_\alpha}\ln X_\alpha .$$

For an arbitrary non-ideal mixture we set

$$\mu_\alpha(T,p,X_\beta)=g_\alpha(T,p)+\frac{RT}{M_\alpha}\ln a_\alpha(T,p,X_\beta) \tag{8.69}$$

and thus we define the *activity* a_α, whose deviation from X_α represents the non-ideal character. Furthermore one defines the *activity coefficient* γ_α by

$$a_\alpha(T,p,X_\beta)=\gamma_\alpha(T,p,X_\beta)X_\alpha \tag{8.70}$$

whose deviation from the value 1 characterizes the deviation from ideal behavior.

Another customary correction factor – especially well-suited for vapors – is the *fugacity* f_α defined by

$$\mu_\alpha(T,p,X_\beta)=g_\alpha(T,p_\alpha(T))+\frac{R}{M_\alpha}T\ln f_\alpha(T,p,X_\beta). \tag{8.71}$$

Here $p_\alpha(T)$ is the saturation pressure of the pure constituent α. If the mixture is a mixture of ideal gases, we have

$$f_\alpha(T,p,X_\beta)=X_\alpha\frac{p}{p_\alpha(T)},$$

because ideal gases are special cases of ideal mixtures, and because $g_\alpha(T,p)$ is a logarithmic function of p in ideal gases, so that

$$g_\alpha(T,p_\alpha(T))+\frac{R}{M_\alpha}T\ln\frac{p}{p_\alpha(T)}=g_\alpha(T,p)$$

holds. One also defines the fugacity coefficient φ_α by the equation

$$f_\alpha(T,p,X_\beta)=\varphi_\alpha(T,p,X_\beta)X_\alpha\frac{p}{p_\alpha(T)} \tag{8.72}$$

so that the deviation of φ_α from the value 1 represents the deviation of the vapor from a mixture of ideal gases.

Note that in a non-ideal mixture we continue to decompose $\mu_\alpha(T,p,X_\beta)$ into $g_\alpha(T,p)$, the chemical potential of pure constituent α, and a logarithmic residue. In a manner of speaking this is purely traditional: What we are doing is replacing one function of $\nu+1$ variables, namely the chemical potential μ_α, by another function of the same variables, either a_α or f_α. We might just as well have stuck to the μ_α's, but convention is strong and a hard-working army of chemical engineers is determined to keep on measuring activities and/or fugacities rather than chemical potentials.

8.7.2 *Raoult's law for non-ideal mixtures*

The equilibrium conditions in a two-phase mixture are always given by Gibbs's phase rule

$$\mu'_\alpha(T,p,X'_\beta)=\mu''_\alpha(T,p,X''_\beta).$$

By (8.69) and (8.71) we may write this in the form

$$a'_\alpha(T,p,X'_\beta)\exp\left[\frac{g'_\alpha(T,p)}{\frac{R}{M_\alpha}T}\right]=f''_\alpha(T,p,X''_\beta)\exp\left[\frac{g''_\alpha(T,p_\alpha(T))}{\frac{R}{M_\alpha}T}\right].$$

For incompressible liquids we have

$$g'_\alpha(T,p)=g'_\alpha(T,p_\alpha(T))+v'_\alpha[p-p_\alpha(T)]$$

and, since $g'_\alpha(T,p_\alpha(T))=g''_\alpha(T,p_\alpha(T))$ holds, we obtain

$$a'_\alpha(T,p,X'_\beta)\exp\left[\frac{v'_\alpha[p-p_\alpha(T)]}{\frac{R}{M_\alpha}T}\right]=f''_\alpha(T,p,X''_\beta).$$

In the exponent we have – by order of magnitude – the ratio of specific volumes of liquid and vapor of constituent α. If this ratio is much smaller than one – as is most often the case* – we may write Gibbs's phase rule in the form

$$a'_\alpha(T,p,X'_\beta)=f''_\alpha(T,p,X''_\beta). \tag{8.73}$$

When expressed by the coefficients γ_α and φ_α of activity and fugacity, respectively, we obtain, *cf.* (8.70) and (8.72)

$$\gamma'_\alpha(T,p,X'_\beta)X'_\alpha p_\alpha(T)=\varphi''_\alpha(T,p,X''_\beta)X''_\alpha p\,. \tag{8.74}$$

This is Raoult`law for non-ideal mixtures. If $\gamma'_\alpha=f''=1$, it falls back to the form (8.48) for ideal mixtures.

If we consider a binary mixture we obtain from (8.74)

$$\begin{aligned}\gamma'_1(T,p,X'_1)\qquad X'_1p_1(T)&=\varphi''_1(T,p,X''_1)\qquad X''_1p\\ \gamma'_2(T,p,X'_1)(1-X'_1)p_2(T)&=\varphi''_2(T,p,X''_1)(1-X''_1)p\,.\end{aligned} \tag{8.75}$$

Just as in ideal mixtures these are two equations for the two variables X'_1 and X''_1, but they are no longer *linear* equations. Therefore we cannot expect that – for given T – the functions $X'_1(p)$ and $X''_1(p)$ are those of Fig. 8.7.

8.7.3 *Determination of the activity coefficient*

Indeed, for non-ideal mixtures the (p,X_1)-phase diagrams exhibit more or less distorted boiling and dew lines as well as distorted curves for the partial pressures p''_α. Thus Fig. 8.16 shows phase diagrams for the mixtures acetone CS_2, and

* It is *not* the case, when we are close to the critical point of one or the other constituent, or of both.

acetone-chloroform, respectively; only the boiling lines and the partial pressures are shown, *not* the dew lines. For ideal mixtures all three curves would be straight lines.

We assume that the vapor phase may be considered as a mixture of ideal gases. In this case the fugacity coefficients φ''_α in (8.74) are equal to 1 and the products $p X''_\alpha$ are equal to the partial vapor pressures. Therefore we may write (8.74) in the form

$$p''_\alpha = \gamma'_\alpha X'_\alpha p_\alpha(T) \text{ and, of course, we have } p''^{\text{id}}_\alpha = X'_\alpha p_\alpha(T),$$

if p''^{id}_α denotes the partial pressures in the ideal case; these are represented by the dashed lines in Fig. 8.16a. Therefore the activity coefficient is given by

$$\gamma'_\alpha = \frac{p''_\alpha}{p''^{\text{id}}_\alpha},$$

so that γ'_1 is given by the length ratios $\frac{AE}{BE}$, while γ'_2 is determined by $\frac{CE}{DE}$, *cf.* Fig. 8.16 a. Obviously these ratios depend on X'_1 and on the pair (p,T). $p = p''_1 + p''_2$ is given by the height of the boiling line, and $T = 32.5°\text{C}$ is the temperature for which the phase diagram was experimentally determined.

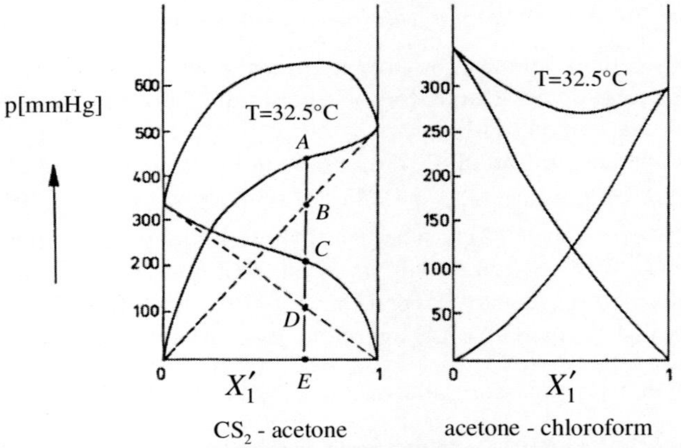

Fig. 8.16 Two realistic phase diagrams. Boiling lines and partial vapor pressures

A significant qualitative difference between the two diagrams of Fig. 8.16 is that on the left hand side the boiling line is higher than the ideal straight line between the ordinate values at $X'_1 = 0$ and $X'_1 = 1$, while on the right hand side the boiling line lies below that straight line. We shall explain in Paragraph 8.7.5 that in the first case unequal next neighbors among the atoms or molecules are energetically unfavorable so that boiling and evaporation can start at high pressure. In the second case unequal next neighbors are tightly bound so that it takes a low pressure to start the evaporation. In the first case mixing of the liquids requires energy – usually by heating – while in the second case mixing is accompanied by a release of heat.

8.7.4 *Determination of fugacity coefficients*

The fugacity coefficients may be determined by volume measurements and by the use of $(8.10)_2$

$$\left(\frac{\partial \mu_\alpha}{\partial p}\right)_{T,m_\beta} = \left(\frac{\partial V}{\partial m_\alpha}\right)_{T,p,m_{\beta\neq\alpha}} \quad \text{hence, by } \mu_\alpha = g_\alpha(T, p_\alpha(T)) + \frac{RT}{M_\alpha}\ln\left[\varphi_\alpha X_\alpha \frac{p}{p_\alpha(T)}\right]$$

$$\left(\frac{\partial \ln \varphi_\alpha}{\partial p}\right)_{T,m_\beta} = \frac{M_\alpha}{RT}\left(\frac{\partial V}{\partial m_\alpha}\right)_{T,p,m_{\beta\neq\alpha}} - \frac{1}{p}.$$

φ_α follows by integration over p, starting with a very small value of p – effectively p=0 –, where the vapor behaves as an ideal gas and φ_α is equal to 1

$$\varphi_\alpha(T, p, X_\beta) = \int_0^p \left[\frac{M_\alpha}{RT}\left(\frac{\partial V}{\partial m_\alpha}\right)_{T,p,m_{\beta\neq\alpha}} - \frac{1}{p}\right] \mathrm{d}p\ .$$

The function $\partial V/\partial m_\alpha$ in the integral must be measured by adding small masses Δm_α to the vapor mixture and registering the volume change. These measurements must be repeated many times for different values of p and X_β.

8.7.5 *Activity coefficient and heat of mixing. Construction of a phase diagram*

In this paragraph we shall demonstrate how non-ideal phase diagrams may be calculated when the energetic conditions for the formation of unequal next neighbors are taken into account in the liquid.

We recall the characterization of ideal mixtures in Paragraph 8.2.3 and consider a non-ideal mixture by putting $H_{\text{mix}} \neq 0$. For simplicity we assume that S_{mix} is still given by the expression $(8.21)_4$. The heat of mixing may be attributed to the phenomenon that the formation of neighbors of molecules of different constituents is not energetically neutral. It may happen that the formation of such a pair costs energy and it may also happen that it is energetically favorable. Purely statistically the expectation value for unequal pairs of particles is equal to $2N_1 \cdot \frac{N_2}{N}$, if N_1 and N_2 are the particle numbers of the constituents and N is their sum. Therefore we formulate the simple ansatz

$$H_{\text{mix}} = 2e\,N_1 \frac{N_2}{N} \quad \text{with } e \begin{cases} > 0: \text{ energetic malus} \\ < 0: \text{ energetic bonus}. \end{cases} \tag{8.76}$$

Expressed in terms of the constituent masses m_1 and m_2 this reads

$$H_{\text{mix}} = 2e\frac{1}{m_0}\frac{m_1 m_2}{m_1 M_2 + m_2 M_1}.$$

And therefore, by (8.2), there must be an additional term of the form $2\frac{e}{m_0}\frac{1}{M_\alpha}(1 - X_\alpha)^2$ in the chemical potential of ideal mixtures. We obtain

$$\mu_\alpha = g_\alpha(T,p) + \frac{R}{M_\alpha} T \ln X_\alpha + 2\frac{e}{m_0}\frac{1}{M_\alpha}(1-X_\alpha)^2 . \tag{8.77}$$

Comparison of this expression with (8.69), (8.70) leads to an activity coefficient of the form

$$\gamma_\alpha = \exp\left[2\frac{e}{kT}(1-X_\alpha)^2\right]. \tag{8.78}$$

Raoult's law (8.75) reads in this case

$$\begin{aligned} \exp\left[\frac{2e}{kT}(1-X_1')^2\right] X_1' p_1(T) &= X_1'' p \\ \exp\left[\frac{2e}{kT}X_1'^2\right](1-X_1') p_2(T) &= (1-X_1'')p , \end{aligned} \tag{8.79}$$

provided that the fugacity coefficients are equal to 1.

Adding the equations we obtain

$$p = \exp\left[\frac{2e}{kT}(1-X_1')^2\right] X_1' p_1(T) + \exp\left[\frac{2e}{kT}X_1'^2\right](1-X_1'') p_2(T). \tag{8.80}$$

This function is shown in Fig. 8.17 for three values of $\frac{e}{kT}$. The ideal linear function follows for $e=0$. For an energetic malus $e>0$ the liquid boils at a higher than the ideal pressure and for an energetic bonus $e<0$ the pressure must be lowered below the ideal case, in order to bring the liquid to the boiling point. The figure also exhibits the partial pressures $p_1'' = X_1'' p$ and $p_2'' = (1-X_1'')p$ which are given by (8.79).

If we increase the energetic malus e, we obtain the diagrams of Fig. 8.18 which shows boiling *and* dew lines – both as numerical solutions of the equations (8.79) for some fixed temperature. The boiling lines are solid and the dew lines are dashed. Of particular interest are the lower graphs of the figure.

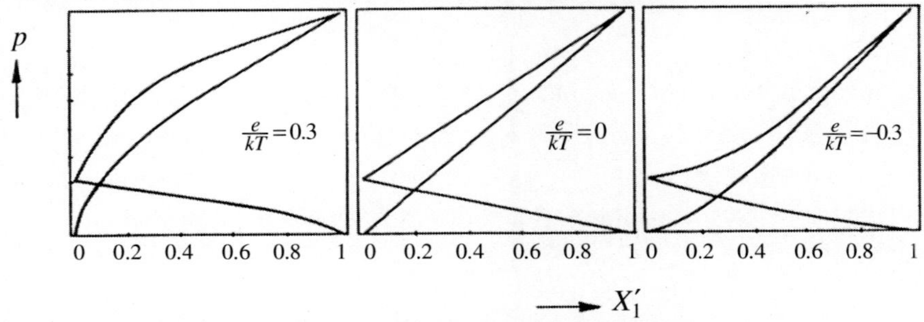

Fig. 8.17 Boiling lines and partial vapor pressures for $e>0$ and $e<0$

The graphs of Fig. 8.18 c exhibit azeotropy (Greek: a "not", and zein "boil", and trop "change", *i.e.* no "change during boiling") in the point, where the boiling line and the dew line touch: The mol fraction corresponding to the point of osculation defines an azeotropic mixture. That mixture evaporates and condenses without change of pressure and at a constant mol fraction, just like a pure vapor.

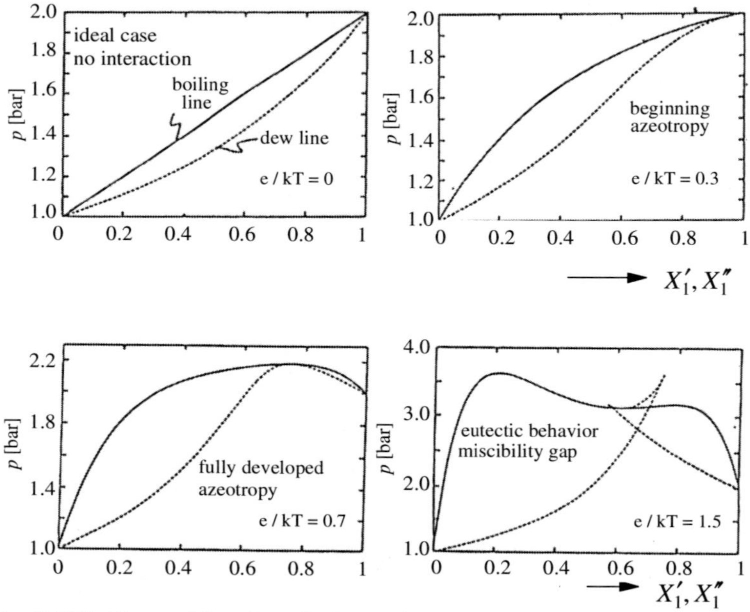

Fig. 8.18 Boiling and dew lines for increasing values of the energetic malus *e*

The graphs of Fig. 8.18 d exhibit a loop in the dew line for large values of *e*. We thus encounter eutectic behavior of the mixture and the phenomenon of a miscibility gap. The interpretation will be deferred to Sect. 8.8 and Fig. 8.23.

8.7.6 *Henry coefficient*

From the *ideal* form (8.48) of Raoult's law we have been able to show that the partial vapor pressure p''_α is proportional to the mol fraction X'_α of constituent α in the liquid. We have

$$p''_\alpha = p_\alpha(T)X'_\alpha$$

so that, of course, for $X'_\alpha = 1$ we obtain $p''_\alpha = p_\alpha(T)$. In a *real* mixture we have seen, *cf.* Figs. 8.16 and 8.17 that p''_α is *not* proportional to X'_α; indeed the graphs are strongly curved. However, for $X'_1 << 1$ the curve p''_1 has a tangent which coincides with the curve in a small interval, so that in this interval we may write

$$p''_1 = H(T)X'_1. \tag{8.81}$$

The factor of proportionality $H(T)$ is *not* $p_1(T)$ – except in an ideal mixture.* It is called the *Henry coefficient* and it can be measured and is tabulated. For dilute solutions of O_2 and CO_2 in water, the Henry coefficient $H(T)$ has values as shown in Table 8.1.

---

* Actually, there may not even *be* a value $p_1(T)$, because T may be bigger than the critical temperature of constituent 1, *cf.* Fig. 8.19.

Table 8.1 Henry coefficients in bar for water as the solvent

T [K] / solute	290	310	330
CO_2	1.28	2.17	3.22
O_2	38	52	61

Inspection shows that H grows with increasing T. Therefore, for a given p''_α, more gas can be dissolved in water at low temperature than at high temperature. For this reason the arctic oceans are much denser populated by fish than the tropical ones. Fish thrive in the oxygen-rich cold sea.

We also understand now that in an absorption refrigerator the absorber must be cooled in order to absorb more ammonia and that the generator must be heated so as to drive the ammonia out of the solution, *cf.* Chap.6.

Another interesting case is CO_2. Mineral water is "carbonated" by filling the bottles in a CO_2-atmosphere under high pressure and then sealing them. The higher the pressure the more CO_2 is dissolved. When we open the bottle, the CO_2 "bubbles out," and when we push it through the narrow ducts of our throat, even more bubbles appear and tickle our palate so that the thirst is quenched. If we let an open bottle stand outside the refrigerator, the high temperature releases much of the CO_2 from the solution: the sparkling water becomes "flat."

In the CO_2-conscious society, in which we live, a few numbers may be interesting. The partial pressure of CO_2 in our atmosphere is $p''_{CO_2} = 3 \cdot 10^{-4}\,\text{bar}$, *cf.* (P.8). Hence follows from (8.81) with a Henry coefficient of $1.28\,\text{bar}$ corresponding to 17°C

$$X'_{CO_2} = 2.3 \cdot 10^{-4} .$$

If we assume that the seven seas cover 70% of the earth's surface and have a mean depth of 3.8 km we calculate the mass of dissolved CO_2 in the sea as

$$m_{CO_2}^{\text{Sea}} = 5.6 \cdot 10^{16}\,\text{kg} .$$

It is instructive to compare this value with the mass of CO_2 in the atmosphere. We obtain

$$m_{CO_2}^{\text{Atmosphere}} = \frac{p''_{CO_2} V}{\frac{R}{M_{CO_2}} T} \approx 2.3 \cdot 10^{15}\,\text{kg} .$$

Here we have used the volume V of a spherical shell of radius $6 \cdot 10^6\,\text{m}$ with a thickness of $8 \cdot 10^3\,\text{m}$, *cf.* Paragraph 1.4.4.

We conclude from this calculation that the sea contains about 24 times as much CO_2 as the atmosphere.

A drastic case where Raoult's ideal law fails, because the conditions for its derivation are violated, is the mixture of sulphur dioxide SO_2, and carbon dioxide CO_2. Fig. 8.19 shows boiling and dew lines in a (p,X)-diagram at different temperatures. Carbon dioxide has the critical temperature of 32.5°C so that beyond that temperature no saturation pressure exists. Therefore, at 70°C (say) the boiling and the due lines meet at some mol fraction $X_{CO_2} \neq 1$. Note also the steep slope of the boiling line near $X_{CO_2} = 0$, which indicates a large Henry coefficient of CO_2 in SO_2.

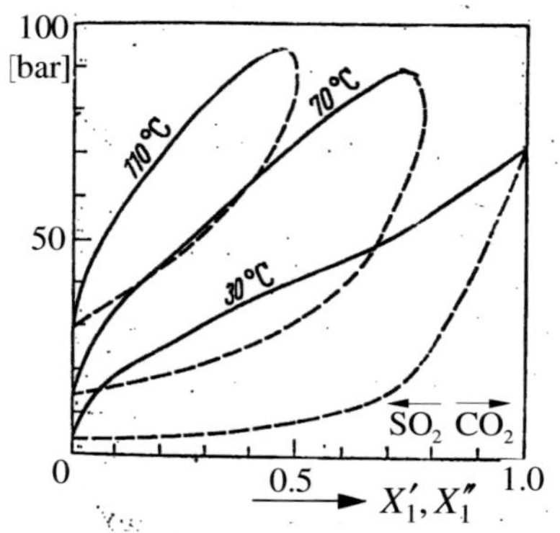

Fig. 8.19 Boiling and dew lines of the solution CO_2 in SO_2 for different temperatures

8.8 Gibbs free energy of a binary mixture in two phases

8.8.1 *Graphical determination of equilibrium states*

So far in the consideration of phase equilibria in mixtures we have made use of Gibbs's phase rule by which the chemical potentials μ_α are all homogeneous throughout the system. We have thus arrived at simple phase diagrams, e.g. in Figs. 8.7, 8.17, and 8.18. However, it is instructive – and useful for the understanding and interpretation of phase diagrams – to construct these diagrams from the consideration of the Gibbs free energies directly, and we proceed to do so. Partly we thus repeat arguments that we have used before, but we also learn new things, or at least new aspects of old results.

We consider a binary mixture of constituents 1 and 2 which are both present in the liquid phase ′ and the vapor phase ″. The total mol number is $\nu = \nu_1 + \nu_2$. The Gibbs free energy G and the mol number ν_1 may be decomposed as

$$G = G'(T, p, \nu_1') + G''(T, p, \nu_1'') \text{ and } \nu_1 = \nu_1' + \nu_1''.$$

Thus with the molar quantities

$$\tilde{g} = \frac{G}{\nu}, \; \tilde{g}' = \frac{G'}{\nu'}, \; \tilde{g}'' = \frac{G''}{\nu''}, \; X_1 = \frac{\nu_1}{\nu}, \; X_1' = \frac{\nu_1'}{\nu'}, \; X_1'' = \frac{\nu_1''}{\nu''}$$

we obtain

$$\tilde{g} = \frac{\nu'}{\nu}\tilde{g}'(T, p, X_1') + \frac{\nu''}{\nu}\tilde{g}''(T, p, X_1'') \text{ and } X_1 = \frac{\nu'}{\nu}X_1' + \frac{\nu''}{\nu}X_1''.$$

With the molar vapor fraction $z = \nu''/\nu$ we may write

$$\tilde{g} = (1-z)\tilde{g}'(T, p, X_1') + z\,\tilde{g}''(T, p, X_1'') \text{ and } X_1 = (1-z)X_1' + z\,X_1''. \quad (8.82)$$

By (8.6) they are related to the chemical potentials, and we have $\tilde{g}' = \sum_{\alpha=1}^{2} X'_\alpha \tilde{\mu}'_\alpha$ and $\tilde{g}'' = \sum_{\alpha=1}^{2} X''_\alpha \tilde{\mu}''_\alpha$. $\tilde{g}'$ and $\tilde{g}''$ are the molar Gibbs free energies of the phases.

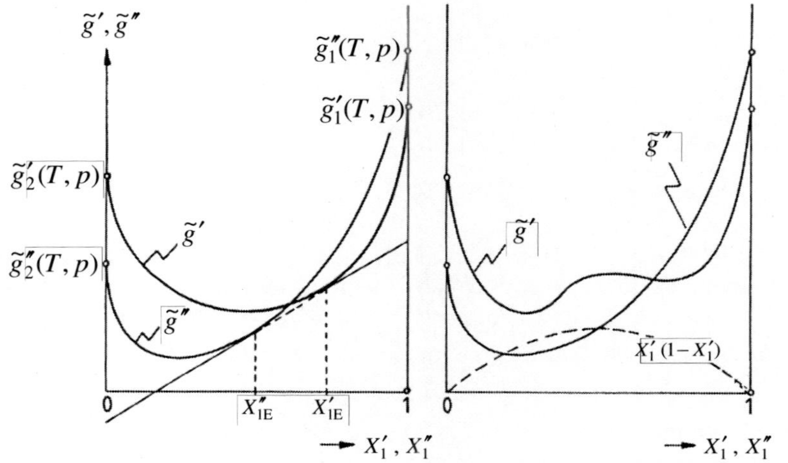

Fig. 8.20 Molar Gibbs free energies g' and g''

a. Both phases are ideal mixtures

b. Phase " is ideal, phase ' has a positive heat of mixing

We consider the case where both phases have chemical potentials of the form (8.77) so that they exhibit heats of mixing determined by the interaction factors e' and e''. Thus $\tilde{g}'$ and $\tilde{g}''$ both have the analytic form

$$\tilde{g}^h = \sum_{\alpha=1}^{2} X_\alpha^h \left[\tilde{g}_\alpha^h(T,p) + RT \ln X_\alpha^h + 2\frac{e^h}{\mu_0}\left(1 - X_\alpha^h\right)^2 \right] \quad (h = ',") \tag{8.83}$$

The only variable is X'_α or X''_α, respectively, and Fig. 8.20 shows qualitative graphs of these functions. In Fig. 8.20 a both graphs are drawn for $e = 0$; the convex character is due to the entropies of mixing. In Fig. 8.20 b we have put $e' > 0$ and $e'' = 0$; the effect of the heat of mixing in the liquid phase is given by

$$2\frac{e'}{m_0}\sum_{\alpha=1}^{2} X'_\alpha(1 - X'_\alpha)^2 = 2\frac{e'}{m_0} X'_1(1 - X'_1),$$

which is represented by the dashed concave parabola in the figure. If e' is large enough that additive term will "push" a concave bulge into the graph for g' as shown in the figure.

The total molar Gibbs free energy $(8.82)_1$ is a function of X'_1, X''_1, and z – for fixed values of T and p – but, by $(8.82)_2$, these variables are not independent.

In equilibrium $\tilde{g}$ must be minimal under the constraint $X_1 =$ const. We take care of the constraint by a Lagrange multiplier λ and minimize the function

$$\phi = (1-z)\tilde{g}'(X_1') + z\,\tilde{g}''(X_1'') - \lambda\left[X_1 - (1-z)X_1' - z\,X_1''\right] \tag{8.84}$$

without constraint. Necessary conditions for a minimum are

$$\frac{\partial \phi}{\partial X_1'} = 0: \quad (1 - z_{\mathrm{E}})\left[\left.\frac{\partial \tilde{g}'}{\partial X_1'}\right|_{\mathrm{E}} + \lambda\right] = 0$$

$$\frac{\partial \phi}{\partial X_1'} = 0: \quad z_{\mathrm{E}}\left[\left.\frac{\partial \tilde{g}''}{\partial X_1''}\right|_{\mathrm{E}} + \lambda\right] = 0$$

$$\frac{\partial \phi}{\partial z} = 0: \quad \tilde{g}''(X_{1\mathrm{E}}'') - \tilde{g}'(X_{1\mathrm{E}}') + \lambda\left[X_{1\mathrm{E}}'' - X_{1\mathrm{E}}'\right] = 0 \tag{8.85}$$

and the constraint $X_1 = (1 - z_{\mathrm{E}})X_{1\mathrm{E}}' + z_{\mathrm{E}}X_{1\mathrm{E}}''$, of course. The index E characterizes equilibrium. These are four equations for the determination of the four unknowns $X_{1\mathrm{E}}'$, $X_{1\mathrm{E}}''$, z_{E}, and λ. An analytical solution is difficult or impossible because of the complicated form of the functions $\tilde{g}'(X_1')$ and $\tilde{g}''(X_1'')$, *cf.* (8.83).

However, there is a simple graphical method for the determination of $X_{1\mathrm{E}}'$ and $X_{1\mathrm{E}}''$ from the graphs of Fig. 8.20. In order to see that, we eliminate λ between the equations (8.85) and obtain

$$\left.\frac{\partial \tilde{g}'}{\partial X_1'}\right|_{\mathrm{E}} = \left.\frac{\partial \tilde{g}''}{\partial X_1''}\right|_{\mathrm{E}} = \frac{\tilde{g}''(X_{1\mathrm{E}}'') - \tilde{g}'(X_{1\mathrm{E}}')}{X_{1\mathrm{E}}'' - X_{1\mathrm{E}}'}. \tag{8.86}$$

It follows that $X_{1\mathrm{E}}'$ and $X_{1\mathrm{E}}''$ are the abscissae of those two points of $\tilde{g}'(X_1')$ and $\tilde{g}''(X_1'')$ in which these graphs

i.) have equal slopes, and
ii.) where the slopes are equal to the difference quotient.

In other words $X_{1\mathrm{E}}'$ and $X_{1\mathrm{E}}''$ are the abscissae of the points where the common tangent of $\tilde{g}'(X_1')$ and $\tilde{g}''(X_1'')$ touches these curves. Fig. 8.20a illustrates the graphical construction of mol fractions.

Once $X_{1\mathrm{E}}'$ and $X_{1\mathrm{E}}''$ are thus known, z_{E} may be determined from $(8.85)_4$

$$z_{\mathrm{E}} = \frac{X_1 - X_{1\mathrm{E}}'}{X_{1\mathrm{E}}'' - X_{1\mathrm{E}}'}.$$

Finally the equilibrium value $\tilde{g}_{\mathrm{E}}$ of the Gibbs free energy follows from $(8.82)_1$

$$\tilde{g}_{\mathrm{E}} = \tilde{g}'(X_{1\mathrm{E}}'') + \frac{X_1 - X_{1\mathrm{E}}'}{X_{1\mathrm{E}}'' - X_{1\mathrm{E}}'}\left[\tilde{g}''(X_{1\mathrm{E}}'') - \tilde{g}'(X_{1\mathrm{E}}')\right]. \tag{8.87}$$

This is a linear function in X_1. It represents the piece of the common tangent between the touching points. In Fig. 8.20 a this piece is dashed.

One may summarize all this by saying that the solution is decomposed into two phases, because in this manner it has a lower free energy along the tangent than in either of the single phases.

8.8.2 *Graphical representation of chemical potentials*

We refer to Fig. 8.21 which shows a molar Gibbs free energy – convex for simplicity – and its tangent in the point X_1^*. The mathematical form of the tangent is

$$t(X_1) = \tilde{g}\left(X_1^*\right) + \left.\frac{\partial \tilde{g}}{\partial X_1}\right|_{X_1^*} \left(X_1 - X_1^*\right).$$

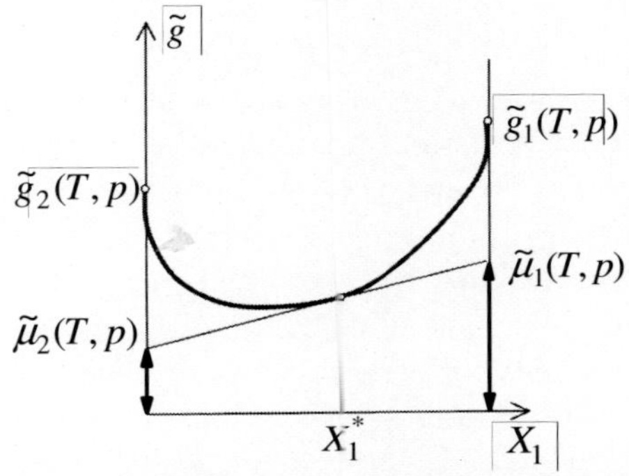

Fig. 8.21 Molar Gibbs free energies and molar chemical potentials

The intercepts of the tangent on the vertical lines $X_1 = 0$ and $X_1 = 1$ are equal to

$$t(0) = \tilde{g}\left(X_1^*\right) - \left.\frac{\partial \tilde{g}}{\partial X_1}\right|_{X_1^*} X_1^* \quad \text{and} \quad t(1) = \tilde{g}\left(X_1^*\right) + \left.\frac{\partial \tilde{g}}{\partial X_1}\right|_{X_1^*} \left(1 - X_1^*\right).$$

Now, since by (8.6) we have $\tilde{g}\left(X_1^*\right) = \tilde{\mu}_1\left(X_1^*\right)X_1^* + \tilde{\mu}_2\left(X_1^*\right)\left(1 - X_1^*\right)$, and hence, by (8.7) $\left.\frac{\partial \tilde{g}}{\partial X_1}\right|_{X_1^*} = \tilde{\mu}_1\left(X_1^*\right) - \tilde{\mu}_2\left(X_1^*\right)$, we conclude that $t(0) = \tilde{\mu}_2\left(X_1^*\right)$ and $t(1) = \tilde{\mu}_1\left(X_1^*\right)$ so that the intercepts represent the chemical potentials.

When $\tilde{g}'$ and $\tilde{g}''$ intersect in the manner of Fig. 8.20a, and if phase equilibrium requires $\mu'_\alpha = \mu''_\alpha$, it is therefore clear that in the equilibrium points the two graphs must have a common tangent. This agrees with the more formal argument of Paragraph 8.8.1 as it must.

8.8.3 *Phase diagram with unrestricted miscibility*

We fix T and X_1 and ask for which pressures the mixture is a liquid, or a vapor, or when both phases exist. The answer may be summarized in a (p, X_1)-diagram which is first constructed – in Fig. 8.22 – for the case that both phases are ideal mixtures. The Gibbs free energies $\tilde{g}'$ and $\tilde{g}''$ are then both convex curves.

The pressure dependence of the two graphs $\tilde{g}'$ and $\tilde{g}''$ is determined by the "anchor points" $\tilde{g}'_2(T,p)$ and $\tilde{g}''_2(T,p)$ at $X_1 = 0$, and $\tilde{g}'_1(T,p)$ and $\tilde{g}''_1(T,p)$ at

$X_1 = 1$. All of these values depend on p, and $\tilde{g}'_\alpha(T,p)$ and $\tilde{g}''_\alpha(T,p)$ grow differently fast with increasing p. For instance:

- When the liquids are both incompressible, the values $\tilde{g}'_\alpha(T,p)$ grow linearly in p with the small molar volumes $\tilde{v}'_\alpha$ as coefficients.
- When the pure vapors are ideal gases, the values $\tilde{g}''_\alpha(T,p)$ grow as $p \ln p$ with the large molar volumes $\tilde{v}''_\alpha$ as coefficients.

Therefore $\tilde{g}''$ is smaller than $\tilde{g}'$ for small pressures in the whole X_1-range, and for large pressures the opposite holds true. For small pressures the mixture is therefore a vapor and for large pressures it is a liquid.

For intermediate pressures the graphs of $\tilde{g}'$ and $\tilde{g}''$ intersect and the common tangent shifts with changing p as illustrated in Fig. 8.22. If the line between the points of contact is projected onto the corresponding isobar in the (p, X_1)–diagram, one obtains one point each on the boiling line and the dew line, see the figure. Connecting these points for all pressures, we obtain the full boiling and dew lines. In this graphical manner one may construct the phase diagram which we have previously – in Paragraph 8.5.2 – determined analytically.

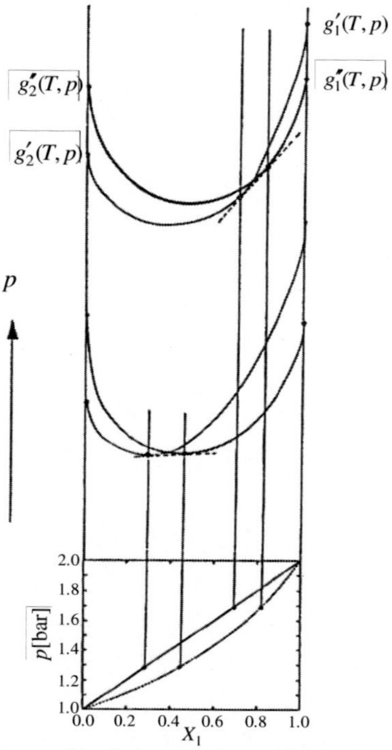

Fig. 8.22 On the graphical constructions of the phase diagram from the Gibbs free energies $\tilde{g}'$ and $\tilde{g}''$.

8.8.4 *Miscibility gap in the liquid phase*

In Fig. 8.23 we have drawn the Gibbs free energies of an ideal vapor phase and of a liquid phase with a positive heat of mixing, *i.e.* $e > 0$. We choose the same parameter values as in Fig. 8.18 d. This will enable us to understand the loop in the dew line whose discussion we have previously postponed. In order to obtain the mol fractions of phase equilibrium we use the method of the common tangent described earlier in Paragraph 8.8.1 through 8.8.3: The tangential lines between the points of contact are projected onto the appropriate isobar in a (p, X_1)–diagram and we thus obtain points on the boiling and the dew line.

In the present case the non-convex shape of $\tilde{g}'$ allows the construction of *two* tangents for intermediate pressures, *cf.* Fig. 8.23 for $p = 2.5\,\text{bar}$.

For $p = 2.83\,\text{bar}$ the two tangents coincide and for still higher pressures the common tangents of $\tilde{g}'$ and $\tilde{g}''$ intersect; this leads to the loop. However, the loop is irrelevant, because there is an energetically more favorable possibility. Indeed from $p = 2.83\,\text{bar}$ upward the tangent of the two convex parts of $\tilde{g}'$ lies lower than the graph $\tilde{g}''$. This means that the solution is entirely in a liquid state but *the liquid decomposes into two liquid phases* whose mol fractions are those of the points of contact of the tangent to $\tilde{g}'$. If, for different pressures, the tangents are projected into the (p, X_1)–diagram, we obtain for various pressures two steep lines which enclose a miscibility gap in the liquid. This is a region of coexistence of a 1-rich and a 2-rich liquid in equilibrium. Invariably one of those liquids is lighter than the other one, so that the liquid solution separates into horizontal layers, one rich in constituent 1 and the other one rich in constituent 2.

The pressure where the vapor phase vanishes is called the eutectic pressure. In the example it has the value $p = 2.83\,\text{bar}$.

8.9 Alloys

8.9.1 (T, c_1)*–diagrams*

Thermodynamics of solid alloys and their melts is largely equivalent to thermodynamics of solutions and their vapors. It is true though that the pressure is less important as a dependent variable for solids and melts, because both are nearly incompressible so that pressure changes cannot achieve much. The temperature is more important. Also metallurgists usually prefer the concentrations c_α for characterizing the composition of an alloy rather than the mol fractions X_α. For these reasons the equilibrium properties of binary alloys are most often laid down in (T, c_1)-diagrams.

Such diagrams may be graphically constructed from known graphs of g' and g'' in a manner analogous to that for solutions. Of course ' and " now characterize the solid and the liquid phase, respectively. Fig. 8.24 a shows the construction for the case of ideal mixtures both in the solid and the liquid phase. And

Fig. 8.24b refers to the case of a positive heat of mixing in the solid phase which creates a miscibility gap in that phase.

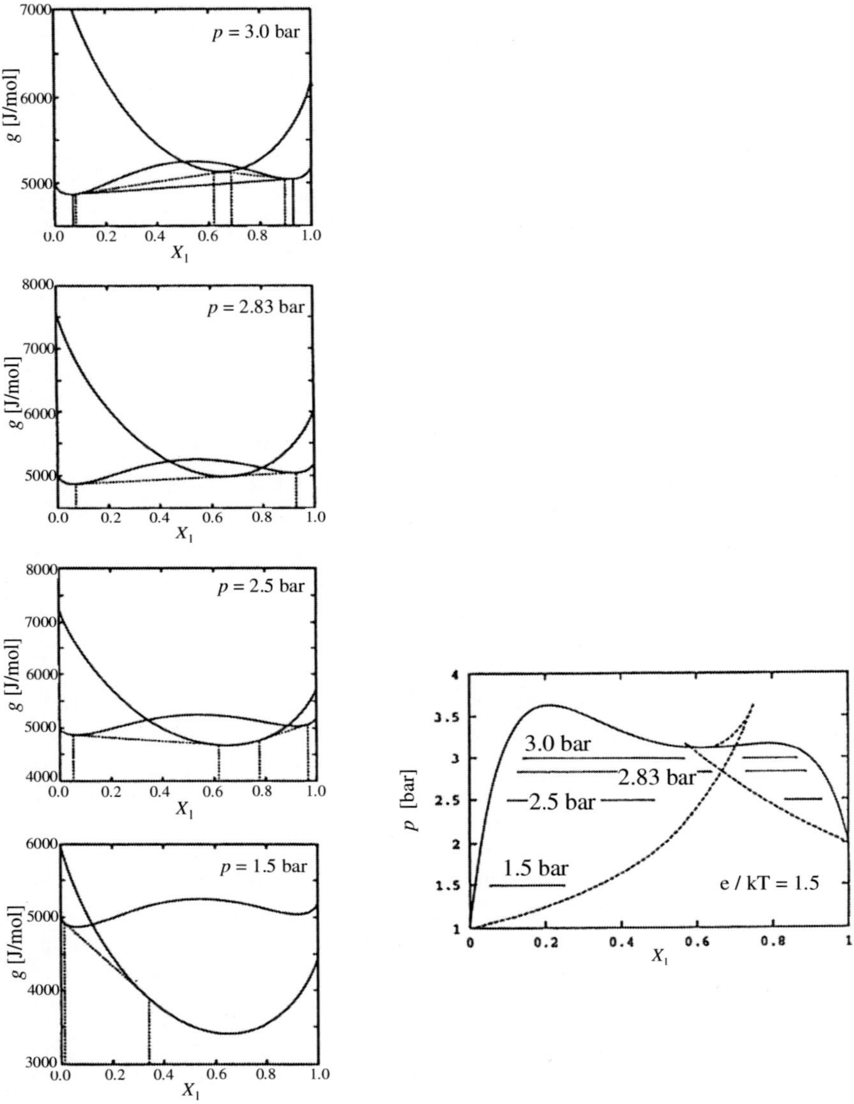

Fig. 8.23 On the graphical constructions of the phase diagram with a miscibility gap in the liquid phase

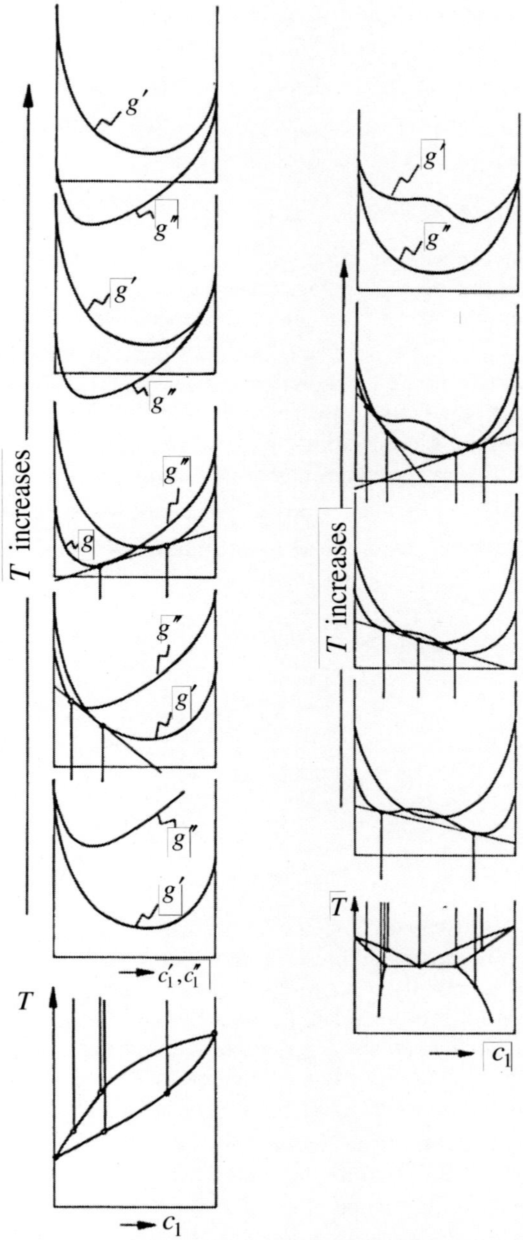

Fig. 8.24 On the graphical constructions of (T, c_1)-phase diagrams
a. Unrestricted miscibility. **b.** Miscibility gap.

For the construction of both diagrams it is important to know how the "anchor points" $g'_\alpha(T,p)$ and $g''_\alpha(T,p)$ depend on temperature. $g'_\alpha(T,p)$ rises faster than $g''_\alpha(T,p)$ for increasing temperature. This is essentially due to the term $-Ts$ in the Gibbs free energy $g = u - Ts + pv$, since s'' is greater than s'. Recall that $s'' - s' = r/T$ holds, where r is the heat of melting.

8.9.2 *Solid solutions and the eutectic point*

Fig. 8.25 shows a typical phase diagram with a miscibility gap in the solid phase. The liquid region is on the top of the diagram and the solid lies under the two "triangles." In the regions denoted by α and β we find *solid solutions*. Here the crystal structure is determined by the neighboring pure solid even though a few atoms of the other constituent are embedded in the crystal. The possibility for such an embedding is quite limited, since it is accompanied by a considerable energetic malus; this is the reason for the miscibility gap, which contains crystallites of type α and β. The miscibility gap becomes narrower for higher temperatures, because the curvature imparted to g' by the entropy of mixing becomes larger for growing T so that the heat of mixing cannot create quite so large a "bulge" in g'.

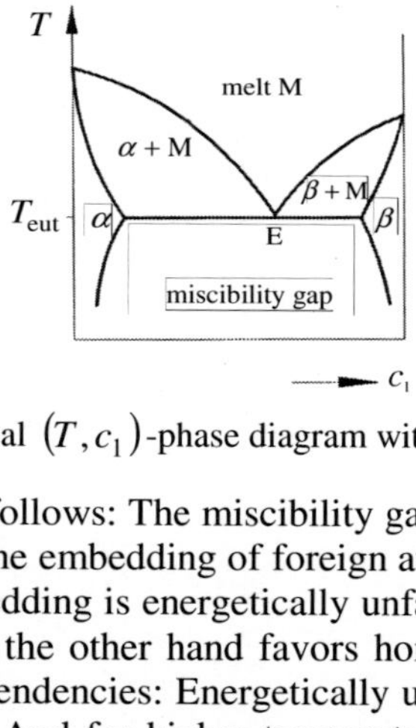

Fig. 8.25 A typical (T, c_1)-phase diagram with a miscibility gap.

One may also argue as follows: The miscibility gap is a consequence of the energetic malus involved in the embedding of foreign atoms into the prevailing crystal structure. Such an embedding is energetically unfavorable for all temperatures. The entropy of mixing on the other hand favors homogeneous mixing. Thus we encounter two competing tendencies: Energetically unmixing is favorable and entropically mixing is good. And for higher temperatures the entropic contribution becomes more important – because of the $T \cdot s$-term in g. Therefore higher temperature favors mixing and diminishes the miscibility gap.

The point E in Fig. 8.25 is called the eutectic point (Greek: eutectos "easy to melt"). The corresponding temperature T_E is the lowest possible melting temperature.

8.9.3 *Gibbs phase rule for a binary alloy*

We consider a point inside one of the "triangles" in Fig. 8.25 where α -crystallites are in equilibrium with the melt. We have $\nu = 2$ constituents and $f = 2$ phases. Therefore, by (8.42) we have $F = 2$ degrees of freedom. However, since in phase diagrams of the type of Fig. 8.25 the pressure is kept constant, there is only one degree of freedom. And indeed, melt and α -crystallite must change their composition along the right or left boundary of the triangle, respectively, with changing temperature. This is to say that both phases have only *one* possibility to change.

In the eutectic point we have three phases: the solid solutions α and β , and the melt. Therefore $F = 1$ holds or, for a constant pressure, $F = 0$. If we allowed for pressure changes, we could plot the pressure variable perpendicular to the page. In this manner we should obtain a eutectic line and F would indeed be equal to one.

8.10 Ternary Phase Diagrams

8.10.1 *Representation*

For a ternary solution – a solution of three constituents – the Gibbs phase rule (8.42) permits a maximum of four degrees of freedom: pressure, temperature, and two mol fractions. Thus a possible equilibrium state may be represented as a point in a four-dimensional space which is difficult to visualize. Therefore it is common practice to fix the pressure at some practical value, – mostly 1 atm – and draw a three-dimensional phase diagram with the temperature on the vertical axis, and the two independent phase fractions on the horizontal base plane, *cf.* Fig. 8.26 a. On that plane the most common representation of the composition of the solution is in the form of an equilateral triangle with side length 1 as in Fig. 8.26 b; that representation treats all three constituents equally. A point P_1 on the side AB of the triangle corresponds to a binary solution of constituents A and B with mol fractions $X_A = P_1B$ and $X_B = P_1A$. And an interior point P of the triangle corresponds to a ternary mixture with mol fractions $X_A = PS$, $X_B = PR$, and $X_C = PP_1$. It is easily confirmed from the geometry of the triangle that the sum of the mol fractions equals 1, as it must be. Thus a straight line through one corner represents states in which two of the constituents are in a fixed proportion to each other.

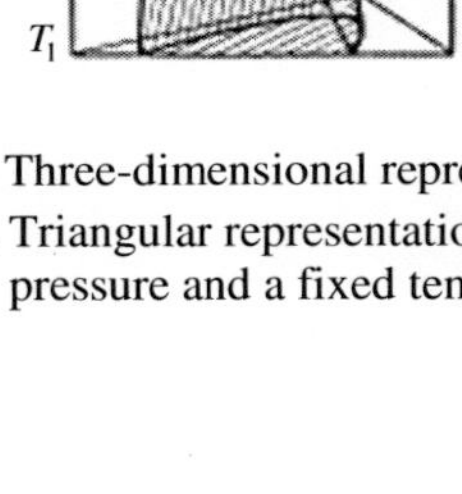

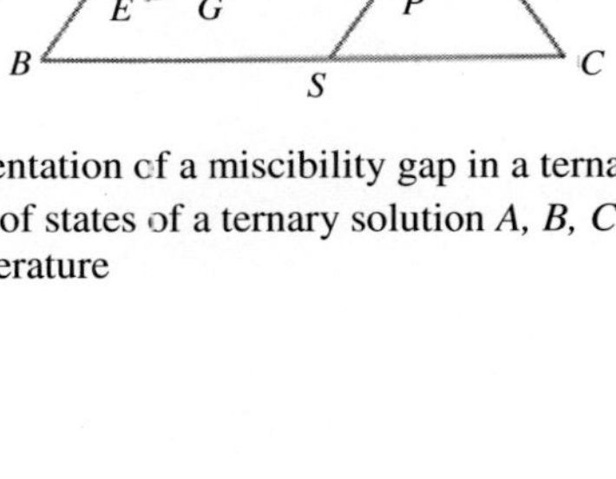

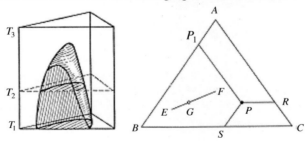

Fig. 8.26 a. Three-dimensional representation of a miscibility gap in a ternary solution
b. Triangular representation of states of a ternary solution *A, B, C* at a fixed pressure and a fixed temperature

When two ternary mixtures of states E and F in Fig. 8.26 b are mixed, the eventual mixed state is found by the "center of gravity rule:" If G represents that mixed state, the proportions of mixtures E and F are given by the ratios $\frac{FG}{EF}$ and $\frac{EG}{EF}$ respectively.

8.10.2 *Miscibility gaps in ternary solutions*

It is clear from the higher variability of ternary solutions that their phase diagrams may exhibit a greater variety than those of binary solutions. Therefore we cannot describe all aspects of the behavior of such solutions in this book. We ignore the vapor-liquid phenomena altogether and concentrate on the occurrence of miscibility gaps in solutions.

Let the constituents B and C in a binary solution have a miscibility gap in the range of mol fractions $BD < X_C < BE$, *cf.* Fig. 8.27. As we have explained before – in Paragraph 8.8.4 – this means that the solution separates into two layers with mol fractions BD and BE which are in phase equilibrium.

The addition of the third constituent A to that solution may decrease the size of the miscibility gap. And usually the addition of A affects the mol fractions of the C-rich solution and the B-rich solution differently so that phase equilibrium lies in the states D' and E'. Those points have to be found experimentally in a laborious process. They vary with X_A and trace out the binodal curve, so-called because it is constructed from data in pairs. The miscibility gap with two layers of fluids is embraced by the binodal curve.

On the straight line between the equilibrium states D' and E' we find the overall phase fractions – averaged over both layers – of the solution. Such a line is called a *tie line,* or sometimes a *conode.* A point F' with a given X_A on the tie line consists of states D' and E' in the proportion $\frac{F'E'}{F'D'}$. In Fig. 8.27 the distance of the equilibrium points D' and E', or D'' and E'' becomes shorter and eventually the two branches of the binodal curve come together in the *plait point* P.

As mentioned before, *cf.* Paragraph 8.9.2, the miscibility gap is likely to increase with decreasing temperature. Thus in Fig. 8.27 b we see a small gap for 45°C and a large one for 25°C. The second one extends through the whole triangle so that the binary solution C_6H_{12}/ C_6H_7N – which had unrestricted miscibility at 45°C – exhibits a miscibility gap at 25°C.

At high temperature it may happen that the ternary solution develops full miscibility of all of its binary sub-solutions. The temperature where the miscibility gap disappears is called the *binary consolute temperature*, *cf.* Fig.8.28 a. It may also happen, that at some higher temperature none of the three binary sub-solutions has a miscibility gap but the ternary system has one, island-like, within the triangle, *cf.* Fig. 8.28 b. For increasing temperature the island shrinks to zero at the *final consolute temperature.*

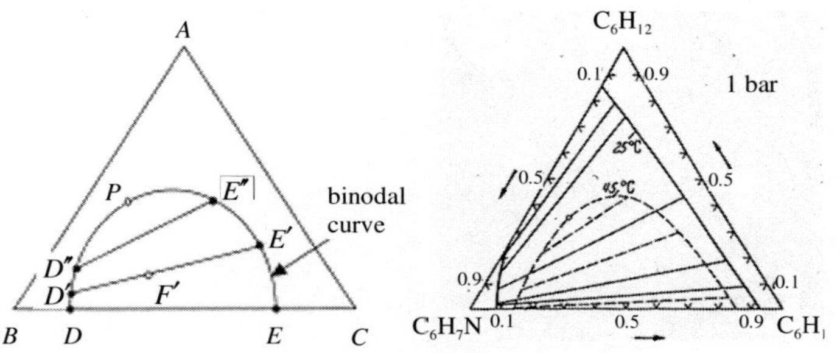

Fig.8.27 a. Binodal curve, tie lines and plait point
b. Phase diagram for aniline C_6H_7N, hexane C_6H_{14}, and methylcyclopentane C_6H_{12}

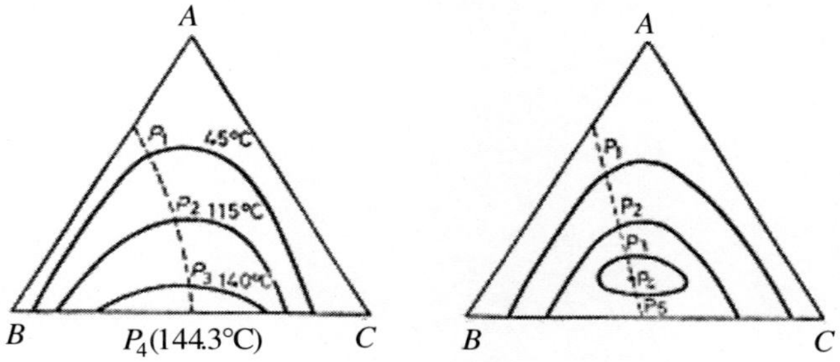

Fig.8.28 a. Vanishing miscibility gap for diphenylhexane (A), furfural (B), and docosane (C)
b. Island-like miscibility gap for acetone (A), water (B), and phenol (C)*

Among a near-infinite variety of possibilities is the interesting one that has miscibility gaps in all three binary subsystems so that there are binodal curves on all three sides of the triangle, *cf.* Fig. 8.29 a. And at decreasing temperatures the miscibility gaps grow and eventually merge to give a phase diagram of the form shown in Fig. 8.29 b with an interior triangle DEF somewhere inside the triangle ABC. In the interior triangle we have three phases – and three layers – so that by Gibbs´s phase rule there are two degrees of freedom or, in actual fact, no degree of freedom, since T and p are fixed. The interior triangle is a new type of miscibility gap in which the three layers have the compositions D, and E, and F, so that the layers vary only by their thickness.

* Comparison with Fig. 8.26 a shows that that qualitative figure represents a miscibility gap which – at high temperature – develops into an "island."

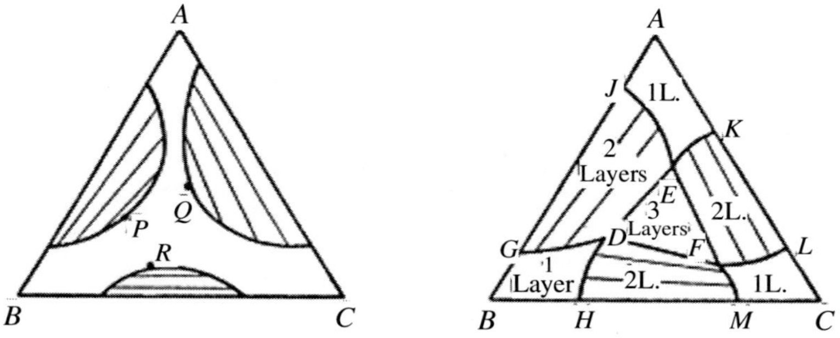

Fig. 8.29 a. Three qualitative miscibility gaps for a ternary solution like nitroethane, glycol, decyl alcohol
b. Phase diagram for the same solution at lower temperature.

9 Chemically reacting mixtures

9.1 Stoichiometry and law of mass action

9.1.1 *Stoichiometry*

We denote the mass of a constituent α in a homogeneous mixture by m_α. The corresponding mass balance reads

$$\frac{\mathrm{d}m_\alpha}{\mathrm{d}t} = \tau_\alpha V \ , \ \alpha = 1, 2, \cdots, \nu \ , \tag{9.1}$$

where τ_α is the density of mass production of constituent α. It is clear that the sum of all the ν values τ_α must be equal to zero, because the total mass is conserved in a chemical reaction. However, this is not the only condition which the production densities τ_α must satisfy. Further conditions follow from the fact that the number of atoms – and their masses – must be conserved in the reaction; the atoms present before the reaction are still present after the reaction, although they may have changed their arrangement in molecules. The investigations of the conditions for this kind of arrangement is the subject of stoichiometry (Greek: stoichos "order" and metron "measure").

The reaction of hydrogen and oxygen when they form water is written in the well-known stoichiometric equation

$$2\,H_2 + O_2 \rightarrow 2\,H_2O \ . \tag{9.2}$$

The indices indicate that hydrogen and oxygen start out in their molecular form, which consists of two atoms each, and H_2O means that the water molecule consists of two H-atoms and one O atom. The coefficients 2, 1 and, again, 2 in (9.2) – in front of the symbols H_2, O_2, and H_2O for the constituents – are called *stoichiometric coefficients*, and they are denoted by γ_{H_2}, γ_{O_2}, and γ_{H_2O}. They indicate how many molecules H_2, O_2, and H_2O participate in the reaction. The constituents on the left hand side of the arrow – here H_2 and O_2 – are called *reactants* and their stoichiometric coefficients are counted as negative so that we have $\gamma_{H_2} = -2$, $\gamma_{O_2} = -1$. The constituents behind the arrow, – here only one –, are called *products* and their stoichiometric coefficients are counted as positive; *i.e.* here $\gamma_{H_2O} = 2$. With this convention the mass conservation in one reaction may be written in the form

$$\sum_{\alpha=1}^{\nu} \gamma_\alpha m_\alpha = 0 \ . \tag{9.3}$$

m_α is the mass of the molecules of constituent α. Among the molecules each reaction always occurs in both directions such that for instance – by (9.2) – water molecules are continuously formed from hydrogen and oxygen molecules, and other water molecules are decomposed; this is so *even in equilibrium*. Per unit time and per volume element the balance of formation and decomposition is described by the *reaction rate density* λ. In equilibrium this quantity is equal to zero; we take λ as positive when the reaction occurs in the direction of the arrow, otherwise λ is negative.

The reaction rate density determines the mass production densities τ_α. We have

$$\tau_\alpha = \gamma_\alpha m_\alpha \lambda \tag{9.4}$$

so that among all τ_α's only one is independent.

In reality the reaction (9.2) occurs very seldom, if at all, because three molecules of the proper type must meet at essentially the same location before a water molecule can be formed. More realistically water forms as the result of a complicated sequence of four reactions which require the meeting of only two molecules. Such a sequence could read for example

$$\begin{array}{llllll}
(1) & -H_2+2H & & & & =0\\
(2) & & -O_2+2O & & & =0\\
(3) & -H & & -O+OH & & =0\\
(4) & -H & & & -OH+H_2O & =0\ .
\end{array}$$

Of course, here we have four reaction rate densities λ^a, $a=1,\cdots,4$, one for each reaction. Chemists have to grapple with this and the determination of λ^a as functions of T, p, and m_α ($\alpha=1,2,\cdots,\nu$) is the difficult subject of *chemical kinetics*. Our gross reaction (9.2) is a linear combination of the partial reactions (1) through (4) in the form $2\times(1)+(2)+2\times(3)+2\times(4)$. For the evaluation of the balance equations of mass, energy, and entropy in a reaction we may often limit the attention to the gross reaction, and that is what we shall do in this book.

From (9.1) and (9.4) we obtain

$$\frac{\mathrm{d}m_\alpha}{\mathrm{d}t}=\gamma_\alpha m_\alpha \lambda V\ ,\ \text{hence}\ \ m_\alpha(t)=m_\alpha(t_0)+\gamma_\alpha m_\alpha \underbrace{\int_{t_0}^{t}\lambda V \mathrm{d}t}_{\boldsymbol{R}(t)-\boldsymbol{R}(t_0)}\ , \tag{9.5}$$

where the brace defines the *extent of reaction*, called $\boldsymbol{R}$, a dimensionless quantity. We thus have:

$$m_\alpha(t)=m_\alpha(t_0)+\gamma_\alpha m_\alpha\left[\boldsymbol{R}(t)-\boldsymbol{R}(t_0)\right]. \tag{9.6}$$

Note that one single $\boldsymbol{R}$ determines all m_α's.

By (9.5) we obtain with $m_\alpha = N_\alpha m_\alpha$

$$\frac{\mathrm{d}}{\mathrm{d}t}\left(\frac{N_\alpha}{\gamma_\alpha}\right)=\lambda V$$

so that the rate of change of N_α/γ_α is independent of α. Therefore we may write

$$\frac{\mathrm{d}}{\mathrm{d}t}\left(\frac{N_\alpha}{\gamma_\alpha}\right)=\frac{\mathrm{d}}{\mathrm{d}t}\left(\frac{N_\beta}{\gamma_\beta}\right)\ ,\ \text{hence}\ \ \frac{N_\alpha}{\gamma_\alpha}-\frac{N_\beta}{\gamma_\beta}=\frac{N_\alpha(t_0)}{\gamma_\alpha}-\frac{N_\beta(t_0)}{\gamma_\beta}\ . \tag{9.7}$$

These are $\nu - 1$ independent equations for the ν particle numbers N_α.

For the water reaction (9.2) – with three constituents – the equations (9.7) represent two equations, *e.g.*

$$\frac{N_{H_2}}{2} - N_{O_2} = \frac{N_{H_2}(t_0)}{2} - N_{O_2}(t_0) \text{ and } N_{H_2} + N_{H_2O} = N_{H_2}(t_0) + N_{H_2O}(t_0). \quad (9.8)$$

9.1.2 *Application of stoichiometry. Respiratory quotient* RQ

The combustion of $1\,\text{mol}$ of sugar – more exactly of glucose $C_6H_{12}O_6$ – according to the stoichiometric formula

$$C_6H_{12}O_6 + 6O_2 \rightarrow 6CO_2 + 6H_2O$$

requires the consumption of $6\,\text{mol}$ of oxygen, and $6\,\text{mol}$ CO_2 are produced. If this happens in an animal or in a human being, oxygen and carbon dioxide are transported by the air they breathe. The amount of inhaled – and consumed – O_2 and exhaled CO_2 is therefore equal, and we say that the respiratory quotient is equal to one. Since this is much the same for all carbohydrates we write

$$\text{RQ}_{\text{Carbohydrates}} = \frac{\text{amount of exhaled } CO_2}{\text{amount of consumed } O_2} = 1.$$

Indeed this is confirmed in experiments for animals that are primarily fed with carbohydrates.

Fat molecules have fewer O-atoms relative to their C- and H-atoms. As a representative for a fat molecule we choose triolein $C_{54}H_{104}O_6$ which reacts in an organism according to the stoichiometric formula

$$C_{54}H_{104}O_6 + 77O_2 \rightarrow 54CO_2 + 52H_2O.$$

For 77 consumed O_2-molecules we have only 54 exhaled CO_2-molecules, and therefore the respiratory quotient is equal to

$$\text{RQ}_{\text{Fat}} = \frac{54}{77} = 0.7.$$

The third essential component of food are the proteins which vary too much in chemical composition to be characterized by a simple formula. However, the oxygen content lies somewhere between fats and carbohydrates and a typical respiratory quotient is $\text{RQ}_{\text{Protein}} \approx 0.8$.

This value is also the RQ of a man with a well-balanced diet. If one eats excessively many carbohydrates or fats, the RQ shifts to higher or lower values, respectively.

9.1.3 *Law of mass action*

From the stability considerations of Paragraph 4.2.7 we conclude that for constant and homogeneous fields T and p the Gibbs free energy tends to a minimum. Therefore, in order to determine the equilibrium for a mixture of chemically reacting constituent, we minimize

$$G=\sum_{\delta=1}^{\nu}\mu_\delta m_\delta .$$

By (9.6) and for fixed T and p this quantity depends only on the extent of reaction and we calculate

$$\frac{\partial G}{\partial \mathcal{R}}=0 \text{ hence } \sum_{\gamma=1}^{\nu}\sum_{\delta=1}^{\nu}\frac{\partial \mu_\delta}{\partial m_\gamma}m_\delta\frac{\mathrm{d}m_\gamma}{\mathrm{d}\mathcal{R}}+\sum_{\delta=1}^{\nu}\mu_\delta\frac{\mathrm{d}m_\delta}{\mathrm{d}\mathcal{R}}=0 .$$

The first term vanishes because of the Gibbs-Duhem relation (8.7) and the second one is simplified by $\mathrm{d}m_\delta/\mathrm{d}\mathcal{R}=\gamma_\delta \mathfrak{m}_\delta$ so that we obtain with $\mathfrak{m}_\delta=\frac{M_\delta}{A}$

$$\sum_{\delta=1}^{\nu}\mu_\delta\big|_{\mathrm{E}}\gamma_\delta M_\delta=0 \text{ or } \sum_{\delta=1}^{\nu}\tilde{\mu}_\delta\big|_{\mathrm{E}}\gamma_\delta=0 , \tag{9.9}$$

where $\tilde{\mu}_\delta$ is the molar chemical potential, *cf.* Paragraph 8.6.1. This is the *law of mass action*. It furnishes one relation which in equilibrium – index E – determines the extent of reaction as a function of T and p, and therefore the masses of the reacting constituents. If a mass of a reacting constituent is changed, the equilibrium shifts, hence the name *mass-action.*

An explicit evaluation of the law requires explicit knowledge of the chemical potentials.

9.1.4 *Law of mass action for ideal mixtures and mixtures of ideal gases*

In an ideal mixture the chemical potentials have the form given by (8.25) so that the law of mass action may be written as

$$\sum_{\delta=1}^{\nu}\gamma_\delta\ln X_\delta=-\sum_{\delta=1}^{\nu}\gamma_\delta\frac{g_\delta(T,p)}{\frac{R}{M_\delta}T}$$

or, after an easy reformulation

$$\prod_{\delta=1}^{\nu}X_\delta^{\gamma_\delta}=\exp\left[-\sum_{\delta=1}^{\nu}\gamma_\delta\frac{g_\delta(T,p)}{\frac{R}{M_\delta}T}\right]. \tag{9.10}$$

This is the law of mass action for ideal mixtures. For a given pair (T,p) it provides a relation between the mol fractions of the constituents. The (T,p)-dependence is through $g_\delta(T,p)$, the specific Gibbs free energy of the pure substance δ.

A still simpler – and instructive – form of the law of mass action results when reactants and products are all ideal gases. In this case we have

$$X_\delta=\frac{n_\delta}{\sum_{\beta=1}^{\nu}n_\beta} \text{ or, with } p_\delta=n_\delta kT: \; X_\delta=\frac{p_\delta}{p} ,$$

and the specific Gibbs free energies read, by (8.17)

$$g_\delta(T,p)=\left(z_\delta\frac{R}{M_\delta}T+\alpha_\delta\right)-T\left[(z_\delta+1)\frac{R}{M_\delta}\ln T-\frac{R}{M_\delta}\ln p+\beta_\delta\right]+\frac{R}{M_\delta}T\,.$$

Insertion into (9.10) shows that the pressure p on both sides cancels and we obtain

$$\prod_{\delta=1}^{\nu}p_\delta^{\gamma_\delta}=T^{\sum_{\delta=1}^{\nu}\gamma_\delta(z_\delta+1)}\exp\left[-\sum_{\delta=1}^{\nu}\gamma_\delta(z_\delta+1)\right]\exp\sum_{\delta=1}^{\nu}\left[-\gamma_\delta\frac{\alpha_\delta-T\beta_\delta}{\frac{R}{M_\delta}T}\right].\qquad(9.11)$$

The right hand side is a function of T only and therefore we may abbreviate (9.11) by writing

$$\prod_{\delta=1}^{\nu}p_\delta^{\gamma_\delta}=K_p(T).\qquad(9.12)$$

The chemical "constant" $K_p(T)$ has been tabulated for many ideal gas mixtures. It depends on the additive constants α and β of the internal energy and entropy of all constituents. We shall discuss later – in Sect. 9.2 – how these constants can be determined so that $K_p(T)$ may be known.

The fact that $K_p(T)$ depends on T only does not mean that the pressure cannot influence the concentrations of ideal gases in equilibrium. Generally such a pressure dependence does exist, because the left hand side of (9.12) contains the pressure implicitly; after all, the partial pressures must add up to the total pressure:

$$\sum_{\beta=1}^{\nu}p_\beta=p\,.$$

9.1.5 *On the history of the law of mass action*

The above derivation of the law of mass action is due to Gibbs. However, even before Gibbs the law was known in its essential features from intuitive and suggestive arguments. The original discoverers were Cato Maximilian GULDBERG (1836-1902) and Peter WAAGE (1833-1900), two Norwegian professors at the University of Christiania, nowadays Oslo. Their work was published in 1863 in Norwegian and therefore it escaped the attention of most chemists. Nor did a French translation find much interest and it was only the German translation in 1879 which was appreciated by VAN'T HOFF who worked on similar problems.

The argument by GULDBERG and WAAGE runs somewhat as follows: We assume that a reaction has the stoichiometric coefficients γ_α, which determine how many molecules of constituent α must meet for the reaction to proceed. The probability for a molecule α to be at a point ought to be proportional to n_α and the probability for γ_α molecules to meet at that point is then determined by $n_\alpha^{\gamma_\alpha}$. The reaction can only occur when the molecules of all reactants come together at the point and the probability for this is given by

$$A\prod_{\alpha^-}n_\alpha^{|\gamma_\alpha|}\,,$$

where the product must be taken over all constituents with negative stoichiometric coefficients and where A is some factor of proportionality. Analogously the probability for the reverse reaction is equal to

$$B\prod_{\alpha^+} n_\alpha^{\gamma_\alpha} ,$$

where we must extend the multiplication over all α's with positive stoichiometric coefficients.

In equilibrium both probabilities must be equal so that we have

$$A\prod_{\alpha^-} n_\alpha^{|\gamma_\alpha|} = B\prod_{\alpha^+} n_\alpha^{\gamma_\alpha} \quad \text{or} \quad \prod_{\alpha=1}^{\nu} n_\alpha^{\gamma_\alpha} = Q \text{, if we put } Q = A/B .$$

Comparison with (9.10) shows agreement, because n_α is proportional to X_α. However, this simple argument did not suffice for the determination of the T- and p-dependence of Q.

9.1.6 *Examples for the law of mass action for ideal gases*

- *From hydrogen and iodine to hydrogen iodide and vice versa*

We consider the reaction

$$H_2 + I_2 \to 2\,HI . \tag{9.13}$$

Iodine is a gray solid at room temperature. Gentle heating produces a violet vapor and above $460\,K$ it exists only in vapor form, and we consider it as an ideal gas. Accordingly we write the law of mass action (9.12) appropriate to this reaction in the form

$$\frac{p_{HI}^2}{p_{H_2} p_{I_2}} = K_p(T). \tag{9.14}$$

We start the reaction with $1\,mol$ of H_2 and I_2 each and no HI, so that the conservation laws (9.7) of atomic numbers read

$$N_{H_2} - N_{I_2} = 0 \quad \text{and} \quad N_{H_2} + \frac{N_{HI}}{2} = A\,mol ,$$

where A is the Avogadro number, the number of particles in $1\,mol$. With the ideal gas law $p_\alpha V = N_\alpha kT$ we thus have (recall that $R = Ak$)

$$p_{H_2} = p_{I_2} \quad \text{and} \quad p_{H_2} + \frac{1}{2} p_{HI} = \frac{RT}{V} mol . \tag{9.15}$$

And, of course, the sum of all partial pressures is the total pressure p

$$p_{H_2} + p_{I_2} + p_{HI} = p . \tag{9.16}$$

Given p and T, with (9.14) through (9.16) we have four relations for the partial pressures and for V. They are easily solved and we obtain

$$p_{HI} = \frac{\sqrt{K_p}}{2+\sqrt{K_p}} p , \quad p_{H_2} = p_{I_2} = \frac{p}{2+\sqrt{K_p}} , \quad V = 2\frac{RT}{p} mol .$$

The values of K_p may be read off from Table 9.1. For $T = 600\,\text{K}$ and $p = 1\,\text{bar}$ we thus obtain

$$p_{\text{HI}} = 0.81\,\text{bar}\ ,\quad p_{\text{H}_2} = p_{\text{I}_2} = 0.095\,\text{bar}\ ,\quad V = 0.1\,\text{m}^3 .$$

For higher values of T, *i.e.* smaller values of K_p, the partial pressure of hydrogen iodide decreases. The volume does not change during the reaction, since the number of molecules is unchanged.

- *Decomposition of carbon dioxide into carbon monoxide and oxygen*

If carbon dioxide is heated, it decomposes into carbon monoxide and oxygen. We shall investigate this decomposition in its dependence on pressure and temperature. The stoichiometric equation for the process may be written in the form

$$CO_2 \rightarrow CO + \frac{1}{2} O_2 . \qquad (9.17)$$

Accordingly the law of mass action (9.12) and the number conservation laws read

$$\frac{p_{CO}\sqrt{p_{O_2}}}{p_{CO_2}} = K_p(T),\ N_{CO} - 2N_{O_2} = 0,\ N_{CO} + N_{CO_2} = A\,\text{mol}, \qquad (9.18)$$

if we start with $1\,\text{mol}$ of CO_2 and no CO, nor O_2.

Table 9.1 Chemical constants $K_p(T)$

T [K]	$K_p = \frac{p_{\text{HI}}^2}{p_{\text{H}_2} p_{\text{I}_2}}$
600	69.4
800	37.2
1000	26.6

T [K]	$K_p = \frac{p_{CO}\sqrt{p_{O_2}}}{p_{CO_2}}$ $[\sqrt{\text{bar}}]$
298	$9.04 \cdot 10^{-45}$
400	$3.89 \cdot 10^{-33}$
600	$8.61 \cdot 10^{-21}$
800	$1.27 \cdot 10^{-14}$
1000	$6.32 \cdot 10^{-11}$
1200	$1.81 \cdot 10^{-8}$
1500	$5.08 \cdot 10^{-6}$
2000	$1.37 \cdot 10^{-4}$
2500	$3.80 \cdot 10^{-2}$
3000	$3.12 \cdot 10^{-1}$
3500	1.59

Values for $K_p(T)$ for the reaction are given in Table 9.1. By use of $p_{CO_2} + p_{CO} + p_{O_2} = p$ we may eliminate p_{CO_2} and p_{O_2} from $(9.18)_1$ and obtain a cubic equation for p_{CO}, *viz.*

$$\sqrt{\frac{p}{2K_p^2}}\left(\frac{p_{CO}}{p}\right)^{\frac{3}{2}}+\frac{3}{2}\frac{p_{CO}}{p}-1=0. \tag{9.19}$$

Fig. 9.1 shows the solution $1-\frac{p_{CO_2}}{p}$ as a function of T and for two values of p.

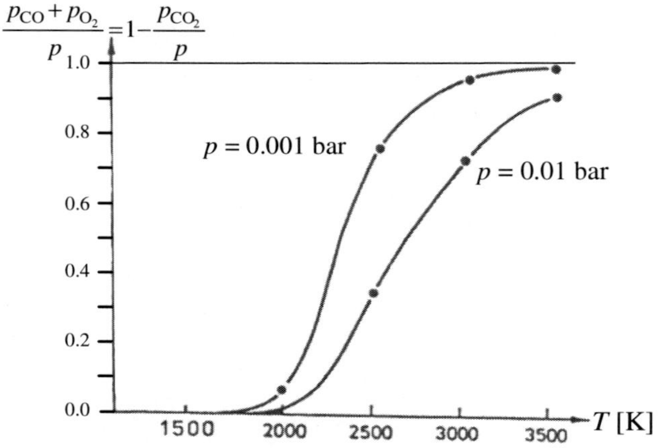

Fig. 9.1 Decomposition of CO_2 into CO and O_2

Inspection of the figure shows that up to 1500 K there is no appreciable decomposition of CO_2. For very large values of T, however, nearly all of the CO_2 is decomposed and the partial pressure p_{CO_2} tends to zero; it reaches this value earlier for a smaller pressure. Thus a decrease of pressure favors the decomposition.

9.1.7 *Equilibrium in stoichiometric mixtures of ideal gases*

The previous examples show that it is a little awkward to exploit the conservation laws (9.7) of atomic numbers for each reaction separately. Therefore we proceed to derive a form of the law of mass action in which the particle conservation laws have been taken care of in advance. We limit the attention to *stoichiometric mixtures* which contain the "right" number of reactants such that all reactants may in fact be fully converted into products, without leaving any residues of reactants. Also we introduce a synthetic description by denoting the reactant constituents by capital letters $A, B, C, \cdots$ and the produced constituents by $E, F, G, \cdots$. The corresponding stoichiometric coefficients are denoted by $a, b, c, \cdots$ and $e, f, g, \cdots$, respectively. The synthetic – single – reaction thus reads

$$a\mathrm{A}+b\mathrm{B}+c\mathrm{C}+\cdots\rightarrow e\mathrm{E}+f\mathrm{F}+g\mathrm{G}+\cdots. \tag{9.20}$$

In a stoichiometric mixture the initial values of the reactants N_A^0, N_B^0, N_C^0, *etc.* have the same ratios as the corresponding stoichiometric coefficients, while the products are absent at the beginning

$$\frac{N_{\rm A}^0}{a}=\frac{N_{\rm B}^0}{b}=\frac{N_{\rm C}^0}{c}=\cdots \quad \text{and} \quad N_{\rm E}^0=N_{\rm F}^0=N_{\rm G}^0=\cdots=0\,.$$

In this case the conservation laws (9.7) imply

$$\frac{N_{\rm A}}{a}=\frac{N_{\rm B}}{b}=\frac{N_{\rm C}}{c}=\cdots \quad \text{and} \quad \frac{N_{\rm E}}{e}=\frac{N_{\rm F}}{f}=\frac{N_{\rm G}}{g}=\cdots\,. \tag{9.21}$$

We choose $N_{\rm A}^0=|a|\,A\,\text{mol}$, where A is the Avogadro number. Thus we start with $|a|\,\text{mol}$ of constituent A, $|b|\,\text{mol}$ of constituent B, $|c|\,\text{mol}$ of constituent C, *etc.* Thus after $A\,\text{mol}$ molecular reactions we arrive at $N_{\rm A}=N_{\rm B}=N_{\rm C}=\cdots=0$.

The relative extent of reaction is defined by

$$r=\frac{R}{A\,\text{mol}}\,, \text{ so that } 0\le r\le 1\,. \tag{9.22}$$

Thus r is a good measure of how far the reaction has proceeded from $\text{A},\text{B},\text{C},\cdots$ at time t_0 to $\text{E},\text{F},\text{G},\cdots$ at time t. There is therefore an easy relation between the partial pressures $\frac{p_{\rm A}}{p},\frac{p_{\rm B}}{p},\frac{p_{\rm C}}{p},\cdots$ or $\frac{p_{\rm E}}{p},\frac{p_{\rm F}}{p},\frac{p_{\rm G}}{p},\cdots$ and r. Indeed, since our constituents are ideal gases, we may write

$$\frac{p_{\rm A}}{p}=\frac{N_{\rm A}}{N_{\rm A}+N_{\rm B}+\cdots+N_{\rm E}+N_{\rm F}+\cdots}\,, \qquad \frac{p_{\rm E}}{p}=\frac{N_{\rm E}}{N_{\rm A}+N_{\rm B}+\cdots+N_{\rm E}+N_{\rm F}+\cdots}$$

$$=\frac{a\frac{N_{\rm A}}{a}}{\frac{N_{\rm A}}{a}(a+b+\cdots)+\frac{N_{\rm E}}{e}(e+f+\cdots)}\,, \qquad =\frac{e\frac{N_{\rm E}}{e}}{\frac{N_{\rm A}}{a}(a+b+\cdots)+\frac{N_{\rm E}}{e}(e+f+\cdots)}\,.$$

Also, by number conservation (9.6) we have

$$\frac{N_{\rm A}}{|a|}=\frac{N_{\rm A}^0}{|a|}-r\,A\,\text{mol} \quad \text{and} \quad \frac{N_{\rm E}}{e}=r\,A\,\text{mol}\,.$$

Hence follows with $\frac{N_{\rm A}^0}{|a|}=A\,\text{mol}$

$$\frac{p_{\rm A}}{p}=|a|\frac{1-r}{(1-r)m+rn}\,, \qquad \frac{p_{\rm E}}{p}=e\frac{r}{(1-r)m+rn}\,, \tag{9.23}$$

where m and n are the absolute values of the sums over all negative and positive stoichiometric coefficients, respectively. Analogous expressions are derived for the other constituents $\text{B},\text{C},\cdots$ and $\text{F},\text{G},\cdots$.

The relations (9.23) and their analoga for $\text{B},\text{C},\cdots$ and $\text{F},\text{G},\cdots$ are always true during the reaction. They represent the partial pressures of the constituents and we recognize that they only depend on the relative extent of reaction. In *equilibrium,* however, the law of mass action (9.12) must be satisfied in addition, and we obtain by (9.23)

$$K_p(T)=\frac{e^e f^f g^g \cdots}{|a|^{|a|}|b|^{|b|}|c|^{|c|}\cdots}\frac{r^n}{(1-r)^m}\left(\frac{p}{m+(n-m)r}\right)^{n-m}\,. \tag{9.24}$$

This is a relation between p, T, and r; it is explicit in r and p, so that we may use it to find out how the pressure affects the equilibrium value of r.

As examples for (9.24) we consider the ammonia synthesis and the decomposition of CO_2

$$N_2 + 3H_2 \rightarrow 2NH_3 \qquad \Bigg\| \qquad 2CO_2 \rightarrow 2CO + O_2$$

and obtain

$$K_p(T) = \frac{2^2}{3^3}\frac{r^2}{(1-r)^4}\frac{(4-2r)^2}{p^2} \qquad \Bigg\| \qquad K_p(T) = \frac{r^3}{(1-r)^2}\frac{p}{2+r}. \tag{9.25}$$

For $r \ll 1$ we conclude

$$K_p(T) \approx \frac{64}{27}\frac{r^2}{p^2} \qquad \Bigg\| \qquad K_p(T) \approx \frac{1}{2}p\,r^3 \tag{9.26}$$

so that for high pressure we obtain proportionally more ammonia, while the decomposition of CO_2 is favored by a small pressure; in the latter case r is proportional to $1/\sqrt[3]{p}$.

Another example is the dissociation of hydrogen molecules according to the stoichiometric condition $H_2 \rightarrow 2H$. By (9.24) we have

$$K_p(T) = 4\frac{r^2}{1-r^2}p \quad \text{hence for } r \ll 1\text{:} \quad K_p(T) \approx 4r^2 p. \tag{9.27}$$

This means that initially, when the decomposition begins, the relative extent is proportional to $1/\sqrt{p}$: More atomic hydrogen appears for lower pressure.

9.2 Heats of reaction, entropies of reaction, and absolute values of entropies

9.2.1 *The additive constants in u and s*

When the masses of the constituents disappear in a reaction, the energy and entropy of the constituents disappear along with them. Also, the reaction products emerge complete with mass, energy, and entropy and these include the additive constants in energy and entropy. So far in this book these constants were quite unimportant, but now – when chemical reactions occur – their values do matter; they play an important role. And so far we could be negligent about the constants – calling them simply α and β – but now it is appropriate to look at them more closely, and to relate them to the absolute values of energy and entropy in a reference state.

We recall that u and s – or also h and s – were determined by integration from

$$
\begin{aligned}
\mathrm{d}h &= c_p(T,p)\mathrm{d}T + \left[\upsilon(T,p) - T\left(\frac{\partial \upsilon}{\partial T}\right)_p (T,p) \right] \mathrm{d}p \\
\mathrm{d}s &= \frac{c_p(T,p)}{T}\mathrm{d}T - \left(\frac{\partial \upsilon}{\partial T}\right)_p (T,p)\mathrm{d}p\,,
\end{aligned}
\tag{9.28}
$$

where $\mathrm{d}s$ has been obtained from the Gibbs equation $T\mathrm{d}s = \mathrm{d}u + p\,\mathrm{d}\upsilon$. Integration between a reference state $(T_\mathrm{R}, p_\mathrm{R})$ and some state (T,p) gives

$$
\begin{aligned}
h(T,p) - h(T_\mathrm{R}, p_\mathrm{R}) &= \int_{T_\mathrm{R}}^{T} c_p(z,p)\mathrm{d}z + \int_{p_\mathrm{R}}^{p} \left[\upsilon(T_\mathrm{R}, z) - T_\mathrm{R}\left(\frac{\partial \upsilon}{\partial T}\right)_p (T_\mathrm{R}, z) \right] \mathrm{d}z + \sum_i r(T_i) \\
s(T,p) - s(T_\mathrm{R}, p_\mathrm{R}) &= \int_{T_\mathrm{R}}^{T} \frac{c_p(z,p)}{T}\mathrm{d}z - \int_{p_R}^{p} \left(\frac{\partial \upsilon}{\partial T}\right)_p (T_\mathrm{R}, z)\mathrm{d}z + .\sum_i \frac{r(T_i)}{T_i}
\end{aligned}
\tag{9.29}
$$

The sum over i occurs, if we pass through a phase transition along the path from $(T_\mathrm{R}, p_\mathrm{R})$ to (T,p). If this is the case at a temperature T_i – and the pressure p_i – the enthalpy is discontinuous and has a jump $r(T_i)$, where $r(T_i)$ is the latent heat of the phase transition; the entropy has a jump $r(T_i)/T_i$.

Chemists usually choose the reference state as

$$p_\mathrm{R} = 1\,\mathrm{atm} \quad \text{and} \quad T_\mathrm{R} = 298\,\mathrm{K}\,. \tag{9.30}$$

The right hand side of (9.29) may obviously be determined from (p,T,υ)-measurements and measurements of specific and latent heats. For ideal gases, where the thermal and caloric equations of state are known, we obtain from (9.29) with $h_R = h(T_\mathrm{R}, p_\mathrm{R})$ and $s_R = s(T_\mathrm{R}, p_\mathrm{R})$

$$
\begin{aligned}
h(T,p) - h_\mathrm{R} &= (z+1)\tfrac{R}{M}(T - T_\mathrm{R}) \quad \text{and} \\
s(T,p) - s_\mathrm{R} &= (z+1)\tfrac{R}{M}\ln\tfrac{T}{T_\mathrm{R}} - \tfrac{R}{M}\ln\tfrac{p}{p_\mathrm{R}}
\end{aligned}
\tag{9.31}
$$

provided that on the whole path between $(T_\mathrm{R}, p_\mathrm{R})$ and (T,p) the ideal gas laws are valid. This, however, is seldom the case.

Consider for example the case of water: Water is liquid in the reference state (9.30), and it is only in the vapor state where we may approximately consider it as an ideal gas. Thus when we wish to calculate $h(T,p)$ and $s(T,p)$ in the vapor state, we must take the changes of enthalpy and entropy during evaporation into account. Fig. 9.2 shows a possible integration path schematically and the corresponding values of $h(T,p)$ and $s(T,p)$. We have assumed that liquid water is incompressible so that $(\partial \upsilon/\partial T)_p = 0$.

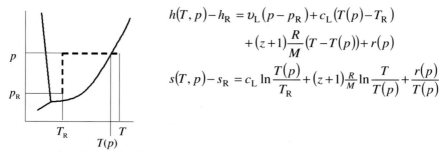

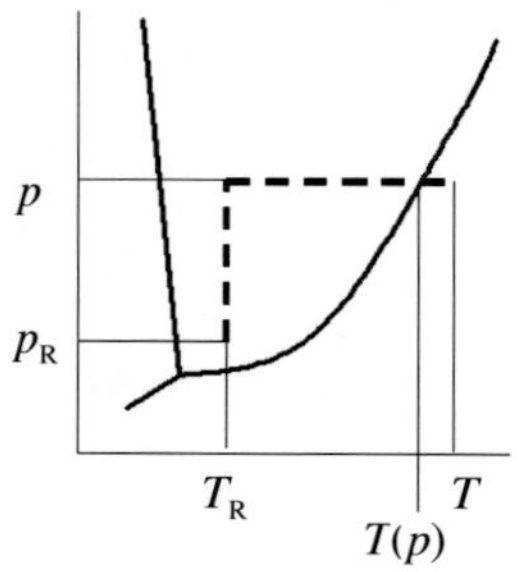

$$h(T,p)-h_R = v_L(p-p_R)+c_L(T(p)-T_R) + (z+1)\frac{R}{M}(T-T(p))+r(p)$$

$$s(T,p)-s_R = c_L \ln\frac{T(p)}{T_R}+(z+1)\tfrac{R}{M}\ln\frac{T}{T(p)}+\frac{r(p)}{T(p)}$$

Fig. 9.2 $h(T,p)$ and $s(T,p)$ for water vapor, if the liquid is considered as incompressible and the vapor as an ideal gas (c_L and v_L are the specific heat and specific volume of the liquid, $T(p)$ is the temperature of evaporation and $r(p)$ the latent heat t pressure p)

9.2.2 *Heats of reaction*

As is well known, chemical reactions are often accompanied by a release of heat, sometimes dramatically so. However, there are also reactions that require a supply of heat. The sign of the heating and its value may be calculated from the First Law in the form

$$\dot{Q}=\frac{dU}{dt}+p\frac{dV}{dt} \quad \text{or} \quad \dot{Q}=\frac{dH}{dt}-V\frac{dp}{dt}.$$

If the reaction between the beginning at time t_0 and the end at time t is isochoric or isobaric, the heats of reaction are accordingly

$$Q^v = U(t)-U(t_0) \text{ or } Q^p = H(t)-H(t_0). \tag{9.32}$$

We shall be interested mostly in the isobaric case. If Q^p is positive, we speak of an endothermic reaction, otherwise the reaction is exothermic.

In order to relate the heat of reaction Q^p to the enthalpies of the pure constituents we write

$$H(t)=\sum_{\alpha=1}^{\nu} m_\alpha(t)h_\alpha(T,p_\beta(t))$$

or, by the formulae of Paragraph 8.2.1, and 8.2.2

$$H(t)=H_{mix}(t)+\sum_{\alpha=1}^{\nu} m_\alpha(t)h_\alpha(T,p), \tag{9.33}$$

where H_{mix} is the heat of mixing at time t. In particular, for an ideal mixture – to which we restrict the attention here – H_{mix} is zero and therefore we obtain by (9.5)

$$Q^p=\sum_{\alpha=1}^{\nu}\gamma_\alpha m_\alpha h_\alpha(T,p)[\mathcal{R}(t)-\mathcal{R}(t_0)] \text{ or } Q^p=\sum_{\alpha=1}^{\nu}\gamma_\alpha \tilde{h}_\alpha(T,p)\frac{\mathcal{R}(t)-\mathcal{R}(t_0)}{A}.$$

Thus the molar heat of reaction is equal to

$$\Delta\tilde{h} = \sum_{\alpha=1}^{\nu} \gamma_\alpha \tilde{h}_\alpha(T,p) \text{ or, in the reference state } \Delta\tilde{h}_R = \sum_{\alpha=1}^{\nu} \gamma_\alpha \tilde{h}_\alpha(T_R,p_R). \tag{9.34}$$

$\Delta\tilde{h}_R$, the reference value of the molar heat of reaction, has been measured for many reactions and the values are listed in handbooks, see Table 9.2.

Table 9.2 Reference values of molar heats of reaction $\Delta\tilde{h}_R$ and molar entropies of reaction $\Delta\tilde{s}_R$.

Reaction	$\Delta\tilde{h}_R$ [kJ/mol]	$\Delta\tilde{s}_R$ [J/(mol K)]
$H_2 \to 2H$	435.8	98.64
$O_2 \to 2O$	494.8	116.76
$H_2 + \frac{1}{2}O_2 \to H_2O$	–285.9	–166.54
$H_2 + I_2 \to 2HI$	51.8	165.90
$C + \frac{1}{2}O_2 \to CO$	–110.5	89.14
$C + O_2 \to CO_2$	–339.5	2.87
$CO_2 + H_2O \to \frac{1}{6}C_6H_{12}O_6 + O_2$	466.3	–40.10
$N_2 + 3H_2 \to 2NH_3$	–92.4	–178.6

We emphasize that the values $\Delta\tilde{h}_R$ in the table are valid only, if both the reactants and the products have the reference state (9.30). Thus in particular, water must be considered as liquid, and carbon, glucose $C_6H_{12}O_6$, and iodine as solid.

9.2.3 *Entropies of reaction*

We recall the law of mass action (9.10) for ideal mixtures which – by $g = h - Ts$ – may be written in the form

$$\prod_{\delta=1}^{\nu} X_\delta^{\gamma_\delta} = \exp\left[-\frac{1}{RT}\sum_{\delta=1}^{\nu}\gamma_\delta \tilde{h}_\delta(T,p) + \frac{1}{R}\sum_{\delta=1}^{\nu}\gamma_\delta \tilde{s}_\delta(T,p)\right]. \tag{9.35}$$

$\sum_{\delta=1}^{\nu}\gamma_\delta \tilde{h}_\delta(T,p)$ may be calculated from the heat of reaction $\Delta\tilde{h}_R$, since $\tilde{h}_\delta(T,p)$ is known as a function of T and p. Therefore the *molar entropy of reaction* $\Delta\tilde{s}(T,p) \equiv \sum_{\delta=1}^{\nu}\gamma_\delta \tilde{s}_\delta(T,p)$ may be obtained from the quantitative determination of mol fractions on the left hand side of (9.35), – in equilibrium of course. For the reference state the molar entropy of reaction

$$\Delta\tilde{s}_R \equiv \sum_{\delta=1}^{\nu}\gamma_\delta \tilde{s}_\delta(T_R,p_R) \tag{9.36}$$

is tabulated, and Table 9.2 shows some values.

9.2.4 *Le Chatelier's principle of least constraint*

We investigate the shift of chemical equilibrium under changes of temperature and pressure. The point of departure is the law of mass action in the form (9.10) for ideal solutions. We differentiate this equation with respect to T and p and use the known relations

$$\widetilde{s}_\delta(T,p)=-\left(\frac{\partial\widetilde{g}_\delta(T,p)}{\partial T}\right)_p \quad\text{and}\quad \widetilde{v}_\delta(T,p)=\left(\frac{\partial\widetilde{g}_\delta(T,p)}{\partial p}\right)_T .$$

Thus we obtain

$$\frac{\partial\ln\left(\prod\limits_{\delta=1}^{\nu}X_\delta^{\gamma_\delta}\right)}{\partial T}=\frac{\sum\limits_{\delta=1}^{\nu}\gamma_\delta\widetilde{h}_\delta(T,p)}{RT^2}\quad\text{and}\quad\frac{\partial\ln\left(\prod\limits_{\delta=1}^{\nu}X_\delta^{\gamma_\delta}\right)}{\partial p}=-\frac{\sum\limits_{\delta=1}^{\nu}\gamma_\delta\widetilde{v}_\delta(T,p)}{RT}. \tag{9.37}$$

Since there is no heat of mixing nor a volume of mixing in an ideal solution, the right hand sides of (9.37) represent the molar heat of reaction $\Delta\widetilde{h}$ and the molar volume change $\Delta\widetilde{v}(T,p)$ during the reaction

$$\frac{\partial\ln\left(\prod\limits_{\delta=1}^{\nu}X_\delta^{\gamma_\delta}\right)}{\partial T}=\frac{\Delta\widetilde{h}(T,p)}{RT^2}\quad\text{and}\quad\frac{\partial\ln\left(\prod\limits_{\delta=1}^{\nu}X_\delta^{\gamma_\delta}\right)}{\partial p}=-\frac{\Delta\widetilde{v}(T,p)}{RT}. \tag{9.38}$$

From this we conclude that

- an exothermal reaction favors the reactants at higher temperature,
- an endothermal reaction favors the products at higher temperature,
- a volume decreasing reaction favors the products at higher pressure,
- a volume increasing reaction favors the reactants at higher pressure.

These are special cases – for ideal mixtures – of the principle of least constraint which was postulated by LE CHATELIER: If one parameter is changed, the equilibrium shifts so as to minimize the effect of the change.

For instance, if the pressure is increased, the reaction moves so as to decrease the volume; in this manner the pressure increase will eventually be less than without the shift of chemical equilibrium.

As a heuristic principle LE CHATELIER's axiom played an important role in the chemistry of the 19th and the 20th century. In our examples of Paragraphs 9.1.6 and 9.1.7 we find some confirmation. Indeed, the volume-increasing decomposition of CO_2 favors CO and O_2 for lower pressure, or the formation of H_2 from H favors H_2 for higher pressure.

9.3 Nernst's heat theorem. The Third Law of thermodynamics

9.3.1 *Third Law in Nernst's formulation*

Hermann Walter NERNST (1864-1941) extrapolated his observations of physical, chemical, and electrochemical phenomena at low temperature, and conjectured that the entropy becomes independent of pressure for $T \to 0$. Also he found that the entropies for different phases of a body are equal for low temperatures – at least when the phases are crystalline. These findings or rather extrapolations became known as the Third Law of thermodynamics.

In order to appreciate the statement about the phases, we need to know that many solids – at the same pressure and temperature – may be present in different crystalline phases, *i.e.* with a different crystal structure. For example tin: At $p = 1\,\text{bar}$ tin is tetragonal above $T = 13.2°\text{C}$; it is then called "white tin." Below this temperature the stable phase is "gray tin" which has a cubic crystal structure. However, white tin may be cooled far below $13.2°\text{C}$. It is then called "metastable" and converts to gray tin very slowly so that measurements of specific heats, *etc.* can be made for both phases at low temperature.

For protracted periods of cold, however, tin plates may suffer from the "tin pest;" the white tin of the plates crumbles into powdery gray tin.

9.3.2 *Application of the Third Law. The latent heat of the transformation gray→white in tin*

Gray tin is stable below $286.3\,\text{K}$, while white tin is metastable. Thus we may calculate the entropies of both phases by measuring specific heats and integrating over them between $0\,\text{K}$ – or as low as we can get – and $T_\text{T} = 286.3\,\text{K}$

$$\begin{aligned} s^\text{w}(T_\text{T}, p) - s^\text{w}(0\,\text{K}) &= \int_0^{T_\text{T}} \frac{c_p^\text{w}(\tau, p)}{\tau} \text{d}\tau , \\ s^\text{g}(T_\text{T}, p) - s^\text{g}(0\,\text{K}) &= \int_0^{T_\text{T}} \frac{c_p^\text{g}(\tau, p)}{\tau} \text{d}\tau \end{aligned} \tag{9.39}$$

The entropies s^w and s^g at $0\,\text{K}$ *are independent of* p *and equal* according to the Third Law. Therefore from (9.39) we obtain

$$s^\text{w}(T_\text{T}, p) - s^\text{g}(T_\text{T}, p) = \int_0^{T_\text{T}} \frac{c_p^\text{w}(\tau, p) - c_p^\text{g}(\tau, p)}{\tau} \text{d}\tau = 62 \frac{\text{J}}{\text{kgK}} \tag{9.40}$$

where the second equation results from measurements of c_p^w and c_p^g at $p = 1\,\text{bar}$.

The latent heat of transformation from gray tin to white tin at T_T may thus be calculated as

$$r^{\text{g}\to\text{w}}(1\,\text{bar}) = T_\text{T}\left[s^\text{w}(T_\text{T}, 1\,\text{bar}) - s^\text{g}(T_\text{T}, 1\,\text{bar})\right] = 17.85 \cdot 10^3 \frac{\text{J}}{\text{kg}} . \tag{9.41}$$

The actually measured value of the latent heat is $18.21 \cdot 10^3\,\text{J/kg}$. The agreement is good enough given the unavoidable errors in the measurements of specific heats. Such calculations and measurements have convinced physicists that the Third Law is valid.

NERNST reassures us concerning the emergence of further thermodynamic laws:

The First Law had three discoverers: Mayer, Joule, Helmholtz.
The Second Law had two discoverers: Carnot and Clausius.
The Third Law has only one discoverer, namely himself: NERNST.
The Fourth Law ⋯ (?)

9.3.3 *Third Law in PLANCK's formulation*

Despite NERNST's claim there were really *two* discoverers of the Third Law. Indeed, Planck strengthened the law considerably. We recall that, according to NERNST, the entropies of a body at $T = 0\,\mathrm{K}$ are independent of pressure and phase. But no statement was made about their values, or about different values – perhaps – for different bodies. PLANCK suggested that all crystalline bodies have zero entropy at $T = 0\,\mathrm{K}$.

$$S \underset{T\to 0}{\rightarrow} 0\,,\ \text{hence follows}\ s(T,p) = \int_0^T \frac{c_p(\tau,p)}{\tau}\mathrm{d}\tau + \sum_i \frac{r(T_i)}{T_i}. \tag{9.42}$$

This means that the entropy has a definite absolute value which may be determined from specific heat measurements and measurements of latent heats $r(T_i)$ which may occur at T_i on the path from 0 to T. In particular, we may thus measure and calculate the entropy of a body at the reference state $T_\mathrm{R} = 298\,\mathrm{K}$ and $p_\mathrm{R} = 1\,\mathrm{atm}$.

$$s(T_\mathrm{R}, p_\mathrm{R}) = \int_0^{T_\mathrm{R}} \frac{c_p(\tau, p_\mathrm{R})}{\tau}\mathrm{d}\tau + \sum_i \frac{r(T_i)}{T_i}. \tag{9.43}$$

Such entropies $\tilde{s}_\mathrm{R} = \tilde{s}(T_\mathrm{R}, p_\mathrm{R})$ are listed in handbooks and some important molar values follow

$$\begin{aligned}
&\tilde{s}_\mathrm{R}^\mathrm{H} = 114.6\frac{\mathrm{J}}{\mathrm{molK}} && \tilde{s}_\mathrm{R}^{\mathrm{H}_2} = 130.6\frac{\mathrm{J}}{\mathrm{molK}} && \\
&\tilde{s}_\mathrm{R}^\mathrm{O} = 160.9\frac{\mathrm{J}}{\mathrm{molK}} && \tilde{s}_\mathrm{R}^{\mathrm{O}_2} = 205.0\frac{\mathrm{J}}{\mathrm{molK}} && \\
&\tilde{s}_\mathrm{R}^\mathrm{C} = 5.7\frac{\mathrm{J}}{\mathrm{molK}} && \tilde{s}_\mathrm{R}^\mathrm{CO} = 197.4\frac{\mathrm{J}}{\mathrm{molK}} && \tilde{s}_\mathrm{R}^{\mathrm{CO}_2} = 213.6\frac{\mathrm{J}}{\mathrm{molK}} \\
&\tilde{s}_\mathrm{R}^\mathrm{N} = 153.1\frac{\mathrm{J}}{\mathrm{molK}} && \tilde{s}_\mathrm{R}^{\mathrm{N}_2} = 191.5\frac{\mathrm{J}}{\mathrm{molK}} && \tilde{s}_\mathrm{R}^{\mathrm{NH}_3} = 192.5\frac{\mathrm{J}}{\mathrm{molK}} \\
&\tilde{s}_\mathrm{R}^{\mathrm{H}_2\mathrm{O}} = 66.6\frac{\mathrm{J}}{\mathrm{molK}} && \tilde{s}_\mathrm{R}^{\mathrm{C}_6\mathrm{H}_{12}\mathrm{O}_6} = 213.0\frac{\mathrm{J}}{\mathrm{molK}} && \\
&\tilde{s}_\mathrm{R}^{\mathrm{I}_2} = 116.1\frac{\mathrm{J}}{\mathrm{molK}} && \tilde{s}_\mathrm{R}^\mathrm{HI} = 206.3\frac{\mathrm{J}}{\mathrm{molK}}. &&
\end{aligned} \tag{9.44}$$

Carbon at T_R, p_R may occur as graphite or diamond. The value given in (9.44) is the value for the stable phase graphite.

PLANCK had recognized that the entropy of an ideal gas can be calculated by the quantization of the phase space of atoms – spanned by their coordinates and momenta – into cells of h^3, where h is the Planck constant. Thus he was able to link the Third Law to the quantum mechanics originated by him. He could calculate the entropy of monatomic gases much as we have done in Chap. 5, *cf.* $(5.14)_1$ but *with a specific value for the constant.* He arrived at

$$s(T,p)=\frac{k}{m}\left[\frac{5}{2}\ln T-\ln p+\ln\left(\frac{\sqrt{2\pi m}}{h^3}k^{3/2}e^{5/2}\right)\right]. \quad (9.45)$$

Starting with this value for a pair (T,p) in the monatomic ideal gas state of a body – *i.e.* at high temperature and low pressure – one may calculate $s(T_R,p_R)$ by integrating over specific heats and, perhaps, by summing over latent heats.

Usually one thus obtains the same value as the one calculated in (9.43), thus confirming PLANCK's version of the Third Law. But sometimes there is a difference which is a clear sign that a body has a non-zero entropy at absolute zero. This occurs for amorphous bodies like glass and some polymers. For these materials neither NERNST's nor PLANCK's version of the Third Law is valid.

9.3.4 *Absolute values of energy and entropy*

In Paragraphs 9.2.2 and 9.2.3 we have shown – for ideal mixtures – that the combinations

$$\Delta\tilde{h}_R=\sum_{\alpha=1}^{\nu}\gamma_\alpha\tilde{h}_\alpha(T_R,p_R) \text{ and } \Delta\tilde{s}_R=\sum_{\alpha=1}^{\nu}\gamma_\alpha\tilde{s}_\alpha(T_R,p_R). \quad (9.46)$$

of the additive constants in energy and entropy could be calculated from thermal and caloric measurements, and from the quantitative analysis of reaction products. The values of the constants themselves were not attainable in this way.

But now, in Paragraph 9.3.3, we have seen that the Third Law – in Planck's form – permits the determination of the absolute value of the entropies, *cf.* (9.45). What about energy? Here too the first decade of the 20th century, and the fast progress of physics at that time, have provided the possibility to determine absolute values of the energies. That was achieved by Einstein's result $E=mc^2$. However, this is not a practical tool for chemists, because the masses m of the atoms and molecules cannot be measured with sufficient accuracy. Still, in principle, the additive constants in both energy and entropy are now identifiable. Note that this is more than the chemists need, because all they require are the combinations (9.46).

9.4 Energetic and entropic contributions to equilibrium

9.4.1 *Three contributions to the Gibbs free energy*

We know that the chemical equilibrium lies in the minimum of the Gibbs free energy, just like phase equilibria do. And it is instructive to illustrate this fact graphically. This has the advantage that we recognize how energy, or enthalpy, and entropy compete in establishing equilibrium. We restrict the attention to reactions between ideal gases, although very similar considerations are valid for arbitrary ideal mixtures, and even for non-ideal mixtures. The advantage of ideal gases is, of course, that all state functions are explicitly known.

We have generally

$$G=\sum_{\alpha=1}^{\nu}\nu_\alpha\tilde{h}_\alpha(T,p_\beta)-T\sum_{\alpha=1}^{\nu}\nu_\alpha\tilde{s}_\alpha(T,p_\beta)$$

and in an ideal mixture according to Paragraph 8.2.2

$$G=\sum_{\alpha=1}^{\nu}\nu_\alpha\tilde{h}_\alpha(T,p)-T\sum_{\alpha=1}^{\nu}\nu_\alpha\left[\tilde{s}_\alpha(T,p)-R\ln\frac{\nu_\alpha}{\sum_{\beta=1}^{\nu}\nu_\beta}\right].$$

Since we are interested in ideal gas mixtures we may set

$\tilde{h}_\alpha(T,p)=\tilde{h}_\alpha(T_R,p_R)+(z_\alpha+1)R(T-T_R)$, and

$$\tilde{s}_\alpha(T,p)=\tilde{s}_\alpha(T_R,p_R)+(z_\alpha+1)R\ln\frac{T}{T_R}-R\ln\frac{p}{p_R}.$$

Furthermore we use (9.6) in the form $\nu_\alpha(t)=\nu_\alpha(t_0)+\gamma_\alpha[\boldsymbol{R}(t)-\boldsymbol{R}(t_0)]/A$ and start the reaction with $\boldsymbol{R}(t_0)=0$, and with a stoichiometric mixture of reactants. Thus we obtain

$$\begin{aligned}G-G(0)=&\left[\Delta\tilde{h}_R+\sum_{\alpha=1}^{\nu}\gamma_\alpha(z_\alpha+1)R(T-T_R)\right]\frac{\boldsymbol{R}}{A}-\\&-T\left[\Delta\tilde{s}_R+\sum_{\alpha=1}^{\nu}\gamma_\alpha(z_\alpha+1)R\ln\frac{T}{T_R}-\sum_{\alpha=1}^{\nu}\gamma_\alpha R\ln\frac{p}{p_R}\right]\frac{\boldsymbol{R}}{A}+\\&+RT\sum_{\alpha=1}^{\nu}\left[\left(\nu_\alpha(t_0)+\gamma_\alpha\tfrac{\boldsymbol{R}}{A}\right)\ln\frac{\nu_\alpha(t_0)+\gamma_\alpha\frac{\boldsymbol{R}}{A}}{\sum_{\beta=1}^{\nu}\left[\nu_\beta(t_0)+\gamma_\beta\frac{\boldsymbol{R}}{A}\right]}-\nu_\alpha(t_0)\ln\frac{\nu_\alpha(t_0)}{\sum_{\beta=1}^{\nu}\nu_\beta(t_0)}\right]\end{aligned}\tag{9.47}$$

and the range of $\boldsymbol{R}$ is given by

$$0\le\frac{\boldsymbol{R}}{A}\le-\frac{\nu_\alpha(t_0)}{\gamma_\alpha}\quad\text{for all }\gamma_\alpha<0.\tag{9.48}$$

Thus G contains three terms which represent

- the enthalpy of the mixture which – for ideal mixtures – is the sum of the enthalpies of the unmixed constituents,
- the entropy of the unmixed constituents, and
- the entropy of mixing.

The first two terms are linear in $\boldsymbol{R}$ so that they do not show minima inside the range (9.48); only end-point minima occur according to these terms. If there is an interior minimum, it must be due to the entropy of mixing. We investigate some instructive examples.

9.4.2 *Examples for minima of the Gibbs free energy*

- $H_2 \rightarrow 2H$

Both molecular and atomic hydrogen are ideal gases in the reference state (9.30) so that for a low pressure p and high temperature $T > T_R$ the equation (9.47) is valid. Data and initial conditions may be read off from Table 9.3.

Table 9.3 Data and initial conditions for $H_2 \rightarrow 2H$

	γ_α	$\nu_\alpha(t_0)$	$\Delta\tilde{h}_R$	$\Delta\tilde{s}_R$	z_α
H_2	-1	1 mol	$435.8\frac{\text{kJ}}{\text{mol}}$	$98.6\frac{\text{J}}{\text{mol}\cdot\text{K}}$	5/2
H	2	0			3/2

Thus we obtain (9.47) in the explicit form with $r = \mathcal{R}/A\,\text{mol}$ so that the range of r is $0 \leq r \leq 1$:

$$G(r) - G(0) = \left[435.8\,\text{kJ} + \frac{3}{2}8.314\tfrac{\text{J}}{\text{K}}(T - T_R)\right]r -$$

$$-T\left[98.6\tfrac{\text{J}}{\text{K}} + \frac{3}{2}8.314\tfrac{\text{J}}{\text{K}}\ln\frac{T}{T_R} - 8.314\tfrac{\text{J}}{\text{K}}\ln\frac{p}{p_R}\right]r +$$

$$+8.314\tfrac{\text{J}}{\text{K}}T\left[(1-r)\ln\frac{1-r}{1+r} + 2r\ln\frac{2r}{1+r}\right]. \tag{9.49}$$

Fig. 9.3 shows plots of this function for $T = 298\,\text{K}$, $p = 1\,\text{bar}$, and $T = 3000\,\text{K}$, $p = 0.01\,\text{bar}$. We conclude that for the low temperature the minimum of G occurs at – or near – $r = 0$ so that H_2 is the stable constituent. At the high temperature there is a mixture of constituents in equilibrium, where G has a minimum at $r \approx 0.7$.

The figures also show the three lines of (9.49) separately. The first line labeled "energy" has an end-point minimum at H_2, while the third one labeled "entropy of mixing" is responsible for the genuine minimum at $r \approx 0.7$, at least for the high temperature. Thus we may say that energy favors the molecules, while entropy favors the atoms and the entropy of mixing favors mixing – naturally. The entropy of mixing, however, cannot achieve much at low temperature, because it is too small. In Fig. 9.3 a it cannot even be seen, because – to the naked eye – its graph coincides with the r-axis.

There is a subtle point concerning the energy of mixing for $r \rightarrow 0$ and $r \rightarrow 1$. In both cases its slope is infinite. Therefore – however dominant the linear curves of energy and entropy may be – the entropy of mixing enforces a minimum close to an end point. This minimum is so close to the end point, and so small that it cannot be seen on the scale of Fig. 9.3 a, but it is there: In Fig. 9.3 a it occurs at

$r \approx 10^{-8000}$ so that – with only 10^{22} molecules – we will not see a single H_2 dissociated into H-atoms at the low temperature.

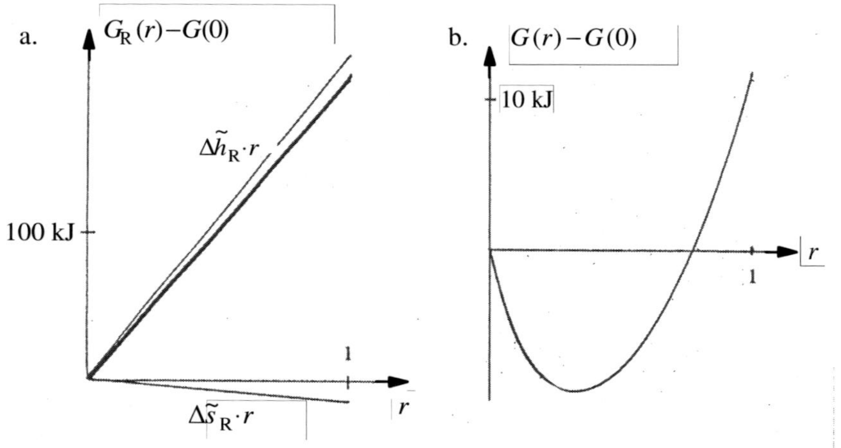

Fig. 9.3 Gibbs free energy as a function of r for $H_2 \to 2H$
The minima of the bold curves define the chemical equilibrium
a. Reference state **b.** $T = 3000$ K, $p = 0.01$ bar

- $N_2 + 3H_2 \to 2NH_3$ *Haber-Bosch synthesis of ammonia*

For the ammonia reaction we have data and initial values as shown in Table 9.4.

Table 9.4 Data and initial conditions for $N_2 + 3H_2 \to 2NH_3$

	γ_α	$\nu_\alpha(t_0)$	$\Delta\tilde{h}_R$	$\Delta\tilde{s}_R$	z_α
N_2	-1	1 mol			5/2
H_2	-3	3 mol	$-92.4\frac{\text{kJ}}{\text{mol}}$	$-178.6\frac{\text{J}}{\text{mol}\cdot\text{K}}$	5/2
NH_3	2	0			3

For this choice the explicit form of G reads, according to (9.47)

$$G(r)-G(0)=\left[-92.4\,\text{kJ}-6\cdot 8.314\tfrac{\text{J}}{\text{K}}(T-T_R)\right]r-$$
$$-T\left[-178.6\tfrac{\text{J}}{\text{K}}-6\cdot 8.314\tfrac{\text{J}}{\text{K}}\ln\frac{T}{T_R}-8.314\tfrac{\text{J}}{\text{K}}\ln\frac{p}{p_R}\right]r+ \quad (9.50)$$
$$+8.314\tfrac{\text{J}}{\text{K}}T\left[(1-r)\ln\frac{1-r}{4-2r}+3(1-r)\ln\frac{3(1-r)}{4-2r}\right.$$
$$\left.+2r\ln\frac{2}{4-2r}+\ln 4-3\ln\frac{3}{4}\right].$$

The range of r is $0 \le r \le 1$, according to (9.48).

We plot this function first for $T = T_R$, $p = p_R$ in Fig. 9.4 and obtain a graph with a minimum very close to $r = 0.98$ so that we might expect that ammonia is easily formed. In reality, however, there is practically no ammonia, because, before the reaction can occur, the tightly bound molecules H_2 and N_2 must be converted into their atomic forms and this decomposition requires a catalyst. In the Haber-Bosch synthesis one uses perforated iron sheets. Their effectiveness as catalysts, however, requires a high temperature of 500°C (say). But at that temperature the entropic part of G – the second line in (9.50) – is emphasized, and we obtain a graph whose minimum lies at $r = 0$, *cf.* Fig. 9.4., so that again no ammonia forms. In order to decrease the entropic influence, one may raise the pressure – see (9.50) – and, indeed, if the pressure is raised to $200\,\mathrm{bar}$, we obtain a graph for G that has a minimum at about $r = 0.48$, *cf.* Fig. 9.4. Thus ammonia can be produced in quantity at high temperature and high pressure.

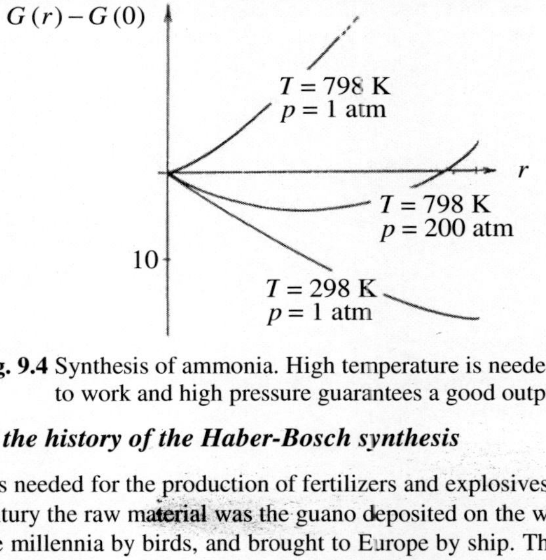

Fig. 9.4 Synthesis of ammonia. High temperature is needed for the catalyst to work and high pressure guarantees a good output

9.4.3 *On the history of the Haber-Bosch synthesis*

Ammonia is needed for the production of fertilizers and explosives. Up to the early years of the 19th century the raw material was the guano deposited on the west coast of South America over the millennia by birds, and brought to Europe by ship. The Haber-Bosch synthesis was developed in 1908, just in time for the First World War. It was clear that in case of war Germany would be cut off from guano imports by a British naval blockade and therefore a huge ammonia plant was built in Saxony. Its output supplied the German army throughout the four years of war easily. The country ran out of men and food and morale, but never of explosives. HABER received the Nobel prize in 1918 and, after the war, he was made director of the Kaiser Wilhelm Institute for physical chemistry.

HABER's patriotism led him to propagate and direct the use of chlorine and mustard gas as a means of warfare on the western front. The opposing troops fled and the result was an unprecedented five-mile gap in the front. The strategic effect, however, was nil, since the German general staff had not really believed that the project would work, and was not prepared for an offensive.

Fritz HABER (1868-1924) continued in what he saw as his patriotic duty after the disastrous war. He attempted to isolate gold from sea water in the hope of helping Germany repay the huge war indemnity demanded after the Versailles treaty. In this effort HABER failed. However, he could have saved himself the effort since in the end the indemnity was never paid.

Only a little later HABER became a tragic figure. He was Jewish and when HITLER came to power he was stripped of all posts and driven into exile. He was not alone in that, of course, but while others were made welcome in Britain by an international initiative of scientists led by Ernest RUTHERFORD, HABER was not, because of his poison gas activity. He left for Italy but died *en route*.

The Haber-Bosch synthesis of ammonia is still in these days one of the big money makers of the chemical industry. Scientifically it was important, because it started the field of catalytic high pressure chemistry by which later the conversion of coal into liquid fuels was achieved.

9.5 The fuel cell

9.5.1 Chemical Reactions

In the fuel cell a cold "combustion" of hydrogen H_2 and oxygen O_2 takes place which together form water. In this reaction energy and entropy are freed according to Table 9.2 and we have

$$\Delta\tilde{g}_R = \Delta\tilde{h}_R - T_R \Delta\tilde{s}_R = -263.3 \frac{\text{kJ}}{\text{mol}} . \qquad (9.51)$$

Therefore this reaction may provide energy. In the fuel cell this energy is obtained by the use of catalysts and an electrolyte.

Fig. 9.5 shows a schematic picture of a fuel cell and we proceed to describe the partial reactions that occur on the side of the anode (left) and of the cathode (right).

anodic catalyst	cathodic catalyst	
$H_2 \to 2H$	$\frac{1}{2}O_2 \to O$.	(9.52)

The dissociations (9.52) occur on the metallic surface which has a very large area in the porous catalysts.

anodic oxidation of H*	cathodic reduction of O	
$2H \to 2H^+ + 2e^-$	$O + H_2O + 2e^- \to 2OH^-$.	(9.53)

e^- denotes the negative elementary charge – the charge of an electron – whose magnitude is equal to $1.60206 \cdot 10^{-19}$ C. The water that participates in the cathodic reduction, is a constituent of the electrolyte. Thus we have

a surplus of $2e^-$ at the anode	a lack of $2e^-$ at the cathode.

* Oxidation is the release of electrons and reduction is the absorption of electrons.

This charge difference leads to a voltage difference between the electrodes – anode and cathode – or, if there is an electrically conducting connection, to a current of $2e^-$ through the external resistor R_e. The exchange of the other charges, the H^+- and the OH^--ions occurs through the electrolyte by ion migration; the details depend on the type of the fuel cell but in any case the final reaction is

$$2H^+ + 2OH^- \rightarrow H_2O\,. \tag{9.54}$$

The over-all reaction results from a summation of the partial reactions (9.52) through (9.54) and we obtain

$$H_2 + \tfrac{1}{2}O_2 \rightarrow H_2O\,. \tag{9.55}$$

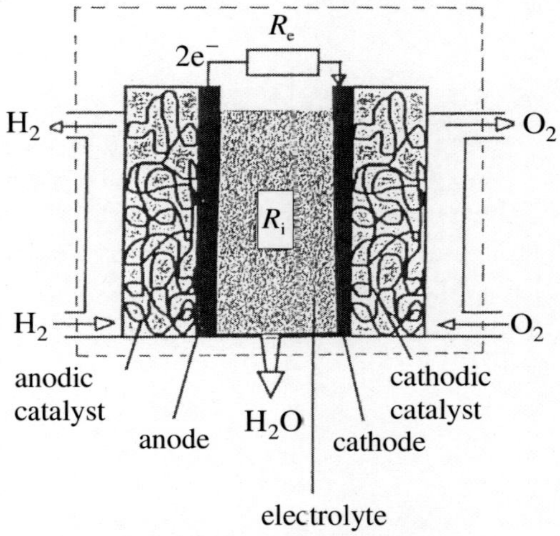

Fig. 9.5 A fuel cell (schematic)

9.5.2 *Various types of fuel cells*

Fuel cells differ by the type of their electrolyte. For the alkaline fuel cell, called AFC, the electrolyte is a base, *e.g.* KOH in water solution, which is fully dissociated into K^+- and OH^--ions. On the side of the anode the H^+-ions – formed according to $(9.53)_1$ – combine with two OH^--ions of the electrolyte and form water as indicated by (9.54). The OH^--deficiency thus created at the anode is compensated for by the migration of the OH^--ions, formed at the cathode, through the electrolyte. Thus the reaction product water appears at the anode. The working temperature of an AFC is about 80°C so that the water is liquid.

For the phosphoric acid fuel cell, called PAFC, the electrolyte is a water solution of phosphoric acid H_3PO_4, which is fully decomposed into H^+- and PO_4^{3-}-ions. The OH^--ions appearing at the cathode according to $(9.53)_2$ combine with the H^+-ions of the acid to form water, which therefore appears at the cathode.

The shortfall of H^+-ions near the cathode is then compensated by the migration of H^+-ions created in the anode through the electrolyte. The working temperature is about 180°C so that the emerging water is in the vapor phase.

Another promising device, called PEMFC, uses a polyelectrolytic membrane for the electrolyte, *cf.* Sect. 11.6 for a treatment of polyelectrolytic gels. Inside the gel-membrane there is a network of long-chain molecules which releases H^+-ions. The exchange of charges occurs much in the same way as for PAFC, although the working temperature is usually below the boiling point of water.

Here is not the proper place to discuss the relative advantages of the various designs. It may suffice to say that an AFC needs very pure hydrogen and oxygen. Even the tiny CO_2-content of air forbids the use of air as a cheap source for the oxygen.

9.5.3 *Thermodynamics*

We consider the open system, whose surface ∂V is shown in Fig. 9.5 by the dashed line. Hydrogen and oxygen flow in and out, and the water flows out; the molar flow rates are $\dot{\nu}_\alpha$. We denote the rate of reactions inside the volume by Λ. In this case, by (9.55), the conservation laws of the number of atoms read

$$\begin{aligned} \dot{\nu}_{H_2}^{out} - \dot{\nu}_{H_2}^{in} &= -\frac{\Lambda}{A} \\ \dot{\nu}_{O_2}^{out} - \dot{\nu}_{O_2}^{in} &= -\frac{\Lambda}{2A} \\ \dot{\nu}_{H_2O}^{out} &= \frac{\Lambda}{A}. \end{aligned} \tag{9.56}$$

We assume that the in- and outflows of H_2 and O_2 occur at the same pressure p and temperature T, *i.e.* with equal specific enthalpies and entropies. Thus we obtain from the energy balance

$$\left(\dot{\nu}_{H_2}^{out} - \dot{\nu}_{H_2}^{in}\right)\tilde{h}_{H_2} + \left(\dot{\nu}_{O_2}^{out} - \dot{\nu}_{O_2}^{in}\right)\tilde{h}_{O_2} + \dot{\nu}_{H_2O}^{out}\tilde{h}_{H_2O} = \dot{Q}\,, \tag{9.57}$$

where $\dot{Q}$ is the heating of the surface ∂V which is needed to maintain a constant temperature.

The entropy balance reads accordingly

$$\left(\dot{\nu}_{H_2}^{out} - \dot{\nu}_{H_2}^{in}\right)\tilde{s}_{H_2} + \left(\dot{\nu}_{O_2}^{out} - \dot{\nu}_{O_2}^{in}\right)\tilde{s}_{O_2} + \dot{\nu}_{H_2O}^{out}\tilde{s}_{H_2O} = \frac{\dot{Q}}{T} + P\,. \tag{9.58}$$

P is the production rate of entropy inside ∂V. By the Second Law it is equal to the Joule heating of the electric current I divided by the temperature. The Joule heating occurs through the motion of the electrons in the external resistor R_e, and through the motion of ions in the electrolyte, whose resistance we denote by R_i. The laws of electrodynamics give the Joule heating as RI^2 so that we may write

$$P = \frac{(R_e + R_i)I^2}{T}. \tag{9.59}$$

Elimination of $\dot{\nu}_\alpha$, $\dot{Q}$, and P between (9.56) through (9.59) gives with $g = h - Ts$

$$\underbrace{\left(-\tilde{g}_{H_2} - \tfrac{1}{2}\tilde{g}_{O_2} + \tilde{g}_{H_2O}\right)}_{\Delta\tilde{g}(T,p)}\frac{\Lambda}{A} = -(R_e + R_i)I^2 . \tag{9.60}$$

Since each single reaction, by Paragraph 9.5.1, leads to the passage of the charge $2e^-$ from the anode to the cathode we have

$$I = 2e^- \Lambda$$

so that one I cancels out from (9.60). If we let the reaction occur in the reference state, we thus obtain

$$R_e I = -\frac{\Delta\tilde{g}_R}{2e^- A} - R_i I . \tag{9.61}$$

By Ohm's law $R_e I$ is the voltage U between the anode and the cathode, *i.e.* the voltage that may be put to use for lighting (say), or to drive an engine. Thus we obtain for that voltage, *cf.* (9.51)

$$U = \underbrace{-\frac{\Delta\tilde{g}_R}{2e^- A}}_{U_0} - R_i I = 1.23\,\mathrm{V} - R_i I . \tag{9.62}$$

The constant $e^- A \approx 96000\,\mathrm{C/mol}$ is often called the Faraday constant; it is usually denoted by *F*. $\Delta\tilde{g}_R$ is equal to –263.3kJ/kg by (9.51), so that $U_0 = 1.23\,\mathrm{V}$ holds. U_0 is called the "zero point" voltage; it represents the voltage between the electrodes when no current is drawn from the cell.

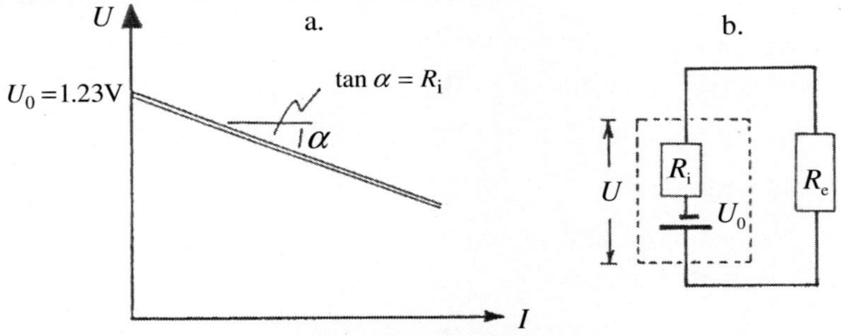

Fig. 9.6 On the fuel cell
a. Ideal (U,I)-characteristic
b. Electric circuit of a fuel cell

Fig. 9.6 shows the graph of this simple (U,I)-relation for an internal resistance $R_i = 4\cdot 10^{-2}\,\Omega$ which is typical for the electrolyte of an alkaline fuel cell. Concerning the simple straight graph of the figure we must say that a real fuel cell deviates from this appreciably: First of all, in the neighborhood of $I = 0$ there is a sharp voltage drop of about 20% before the graph turns linear. And for large values of I the graph turns downward. All of this is due to secondary effects: Shielding of the electrons by spatial charge distributions, emergence of other constituents, such as H_2O_2, *etc.* We do not discuss these, they are difficult to describe quantitatively.

9.5.4 *Effects of temperature and pressure*

The value $U_0 = 1.23\,\mathrm{V}$ of the voltage changes a little, if the fuel cell works at values of (T, p) other than the reference values. As long as the emerging water is still liquid, we have

$$\Delta\tilde{h}(T,p) = \Delta\hat{h}_{\mathrm{R}} + \left[-\left(z_{\mathrm{H}_2}+1\right) - \tfrac{1}{2}\left(z_{\mathrm{O}_2}+1\right)\right]R(T-T_{\mathrm{R}}) + \tilde{c}_{\mathrm{LW}}(T-T_{\mathrm{R}})$$

$$\Delta\tilde{s}(T,p) = \Delta\tilde{s}_{\mathrm{R}} + \left[-\left(z_{\mathrm{H}_2}+1\right) - \tfrac{1}{2}\left(z_{\mathrm{O}_2}+1\right)\right]R\ln\frac{T}{T_{\mathrm{R}}} + \tilde{c}_{\mathrm{LW}}\ln\frac{T}{T_{\mathrm{R}}} -$$

$$\vdots \qquad\qquad - R\left[-\ln\frac{p_{\mathrm{H}_2}}{p_{\mathrm{R}}} - \tfrac{1}{2}\ln\frac{p_{\mathrm{O}_2}}{p_{\mathrm{R}}}\right],$$

where $z_{\mathrm{H}_2} = z_{\mathrm{O}_2} = \frac{5}{2}$ and $\tilde{c}_{\mathrm{LW}} = 75.25\,\mathrm{J/(mol\cdot K)}$ is the molar specific heat of liquid water. Therefore we obtain

$$\Delta\tilde{g}(T,p) = \Delta\tilde{g}_{\mathrm{R}} - (T-T_{\mathrm{R}})\Delta\tilde{s}_{\mathrm{R}} + RT\left[-\ln\frac{p_{\mathrm{H}_2}}{p_{\mathrm{R}}} - \tfrac{1}{2}\ln\frac{p_{\mathrm{O}_2}}{p_{\mathrm{R}}}\right] + \left[-\tfrac{21}{4}R + 75.24\frac{\mathrm{J}}{\mathrm{mol\cdot K}}\right]\left[(T-T_{\mathrm{R}}) - T\ln\frac{T}{T_{\mathrm{R}}}\right]. \tag{9.63}$$

For a temperature of 75°C and $p_{\mathrm{H}_2} = p_{\mathrm{O}_2} = p_{\mathrm{R}}$ we obtain $\Delta\tilde{g}(T,p) - \Delta\tilde{g}_{\mathrm{R}} = 8.44\,\mathrm{J/mol}$ so that the voltage U_0 is *reduced* by 3.6%. If the pressure of both gases is equal to $p = 10\,\mathrm{bar}$ and the temperature $T = T_{\mathrm{R}}$ we obtain $\Delta\tilde{g}(T,p) - \Delta\tilde{g}_{\mathrm{R}} = -8.56\,\mathrm{J/mol}$ so that U_0 *grows* by 3.6%.

9.5.5 *Power of the fuel cell*

The electric circuit with a fuel cell of voltage U_0 and resistances R_{i} and R_{e} is shown in Fig. 9.6 b. By the laws of electrodynamics the working of the fuel cell is equal to $\dot{W} = U\,I$, where U is the voltage drop across the external resistor. Kirchhoff's law implies

$$U = U_0 - R_{\mathrm{i}}I \quad \text{and} \quad U_0 - R_{\mathrm{i}}I - R_{\mathrm{e}}I = 0\,.$$

Hence follows

$$\dot{W} = (U_0 - R_{\mathrm{i}}I)I \quad \text{or, with } I = \frac{U_0}{R_{\mathrm{i}} + R_{\mathrm{e}}}$$

$$\dot{W} = \frac{R_{\mathrm{e}}}{(R_{\mathrm{i}} + R_{\mathrm{e}})^2}U_0^2\,. \tag{9.64}$$

We conclude that $\dot{W}$ is zero for $R_e = 0$ and $R_e \to \infty$, because either U is zero, or $I = 0$ in these cases. In-between, for $R_e = R_i$, the working has a maximum, *viz.*

$$\dot{W}_{max} = \frac{U_0^2}{4R_i} \quad \text{with} \quad I_{max} = \frac{U_0}{2R_i} \quad \text{and} \quad U_{max} = \frac{U_0}{2}. \tag{9.65}$$

For $R_i = 10^{-2}\,\Omega$ as a typical internal resistance of a PAFC and for $U_0 = 1.23\,\text{V}$, *cf.* (9.62), we obtain

$$\dot{W}_{max} \approx 40\,\text{W} \quad \text{and} \quad I_{max} = 60\,\text{A}.$$

It is therefore clear that *one single* fuel cell cannot power a car. Even small cars have nowadays a power of $40\,\text{kW}$ and for this purpose one needs a "stack" of 1000 fuel cells – less when R_i is smaller. Each single cell must be equipped with pipes for the in- and outflow of H_2 and O_2 and for the draining of water. This presents a considerable challenge for the engineers and even the sheer weight of the stack is a problem.

9.5.6 *Efficiency of the fuel cell*

Scientific folklore has it that the efficiency of a fuel cell is very high. Not being a thermal engine, the Carnot efficiency-limit is not relevant for a fuel cell and people say that the efficiency might therefore be equal to one. It is indeed true that the fuel cell converts chemical energy into electric energy. We must therefore redefine the notion of efficiency; or rather, we must fall back to the generic notion of the efficiency as the quotient of the desired output and the required input. The required input may be defined as $|\Delta \tilde{g}_R|$, the change of the Gibbs free energy when water is split into hydrogen and oxygen, which are our fuels.* Thus we have, *cf.* (9.62)

$$\Delta \tilde{g}_R = 2e^- A U_0.$$

The desired output is the molar electric energy $U 2e^- A$ passing through R_e for each mol of water produced. Thus the efficiency comes out as

$$e = \frac{U 2e^- A}{|\Delta g_R|} = \frac{U}{U_0} = 1 - \frac{R_i I}{U_0}. \tag{9.66}$$

So, indeed, for very small currents the efficiency is equal to one but, of course, in this case we get very little power. If we withdraw the current $I_{max} = U_0/(2R_i)$ for which, by (9.65), the power has a maximum, we obtain $e = 1/2$. This is still a good efficiency in comparison with heat engines and internal combustion engines. However, in practice the efficiency of fuel cell is halved by the secondary effects mentioned above.

---

* Of course, if we obtain the reactants from natural gas or from air, another input may be relevant.

9.6 Thermodynamics of photosynthesis

9.6.1 *The dilemma of glucose synthesis*

Plants produce glucose $C_6H_{12}O_6$ from the carbon dioxide in the air and from the water in the soil. In the process they set oxygen free. The process is called photosynthesis, since it occurs only under light. The multiple partial steps involved in photosynthesis are still not fully understood by biophysicists, although they are getting close. Here, however, we ignore all detail; we consider the process thermodynamically and balance in- and effluxes. The stoichiometric equation reads

$$CO_2 + H_2O \rightarrow \tfrac{1}{6} C_6H_{12}O_6 + O_2 . \tag{9.67}$$

Like all other processes the glucose synthesis must satisfy the First and Second Law. If the process occurs at fixed values of p and T, *e.g.* at $p = p_R$ and $T = T_R$, this means that the molar heat of reaction must be given by

$$\begin{aligned} \tilde{q}^{\mathrm{p}} &= \Delta \tilde{h}_R \\ \tilde{q}^{\mathrm{p}} &\le T_R \Delta \tilde{s}_R \end{aligned} \quad \text{where by Table 9.2 we have} \quad \begin{aligned} \Delta \tilde{h}_R &= 466.3 \frac{\mathrm{kJ}}{\mathrm{mol}} \\ \Delta \tilde{s}_R &= -40.1 \frac{\mathrm{J}}{\mathrm{mol \cdot K}} \end{aligned} . \tag{9.68}$$

Therefore the two laws of thermodynamics impose conflicting demands: The First Law requires that we provide heat and the Second Law demands that we withdraw heat.

An alternative view of the dilemma is offered by elimination of $\tilde{q}^{\mathrm{p}}$ between the two relations (9.68). Thus the two laws of thermodynamics imply that the Gibbs free energy should decrease

$$\Delta \tilde{g}_R = \Delta \tilde{h}_R - T_R \Delta \tilde{s}_R < 0 . \tag{9.69}$$

On the other hand, the numbers in (9.68) imply that $\Delta \tilde{g}_R$ is equal to $\Delta \tilde{h}_R = 478.3 \frac{\mathrm{kJ}}{\mathrm{mol}}$; it is *positive* !

We should therefore conclude that the process is impossible. However, of course, we know better: the plants *do produce* glucose and the question is how. The only possibility is that the reaction in the plants is accompanied by an entropy increasing process. And the entropy increase in that accompanying process must be so large, that the entropy decrease of the glucose formation can be compensated, and – what is more – that the inequality can be satisfied.

In a manner of speaking the numbers in (9.68) indicate the worst possible case: The energy increases and the entropy decreases when in equilibrium the energy *wants* to be minimal and the entropy *wants* to be maximal, *cf.* Paragraph 4.2.7.

The key to the resolution of the dilemma lies in the observation that a plant uses much more water than is required for the glucose formation, according to the stoichiometric equation (9.67).

Indeed the plant uses 100 to 1000 times more water than that. The water evaporates and the resulting water vapor mixes with the surrounding air. But evaporation alone does not help. It is true that the entropy grows in evaporation by the amount $r(T_R)/T_R$, but the enthalpy also grows by $r(T_R)$, so that the Gibbs free energy remains unchanged. However, if the entropy of mixing is taken into consideration, both laws of thermodynamics can be satisfied. We investigate this proposition in detail. Throughout this section we choose $T = T_R$ and $p = p_R$, because roughly these are the conditions under which photosynthesis takes place.

9.6.2 *Balance of particle numbers*

We consider the open system of Fig. 9.7 around a growing leaf. Dry air and liquid water enter the system and moist air – with a changed proportion of CO_2 and O_2 – leaves it. The influx of liquid water is adjusted so that only water vapor leaves the system. Inside the system a plant leaf grows, *i.e.* glucose is formed and the newly generated glucose is pulled out so as to guarantee stationary conditions. Temperature and pressure are homogeneous.

The balance equations for the particles read

$$\dot{\nu}_\alpha^{\text{out}} - \dot{\nu}_\alpha^{\text{in}} = \gamma_\alpha \frac{\Lambda}{A}. \tag{9.70}$$

Λ is the reaction rate and $\dot{\nu}_\alpha$ are the molar flow rates. Obviously glucose does not enter, so that $\dot{\nu}_{\text{G}}^{\text{in}} = 0$ holds. The entering flow rates of CO_2 and water must at least be equal to Λ/A. If they are equal to Λ/A, all incoming CO_2 – and all water molecules – are used up for the formation of glucose. But we shall provide *more* CO_2 and *more* water. Thus we have

$$\dot{\nu}_{\text{G}}^{\text{in}} = 0 \text{ and } \dot{\nu}_{\text{CO}_2}^{\text{in}} = y\frac{\Lambda}{A},\ \dot{\nu}_{\text{W}}^{\text{in}} = x\frac{\Lambda}{A}, \tag{9.71}$$

where $(x-1)\Lambda/A$ and $(y-1)\Lambda/A$ are the excess flows, which run through the system without participating in the chemical reaction. This does not mean that the excess flows are not necessary, because the excess water evaporates and the excess air provides a greater volume for the evaporated water to mix with.

We assign the partial pressures of the incoming air as*

$$\frac{p_{\text{N}_2}^{\text{in}}}{p} = 0.7897,\ \frac{p_{\text{O}_2}^{\text{in}}}{p} = 0.21,\ \frac{p_{\text{CO}_2}^{\text{in}}}{p} = 0.0003, \tag{9.72}$$

so that they are close to the natural mix, *cf.* (P.8). By use of $p_\alpha V = \nu_\alpha RT$ we obtain from (9.71)₂

$$\dot{\nu}_{\text{CO}_2}^{\text{in}} = \frac{p_{\text{CO}_2}^{\text{in}} \dot{V}^{\text{in}}}{RT} = 3\cdot 10^{-4}\frac{p\dot{V}^{\text{in}}}{RT} = y\frac{\Lambda}{A},\ \text{hence}\ \dot{V}^{\text{in}} = \frac{RT}{p}\frac{y}{3\cdot 10^{-4}}\frac{\Lambda}{A}. \tag{9.73}$$

It follows that the volume of air needed for the production of 1g of glucose must at least – *i.e.* for $y = 1$ – be equal to $2.7\,\text{m}^3$.

The other gaseous inflow rates are therefore given by

---

* We ignore argon and choose p_α so that $p_{\text{N}_2} + p_{\text{O}_2} + p_{\text{CO}_2} = 1$ holds.

$$\dot{\nu}_{N_2}^{in} = p_{N_2}^{in} \frac{\dot{V}^{in}}{RT} = 0.7897 \frac{p\dot{V}^{in}}{RT} = \frac{0.7897}{3\cdot 10^{-4}} y \frac{\Lambda}{A}$$
$$\dot{\nu}_{O_2}^{in} = p_{O_2}^{in} \frac{\dot{V}^{in}}{RT} = 0.21 \frac{p\dot{V}^{in}}{RT} = \frac{0.21}{3\cdot 10^{-4}} y \frac{\Lambda}{A}. \tag{9.74}$$

Hence follow the outflow rates from (9.70)

$$\dot{\nu}_{CO_2}^{out} = (y-1)\frac{\Lambda}{A}, \quad \dot{\nu}_{N_2}^{out} = \frac{0.7897}{3\cdot 10^{-4}} y \frac{\Lambda}{A}, \quad \dot{\nu}_{O_2}^{out} = \left(1 + \frac{0.21}{3\cdot 10^{-4}} y\right)\frac{\Lambda}{A}. \tag{9.75}$$

The glucose produced and the outflowing water vapor correspond to the molar flow rates

$$\dot{\nu}_{G}^{out} = \frac{1}{6}\frac{\Lambda}{A} \text{ and } \dot{\nu}_{W}^{out} = (x-1)\frac{\Lambda}{A}. \tag{9.76}$$

Therefore by (9.71) and (9.73) through (9.76) all in- and outflows are expressed by the reaction rate Λ.

We consider the water vapor as an ideal gas. Therefore the partial pressures of the outflowing gases all follow from $p_\alpha^{out}\dot{V}^{out} = \dot{\nu}_\alpha^{out} RT$. Hence, by summation we have

$$\dot{V}^{out} = \frac{RT}{p}\left[\frac{y}{3\cdot 10^{-4}} + (x-1)\right]\frac{\Lambda}{A} \tag{9.77}$$

and, finally, we obtain the partial pressures of the outflowing gases

$$\frac{p_{CO_2}^{out}}{p} = \frac{3\cdot 10^{-4}(y-1)}{y + 3\cdot 10^{-4}(x-1)} \quad \frac{p_{N_2}^{out}}{p} = \frac{0.7879 y}{y + 3\cdot 10^{-4}(x-1)}$$
$$\frac{p_{O_2}^{out}}{p} = \frac{3\cdot 10^{-4} + 0.21 y}{y + 3\cdot 10^{-4}(x-1)} \quad \frac{p_{W}^{out}}{p} = \frac{3\cdot 10^{-4}(x-1)}{y + 3\cdot 10^{-4}(x-1)}. \tag{9.78}$$

9.6.3 *Balance of energy. Why a plant needs lots of water*

We continue to consider the open system of Fig. 9.7 and balance the in- and outgoing energy fluxes as follows

$$\left(650 \frac{\text{W}}{\text{m}^2} - J\right) a = \sum_\alpha \tilde{h}_\alpha^{out} \dot{\nu}_\alpha^{out} - \sum_\alpha \tilde{h}_\alpha^{in} \dot{\nu}_\alpha^{in}. \tag{9.79}$$

$\tilde{h}_\alpha$ denote the molar enthalpies of the in- and outgoing substances. The left hand side recognizes the heating of the leaf by the sun and the cooling of the leaf by radiation; a is the surface area of the leaf. We recall the considerations of Chap. 7 about heat radiation. There – in Paragraph 7.4.1 – we had stated that about half of the solar constant reaches the surface of the earth; hence the value of

$650\,\mathrm{W/m^2}$ for the influx in (9.79). If we consider the leaf as a black body,* we obtain by the T^4-law and by use of (9.70)

$$\left(650\frac{\mathrm{W}}{\mathrm{m}^2}-5.66\cdot10^{-8}\frac{T^4}{\mathrm{K}^4}\right)\frac{\mathrm{W}}{\mathrm{m}^2}=\sum_\alpha \tilde{h}_\alpha^{\mathrm{out}}\gamma_\alpha\frac{\Lambda}{A}-\sum_\alpha\left(\tilde{h}_\alpha^{\mathrm{out}}-\tilde{h}_\alpha^{\mathrm{in}}\right)\dot{\nu}_\alpha^{\mathrm{in}}\,. \quad (9.80)$$

The only constituent for which $\tilde{h}_\alpha^{\mathrm{out}}$ differs from its reference value $\tilde{h}_\alpha^{\mathrm{R}}$ is water, since water is released as a vapor; therefore we have to set $\tilde{h}_\mathrm{W}^{\mathrm{out}}=\tilde{h}_\mathrm{W}^{\mathrm{R}}+\tilde{r}(T)$, where $\tilde{r}(T)$ is the molar heat of evaporation of water. Furthermore, only water contributes a difference $\tilde{h}_\alpha^{\mathrm{out}}-\tilde{h}_\alpha^{\mathrm{in}}$ in the sum, because the enthalpies of the gases N_2, O_2, and CO_2 depend only on the temperature which is equal for the in- and outflows. Considering all this we obtain an equation for x from $(9.71)_3$

$$\left(650\frac{\mathrm{W}}{\mathrm{m}^2}-5.66\cdot10^{-8}\frac{T^4}{\mathrm{K}^4}\right)\mathrm{W}=\left[\Delta\tilde{h}_\mathrm{R}+\tilde{r}(T)(x-1)\right]\frac{\Lambda/A}{a/\mathrm{m}^2}\,. \quad (9.81)$$

Under favorable conditions a leaf produces 20 g glucose per day and per m^2, *i.e.* $\frac{1}{9}$ mol so that Λ/A equals $0.67\,\mathrm{mol/day}$. Thus from $\Delta\tilde{h}_\mathrm{R}=466.3\,\mathrm{kJ/mol}$, $\tilde{r}(T)=44.0\,\mathrm{kJ/mol}$ and $T=298\,\mathrm{K}$ we obtain $x=600$! In other words the plant needs 600 times as much water than would be required for the formation of glucose according to the stoichiometric reaction. Of course, farmers, gardeners, and housewives know about this, and water their plants sufficiently.

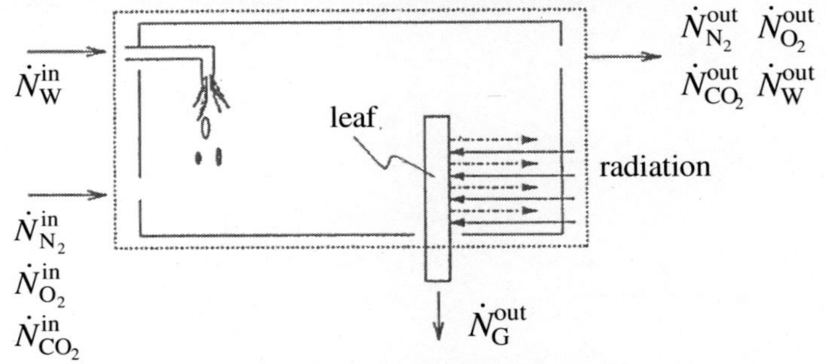

Fig. 9.7 Control volume for stationary growth of a plant

The additional mass of water is evaporated and one may say that the plant cools itself by evaporation – just as animals do. Indeed, if x were equal to one we should obtain $T=327\,\mathrm{K}$ from (9.81), *i.e.* 54°C, and no plant can perform photosynthesis at such a high temperature.

---

* In the visible part of the spectrum the leaf is *not* a black body; it absorbs primarily the red and yellow light and uses it for photosynthesis. This is why the leaf looks green. But this concerns only the small visible part of the total spectrum.

9.6.4 *Balance of entropy. Why a plant needs air*

The balance of entropy for the open system of Fig. 9.8 reads

$$\frac{\left(650\frac{\mathrm{W}}{\mathrm{m}^2}-J\right)a}{T}\leq\sum_\alpha \tilde{s}_\alpha^{\mathrm{out}}\dot{\nu}_\alpha^{\mathrm{out}}-\sum_\alpha \tilde{s}_\alpha^{\mathrm{in}}\dot{\nu}_\alpha^{\mathrm{in}}\,. \tag{9.82}$$

The left hand side represents the heat exchanged by the system with the surroundings by radiation, divided by T. According to the Second Law this quantity must be smaller than the difference of the entropic in- and outflow.

Between (9.82) and (9.79) we eliminate the heat exchanged, and obtain with

$$\sum_\alpha\left(\tilde{h}_\alpha^{\mathrm{out}}-T\tilde{s}_\alpha^{\mathrm{out}}\right)\gamma_\alpha\frac{\Lambda}{A}+\sum_\alpha\left[\tilde{h}_\alpha^{\mathrm{out}}-\tilde{h}_\alpha^{\mathrm{in}}-T\left(\tilde{s}_\alpha^{\mathrm{out}}-\tilde{s}_\alpha^{\mathrm{in}}\right)\right]\dot{\nu}_\alpha^{\mathrm{in}}\leq 0\,. \tag{9.83}$$

The terms with $\tilde{h}_\alpha^{\mathrm{out}}$ and $\tilde{h}_\alpha^{\mathrm{in}}$ have been discussed before, in Paragraph 9.6.3. For $T=T_{\mathrm{R}}$ and $p=p_{\mathrm{R}}$ we have

$$\begin{aligned}
\tilde{h}_{\mathrm{CO_2}}^{\mathrm{in}}&=\tilde{h}_{\mathrm{CO_2}}^{\mathrm{R}} & \tilde{h}_{\mathrm{CO_2}}^{\mathrm{out}}&=\tilde{h}_{\mathrm{CO_2}}^{\mathrm{R}}\\
\tilde{h}_{\mathrm{W}}^{\mathrm{in}}&=\tilde{h}_{\mathrm{W}}^{\mathrm{R}} & \tilde{h}_{\mathrm{W}}^{\mathrm{out}}&=\tilde{h}_{\mathrm{W}}^{\mathrm{R}}+\tilde{r}_{\mathrm{W}}(T)\\
\tilde{h}_{\mathrm{O_2}}^{\mathrm{in}}&=\tilde{h}_{\mathrm{O_2}}^{\mathrm{R}} & \tilde{h}_{\mathrm{O_2}}^{\mathrm{out}}&=\tilde{h}_{\mathrm{O_2}}^{\mathrm{R}}\\
\tilde{h}_{\mathrm{N_2}}^{\mathrm{in}}&=\tilde{h}_{\mathrm{N_2}}^{\mathrm{R}} & \tilde{h}_{\mathrm{N_2}}^{\mathrm{out}}&=\tilde{h}_{\mathrm{N_2}}^{\mathrm{R}}\\
& & \tilde{h}_{\mathrm{G}}^{\mathrm{out}}&=\tilde{h}_{\mathrm{G}}^{\mathrm{R}}\,.
\end{aligned} \tag{9.84}$$

The entropy terms are more complex than the enthalpies, because for the gaseous constituents the entropies depend also on the partial pressures p_α and not only on temperature. We obtain

$$\begin{aligned}
\tilde{s}_{\mathrm{CO_2}}^{\mathrm{in}}&=\tilde{s}_{\mathrm{CO_2}}^{\mathrm{R}}-R\ln\frac{p_{\mathrm{CO_2}}^{\mathrm{in}}}{p} & \tilde{s}_{\mathrm{CO_2}}^{\mathrm{out}}&=\tilde{s}_{\mathrm{CO_2}}^{\mathrm{R}}-R\ln\frac{p_{\mathrm{CO_2}}^{\mathrm{out}}}{p}\\
\tilde{s}_{\mathrm{W}}^{\mathrm{in}}&=\tilde{s}_{\mathrm{W}}^{\mathrm{R}} & \tilde{s}_{\mathrm{W}}^{\mathrm{out}}&=\tilde{s}_{\mathrm{W}}^{\mathrm{R}}-R\ln\frac{p_{\mathrm{W}}^{\mathrm{out}}}{p(T)}+\frac{\tilde{r}_{\mathrm{W}}(T)}{T}\\
\tilde{s}_{\mathrm{O_2}}^{\mathrm{in}}&=\tilde{s}_{\mathrm{O_2}}^{\mathrm{R}}-R\ln\frac{p_{\mathrm{O_2}}^{\mathrm{in}}}{p} & \tilde{s}_{\mathrm{O_2}}^{\mathrm{out}}&=\tilde{s}_{\mathrm{O_2}}^{\mathrm{R}}-R\ln\frac{p_{\mathrm{O_2}}^{\mathrm{out}}}{p}\\
\tilde{s}_{\mathrm{N_2}}^{\mathrm{in}}&=\tilde{s}_{\mathrm{N_2}}^{\mathrm{R}}-R\ln\frac{p_{\mathrm{N_2}}^{\mathrm{in}}}{p} & \tilde{s}_{\mathrm{N_2}}^{\mathrm{out}}&=\tilde{s}_{\mathrm{N_2}}^{\mathrm{R}}-R\ln\frac{p_{\mathrm{N_2}}^{\mathrm{out}}}{p}\\
& & \tilde{s}_{\mathrm{G}}^{\mathrm{out}}&=\tilde{s}_{\mathrm{G}}^{\mathrm{R}}\,.
\end{aligned} \tag{9.85}$$

$p(T)$ in $\tilde{s}_{\mathrm{W}}^{\mathrm{out}}$ is the saturated vapor pressure of water at $298\,\mathrm{K}$ so that $p(T)=0.032\,\mathrm{bar}$ holds according to Table 2.4. One has to consider that the outflowing water is in the vapor phase, whereas water in the reference state is liquid. In order to calculate $\tilde{s}_{\mathrm{W}}^{\mathrm{out}}$, we therefore begin with $\tilde{s}_{\mathrm{W}}^{\mathrm{R}}$ and then add the entropy of evaporation $\tilde{r}_{\mathrm{W}}(T)/T$ at T, and then we let the vapor expand to $p_{\mathrm{W}}^{\mathrm{out}}$.

We insert (9.84), (9.85) in the entropy balance (9.83) and use (9.71) and (9.74) to replace the inflowing molar flow rates. The heat of evaporation $\tilde{r}_W(T)$ drops out everywhere – always for $T = T_R$ – and we obtain the inequality

$$\Delta\tilde{h}_R - T\Delta\tilde{s}_R + RT\left(-\ln\frac{p_{CO_2}^{out}}{p} + \ln\frac{p_{O_2}^{out}}{p} - \ln\frac{p_W^{out}}{p(T)}\right) + \tag{9.86}$$

$$+RT\left(y\ln\frac{p_{CO_2}^{out}}{p} + x\ln\frac{p_W^{out}}{p(T)} + \frac{0.21}{3\cdot 10^{-4}}y\ln\frac{p_{O_2}^{out}}{p_{O_2}^{in}} + \frac{0.7897}{3\cdot 10^{-4}}y\ln\frac{p_{N_2}^{out}}{p_{N_2}^{in}}\right) \le 0\,.$$

The pressures of the in- and outflowing gases are known from (9.72) and (9.78). Also we know $\Delta\tilde{h}_R - T\Delta\tilde{s}_R = 487.3\,\mathrm{kJ/mol}$, *cf.* Paragraph 9.6.1. Therefore the left-hand side of the inequality (9.84) is an explicit function of x and y. However, we also know x from the considerations of Paragraph 9.6.3, where we have obtained $x = 600$. Therefore the left-hand side of the inequality (9.86) depends only on y and we may ask for *the* values of y that satisfy the inequality.

We recall that y characterizes the inflowing amount of air. When $y = 1$ holds, the air carries just enough CO_2 to make the chemical reaction possible. However, for $y = 1$ the inequality (9.86) cannot be satisfied, and we conclude that *more air* needs to be supplied. All logarithmic terms in (9.86) depend on y, but most of them are very small. The leading term, – the one that is by far the largest –, is the term with x. Therefore we may write the inequality in the shortened approximate form

$$\Delta\tilde{h}_R - T\Delta\tilde{s}_R + RT\,x\ln\frac{p_W^{out}}{p(T)} \le 0\,. \tag{9.87}$$

With $(9.78)_4, T = 298\,\mathrm{K}$, $p = 1\,\mathrm{bar}$, $p(T) = 0.032\,\mathrm{bar}$, $x = 600, \Delta\tilde{g}_R = 478.3\frac{\mathrm{kJ}}{\mathrm{mol}}$ we obtain

$$478.3 + 2.48\cdot 600\ln\frac{3\cdot 10^{-4}\cdot 599}{0.032\cdot(y + 3\cdot 10^{-4}\cdot 599)} \le 0\,. \tag{9.88}$$

The smallest value of y that satisfies this inequality is $y = 7.5$. Therefore we must supply seven to eight times as much air to the plant as required by the stoichiometric equation.

9.6.5 *Discussion*

Photosynthesis of glucose is a thermodynamically precarious process: The energy – or enthalpy – grows and the entropy decreases. We recall from Paragraph 4.2.7 that an energy *decrease* and an *increase* of entropy are conducive to equilibrium. The growth of energy is the smaller problem, because the plant may use the solar energy which is abundantly available – actually it is too abundant.

Indeed, we have seen that the energy transmitted to the plant by radiation would heat the plant to such a degree that photosynthesis is not possible. The plant must cool itself by the evaporation of water. From the First Law we have determined the appropriate water supply: The plant needs 600 times more water than it uses for the fabrication of glucose.

But watering is not enough: A plant must also have sufficient air. Indeed, the Second Law demands seven to eight times more air than needed for the glucose formation. The additional air lowers the partial pressure of the evaporated water and thus gives the water vapor sufficient volume to increase its entropy by mixing.

10 Moist air

10.1 Characterization of moist air

10.1.1 *Moisture content*

Unsaturated moist air is a mixture of air and water vapor. Both are considered as ideal gases. Indeed, we use the ideal gas laws for water vapor all the way down to the state of saturation, where condensation occurs. For a given temperature this occurs at the pressure $p' = p(T)$ which may be read off from Table 2.4.

Saturated moist air contains air and saturated water vapor, and may contain liquid water in the form of droplets as in a fog or a cloud.

The total pressure of moist air will be denoted by p_0 in this chapter, the partial pressure of water is p or, in the case of saturation $p' = p(T)$. A volume containing air and water, either as a vapor, or as saturated vapor and liquid – is characterized by the moisture content x

$$x = \frac{m_{\mathrm{W}}}{m_{\mathrm{A}}} \tag{10.1}$$

which gives the ratio of the masses of water and air. If there is no liquid water in the volume, but the vapor is just saturated we write the moisture content as

$$x' = \frac{m'_{\mathrm{V}}(T)}{m_{\mathrm{A}}}, \tag{10.2}$$

where $m'_{\mathrm{V}}(T)$ is the maximal mass of vapor which the air may contain at the given temperature and a given volume.

10.1.2 *Enthalpy of moist air*

We recall the First Law of thermodynamics in the form

$$\dot{Q} = \frac{\mathrm{d}U}{\mathrm{d}t} + p_0 \frac{\mathrm{d}V}{\mathrm{d}t} = \frac{\mathrm{d}H}{\mathrm{d}t} - V\frac{\mathrm{d}p_0}{\mathrm{d}t}. \tag{10.3}$$

It follows that heating at constant p_0 changes the enthalpy $H = U + p_0 V$. Many processes in moist air occur at a fixed pressure p_0, and therefore the enthalpy is the most important caloric quantity in this field.

Since the specific internal energy u contains an arbitrary additive constant, so does the specific enthalpy h. In the field of moist air it is common practice to choose these constants for air and liquid water such that air and water of 0°C have the enthalpies zero. We denote the Celsius temperature by τ and obtain for the enthalpies H and H^{S} of non-saturated and saturated moist air, respectively

$$\begin{aligned} H &= m_{\mathrm{A}} c_p^{\mathrm{A}} \tau + m_{\mathrm{V}} \left(r_0 + c_p^{\mathrm{V}} \tau\right) \\ H^{\mathrm{S}} &= m_{\mathrm{A}} c_p^{\mathrm{A}} \tau + m'_{\mathrm{V}} \left(r_0 + c_p^{\mathrm{V}} \tau\right) + m_{\mathrm{L}} c^{\mathrm{L}} \tau . \end{aligned} \tag{10.4}$$

Here m_{A}, m_{V}, and m_{L} are the masses of air, vapor, and liquid water, respectively, and c_p^{A}, c_p^{V}, and c^{L} are the corresponding specific heats. r_0 is the heat of evaporation of liquid water at 0°C. The numerical values are

$$c_p^{\mathrm{A}} = 1.004 \tfrac{\mathrm{kJ}}{\mathrm{kgK}}\ ,\quad c_p^{\mathrm{V}} = 1.86 \tfrac{\mathrm{kJ}}{\mathrm{kgK}}\ ,\quad c^{\mathrm{L}} = 4.18 \tfrac{\mathrm{kJ}}{\mathrm{kgK}}\ ,\quad r_0 = 2500 \tfrac{\mathrm{kJ}}{\mathrm{kg}}\ . \tag{10.5}$$

In thermodynamics of moist air it is common practice to relate the enthalpy to the mass of air m_{A} rather than to the total mass. This is convenient since there are many processes where the mass of air remains constant. Therefore a new specific enthalpy is defined, namely

$$h_{1+x} = \frac{H}{m_{\mathrm{A}}} = \frac{m}{m_{\mathrm{A}}}\frac{H}{m} = \frac{m_A + m_W}{m_A} h = (1+x)h\ ;$$

this is obviously the enthalpy of $(1+x)\mathrm{kg}$ of moist air. From (10.4) we obtain

$$\begin{aligned} h_{1+x} &= c_p^{\mathrm{A}}\tau + x\left(r_0 + c_p^{\mathrm{V}}\tau\right) \\ h_{1+x}^{\mathrm{S}} &= c_p^{\mathrm{A}}\tau + x'\left(r_0 + c_p^{\mathrm{V}}\tau\right) + (x - x')c^{\mathrm{L}}\tau \end{aligned} \tag{10.6}$$

for unsaturated and saturated moist air. We may combine these relations and write

$$h_{1+x}^{\mathrm{S}} = h_{1+x'} + (x - x')c^{\mathrm{L}}\tau\ , \tag{10.7}$$

where $h_{1+x'}$ is the value of h_{1+x} for $x = x'$.

10.1.3 *Table for moist air*

Table 10.1 presents properties of saturated moist air as functions of τ for $1\,\mathrm{bar}$. The entries for $p' = p(T)$ are those from Table 2.4. The values for the saturated moisture content x' are calculated from

$$m_{\mathrm{V}}' = \frac{p'V}{\frac{R}{M_{\mathrm{W}}}T} \quad \text{and} \quad m_{\mathrm{A}}' = \frac{(p_0 - p')V}{\frac{R}{M_{\mathrm{A}}}T} \quad \text{as the quotient} \quad x' = \frac{M_{\mathrm{W}}}{M_{\mathrm{A}}}\frac{p'}{p_0 - p'} \tag{10.8}$$

and $h_{1+x'}$ follows from (10.6) for $x = x'$.

From the table we may learn that for 0°C the moist air may contain a maximum of $3.8\,\mathrm{g}$ water in $1\,\mathrm{kg}$ air without condensation; at 20°C this value has risen to $14.9\,\mathrm{g}$, and for 70°C we have 281.5 g. So, the water carrying capacity of air increases drastically with temperature. The table may also be used to answer the question for the temperature at which a given mass m_{W} of liquid water fully evaporates in a given mass m_{A} of air and for a pressure $p_0 = 1\mathrm{bar}$. According to (10.8) we have

$$\frac{m_{\mathrm{V}}'}{m_{\mathrm{A}}} = \frac{M_{\mathrm{W}}}{M_{\mathrm{A}}}\frac{p(T)}{p_0 - p(T)}$$

and, if m_{V}' equals m_{W}, the water is fully evaporated. The temperature at which this happens is given implicitly by

$$p(T) = p_0 \frac{x'}{\frac{M_W}{M_A} + x'}.$$

Thus for instance for $x' = 0.01$ we obtain $p(T) = 0.016\,\text{bar}$, hence $T \approx 14°\text{C}$. And for $x' = 0.2$ we obtain $p(T) = 0.24\,\text{bar}$, hence $T \approx 64°\text{C}$. Those temperatures are called *dew points*.

Table 10.1 Properties of moist air at $p_0 = 1$ bar

°C	p' mbar	x' g/kg	$h_{1+x'}$ kJ/kg	°C	p' mbar	x' g/kg	$h_{1+x'}$ kJ/kg
0	6.107	3.822	9.56	35	25.22	37.05	130.3
1	6.566	4.111	11.29	36	59.40	39.28	137.1
2	7.054	4.419	13.06	37	62.74	41.64	144.2
3	7.575	4.747	14.91	38	66.24	44.12	151.7
4	8.129	5.097	16.81	39	69.91	46.75	159.6
5	8.719	5.471	18.76	40	73.75	49.52	167.8
6	9.346	5.868	20.77	41	77.77	52.45	176.4
7	10.012	6.290	22.85	42	81.98	55.55	185.5
8	10.721	6.740	25.00	43	86.39	58.81	195.1
9	11.743	7.219	27.22	44	91.00	62.27	205.1
10	12.271	7.727	29.52	45	95.82	65.91	215.6
11	13.118	8.267	31.90	46	100.86	69.77	226.7
12	14.015	8.841	34.37	47	106.12	73.84	238.4
13	14.967	9.450	36.93	48	111.62	78.14	250.7
14	15.974	10.097	39.59	49	117.36	82.70	263.7
15	17.041	10.783	42.35	50	123.35	87.52	277.3
16	18.17	11.51	45.22	51	129.60	92.61	291.7
17	19.36	12.28	48.20	52	136.12	98.00	306.9
18	20.63	13.10	51.30	53	142.92	103.71	322.9
19	21.96	13.97	54.52	54	150.01	109.77	339.9
20	23.37	14.88	57.88	55	157.40	116.19	357.8
21	24.86	15.85	61.38	56	165.10	122.99	376.8
22	26.42	16.88	65.03	57	173.11	130.21	396.8
23	28.08	17.97	68.84	58	181.46	137.88	418.1
24	29.82	19.12	72.81	59	190.15	146.04	440.6
25	31.66	20.34	76.95	60	199.1	154.71	464.6
26	33.60	21.63	81.28	61	208.6	163.9	490.0
27	35.64	22.99	85.80	62	218.4	173.8	517.0
28	37.97	24.42	90.52	63	228.5	184.3	545.8
29	40.04	25.94	95.45	64	239.1	195.4	576.5
30	42.42	27.55	100.62	65	250.1	207.4	609.3
31	44.91	29.23	106.02	66	261.5	220.2	644.2
32	47.54	31.04	111.67	67	273.3	233.9	681.7
33	50.29	32.94	117.59	68	285.6	248.7	721.8
34	53.18	34.94	123.79	69	298.4	264.5	764.9
				70	311.6	281.5	811.3

10.1.4 *The* (h_{1+x}, x)*-diagram*

The (h_{1+x}, x)-diagram may serve for an easy evaluation of the effect • of heating moist air, or • mixing of two masses of moist air, or • of the supply of water to moist air. The origin of the diagram lies at $(h_{1+x}, x) = (0,0)$ and the oblique axes are shown in Fig. 10.1. We find h_{1+x} and x of a generic point by projecting this point perpendicularly to the axes. The h_{1+x}-axis is scaled so that the point $x = 0.01$ on the x-axis has an enthalpy of $h_{1+x} = 25\,\text{kJ/kg}$. Thus we conclude from $(10.5)_4$ and $(10.6)_1$ that the isotherm $\tau = 0$ of unsaturated moist air coincides with the x-axis.

Fig. 10.1 also shows the dew line, or fog line; it consists of the points $(h_{1+x'}, x')$ that may be read off from Table 10.1. Below the dew line lies the fog region, where we have states with saturated moist air and liquid water. Above the dew line no liquid water exists, this is the region of unsaturated moist air.

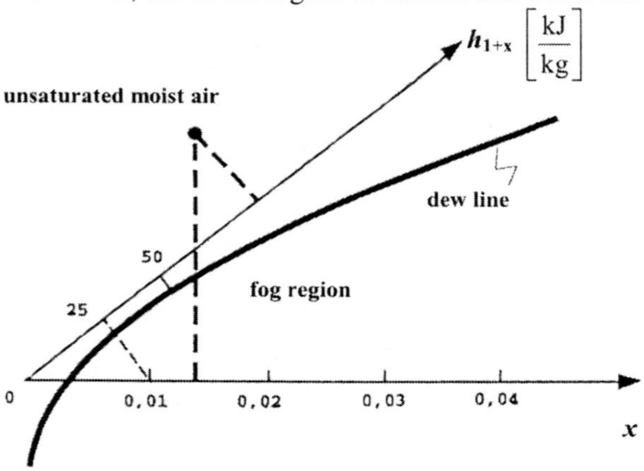

Fig. 10.1 (h_{1+x}, x)-diagram for $p_0 = 1$ bar

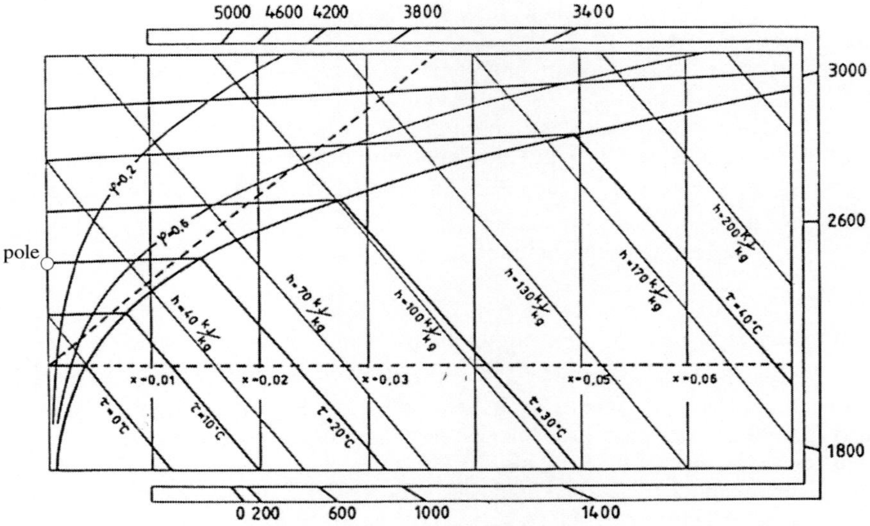

Fig. 10.2 (h_{1+x}, x)-diagram for moist air $p_0 = 1\,\text{bar}$. The axes are dashed. See text for the explanation of the marginal scale.

From (10.5) and $(10.6)_1$ we conclude that the isotherms $\tau > 0$ intersect the vertical line $x = 0$ in ever higher points for growing τ, and that their slope grows slowly. In the fog region the isotherm $\tau = 0$ coincides with the isenthalpic line $h_{1+x'(0)} = 9.56\frac{\text{kJ}}{\text{kg}}$ according to (10.7). All isotherms have a kink on the dew line. Fig. 10.2 shows a realistic (h_{1+x}, x)-diagram. The lines $\varphi = \text{const}$ are lines of a constant *degree of saturation* $x/x(\tau)$.

10.2 Simple processes in moist air

10.2.1 *Supply of water*

If one adds pure water with the specific enthalpy h – either as a liquid, or as wet steam, or as superheated steam – to moist air in the state (h_{1+x}, x), the enthalpy changes by $\Delta H = h\Delta m_W$. The moisture content x changes by $\Delta h_{1+x} = h\Delta x$. It is thus clear that h determines the direction along which the state changes away from its original position. For instance, if we supply ice water with $h = 0$, we move in the direction of the isenthalpic lines in the (h_{1+x}, x)-diagram. Or, if we supply saturated vapor of 100°C with $h = 2676\,\text{kJ/kg}$, we move into a direction with a slightly positive slope.

The (h_{1+x}, x)-diagram of Fig. 10.2 has a marginal scale which facilitates the determination of the direction of motion upon a supply of pure water. The lateral scale shows short lines with values of h in kJ/kg. These short lines must be connected to the pole on the left-hand side of the diagram, and the connecting line gives the direction of motion.

Thus we see that the addition of saturated vapor of 100°C to cold unsaturated moist air will always lead to fog. Also, the addition of liquid water to unsaturated moist air is always accompanied by cooling, even when the temperature of the newly supplied water is higher than the temperature of the air; indeed, the supply line points steeply downward for all values of h between $0\,\mathrm{kJ/kg}$ and $418\,\mathrm{kJ/kg}$. The reason for the cooling effect is clear: The moist air provides some of the heat needed for the evaporation of the supplied liquid and therefore it cools.

10.2.2 *Heating*

When moist air is heated, the moisture content remains unchanged and the state changes along a vertical line in the (h_{1+x}, x)-diagram. Knowing this we conclude that the heating of saturated moist air in the fog region, if sustained, will eventually lead to unsaturated moist air. Thus the early mist in river valleys quickly dissolves under the heating of the morning sun.

The opposite happens when ice cream "steams" on a sultry summer day: The nearly saturated air close to the ice cream is cooled and enters the fog region.

10.2.3 *Mixing*

If two masses of moist air with m_A^1, x^1, τ^1 and m_A^2, x^2, τ^2 are mixed, we obtain a state m_A, x, τ which – in the (h_{1+x}, x)-diagram – lies on a straight line between the states 1 and 2. Indeed, since the masses and the enthalpies add up, we have

$$m_\mathrm{A}(1+x) = m_\mathrm{A}^1(1+x^1) + m_\mathrm{A}^2(1+x^2) \quad \text{and} \quad m_\mathrm{A} h_{1+x} = m_\mathrm{A}^1 h_{1+x^1} + m_\mathrm{A}^2 h_{1+x^2}\,.$$

and hence follows

$$x = \frac{m_\mathrm{A}^1}{m_\mathrm{A}} x^1 + \frac{m_\mathrm{A}^2}{m_\mathrm{A}} x^2 \quad \text{and} \quad h_{1+x} = \frac{m_\mathrm{A}^1}{m_\mathrm{A}} h_{1+x^1} + \frac{m_\mathrm{A}^1}{m_\mathrm{A}} h_{1+x^2}\,. \tag{10.9}$$

Therefore the mixed state divides the connecting line of the initial states in the ratio $m_\mathrm{A}^1 / m_\mathrm{A}^2$ in the same way in which the center of mass between two masses divides the line between them. Students speak of the "center of gravity rule."

One observation to be drawn from this rule of mixing is the fact that the mixing of two saturated masses of moist air always leads to fog. This follows from the observation that the dew line is concave so that the connecting line of the states of the two masses lies entirely in the fog region. It is for this reason that the coast of Newfoundland frequently has foggy weather since that is the area where the saturated air masses of the warm Gulf stream and of the cold Labrador stream mix.

Even the mixing of saturated warm air with unsaturated cold air may produce fog as we see by inspection of the (h_{1+x}, x)-diagram. Thus in winter time, when we exhale the saturated moist air from our lungs and let it mix with the cold dry air from outside, we can see our breath, because small water droplets form. Or else: The steaming of a hot cup of coffee comes about, because the saturated air near the surface mixes with the colder unsaturated air of the room.

10.2.4 *Mixing of moist air with fog*

When we mix fog in the state m_A^1, x^1, τ^1 with unsaturated moist air in the state m_A^2, x^2, τ^2, we know that the state of the mixture lies on a straight line between the states 1 and 2 in the (h_{1+x}, x)-diagram.

It follows that the unsaturated partner becomes colder in the mixing process, independent of its original position. But the fog may become colder or warmer, or it may not change its temperature. This depends on the position of point 1 in relation to point 2, *cf.* Fig. 10.3. If state 2 – state 2´ in the figure – lies on the extension of the fog isotherm τ^1, the fog does not change its temperature. On the other hand when state 2 lies to the right or left of the extended isotherm, the fog becomes warmer or colder, respectively.

Note that we may thus obtain the surprising result that the mixing of two masses with the same temperature leads to a smaller temperature of the mixture. The explanation of such unexpected effects is always the same: There is at least partial evaporation going on and the heat of evaporation cools its environment.

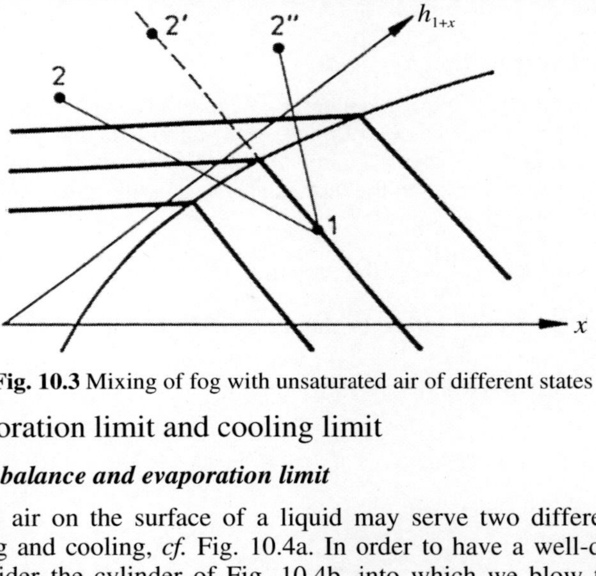

Fig. 10.3 Mixing of fog with unsaturated air of different states

10.3 Evaporation limit and cooling limit

10.3.1 *Mass balance and evaporation limit*

Blowing with air on the surface of a liquid may serve two different purposes, namely drying and cooling, *cf.* Fig. 10.4a. In order to have a well-defined problem, we consider the cylinder of Fig. 10.4b, into which we blow the moist air stream $\dot{m}_I = (1 + x_I)\dot{m}_A$ with temperature τ_I from the left. On the right-hand side the saturated air stream $\dot{m}_{II} = (1 + x'(\tau))\dot{m}_A$ leaves the cylinder. The air stream $\dot{m}_A$ is the same at the entry and at the exit so that the mass m_A is constant inside the cylinder. and $x = m_W / m_A$ are the temperature and the moisture content in the cylinder. Generally both are functions of time during sustained blowing.

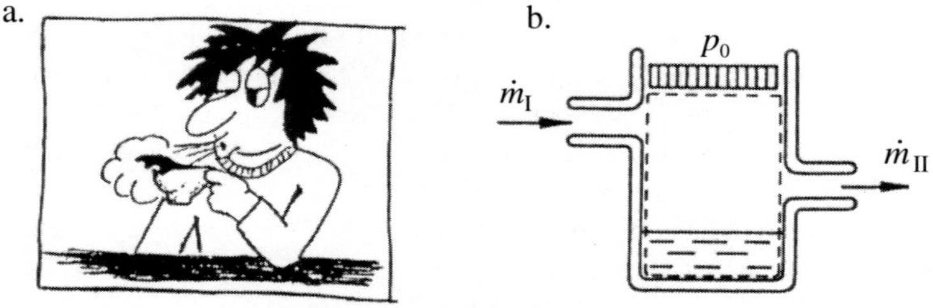

Fig. 10.4 a. Cooling coffee
b. On the balance equations of mass and energy.

The mass balance for the mass $x m_{\mathrm{A}}$ of water in the cylinder reads

$$\frac{\mathrm{d}\, x m_{\mathrm{A}}}{\mathrm{d}t} = \dot{m}_{\mathrm{A}}\left[x_{\mathrm{I}} - x'(\tau)\right] \text{, hence } \frac{\mathrm{d}\, x}{\mathrm{d}t} = \frac{\dot{m}_{\mathrm{A}}}{m_{\mathrm{A}}}\left[x_{\mathrm{I}} - x'(\tau)\right]. \tag{10.10}$$

It follows that water evaporates when the fresh air is dryer than the saturated air above the liquid. In the reverse case, *i.e.* for $x_{\mathrm{I}} > x'(\tau)$, some part of the entering vapor condenses. The areas of the (h_{1+x}, x)-diagram in which the fresh air must lie for evaporation or condensation to occur are thus separated by the vertical evaporation limit, *cf.* Fig. 10.5.

10.3.2 *Energy balance and cooling limit*

We recall the general equation of balance (1.2) and apply it to the energy inside the volume with the dashed surface in Fig. 10.4 b. The kinetic energy is neglected as well as gravitational and frictional forces. Also the cylinder is adiabatically isolated so that no heat flux occurs on the dashed surface. In this case the internal energy U inside the cylinder changes

- by the power $p_0 \frac{\mathrm{d}V}{\mathrm{d}t}$ of the pressure on the piston
- by the convective influx on the left and the power of the pressure p_0 at the entry point $\left(u_{\mathrm{I}} + \frac{p_0}{\rho_{\mathrm{I}}}\right)\dot{m}_{\mathrm{I}} = h_{\mathrm{I}}(1 + x_{\mathrm{I}})\dot{m}_{\mathrm{A}} = h_{1+x_{\mathrm{I}}}\dot{m}_{\mathrm{A}}$
- by the corresponding term at the outlet $h_{1+x'}(\tau)\dot{m}_{\mathrm{A}}$.

Therefore the energy balance reads

$$\frac{\mathrm{d}U}{\mathrm{d}t} = -p_0 \frac{\mathrm{d}V}{\mathrm{d}t} + \dot{m}_{\mathrm{A}}\left[h_{1+x_{\mathrm{I}}} - h_{1+x'}(\tau)\right] \text{, hence } \frac{\mathrm{d}H}{\mathrm{d}t} = \dot{m}_{\mathrm{A}}\left[h_{1+x_{\mathrm{I}}} - h_{1+x'}(\tau)\right]$$

or, by $H = m_{\mathrm{A}} h^{\mathrm{S}}_{1+x}$

$$\frac{\mathrm{d}h^{\mathrm{S}}_{1+x}}{\mathrm{d}t} = \frac{\dot{m}_{\mathrm{A}}}{m_{\mathrm{A}}}\left[h_{1+x_{\mathrm{I}}} - h_{1+x'}(\tau)\right]. \tag{10.11}$$

We insert h^{S}_{1+x}, $h_{1+x_{\mathrm{I}}}$, and $h_{1+x'}$ from (10.6) and (10.7 and obtain the differential equation

$$\left(c_p^{\mathrm{A}} + x'(\tau)c_p^{\mathrm{V}} + [x - x'(\tau)]c^{\mathrm{L}} + \left(r_0 + c_p^{\mathrm{V}}\tau - c^{\mathrm{L}}\tau\right)\frac{\mathrm{d}x'}{\mathrm{d}t}\right)\frac{\mathrm{d}\tau}{\mathrm{d}t} + c^{\mathrm{L}}\tau\frac{\mathrm{d}x}{\mathrm{d}t} =$$

$$= \frac{\dot{m}_{\mathrm{A}}}{m_{\mathrm{A}}}\left[c_p^{\mathrm{A}}\tau_{\mathrm{I}} + x_{\mathrm{I}}\left(r_0 + c_p^{\mathrm{V}}\tau_{\mathrm{I}}\right) - c_p^{\mathrm{A}}\tau - x'(\tau)\left(r_0 + c_p^{\mathrm{V}}\tau\right)\right]. \qquad (10.12)$$

The two equations (10.10) and (10.12) represent a system of two ordinary differential equations for the determination of $x(t)$ and $\tau(t)$ for sustained blowing of air into the cylinder. A solution must be determined numerically, since $x'(\tau)$ is not known analytically. We leave that solution to the interested reader who, of course, must first specify the parameters of the system, namely τ_{I}, x_{I}, $\dot{m}_{\mathrm{A}}/m_{\mathrm{A}}$, and the initial conditions $\tau(0)$, $x(0)$.

Here we limit the attention to the stationary case, where τ has eventually reached a constant value.* In this case we obtain from (10.10) and (10.12)

$$\underbrace{c_p^{\mathrm{A}}\tau + x'(\tau)\left(r_0 + c_p^{\mathrm{V}}\tau\right) + [x_{\mathrm{I}} - x'(\tau)]c^{\mathrm{L}}\tau}_{h^{\mathrm{S}}_{1+x_{\mathrm{I}}}(\tau)\text{ by }(10.6)_2} = \underbrace{c^{\mathrm{L}}\tau_{\mathrm{I}} + x_{\mathrm{I}}\left(r_0 + c_p^{\mathrm{V}}\tau_{\mathrm{I}}\right)}_{h_{1+x_{\mathrm{I}}}(\tau_{\mathrm{I}})\text{ by }(10.6)_1}. \qquad (10.13)$$

This is an equation for the stationary value τ which, again, cannot be solved analytically. But graphically – in the (h_{1+x}, x)-diagram – the solution is easy to find, because the left hand side of (10.13) is the equation for the fog isotherm, as indicated in (10.13). It follows that the stationary temperature τ is *the* value of τ, whose extended fog isotherm passes through the state $(x_{\mathrm{I}}, \tau_{\mathrm{I}})$ of the fresh air in the unsaturated region. In Fig. 10.5 a this extended fog isotherm is shown; it is called the cooling limit. Rather obviously, if the fresh air state $(x_{\mathrm{I}}, \tau_{\mathrm{I}})$ is below the cooling limit, the liquid is cooled by sustained blowing, otherwise it is heated.

Note that the liquid is cooled even if the fresh air is hotter than the liquid, it only needs to be dry enough for high temperature. The cooling effect in such cases is due to the partial evaporation of the liquid and the demand for the latent heat of evaporation.

Note also that the result about the cooling limit fits well into the earlier consideration of mixing unsaturated moist air and fog, *cf.* Paragraph 10.2.4. Indeed, the point $2'$ in Fig. 10.3 lies on the cooling limit of the fog in point 1; therefore the fog does not change its temperature when we mix masses of moist air of the states 1 and $2'$.

---

* Note that x continues to change when τ is constant.

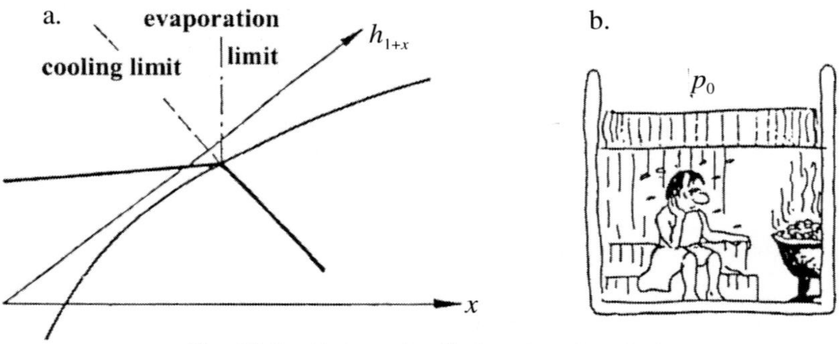

Fig. 10.5 a. Evaporation limit and cooling limit
b. Sauna

10.4 Two Instructive Examples: Sauna and Cloud Base

10.4.1 *A sauna is prepared*

We prepare a sauna by mixing dry air of temperature $\tau_1 = 10°C$ and saturated moist air of temperature $\tau_2 = 24°C$ in such a manner that the mixed air has the moisture content $x_3 = 0.01$. Afterwards we heat the mixture until the cooling limit of liquid of 40°C is reached. We ask for the corresponding temperature τ_4 of the air in the sauna.*

Before we answer the question we shall discuss its significance: The skin of the user of the sauna should not become hotter than 40°C and it will be kept at this "low" temperature by the evaporation of sweat. This means that the surrounding moist air must lie on the cooling limit of the liquid of 40°C. Therefore the sauna should not be heated above this limit. We determine the appropriate temperature in the sauna.

In order to solve the problem we might try to use the (h_{1+x}, x)-diagram of Fig. 10.2, which is repeated in Fig. 10.6 for the present problem. The states 1, 2 , and 3 are marked there by small circles. After all what we have learned, state 3 should lie on a straight line between states 1 and 2 and it should have the moisture content $x_3 = 0.01$; the corresponding temperature is $\tau_3 \approx 17°C$. During the subsequent heating the state of the mixed air moves upwards from point 3 along a vertical line until it reaches the cooling limit for 40°C. Unfortunately the intersection of these lines lies outside the diagram**. Therefore we must determine the temperature τ_4 by calculation, using the equation (10.13). In that equation we set $\tau = 40°C$,

---

* The pressure is $p_0 = 1\text{bar}$ and it remains constant. Therefore air must escape from the sauna through cracks in the door or the walls. In this way an ever smaller mass of air is heated as the heating proceeds. This fact would be a complication for our calculation which we avoid by looking at a situation as shown in Fig. 10.5, in which pressure *and mass* remain unchanged.

** The diagram *is* used in climate control but not under such extreme conditions.

$x_I = 0.01$, and solve for τ_I which is the desired temperature, here called τ_4. We obtain

$$\tau_4 = \tau_I = 136°\text{C}.$$

The user of the sauna survives this high temperature only because his sweat evaporates and thereby cools the skin by withdrawing the heat of evaporation from it.

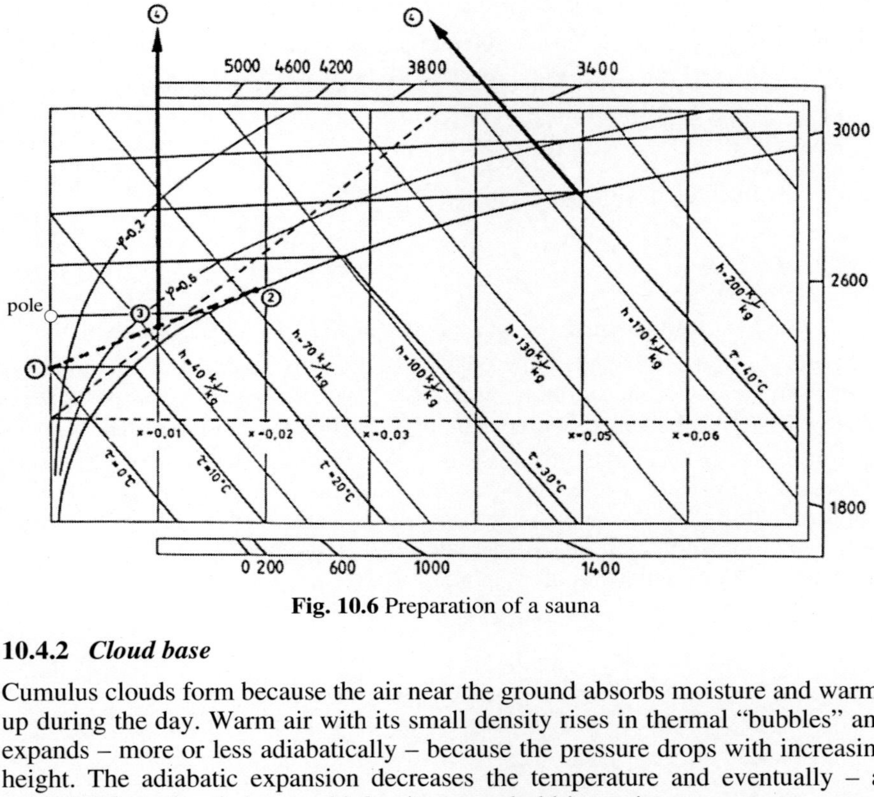

Fig. 10.6 Preparation of a sauna

10.4.2 *Cloud base*

Cumulus clouds form because the air near the ground absorbs moisture and warms up during the day. Warm air with its small density rises in thermal "bubbles" and expands – more or less adiabatically – because the pressure drops with increasing height. The adiabatic expansion decreases the temperature and eventually – at some height – the air is so cold that it cannot hold its moisture content as vapor; droplets form and a cumulus cloud forms.

In Paragraph 1.4.5 we have given the measured average temperature in the atmosphere and we have calculated the pressure as a function of height, *cf.* (1.26). That formula also applies – to a very good approximation – to moist air and we write

$$p_0 = p_{0G}\left(1 - \frac{\gamma}{T_G} z\right)^{\frac{g/\gamma}{R/M_A}}, \quad \text{where} \quad \begin{array}{ll} p_{0G} = 1\,\text{bar} & \gamma = \frac{0.65\,\text{K}}{100\ \text{m}} \\ T_G = 15°\text{C} & M_A = 29 \end{array}. \tag{10.14}$$

The adiabatic cooling of the rising moist air is described by the First Law with $\dot{Q} = 0$, *i.e.*

$$\mathrm{d}U = -p_0 \mathrm{d}V \quad \text{or} \quad \mathrm{d}H = V\mathrm{d}p_0 \,. \tag{10.15}$$

The constitutive equations for H and V are

$$H = m_{\mathrm{A}}\left(c_p^{\mathrm{A}}(T_{\mathrm{A}} - T_{\mathrm{S}}) + x\left[r_0 + c_p^{\mathrm{V}}(T_{\mathrm{A}} - T_{\mathrm{S}})\right]\right)$$
$$V = m_{\mathrm{A}}\left(\frac{R}{M_{\mathrm{A}}} + x\frac{R}{M_{\mathrm{W}}}\right)\frac{T_{\mathrm{A}}}{p_0}\,, \tag{10.16}$$

where $(10.16)_2$ is the ideal gas law for the air-vapor mixture, T_{A} is the temperature of the rising air, and T_{S} is our reference temperature; in meteorology we choose $T_{\mathrm{S}} = 273.15\,\mathrm{K}$. Insertion into $(10.15)_2$ gives the differential equation

$$\left(c_p^{\mathrm{A}} + x c_p^{\mathrm{V}}\right)\mathrm{d}T_{\mathrm{A}} = \left(\frac{R}{M_{\mathrm{A}}} + x\frac{R}{M_{\mathrm{W}}}\right)\frac{T_{\mathrm{A}}}{p_0}\mathrm{d}p_0\,,$$

which is easily integrated with the result

$$T_{\mathrm{A}} = T_{\mathrm{E}}\left(\frac{p_0}{p_{0\mathrm{G}}}\right)^{\frac{R/M_{\mathrm{A}}}{c_p^{\mathrm{A}}}\frac{1+xM_{\mathrm{A}}/M_{\mathrm{W}}}{1+xc_p^{\mathrm{V}}/c_p^{\mathrm{A}}}}\,. \tag{10.17}$$

Here T_{E} is the temperature which the "air bubble" had reached by contact with the ground before it rose. The temperature of the rising thermal bubble may be quite different from the one of the surrounding air; that air is at rest and its temperature is given by $T_{\mathrm{G}} - \gamma z$, the temperature profile of the standard atmosphere. The temperature exchange between the air masses is slow and little effective. This is not true, however, for the pressure exchange and we may assume that the pressure in the rising bubble is equal to that in the surrounding air so that it is given by (10.14). Therefore we may insert (10.14) into (10.17) and obtain the temperature T_{A} of the rising air as a function of z, namely

$$T_{\mathrm{A}} = T_{\mathrm{E}}\left(1 - \frac{\gamma}{T_{\mathrm{G}}}z\right)^{\frac{g}{\gamma c_p^{\mathrm{A}}}\frac{1+xM_{\mathrm{A}}/M_{\mathrm{W}}}{1+xc_p^{\mathrm{V}}/c_p^{\mathrm{A}}}}\,. \tag{10.18}$$

This temperature determines the vapor pressure of saturation $p'(z) = p'(T_{\mathrm{A}}(z))$ which may be read off from Table 10.2.

Table 10.2 Temperature T_{A} of the rising air. Saturation vapor pressure $p'(T_{\mathrm{A}})$ and actual vapor pressure p; all as functions of height z

z [m]	T_A[K]	$p'(T_A)$[hPa]	p [hPa]
0	298	31.7	16.0
500	293	23.4	15.1
1000	288	17.0	14.2
1500	283	12.3	13.4
2000	278	8.7	12.6

The actual vapor pressure in the rising air – not the saturation vapor pressure – is given by the ideal gas law for the vapor

$$p = \frac{1}{V} m_{\mathrm{W}} \frac{R}{M_{\mathrm{W}}} T_{\mathrm{A}} \text{ , hence by } (10.16)_2 \text{ and } (10.14)$$

$$p(z) = \frac{x\, p_{0\mathrm{G}}}{\frac{M_{\mathrm{W}}}{M_{\mathrm{A}}} + x} \left(1 - \frac{\gamma}{T_{\mathrm{G}}} z\right)^{\frac{g/\gamma}{R/M_{\mathrm{A}}}} . \tag{10.19}$$

Clouds form at a height where $p(z)$ becomes larger than $p'(z)$. Table 10.2 and Fig. 10.7 show that this happens at

$$z_{\mathrm{B}} \approx 1400\,\mathrm{m} .$$

The parameters for which the table and figure are drawn are $p_{\mathrm{B}} = 1\,\mathrm{bar}$, $T_{\mathrm{E}} = 298\,\mathrm{K}$, and $x = 0.01$.

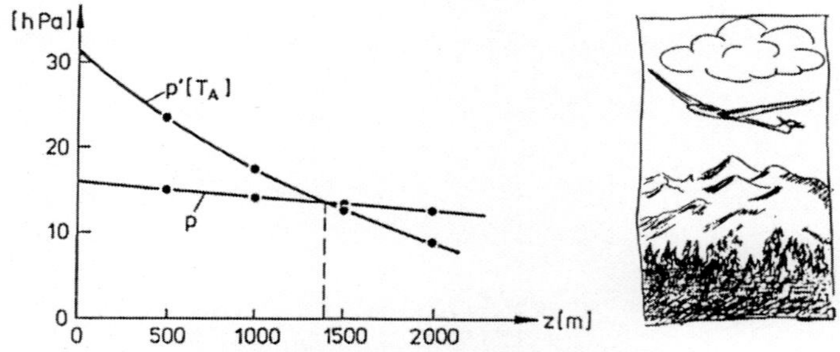

Fig. 10.7 On the height of the cloud base

10.5 Rules of thumb

10.5.1 *Alternative measures of moisture*

When we know $x'(\tau)$ – *e.g.* from Table 10.1 – we may express the moisture content in an alternative manner to $x = m_{\mathrm{W}} / m_{\mathrm{L}}$. Meteorologists prefer the *degree of saturation*, defined as

$$\varphi(x,\tau) = \frac{x}{x'(\tau)} 100\,\% .$$

This quantity measures the amount of water in unsaturated air relative to the maximal amount of water which the air can contain as vapor at the prevailing temperature. It is clear that, for a given x, the degree of saturation depends on temperature. For instance for $x = 0.01$ and $\tau = 20°\mathrm{C}$ we have $\varphi = 67\,\%$ and for the same value of x at $\tau = 25°\mathrm{C}$ we have $\varphi = 49\,\%$.

The *relative humidity* is another measure for the moisture content of air. It is often used by meteorologists and it is defined as

$$\psi(x,\tau) = \frac{p}{p'(\tau)} 100\,\% .$$

This quotient of water vapor pressure and saturation pressure is numerically nearly equal to the degree of saturation, at least if p and $p'(\tau)$ are small compared to p_0.

Another measure for the moisture content – independent of τ – is the dew-point τ_D. This quantity has the dimension of a temperature and is defined as *the* temperature at which water vapor of the moisture content x condenses. From Table 10.1 we see that for $x = 0.01$ the dew point has the value $\tau_D = 14°C$ and for $x = 0.005$ we have $\tau_D = 4°C$.

10.5.2 *Dry adiabatic temperature gradient*

In (10.18) we have calculated the temperature of adiabatically rising air as a function of height. Since x is usually very small – of the order of magnitude $x = 0.01$ – we may neglect the x-dependence in the exponent. Also we have $\frac{\gamma}{T_G} = \frac{1}{288\,\text{K}} \cdot \frac{0.65\,\text{K}}{100} = 0.23 \cdot 10^{-4} \frac{1}{\text{m}}$, so that $\frac{\gamma}{T_G} z << 1$ holds even up to a height of $2000\,\text{m}$. This fact justifies an expansion of the right hand side of (10.18) and we obtain

$$T_A = T_E - \frac{g}{c_p^A} \frac{T_E}{T_G} z .$$

With $g = 9.81\,\text{m/s}^2$, $c_p^A = 1\,\text{kJ/(kg} \cdot \text{K)}$, $T_G = 288\,\text{K}$, $T_E \approx T_G$ we thus obtain

$$T_A = T_E - \gamma_{ad} z \quad \text{where} \quad \gamma_{ad} = \frac{1\,\text{K}}{100\,\text{m}} . \tag{10.20}$$

This means that rising air cools at a rate of $1\,\text{K}$ per $100\,\text{m}$ height. γ_{ad} is called the *dry adiabatic temperature gradient.* "Dry" as long as no droplets form. As soon as this happens, the heat of condensation is released and the rising air cools more slowly than predicted by (10.20), namely with approximately 0.6 K per 100 m gain in height. This is the *moist adiabatic temperature gradient.*

The mixing of the air in the lower layers of the atmosphere is to a large extent due to thermal updrafts, *i.e.* the adiabatic rise of air that was heated up by contact with the soil. One must realize that such a rise does not occur for small temperature gradients in the atmosphere; the temperature gradient must be larger than $1\,\text{K}/100\,\text{m}$. Thus, for instance, the standard atmosphere with the gradient $0.65\,\text{K}/100\,\text{m}$ is perfectly stable. And *inversions* are particularly stable; these are weather conditions in which the temperature *rises* with growing height. Inversions occur by the nightly cooling of the ground and in wintertime they prevent the mixing of the air, sometimes for days, so that fumes from the heating of houses and from the exhaust of cars provide "smog" conditions.

The "Föhn" in the Alps – a warm dry wind "falling" down the northern side of the mountains – occurs when a high pressure region exists in north-eastern Italy and a low pressure in the north-west, both south of the Alps. This creates a south easterly wind of moist air in-between.* The mountains force the air to climb and during that rise it rains off its moisture and cools slowly, *i.e.* with the moist adiabatic temperature gradient. Once arrived at the top, the air is dry. Proceeding north it comes down and heats up with the dry adiabatic temperature gradient. Thus the air arrives in southern Germany considerable warmer than it started out in northern Italy.

For meteorologists, pilots, and especially glider pilots there is a need to perform a quick estimate of the base of cumulus clouds. A rule-of-thumb is used which reads

$$z_{\text{Cloud base}} = 123(\tau_\text{E} - \tau_\text{D})\frac{\text{m}}{\text{K}}. \tag{10.21}$$

For $\tau_E = 25°\text{C}$, $\tau_D = 14°\text{C}$ we obtain 1353 m which checks well with the earlier results of 1400 m read off from the graphs of Fig. 10.7.

10.6 Pressure of saturated vapor in the presence of air

So far in this chapter we have always assumed that the condensation of water vapor occurs at the pressure $p' = p'(T)$ which we may read off from Tables 2.4 or 10.1. This, however, cannot be entirely true, since the presence of air is ignored. We shall now investigate how the presence of air changes the condition for condensation. Beforehand we remark that the change will be minimal so that all previous results in this chapter remain valid in an excellent approximation.

To fix the ideas we consider a situation in which moist air is in equilibrium with liquid water. The equilibrium condition states that the specific Gibbs free energies of liquid water and water vapor are equal**

$$g^\text{L}(T, p_0) = g^\text{V}(T, p_\text{E}), \tag{10.22}$$

where p – as always in this chapter – is the partial pressure of the vapor; accordingly p_E is the partial pressure in equilibrium.

In order to solve (10.22) for p_E we use the fact that for water and pure vapor we have

$$g^\text{L}(T, p(T)) = g^\text{V}(T, p(T)). \tag{10.23}$$

We recall that $\partial g^\text{L}/\partial p = \upsilon^\text{L}$ holds, and that υ^L is constant, if we assume liquid water to be incompressible. Therefore $g^\text{L}(T, p_0)$ is a linear function of p_0 and we may write

$$g^\text{L}(T, p_0) = g^\text{L}(T, p(T)) + \upsilon^\text{L}[p_0 - p(T)]. \tag{10.24}$$

---

* One might have thought that the wind comes from the east under the conditions, but the Coriolis force turn it around so that it comes from the south.

** To be sure, it should be the chemical potentials which are equal, *cf.* Chap. 8. However, in the present case this condition is equivalent to (10.22), since the liquid is considered as pure water, and water vapor is considered as an ideal gas.

On the other hand $g^{\mathrm{V}}(T, p_{\mathrm{E}})$ is the Gibbs free energy of an ideal gas so that we have

$$g^{\mathrm{V}}(T, p_{\mathrm{E}}) = g^{\mathrm{V}}(T, p(T)) + \frac{R}{M_{\mathrm{W}}} T \ln \frac{p_{\mathrm{E}}}{p(T)}. \tag{10.25}$$

We combine (10.22), (10.24), and (10.25) and use (10.23) to obtain

$$p_{\mathrm{E}} = p(T) \exp\left[\frac{\upsilon^{\mathrm{L}}[p_0 - p(T)]}{\frac{R}{M_{\mathrm{W}}} T}\right]. \tag{10.26}$$

It follows that there is indeed a difference between the saturation pressures of the vapor alone and of the vapor, if it is mixed with air. However, the difference is small, since the exponent is usually very much smaller than one. For instance for $T = 323\,\mathrm{K}$ with $p(T) = 0.123\,\mathrm{bar}$, $p_0 = 1\,\mathrm{bar}$ and $\upsilon^{\mathrm{L}} = 1\,\mathrm{cm}^3/\mathrm{g}$ the exponent has the value $0.6 \cdot 10^{-3}$. Therefore we have made a minimal mistake when we set $p_{\mathrm{E}} = p(T)$ in previous sections of this chapter.

11 Selected problems in thermodynamics

11.1 Droplets and bubbles

11.1.1 *Available free energy*

We investigate the equilibrium of a liquid droplet in its vapor, and – at the same time – we discuss the equilibrium conditions for a vapor bubble surrounded by liquid. Fig. 11.1 shows the systems to be considered and introduces some notation. Droplets and bubbles are considered as spherical, p is the pressure on the outside and T the temperature. In some instances we divide the page along a vertical line; in such cases the formulae to the left of the line refer to droplets, those to the right refer to bubbles.

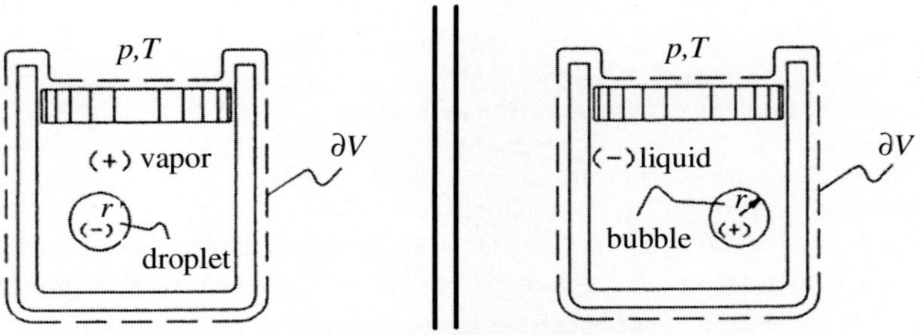

Fig. 11.1 Liquid droplet in vapor and vapor bubble in liquid

After the considerations of Paragraph 4.2.7 – applied to the volume inside ∂V in Fig. 11.1 – the available free energy reads

$$A = E - TS + pV \tag{11.1}$$

where E is the energy. This quantity must have a minimum in equilibrium. The potential energy of the piston is irrelevant here and it is ignored. Also we neglect the kinetic energy and therefore the available free energy assumes the form

$$A = F + pV \,, \tag{11.2}$$

where F is the free energy.

V is the sum of the two parts V^+ and V^-, while F consists of three parts, $F^+ + F^- + F^S$, where F^S is the surface energy of a droplet or a bubble. We assume that this surface energy is proportional to the surface area $4\pi r^2$ of the phase boundary and obtain – with σ as the factor of proportionality –

$$A = m^+ f^+\left(T, \upsilon^+\right) + m^- f^-\left(T, \upsilon^-\right) + 4\pi r^2 \sigma(T) + p\left(V^+ + V^-\right). \tag{11.3}$$

$\sigma(T)$ is called the surface tension ; it is positive and depends on temperature. Among the four variables m^+, m^-, $V^+ = \upsilon^+ m^+$, $V^- = \upsilon^- m^-$ only three are independent, because the total mass $m = m^+ + m^-$ is constant. We choose V^+, V^-, and m^- as independent variables.

11.1.2 *Necessary and sufficient conditions for equilibrium*

Necessary conditions for equilibrium are obtained by setting the derivatives of $\mathcal{A}$ with respect to the three independent variables equal to zero. By

$$p = -\frac{\partial f}{\partial \upsilon}\ , \quad V^{\pm} = \frac{4\pi}{3} r^3\ , \quad \text{and} \quad g = f + p\upsilon$$

we obtain easily

$p^+ = p$	$p^+ = p + \frac{2\sigma}{r}$
$p^- = p + \frac{2\sigma}{r}$	$p^- = p$

$$g^-\left(p^-, T\right) = g^+\left(p^+, T\right). \tag{11.4}$$

We conclude that in this system the pressure is not homogeneous. Indeed, the pressure inside the droplet or bubble is greater than the outside pressure and the pressure difference is inversely proportional to r.

The first two lines of the necessary equilibrium conditions (11.4) represent the conditions of dynamic equilibrium, or equilibrium of forces, while the third line represents phase equilibrium between liquid and vapor.

Sufficient for equilibrium – or for the stability of the state characterized by (11.4) – is the requirement that the matrix of second derivatives of $\mathcal{A}$ with respect to V^+, V^-, and m^- be positive definite. This is a cumbersome condition and we shall not exploit it in any generality. Instead, we shall make the realistic assumption that in the systems of Fig. 11.1 the dynamic equilibrium is more or less instantaneously assumed, while phase equilibrium is slowly established afterwards. In this manner the first two of the three lines of (11.4) become conditions for the determination of two out of the three variables and $\mathcal{A}$ becomes a function of *one* variable only, namely r. This function of *one* variable can be judged most easily with respect to stability: It defines stable equilibria where minima occur, and unstable equilibria where maxima occur.

11.1.3 *Available free energy as a function of radius*

We take the dynamic equilibrium into account in the available free energy $\mathcal{A}$ by writing

$p\left(V^+ + V^-\right) = pV^+ + p^-V^- - \left(p^- - p\right)V^-$	$p\left(V^+ + V^-\right) = p^+V^+ + pV^- - \left(p^+ - p\right)V^+$
and with	and with
$p = p^+\ ,\quad p^- = p + \frac{2\sigma}{r}$ by $(11.4)_1$	$p = p^-\ ,\quad p^+ - p = \frac{2\sigma}{r}$ by $(11.4)_2$

$$p\left(V^+ + V^-\right) = p^+V^+ + p^-V^- - \frac{8\pi}{3} r^2 .$$

Thus we obtain the available free energy (11.3) with $g = f + p\upsilon$

$$\mathcal{A} = m^+ g^+\left(p^+, T\right) + m^- g^-\left(p^-, T\right) + \frac{4\pi}{3} r^2 . \tag{11.5}$$

We refer $\mathcal{A}$ to the state where $r=0$ holds, *i.e.* the pure vapor pressure or the pure liquid. Therefore we subtract $mg^{+}(p^{+},T)$ – or $mg^{-}(p^{-},T)$ for the bubbles – from both sides of (11.5) and obtain, suppressing the T-dependence in $g(p,T)$,

$$\bar{\mathcal{A}}=-m^{-}\left[g^{+}(p^{+})-g^{-}(p^{-})\right]+\frac{4\pi}{3}\sigma r^{2}$$

$$\bar{\mathcal{A}}=m^{+}\left[g^{+}(p^{+})-g^{-}(p^{-})\right]+\frac{4\pi}{3}\sigma r^{2}. \tag{11.6}$$

We assume that the liquid is incompressible, whereas the vapor is an ideal gas. In this case we have

$$m^{-}=\frac{1}{\upsilon^{-}}\frac{4\pi}{3}r^{3}\quad\text{with}\quad \upsilon^{-}=\text{const}$$

$$m^{+}=\frac{p^{+}\frac{4\pi}{3}r^{3}}{\frac{k}{m}T}$$

and $\bar{\mathcal{A}}$ from (11.6) reads with $(11.4)_{1,2}$

$$\bar{\mathcal{A}}(r)=-\frac{1}{\upsilon^{-}}\frac{4\pi}{3}r^{3}\left[g^{+}(p)-g^{-}\left(p+\frac{2\sigma}{r}\right)\right]+\frac{4\pi}{3}\sigma r^{2}$$

$$\bar{\mathcal{A}}(r)=\frac{\frac{4\pi}{3}r^{3}}{\frac{k}{m}T}\left(p+\frac{2\sigma}{r}\right)\left[g^{+}\left(p+\frac{2\sigma}{r}\right)-g^{-}(p)\right]+\frac{4\pi}{3}\sigma r^{2} \tag{11.7}$$

As indicated $\bar{\mathcal{A}}$ is a function of the single variable r. To be sure, the dependence on r is not explicit, since we do not know the functions $g^{\pm}$ explicitly. But, as in Paragraph 8.5.1, we may use the condition $g^{+}(p(T))=g^{-}(p(T))$ of phase equilibrium *without* surface energy and write

$$g^{+}(p)=g^{+}(p(T))+\frac{k}{m}T\ln\frac{p}{p(T)}$$
$$g^{-}\left(p+\tfrac{2\sigma}{r}\right)=g^{-}(p(T))+\upsilon^{-}\left[p-p(T)+\tfrac{2\sigma}{r}\right]$$

$$g^{+}\left(p+\tfrac{2\sigma}{r}\right)=g^{+}(p(T))+\frac{k}{m}T\ln\frac{p+\frac{2\sigma}{r}}{p(T)}$$
$$g^{-}(p)=g^{-}(p(T))+\upsilon^{-}\left[p-p(T)\right]$$

Inserting this into (11.7), we get rid of the Gibbs free energy altogether and obtain

$$\bar{\mathcal{A}}(r)=4\pi\sigma r^{2}-\frac{4\pi}{3}r^{3}p\cdot\left[\frac{\frac{k}{m}T}{p\upsilon^{-}}\ln\frac{p}{p(T)}-\left(1-\frac{p(T)}{p}\right)\right]$$

$$\bar{\mathcal{A}}(r)=\frac{4\pi}{3}\sigma r^{2}+\frac{4\pi}{3}r^{3}\left(p+\frac{2\sigma}{r}\right)\cdot\left[\ln\frac{p+\frac{2\sigma}{r}}{p(T)}-\frac{p\upsilon^{-}}{\frac{k}{m}T}\left(1-\frac{p(T)}{p}\right)\right].$$

In both formulae $p\upsilon^{-}/\left(\frac{k}{m}T\right)$ is a small factor, since the specific volume of the vapor is much larger than that of the liquid. Therefore the second terms in the square brackets may be neglected and we have

$$\bar{\mathcal{A}}(r)=4\pi\sigma r^2-\frac{4\pi}{3}r^3 p\frac{\frac{k}{m}T}{p\upsilon^-}\ln\frac{p}{p(T)} \qquad \Bigg\| \qquad \bar{\mathcal{A}}(r)=\frac{4\pi}{3}\sigma r^2+\frac{4\pi}{3}r^3\cdot\left(p+\frac{2\sigma}{r}\right)\ln\frac{p+\frac{2\sigma}{r}}{p(T)}. \tag{11.8}$$

These are now two simple analytic functions of r. The dependence on r is a little simpler for droplets than for bubbles.

11.1.4 *Nucleation barrier for droplets*

We consider the r-dependence of the available free energy $\bar{\mathcal{A}}$ in $(11.8)_1$ for droplets and, in particular, its extrema. The function is the sum of a parabola with positive curvature and a cubic function whose contribution depends on p. If $p < p(T)$ holds, the cubic term – including its minus sign – is positive and the only extremum lies at $r = 0$. This means that the system has a stable equilibrium in the vapor phase. For $p > p(T)$ the situation is more complex: We still have the minimum at $r = 0$ which is dictated by the surface tension. But for large r we have another – possibly very deep – end-point minimum which corresponds to the state of complete condensation. In-between these minima there is a maximum at the extremal – so-called *critical* -- radius

$$r_C=\frac{2\sigma\upsilon^-}{\frac{k}{m}T\ln\frac{p}{p(T)}}\,,\quad\text{hence}\quad \bar{\mathcal{A}}(r_C)=\frac{4\pi}{3}\sigma r_C^2\,. \tag{11.9}$$

This maximum represents an unstable equilibrium. Fig. 11.2a shows some graphs $\bar{\mathcal{A}}(r)$ – for different values of the parameter $p/p(T)$ – for water with $\sigma = 7.5\cdot 10^{-2}\,\mathrm{N/m}$, $T = 5°\mathrm{C}$.

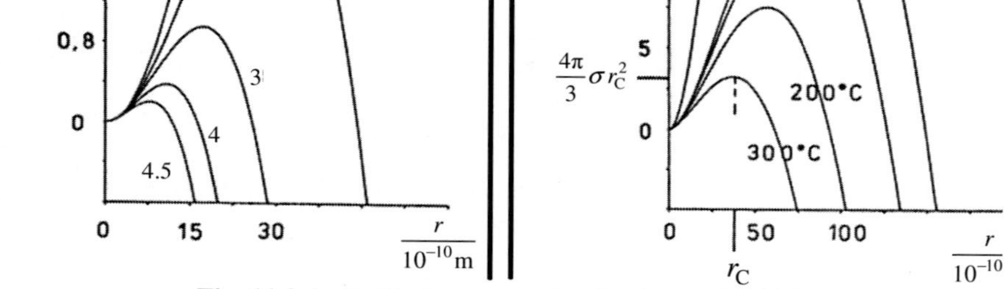

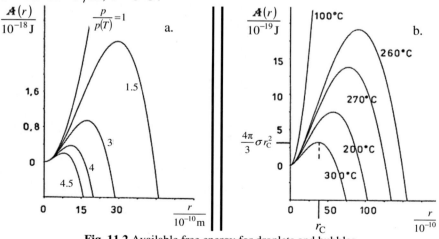

Fig. 11.2 Available free energy for droplets and bubbles
a. Droplet in vapor at T=298 K
b. Bubble in liquid for p=1 atm

Note that the "normal" case, where $\sigma = 0$ holds, has no maximum. For $p < p(T)$ and $p > p(T)$ there are minima for $r = 0$ or for large values of r respectively. And for $p = p(T)$ we have $\overline{\mathbf{A}}(r) \equiv 0$, which means that an indifferent equilibrium exists: It is then energetically irrelevant how much fluid has evaporated.

The interpretation of the maximum is clear: When we have a droplet with $r < r_C$, the system can lower its free energy by evaporating the droplet; the droplet will vanish. On the other hand, for $r > r_C$ the droplet will increase until the whole system is liquid, because in this manner the free energy decreases.

r_C is called the nucleation radius and $\overline{\mathbf{A}}(r_C)$ is the nucleation barrier. This means that a droplet must first reach the size r_C – by a fluctuation (say) or, more probably by attaching itself to a dust particle – before it can serve as a nucleus for condensation.

The existence of the nucleation barrier has many consequences and applications. The barrier permits a vapor to exist, although the pair (p,T) of variables lies firmly inside the liquid range of a (p,T)-diagram, *cf.* Fig. 2.5. Examples include

- the air saturated with moisture over the pacific is so free of dust when it enters northwestern Australia that it cannot nucleate. Farmers force it to "rain-off" by sending weather pilots to a great height who spread some substance on which nuclei can be formed;
- closer at home we notice the same effect in the formation of condensation trails of high flying airplanes. The hot, partially ionized exhaust gases serve as nuclei in this case;
- on a smaller scale the same thing happens in the Wilson chamber which is used for the detection of elementary particles. By adiabatic expansion of dust-free moist air the temperature is decreased so that it is smaller than the temperature $T(p)$ of the saturation curve. Yet no nucleation occurs until elementary particles enter the chamber and ionize air molecules which in turn may serve as nucleation sites. Thus a sequence of droplets identifies the path of an elementary particle.

For clean vapors, which are shielded from all kinds of vibration so that no appreciable density fluctuations occur, p may be four (!) times larger than $p(T)$ and yet no condensation occurs.

11.1.5 *Nucleation barrier for bubbles*

Conditions for bubbles are very similar to those for droplets in principle, although different in detail. The difference is inherent in the different functions $(11.8)_{1,2}$. Fig. 11.2 b shows graphs of $\overline{\mathbf{A}}(r)$ for water vapor at $p = 1\,\text{atm}$ and for different temperatures.

If σ were zero, the graph $\overline{\mathbf{A}}(r)$ would have a minimum at $r=0$ for all $T<100°C$, so that the whole system would be in the liquid phase. And for higher temperatures there would be an end-point-minimum for some large r when the whole system is in the vapor phase.

However, for $\sigma>0$ there is also a maximum at the critical bubble radius

$$r_C = \frac{2\sigma}{p(T)-p}\,, \quad \text{hence} \quad \overline{\mathbf{A}}(r_C) = \frac{4\pi}{3}\sigma r_C^2\,. \tag{11.10}$$

The interpretation of the maximum as a nucleation barrier is analogous to the discussion in the case of droplets. It is clear what $(11.10)_1$ means: The evaporation happens for *the* temperature for which $p(T)$ is equal to the pressure inside the bubble.

The existence of the nucleation barrier for bubbles makes it possible to have overheated liquids, *i.e.* liquids that do not boil although their "normal" boiling temperature $T(p)$ is surpassed. In such liquids the fluctuations or impurities that could give rise to nuclei are lacking. An application is the bubble chamber for the detection of elementary particles. The ions created by such particles may serve as nucleation sites and thus mark the orbits of the particles. In comparison with the Wilson chamber the bubble chamber has an advantage, because of the greater density of the liquid ensuring that even particles which have little interaction with matter – *e.g.* neutrinos – may leave their trace.

11.1.6 *Discussion*

The arguments above about the nucleation barrier of droplets and bubbles go back to W. THOMSON (Lord KELVIN). Therefore the two equations $(11.9)_1$ and $(11.10)_1$ for the critical radii are often called Thomson formulae.

The recognition of a nucleation barrier is important, because it explains the existence of overheated liquids and undercooled vapors. However, the arguments above do not add up to a nucleation theory. Indeed, the question of how the barrier is overcome – and how the nucleation rate depends on r_C and $\overline{\mathbf{A}}(r_C)$ – is not easy to answer and we do not discuss it.

11.2 Fog and clouds. Droplets in moist air

11.2.1 *Problem*

Now that we know – from Sect. 11.1 – that droplets are unstable in their vapor, a person might wonder, why fog and clouds are such stable phenomena that they may persist for days. The answer is that the droplets in a cloud find themselves in, or near, the stable end-point-minima, *cf.* Fig. 11.2 a. And yet, they do not fill the whole system. The reason is air, of course, the main constituent of a cloud.

We discuss the phenomenon by discussing the system of Fig. 11.3 a, where for temperature T, and outside pressure p_0, a water droplet floats, which is surrounded by water vapor *and air*. We investigate the stability of this system in the same manner as previously when we studied the stability – or instability – of the situation drawn in Fig. 11.1 a, where no air was present.

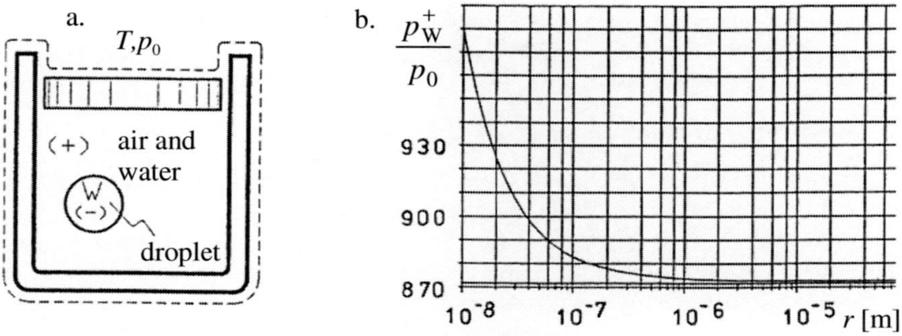

Fig. 11.3 Droplet in water vapor and air
a. System of one droplet
b. Water vapor pressure as a function of the droplet radius at T=5°C

11.2.2 *Available free energy. Equilibrium conditions*

Just like in Paragraph 11.1.1 – or in Paragraph 4.2.7 – we conclude that the available free energy

$$\mathcal{A} = F + p_0 V$$

assumes a minimum in a (stable) equilibrium. We suppose that air and water vapor form a mixture of ideal gases and introduce the following notation

gas space:	masses	m_A , m_W^+	volume V^+
droplet:	mass	$m_W^- = m_W - m_W^+$	volume $V^- = \dfrac{4\pi}{3} r^3$.

The available free energy reads

$$\mathcal{A} = F^+ + F_W^- + 4\pi r^2 \sigma + p_0 \left(V^+ + V^-\right), \tag{11.11}$$

where $F^+ = F_W^+\left(T, m_W^+, V^+\right) + F_A\left(T, m_A, V^+\right)$ and $F_W^- = F_W^-\left(T, m_W^-, V^-\right)$. Therefore we have the four variables m_W^-, m_W^+, V^-, V^+ of which only m_W^-, V^-, V^+ are independent, since $m_W^- + m_W^+ = m_W = \text{const}$ holds.

The necessary equilibrium conditions result from setting the derivatives of $\mathcal{A}$ with respect to the independent variables equal to zero. Thus we obtain with $\mu_\alpha = \partial F / \partial m_\alpha$, $p = -\partial F / \partial V$.

$$\frac{\partial \mathcal{A}}{\partial m_W^-} = 0 \text{ , hence } \frac{\partial F_W^-}{\partial m_W^-} - \frac{\partial F_W^+}{\partial m_W^+} = 0 \text{ , or } \mu_W^- = \mu_W^+ \text{ ,} \tag{11.12}$$

$$\frac{\partial \mathcal{A}}{\partial V^-} = 0 \text{ , hence } \frac{\partial F_W^-}{\partial V^-} + \frac{2\sigma}{r} + p_0 = 0 \text{ , or } p^- - \frac{2\sigma}{r} = p_0 \text{ ,}$$

$$\frac{\partial \mathcal{A}}{\partial V^+} = 0 \text{ , hence } \frac{\partial F_W^+}{\partial V^+} + \frac{\partial F_A}{\partial V^+} + p_0 = 0 \text{ , or } p_W^+ + p_A = p_0 . \tag{11.13}$$

The two equations (11.13) are the dynamic equilibrium conditions, whereas (11.12) represents phase equilibrium. We note that, once again, the pressure inside the system is not homogeneous.

11.2.3 *Water vapor pressure in phase equilibrium*

Before entering into the stability analysis we consider the condition (11.12) a little closer, *i.e.* for specific choices of the chemical potentials. The goal is the determination of the water vapor pressure as a function of the radius of the droplet. Since the liquid phase is pure water, considered incompressible, and since the + phase is a mixture of ideal gases we may write as often before, *cf.* Chap. 8

$$\mu_{\mathrm{W}}^{-} = g_{\mathrm{W}}^{-}\left(T, p^{-}\right) \underset{\text{by (11.13)}_1}{=} g_{\mathrm{W}}^{-}\left(T, p_0 + \tfrac{2\sigma}{r}\right) = g_{\mathrm{W}}^{-}\left(T, p(T)\right) + \upsilon^{-}\left(p_0 + \tfrac{2\sigma}{r} - p(T)\right),$$

$$\mu_{\mathrm{W}}^{+} = g_{\mathrm{W}}^{+}\left(T, p^{+}\right) + \frac{R}{M_{\mathrm{W}}} T \ln \frac{p_{\mathrm{W}}^{+}}{p^{+}} = g_{\mathrm{W}}^{+}\left(T, p(T)\right) + \frac{R}{M_{\mathrm{W}}} T \ln \frac{p_{\mathrm{W}}^{+}}{p(T)}. \tag{11.14}$$

As always the reference to the saturated vapor pressure $p(T)$ is chosen in order to get rid of the specific Gibbs free energies g_{W}; this is possible, since $g_{\mathrm{W}}^{+}(T, p(T)) = g_{\mathrm{W}}^{-}(T, p(T))$ holds. Insertion of (11.14) into (11.12) provides after an easy reformulation

$$p_{\mathrm{W}}^{+} = p(T) \exp\left(\frac{\upsilon^{-}\left[p_0 - p(T) + \frac{2\sigma}{r}\right]}{\frac{R}{M_{\mathrm{W}}} T} \right). \tag{11.15}$$

For the case $r \to \infty$ we recognize this formula from (10.26) which showed that the water vapor pressure – over a plane liquid surface – is not exactly equal to $p(T)$ in the presence of air. In that case the difference was very small. For a finite r, however, the exponential term in (11.15) may lead to a sizable difference of p_{W}^{+} and $p(T)$. Fig. 11.3 b shows p_{W}^{+} as a function of r with $\sigma = 7.5 \cdot 10^{-2}\,\mathrm{N/m}$ – appropriate for water –, $p_0 = 10^5\,\mathrm{Pa}$, and $T = 278\,\mathrm{K}$, corresponding to $p(T) = 872\,\mathrm{Pa}$.

Inspection shows that the water vapor pressure surrounding a small droplet is considerably larger in equilibrium than around a large droplet. Hence we may conclude that many small droplets evaporate more quickly than a single large one. Also a small droplet next to a large one will vanish, while the big droplet grows.

11.2.4 *The form of the available free energy*

As in Sect. 11.1 we now assume that the dynamic equilibrium conditions (11.13) are already satisfied while phase equilibrium is still in the process of adjusting to the condition (11.12). In this case we may use the two equations (11.13) to convert $\mathcal{A}$ into a function of the single variable r. We write

$$p_0\left(V^{+} + V^{-}\right) = p^{+}V^{+} + p^{-}V^{-} + \left(p_0 - p^{+}\right)V^{+} + \left(p_0 - p^{-}\right)V^{-}$$

and, by (11.13)

$$p_0\left(V^+ + V^-\right) = p^+V^+ + p^-V^- - \frac{2\sigma}{r}V^- .$$

If this is introduced into (11.11), we obtain with $G = F + pV$

$$\mathcal{A} = G^+ + G^- + \frac{4\pi}{3}\sigma r^2 .$$

By (8.6) we have $G^+ = m_W^+ \mu_W^+ + m_A \mu_A$ and $G^- = m_W^- g_W^-$. If the chemical potentials are those of an ideal mixture we have

$$\mu_\mathrm{W}^+ = g_\mathrm{W}^+\left(T, p^+\right) + \frac{R}{M_\mathrm{W}} T \ln \frac{N_\mathrm{W}^+}{N_\mathrm{W}^+ + N_\mathrm{A}} ,$$

$$\mu_\mathrm{A}^+ = g_\mathrm{A}^+\left(T, p^+\right) + \frac{R}{M_\mathrm{A}} T \ln \frac{N_\mathrm{A}}{N_\mathrm{W}^+ + N_\mathrm{A}} ,$$

where N denotes particle numbers. Thus it follows

$$\mathcal{A} = \left(m_\mathrm{W} - m_\mathrm{W}^-\right)\left(g_\mathrm{W}^+\left(T, p^+\right) + \frac{R}{M_\mathrm{W}} T \ln \frac{N_\mathrm{W}^+}{N_\mathrm{W}^+ + N_\mathrm{A}} \right) +$$

$$+ m_\mathrm{A}\left(g_\mathrm{A}\left(T, p^+\right) + \frac{R}{M_\mathrm{A}} T \ln \frac{N_\mathrm{A}}{N_\mathrm{W}^+ + N_\mathrm{A}} \right) + m^- g_\mathrm{W}^-\left(T, p^-\right) + \frac{4\pi}{3}\sigma r^2 .$$

With $p^+ = p_0$, by $(11.13)_2$, we obtain a specific form of the reduced available free energy $\overline{\mathcal{A}} = \mathcal{A} - m_\mathrm{W} g_\mathrm{W}^-\left(T, p^-\right) - m_\mathrm{A} g_\mathrm{A}\left(T, p_0\right)$ which is referred to the droplet-free state

$$\overline{\mathcal{A}} = 4\pi\sigma r^2 \qquad \text{surface energy} \qquad (11.16)$$

$$+ kT\left(N_\mathrm{W}^+ \ln \frac{N_\mathrm{W}^+}{N_\mathrm{W}^+ + N_\mathrm{A}} + N_L \ln \frac{N_\mathrm{A}}{N_\mathrm{W}^+ + N_\mathrm{A}} \right) \qquad \text{Gibbs free energy of mixing}$$

$$- m_\mathrm{W}^-\left[g_\mathrm{W}^+\left(T, p_0\right) - g_\mathrm{W}^-\left(T, p_0\right)\right] \qquad \text{Gibbs free energy of condensation}$$

The labels on the right hand side of (11.16) identify the physical meaning of the different contributions to $\mathcal{A}$. The "Gibbs free of condensation" is due to the molecular interaction which forces the vapor to become a liquid. The equation as such does not exhibit explicitly that $\overline{\mathcal{A}}$ depends only on one variable; that *is* the case, however, because we have

$$N_\mathrm{W}^+ = N_\mathrm{W} - N_\mathrm{W}^- , \quad N_\mathrm{W}^- = \frac{m_\mathrm{W}^-}{m_\mathrm{W}} , \quad \text{and} \quad m_\mathrm{W}^- \upsilon^- = \frac{4\pi}{3} r^3$$

so that $\overline{\mathcal{A}}$ depends only on r. There are parameters, of course, namely p_0, T, and N_W, N_A.

A simplified version of (11.16) is obtained, if we replace the Gibbs free energies by reference to $p(T)$

$$g_{\mathrm{W}}^{+}(T,p_0)=g_{\mathrm{W}}^{+}(T,p(T))+\frac{R}{M_{\mathrm{W}}}T\ln\frac{p_0}{p(T)}$$

$$g_{\mathrm{W}}^{-}(T,p_0)=g_{\mathrm{W}}^{-}(T,p(T))+\upsilon^{-}[p_0-p(T)]$$

so that

$$m^{-}\left[g_{\mathrm{W}}^{+}(T,p_0)-g_{\mathrm{W}}^{-}(T,p_0)\right]=\frac{4\pi}{3}r^3\left\{\frac{1}{\upsilon^{-}}\frac{R}{M_{\mathrm{W}}}T\ln\frac{p_0}{p(T)}-[p_0-p(T)]\right\}. \quad (11.17)$$

We insert this into (11.16) and introduce the dimensionless radius x defined by

$$x^3=\frac{N_{\mathrm{W}}^{-}}{N_{\mathrm{W}}}=\left(\frac{r}{a}\right)^3.$$

This is the only variable in $\overline{\mathcal{A}}$; it varies between 0 and 1, where 0 corresponds to the droplet-free state and 1 to the state where all the water is condensed. a is the maximal possible radius of the droplet. An easy reformulation provides

$$\begin{aligned}\overline{\mathcal{A}}&=4\pi\sigma a^2x^2 \\ &+N_{\mathrm{W}}kT\left[\left(1-x^3\right)\ln\frac{1-x^3}{1-x^3+N_{\mathrm{A}}/N_{\mathrm{W}}}+\frac{N_{\mathrm{A}}}{N_{\mathrm{W}}}\ln\frac{1}{1-x^3+N_{\mathrm{A}}/N_{\mathrm{W}}}+\frac{N_{\mathrm{A}}}{N_{\mathrm{W}}}\ln\frac{N_{\mathrm{A}}}{N_{\mathrm{W}}}\right] \\ &-\frac{4\pi}{3}a^3x^3\left\{\frac{1}{\upsilon^{-}}\frac{R}{M_{\mathrm{W}}}T\ln\frac{p_0}{p(T)}-[p_0-p(T)]\right\}.\end{aligned} \quad (11.18)$$

In the first and the last line of (11.18) we recognize the available free energy of the droplet without air, *cf.* Paragraph 11.1.3. The presence of air makes itself known in the entropy of mixing shown in the second line. This expression prevents complete un-mixing of water and air, *i.e.* complete condensation.

In order to give an impression of the relative sizes of the contributions to $\overline{\mathcal{A}}$, we consider a special case with

$$p_0=10^5\,\mathrm{Pa}\ ,\quad T=283\,\mathrm{K}\ ,\quad \text{hence}\quad p(T)=12.3\cdot10^{-2}\,\mathrm{Pa} \quad (11.19)$$

$$N_{\mathrm{W}}=2\cdot10^{15}\ ,\quad \text{hence}\quad a=25\cdot10^{-6}\,\mathrm{m}\quad\text{and}\quad N_{\mathrm{A}}=zN_{\mathrm{W}}\,.$$

We leave the mass of air – or z – as a parameter. In a cloud z, the ratio of air and water, has values between 10 and 100. $\overline{\mathcal{A}}$ thus assumes the form

$$\begin{aligned}\frac{\overline{\mathcal{A}}}{10^{-5}\,\mathrm{J}}&=5.89\cdot10^{-5}x^2 \\ &+0.78\left[\left(1-x^3\right)\ln\frac{1-x^3}{1-x^3+z}+z\ln\frac{1}{1-x^3+z}+(1+z)\ln(1+z)\right] \\ &-3.76\,x^3.\end{aligned} \quad (11.20)$$

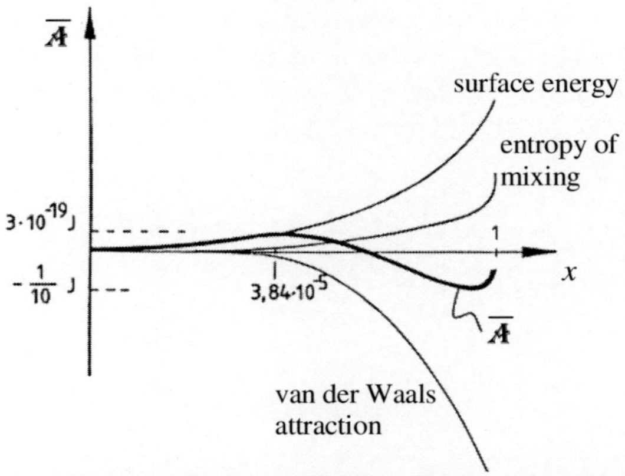

Fig. 11.4 Contributions to the available free energy of a droplet in air (Schematic drawing. Note the scale distortion on both axes)

Fig. 11.4 gives a qualitative view of this function which has minima at $x=0$ and close to $x=1$. In-between there is a maximum, the nucleation barrier. The values given in the figure hold for $z=30$. The figure shows – again only schematically – the contributions of surface tension, the entropy of mixing, and the intermolecular attraction; here labeled as van der Waals attractions. We conclude that now – in the presence of air – the end-point minimum is converted into a "true minimum" as a result of the entropy of mixing. This minimum defines the size of the droplets in the cloud.

11.2.5 *Nucleation barrier and droplet radius*

For $x \ll 1$ the expression (11.20) may be written in the form

$$\frac{\overline{A}}{10^{-5}\,\mathrm{J}} \underset{x\to 0}{\to} 5.89\cdot 10^{-5}x^2 + [0.78\ln(1+z) - 3.76]x^3 .$$

From this, the position and height of the nucleation barrier results as

$$x_{\max} = \frac{11.78\cdot 10^{-5}}{11.1 - 2.34\ln(1+z)} \underset{z=30}{\approx} 4\cdot 10^{-5} ,$$

$$\frac{\overline{A}_{\max}}{10^{-5}\,\mathrm{J}} = \frac{3.03\cdot 10^{-14}}{[0.78\ln(1+z) - 3.96]^2} \underset{z=30}{\approx} 3\cdot 10^{-14} .$$

Hence we conclude that

- the critical radius is of the order of magnitude $10^{-9}\,\mathrm{m}$; it grows with increasing z, the mass of air,
- the height of the nucleation barrier is of the order of magnitude $5\cdot 10^{-19}\,\mathrm{J} \approx 100\,kT$; it grows with increasing z.

Note that the barrier is about 100 times higher than the mean kinetic energy – roughly kT – of the molecules. Therefore the barrier is quite effective in preventing condensation.

The stable minimum close to $x = 1$ occurs, because two opposing tendencies compensate each other, namely the tendency for condensation and the tendency to mix. The latter grows with increasing T and with increasing z so that the droplets in a cloud become smaller, *cf.* Fig. 11.5.

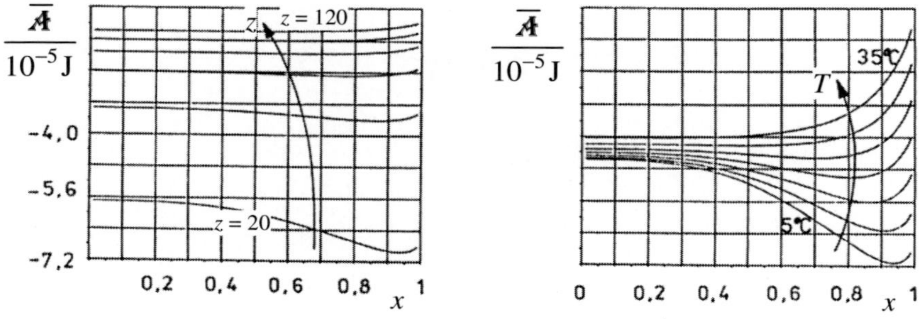

Fig. 11.5 The influence of mass of air and of temperature on the droplet radius in a cloud

11.3 Rubber balloons

11.3.1 *Pressure-radius relation*

We consider spherical balloons of radius r and membrane thickness d. In the deflated state these values are R and D, respectively. The derivation of the pressure-radius relation for a rubber balloon is an instructive example for non-linear elasticity. We cannot present this topic here in any systematic manner for lack of space. Therefore we present a "poor man's approach" which leads us nearly to the correct result.

We focus the attention on an infinitesimally small square cut from the balloon membrane, *cf.* Fig. 11.6. This piece we consider to be loaded only with tangential loads P_1 and P_2. The difference of the normal loads on the large areas of the cut at the base and on the top is so small – in comparison to the tangential loads – that we ignore it. From the First and the Second Laws in the form

$$\frac{\mathrm{d}U}{\mathrm{d}t} = \dot{Q} + \dot{W} \quad \text{and} \quad \frac{\mathrm{d}S}{\mathrm{d}t} = \frac{\dot{Q}}{T}$$

we obtain with $\dot{W} = P_1 \frac{\mathrm{d}L_1}{\mathrm{d}t} + P_2 \frac{\mathrm{d}L_2}{\mathrm{d}t}$ and by elimination of $\dot{Q}$

$$\mathrm{d}F = -S\mathrm{d}T + P_1 \mathrm{d}L_1 + P_2 \mathrm{d}L_2 ,$$

where $F = U - TS$ is the free energy. Since, by Sect. 5.6, the elasticity of rubber is entropy-induced, we have

$$P_1 = -T\frac{\partial S}{\partial L_1} \quad \text{and} \quad P_2 = -T\frac{\partial S}{\partial L_2} . \tag{11.21}$$

In order to calculate P_1 and P_2 as functions of L_1, L_2, and T we may calculate S from the kinetic theory of rubber, *cf.* Sect. 5.5. The entropies in the unloaded and loaded states read, by (5.14) and (5.17)

$$S_0 = k \int_{-\infty}^{+\infty} \left(N \ln 4\pi Z - \frac{\vartheta_1^2 + \vartheta_2^2 + \vartheta_3^2}{\frac{1}{3} N b^2} \right) z_0(\vartheta_1, \vartheta_2, \vartheta_3) \mathrm{d}\vartheta_1 \mathrm{d}\vartheta_2 \mathrm{d}\vartheta_3$$

$$S = k \int_{-\infty}^{+\infty} \left(N \ln 4\pi Z - \frac{\vartheta_1^2 + \vartheta_2^2 + \vartheta_3^2}{\frac{1}{3} N b^2} \right) z(\vartheta_1, \vartheta_2, \vartheta_3) \mathrm{d}\vartheta_1 \mathrm{d}\vartheta_2 \mathrm{d}\vartheta_3 . \tag{11.22}$$

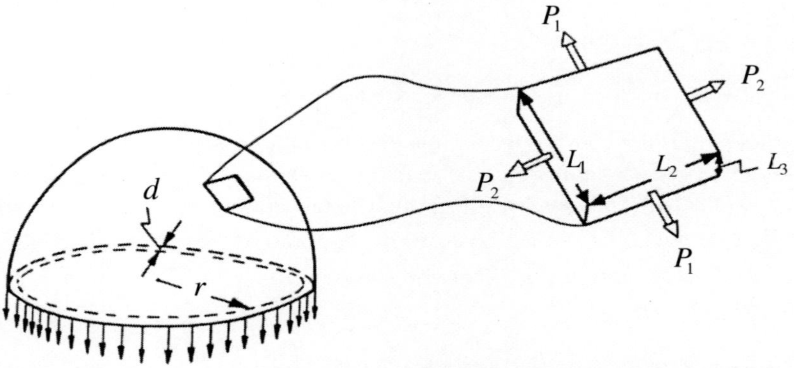

Fig. 11.6 Force balance on the half-balloon. Tangential forces on an infinitesimal square

The distribution functions z_0 and z are given by

$$z_0 = \frac{n}{\sqrt{\frac{1}{3}\pi N b^2}} \exp\left(-\frac{\vartheta_1^2 + \vartheta_2^2 + \vartheta_3^2}{\frac{1}{3}\pi N b^2} \right) \quad \text{and} \tag{11.23}$$

$$z = \frac{n}{\sqrt{\frac{1}{3}\pi N b^2}} \exp\left(-\frac{\frac{1}{\lambda_1^2}\vartheta_1^2 + \frac{1}{\lambda_2^2}\vartheta_2^2 + \lambda_1^2 \lambda_2^2 \vartheta_3^2}{\frac{1}{3}\pi N b^2} \right).$$

The derivation of these expressions makes use of the same arguments as in Sect. 5.5. The only difference is that now the deformation gradient $\underline{\underline{F}}$ reads

$$\underline{\underline{F}} = \begin{bmatrix} \lambda_1 & 0 & 0 \\ 0 & \lambda_2 & 0 \\ 0 & 0 & 1/(\lambda_1 \lambda_2) \end{bmatrix} \quad \text{instead of} \quad \underline{\underline{F}} = \begin{bmatrix} \lambda_1 & 0 & 0 \\ 0 & 1/\sqrt{\lambda_1} & 0 \\ 0 & 0 & 1/\sqrt{\lambda_1} \end{bmatrix}.$$

λ_1 and λ_2 are given as L_1/L_1^0 and L_2/L_2^0, where L_1^0 and L_2^0 are the lengths of the body in the un-deformed state. $\lambda_3 = L_3/L_3^0$ is given by the incompressibility condition $\lambda_1 \lambda_2 \lambda_3 = 1$. Insertion of (11.23) into (11.22) gives after integration

$$S - S_0 = \frac{nk}{2} \left[\lambda_1^2 + \lambda_2^2 + \frac{1}{\lambda_1^2 \lambda_2^2} - 3 \right]. \tag{11.24}$$

Hence follows by insertion into (11.21)

$$t_{11} = \frac{P_1}{L_2 L_3} = \lambda_1 \frac{P_1}{L_2^0 L_3^0} = \frac{nkT}{V}\left[\lambda_1^2 - \frac{1}{\lambda_1^2\lambda_2^2}\right] = \rho\frac{k}{m}T\left[\lambda_1^2 - \frac{1}{\lambda_1^2\lambda_2^2}\right] \tag{11.25}$$

$$t_{22} = \frac{P_2}{L_2 L_3} = \lambda_2 \frac{P_2}{L_1^0 L_3^0} = \frac{nkT}{V}\left[\lambda_2^2 - \frac{1}{\lambda_1^2\lambda_3^2}\right] = \rho\frac{k}{m}T\left[\lambda_2^2 - \frac{1}{\lambda_1^2\lambda_3^2}\right].$$

V is the volume of the membrane square, ρ its density and m the mass of a chain molecule.

In the present case we have

$$\lambda_1 = \lambda_2 = \frac{r}{R}\,,\quad \lambda_3 = \frac{d}{D} \quad \text{so that} \quad dr^2 = DR^2\,; \tag{11.26}$$

the latter relation is the consequence of incompressibility.

The desired pressure-radius relation is obtained as the mechanical equilibrium condition for the half-balloons, *cf.* Fig. 11.6: The tangential force $t_{11} 2\pi r d$ on the circular ring on the equator must be equal to the force of the pressure difference $[p] = p_i - p_o$ inside and outside. This force is equal to $[p]\pi r^2$. Thus we obtain with (11.25), (11.26)

$$[p]\pi r^2 = t_{11} 2\pi r d \quad \text{hence} \quad [p] = 2\underbrace{\rho\frac{k}{m}T}_{s_1 = 3\cdot 10^5 \frac{N}{m^2}}\frac{D}{R}\left[\frac{R}{r} - \left(\frac{R}{r}\right)^7\right]. \tag{11.27}$$

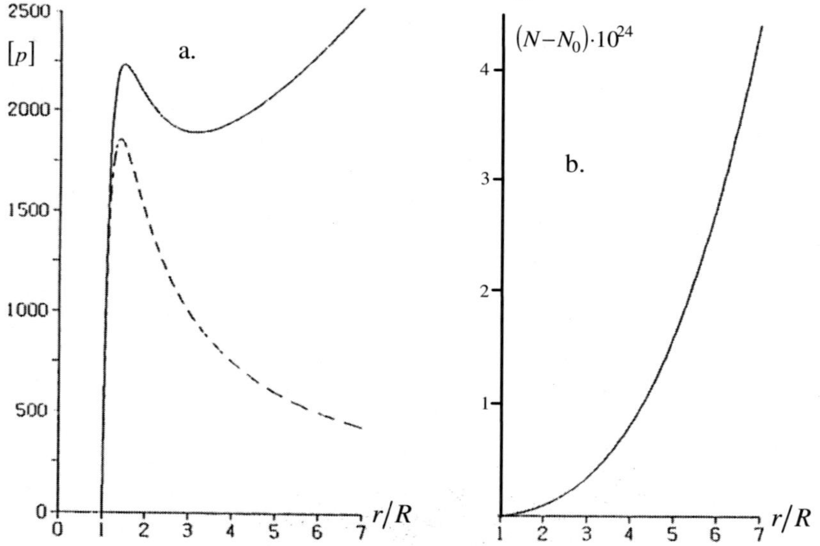

Fig. 11.7 a. Pressure-radius relation of a balloon
dashed: According to the kinetic theory of rubber
solid: Actual relation according to non-linear elasticity
b. Particle number in the balloon as a function of radius

The coefficient s_1, indicated by the brace in (11.27), has a value of 3bar for a typical rubber at room temperature. D/r is roughly equal to $0.5\cdot 10^{-2}$ for a typical balloon. For these values the pressure-radius relation $(11.27)_2$ is shown in Fig. 11.7 by the dashed line. The graph has a maximum at $[p]=0.025\text{bar}$. Behind the maximum the pressure drops which is the reason why – upon inflating a balloon by mouth – a person experiences a certain relief in the lungs after the barrier has been passed. Note, that the maximum of $[p]$ amounts to less than 1% of t_{11}.

However, (11.27) does not represent the correct pressure-radius relation of the balloon. This is due to the fact that the arguments of the kinetic theory of rubber, while instructive and heuristically valuable, are not quite satisfactory for bi-axial deformation. In non-linear elasticity one obtains an additional term such that (11.27) must be replaced by

$$[p]=2s_1\frac{D}{R}\left[\frac{R}{r}-\left(\frac{R}{r}\right)^7\right]\left[1-\frac{s_{-1}}{s_1}\left(\frac{R}{r}\right)^2\right] \tag{11.28}$$

with an additional coefficient s_{-1} which is equal to $-0.1s_1$ for a typical rubber. This is the relation which we shall use. It is shown by the solid graph in 11.7 a.

The new $([p],r)$-relation is non-monotone with two ascending branches at small and large radii and a descending branch in-between. We expect interesting stability properties of a balloon because of this; after all an *increase* of pressure leads to a *decrease* of volume in the descending range. We shall investigate some consequences of this behavior in the following paragraphs.

The number $N-N_0$ of particles blown into the balloon in the process of inflation may be calculated from the ideal gas law

$$(N-N_0)kT=p_i\frac{4\pi}{3}r^3-p_o\frac{4\pi}{3}R^3=\frac{4\pi}{3}\left\{p_o\left(r^3-R^3\right)+[p]r^3\right\}. \tag{11.29}$$

For $p_o=1\text{bar}$ this function is plotted in Fig. 10.7 b. The (N,r)-relation is monotone. Therefore we may eliminate the variable r between the $([p],r)$-relation and the (N,r)-relation and obtain a $([p],N)$-relation which is schematically plotted in Fig. 11.9b. It is non-monotone, of course.

11.3.2 *Stability of a balloon*

We consider the system shown in Fig. 11.8, in which the balloon pressure is balanced by the weight of a piston and the force in an elastic spring. The total mass $m=m_1+m_2$ of air may be changed by letting air enter the cylinder through a valve which is closed afterwards. We ask for the stability of the system for a fixed mass m. From the considerations of Paragraph 4.2.7 we conclude that the available free energy is given by

$$\mathcal{A}=F_{\text{Air}}+F_{\text{Balloon}}+E_{\text{Piston}}^{\text{Pot}}+E_{\text{Spring}}+p_0\left(V_1+V_2\right)\quad\text{with} \tag{11.30}$$

$$F_{\text{Air}}=m_1 f\left(\frac{V_1}{m_1}\right)+m_2 f\left(\frac{V_2}{m_2}\right)=m_1\tfrac{R}{M_A}\ln p_1+m_2\tfrac{R}{M_A}\ln p_2+mc(T)$$

$$F_{\text{Balloon}} = \int_R^r [p] 4\pi r^2 \mathrm{d}r \tag{11.31}$$

$$E_{\text{Piston}}^{\text{Pot}} = m_P g \frac{V_1 + V_{10}}{a}$$

$$E_{\text{Spring}} = \frac{1}{2}\lambda \left(\frac{V_1 + V_{10}}{a}\right)^2 .$$

a is the cross-section of the cylinder and V_{10} is its volume when the spring is unloaded. Thus $(V_1 - V_{10})/a$ is the elastic extension of the spring with the stiffness λ. The specific free energy f of the air depends on $\upsilon = V/m$, or on $p = p\left(\frac{m}{V}\right)$; the dependence on p is logarithmic as shown in (11.31). Therefore the available free energy $\mathcal{A}$ depends on the four variables m_1, m_2, V_1, V_2 of which three are independent; we assume these to be m_2, V_1, V_2. The function $c(T)$ is unimportant for our argument.

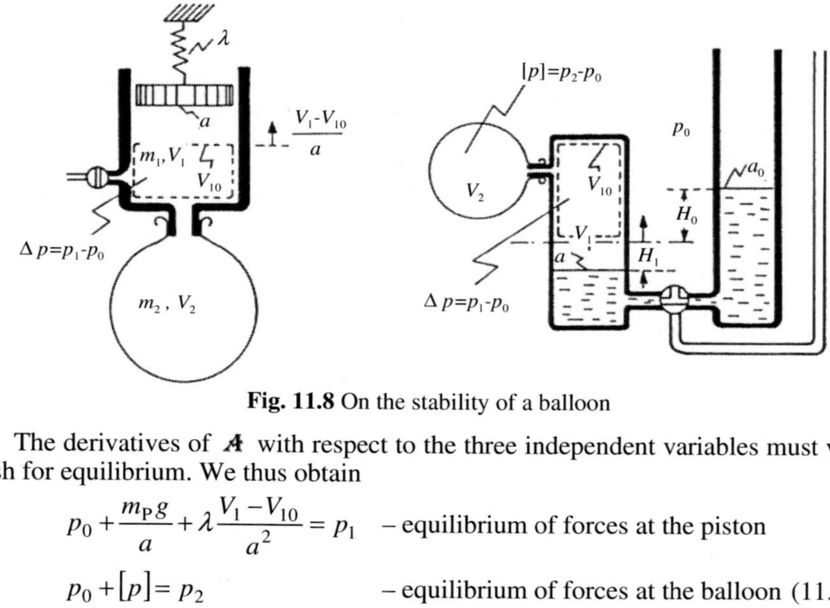

Fig. 11.8 On the stability of a balloon

The derivatives of $\mathcal{A}$ with respect to the three independent variables must vanish for equilibrium. We thus obtain

$$p_0 + \frac{m_P g}{a} + \lambda \frac{V_1 - V_{10}}{a^2} = p_1 \quad \text{– equilibrium of forces at the piston}$$

$$p_0 + [p] = p_2 \quad \text{– equilibrium of forces at the balloon} \tag{11.32}$$

$$g(T, p_1) = g(T, p_2) \quad \text{– equal pressures in balloon and piston.}$$

All of these conditions are somewhat trivial. More interesting are the second derivatives of $\mathcal{A}$ which must form a positive-definite matrix for stability. The exploitation of this condition is straightforward but cumbersome and therefore we present only the result. Two of the three conditions are identically satisfied and the third one reads

$$\frac{1}{4\pi r^2}\frac{\partial[p]}{\partial r} > -\frac{1}{\frac{a^2}{\lambda}+\frac{NkT}{p^2}}. \tag{11.33}$$

From this we conclude first of all that all states are stable for which the balloon finds itself on one of the ascending branches of the $([p],r)$-relation. However, even states on the descending branch *may* be stable. Whether they are, or not, depends on the value of the spring constant λ. Indeed

- for $\lambda=0$, *i.e.* without effective spring, all points on the descending branch of the $([p],r)$-relation are *unstable*,
- for $\lambda\to\infty$, *i.e.* a fixed piston, all these points are *stable*, because even the steepest descending slope still has a larger value than $-p^2/NkT$; at least this is true, if the volume of the cylinder has the same order of magnitude as that of the balloon,
- for intermediate values of λ only the part of steepest descent of the $([p],r)$-relation is unstable.

This result shows clearly – once again – the fact that stability or instability in mechanics is never absolute; it depends on the type of the loading.

11.3.3 *A suggestive argument for the stability of a balloon*

It is all very well to deduce the condition of stability from the positive definiteness of a 3×3 matrix, but it is hardly suggestive. Therefore we consider the problem of stability of the device in Fig. 11.8 a again, but now in a more plausible manner. We assume that the equilibrium at the piston and the equilibrium at the balloon are quickly established, while the equilibrium between the pressures of cylinder and balloon lags behind.

In this case the jump of the pressure across the wall of the cylinder, which we denote by Δp is given by the equilibrium condition $(11.32)_1$

$$\Delta p = \frac{m_P g}{a} + \lambda\frac{V_1-V_{10}}{a^2} \quad \text{or, by } V_1 = \frac{N_1 kT}{p_1} = \frac{(N-N_2)kT}{p_0\left(1+\frac{\Delta p}{p_0}\right)} \approx \frac{(N-N_2)kT}{p_0}$$

$$\Delta p = -\lambda\frac{kT}{p_0 a^2}N_2 + \frac{m_P g}{a} + \frac{\lambda}{a^2}\left(\frac{NkT}{p_0} - V_{10}\right).$$

Thus Δp is a linearly decreasing function of N_2 whose slope is proportional to the stiffness λ of the spring, *cf.* Fig. 11.9 a.*

Similarly we assume that $(11.32)_2$ is satisfied, and while we do not have an analytic form for $[p]$ as a function of N_2, we do know its graph, *cf.* Fig. 11.9 b. If we plot both Δp and $[p]$ as functions of N_2 in a single diagram, the point of

---

* For simplicity m_P has been set equal to zero in Fig. 11.9 a.

intersection determines an equilibrium between cylinder and balloon. The abscissa is denoted by N_2^{E} in Fig. 11.10.

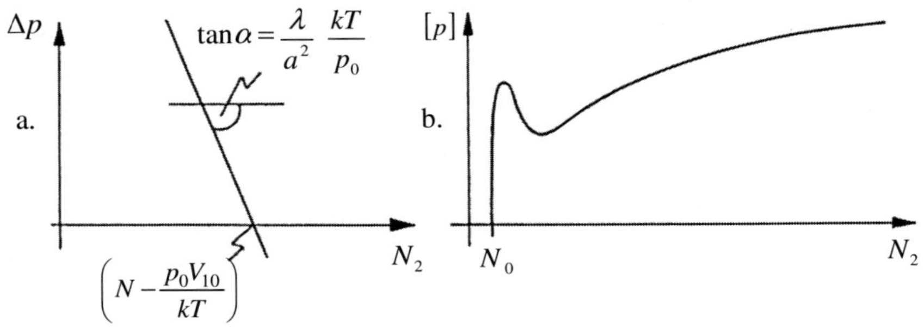

Fig. 11.9 a. Pressure Δp in cylinder as a function of N_2
b. Pressure $[p]$ in balloon as a function of N_2

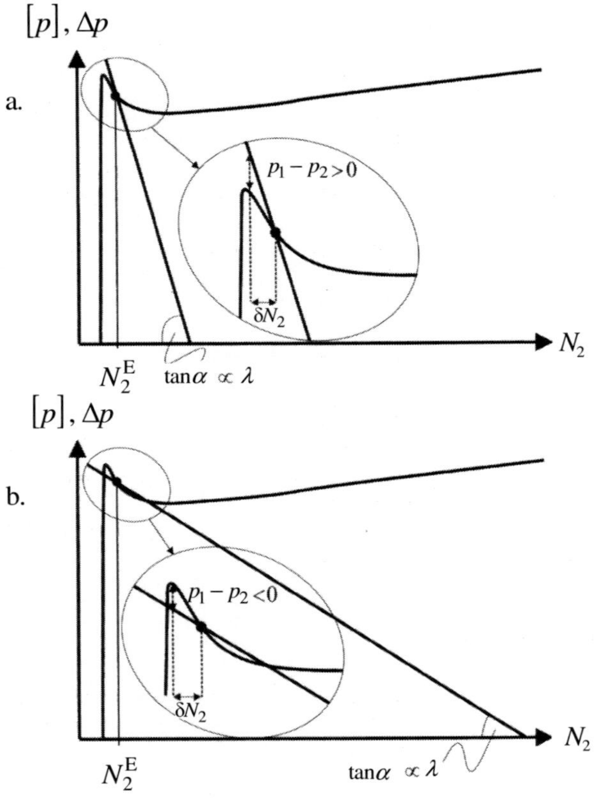

Fig. 11.10 Stability and instability of a spring-cylinder-balloon system
a. stiff spring, **b.** soft spring.

From the graphs of the figure it is particularly easy to evaluate the stability of an equilibrium. Consider Fig. 11.10 a which is appropriate for a stiff spring: We disturb the equilibrium at N_2^{E} by squeezing a little air, δN_2 (say) , from the balloon into the cylinder. Inspection shows that the squeezing increases Δp – or p_1 – and also [p] – or p_2. But Δp increases more than $[p]$. The pressure difference $p_1 - p_2$ will then tend to undo the disturbance: the equilibrium at N_2^{E} is therefore stable for a stiff spring. Fig. 11.10 b shows that the opposite result is obtained for a soft spring.

In summary we conclude that it is the relative slope of the pressure curves of the loading device – here spring and piston – and of the balloon that determines stability. We shall have the opportunity to exploit this conclusion when we consider more than just one balloon.

The device of Fig. 11.8 a consisting of spring, piston, cylinder, and balloon is instructive but impractical, because the seals on the piston present a problem. An equivalent system – without moving parts – is shown in Fig. 11.8 b. In this system an open tube of cross-section a_0 is in contact with another tube of cross-section a on whose only opening the balloon is fixed. It is clear that filling water into the open tube will tend to inflate the balloon. V_{10} is the volume of air in the left tube, when there is no balloon. We refer to Fig. 11.8 and calculate Δp for the devices on the left and on the right. For simplicity we now ignore the weight of the piston.

On the left we have

$$\Delta p = \lambda \frac{V_1 - V_{10}}{a^2}$$

On the right we have

$$\Delta p = \rho_{\mathrm{W}} g (H_1 + H_0)$$

or, by $aH_1 = a_0 H_0$ and $aH_1 = V_1 - V_{10}$

$$\Delta p = a \rho_{\mathrm{W}} g \left(1 + \frac{a}{a_0}\right) \frac{V_1 - V_{10}}{a^2} .$$

Comparison shows that there is an equivalence between the systems, when on the right hand side we interpret the expression

$$a \rho_{\mathrm{W}} g \left(1 + \frac{a}{a_0}\right)$$

as the stiffness. Therefore the "stiffness" of device on the right hand side of Fig. 11.8 may be increased by decreasing a_0. This will happen if the three-way-valve is turned counterclockwise by 90° so that it brings the balloon and *its* tube in contact with the *narrow* tube on the right. We have built that device and it was instructive to see how the balloon, when traversing its unstable range, could be stabilized at any radius by turning the valve.*

---

* Such considerations and many others concerning balloons may be found in a book by one of the authors: I. MÜLLER, P. STREHLOW "Rubber and Rubber Ballooons," Lecture Notes in Physics, Springer Verlag, Heidelberg (2004).

11.3.4 *Equilibria between interconnected balloons*

Two identical balloons, if connected by a thin tube, are in equilibrium with each other, when their pressures are equal. Because of the non-monotone $([p],r)$-relation this does not necessarily mean that the radii are also equal. In the interesting range of pressures there are six possible equilibria – a through f – which are indicated in Fig. 11.11, and which differ by N, the total number of particles. We may ask which of these equilibria are stable.

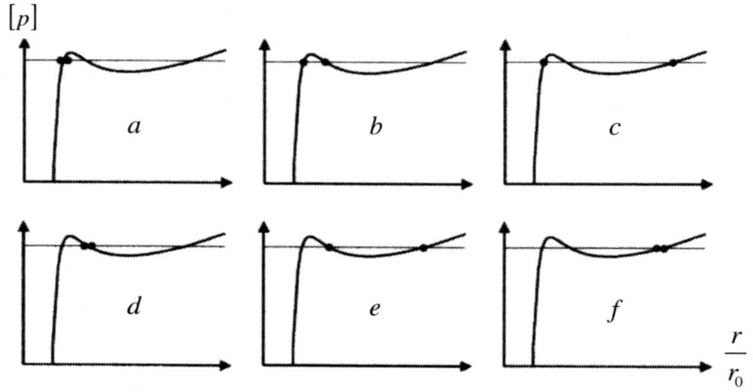

Fig. 11.11 Two balloons and their possible equilibria in the intermediate range of pressures.

This question is easy to answer: The balloons both have a $([p],N)$-relation of the form drawn in Fig. 11.9 b. If we plot $[p](N_1)$ for the first balloon and $[p](N_2)=[p](N-N_1)$ for the second one – both as functions of N_1 – we obtain the curves shown in Fig. 11.12 a, where an equilibrium of the type e is considered. If this equilibrium is disturbed by squeezing δN_1 particles out of balloon 1, the pressure in balloon 1 drops in both balloons. However, the pressure drop in balloon 1 is less than the one in balloon 2, so that the disturbance is not compensated; rather it is enhanced. This means that an equilibrium of type e is unstable. The stability of the other five equilibrium situations is tested in an analogous manner and one finds that only the situations d and e are unstable.

In particular, situation b is stable, although one balloon is resting on the "unstable" branch. One might say in this case *that balloon* 1 *– on the ascending branch – is more stable than balloon* 2 *– on the descending branch – is unstable*, so that the system as a whole is stable. The reason is clear: $N=N_1+N_2$ is constant and balloon 1 cannot provide enough air to assist balloon 2 in moving away from the precarious descending branch.

Fig. 11.12 b represents the pressure in both balloons as a function of the equivalent radius

$$\lambda = \sqrt[3]{\frac{r_1^3 + r_2^3}{2R^3}} \tag{11.34}$$

for different inflation paths of the two balloons. The branches denoted a through f correspond to the situations marked in Fig. 11.11. And the branches d and e are dashed, indicating that they are unstable.

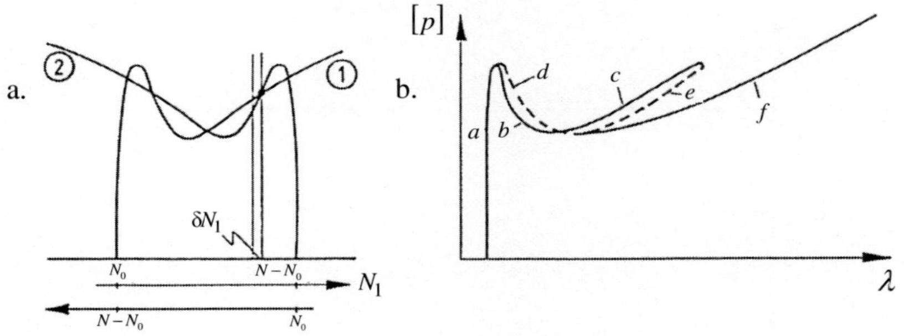

Fig. 11.12 a. On the stability of the situation e
b. Equal pressures in two balloons as a function of the equivalent radius λ

A popular and reliable little experiment in physics classes is performed by connecting two inflated balloons of slightly different size by a pipe. The uninformed students usually expects that the balloons exchange air so that their sizes become equal; instead as they watch, the smaller balloon drops in size to a very small sphere – barely larger than R – while the larger balloon increases its size. It is true that in these experiments the initial state is not an equilibrium state with equal pressures in both balloons. But the same phenomenon occurs when initially the balloons have equal pressures, *i.e.* when we start in situation e. In these cases the two balloons quickly increase their size-difference and tend to assume the situation of types b or c, with one balloon being very small. Often the experimenters do not know that with more air between both balloons the radii do indeed become equal; in such a case they tend to a situation of the type f.

When both balloons are inflated together – through one pipe with two outlets – the state passes through the stable branches in the sequence $abcf$, *cf.* Fig. 11.13, while the branch d is ignored and e is bypassed irreversibly in the manner indicated by the dashed line in the Fig. 11.13 a. Similarly during deflation the balloons pass through $fcba$, ignoring e and bypassing d irreversibly by jumping upward from f to c. The downward jump occurs when the first, already fully inflated balloon *pulls* the second one over the pressure maximum. The upward jump occurs during deflation when one of the balloons on branch f *pushes* the other one back across the pressure maximum. Thus in an inflation-deflation cycle we observe a distinct hysteresis.

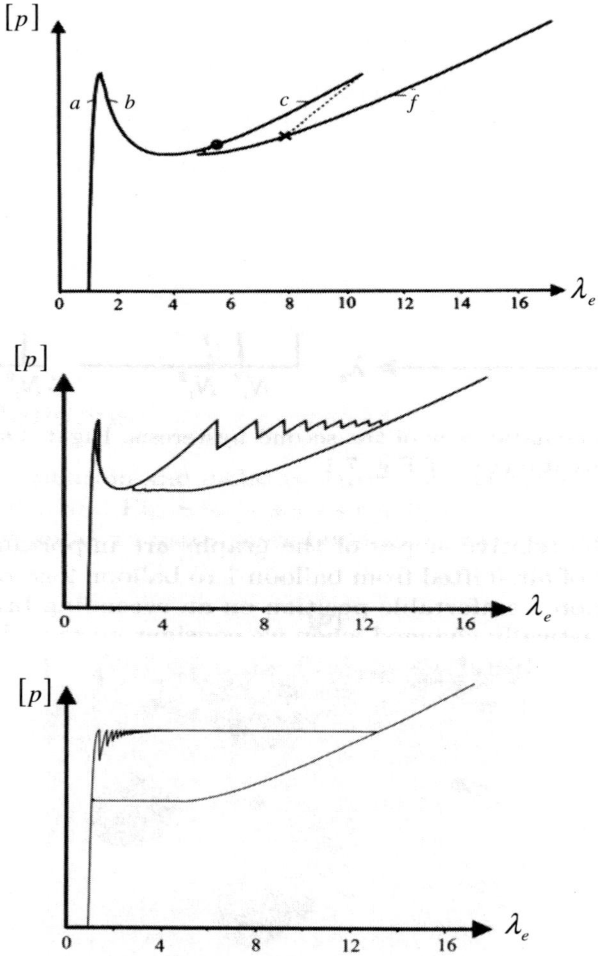

Fig. 11.13 Simultaneous inflation and deflation of two, ten and one thousand balloons

Figs. 11.13 b and c show the corresponding jumps for 10 balloons and for 1000 balloons. Here, for simplicity we have represented the jumps by vertical lines. The hysteresis becomes more and more pronounced for more and more balloons, whereas the spikes in the inflation- and deflation curves develop into a proper pseudo-elastic hysteresis, similar to the one that occurs in shape memory alloys, *cf.* Sect. 12.5.

11.4 Sound

11.4.1 *Wave equation*

Sound is the propagation of disturbances of pressure, density, temperature, and velocity. The propagation is described by the equation of balance of mass, momentum, and (internal) energy, *cf.* (2.2). Gravitation and radiation are not of interest in this context, and we also ignore friction and heat conduction in this section. In this case the equations of balance read

$$\frac{\partial\rho}{\partial t}+\frac{\partial\rho w_i}{\partial x_i}=0\,,$$
$$\frac{\partial\rho w_i}{\partial t}+\frac{\partial\rho w_i w_j+p\delta_{ij}}{\partial x_j}=0\,, \tag{11.35}$$
$$\frac{\partial\rho u}{\partial t}+\frac{\partial\rho u w_i}{\partial x_i}=-p\frac{\partial w_i}{\partial x_i}\,.$$

The thermal and caloric equations of state $p=p(\rho,T)$ and $u=u(\rho,T)$ close the system.

For simplicity we consider the one-dimensional case, in which ρ, T and w_1 depend only on x_1 and t, and where $w_2=w_3=0$ holds. In this manner (11.35) reduces to – with $w_1=w$ and $x_1=x$ –

$$\frac{\partial\rho}{\partial t}+\frac{\partial\rho w}{\partial x}=0\,,$$
$$\frac{\partial\rho w}{\partial t}+\frac{\partial\rho w^2}{\partial x}+\frac{\partial p}{\partial x}=0\,, \tag{11.36}$$
$$\frac{\partial\rho u}{\partial t}+\frac{\partial\rho u w}{\partial x_i}=-p\frac{\partial w}{\partial x}\,.$$

In addition we consider processes $\rho=\rho(x,t)$, $T=T(x,t)$, $w=w(x,t)$ in which the fields deviate only slightly from constant values $\rho=\rho_0$, $T=T_0$, and $w=0$. Therefore we write

$$\rho(x,t)=\rho_0+\overline{\rho}(x,t)\,,$$
$$w(x,t)=\qquad\overline{w}(x,t)\,, \tag{11.37}$$
$$T(x,t)=T_0+\overline{T}(x,t)\,,$$

and we neglect all products of $\overline{\rho}$, $\overline{w}$, and $\overline{T}$ and of their derivatives. Insertion of these decompositions into (11.36) will then provide a linear system of equations of the form

$$\frac{\partial\overline{\rho}}{\partial t}+\rho_0\frac{\partial\overline{w}}{\partial x}=0\,,$$
$$\rho_0\frac{\partial\overline{w}}{\partial t}+\left(\frac{\partial p}{\partial\rho}\right)_0\frac{\partial\overline{\rho}}{\partial x}+\left(\frac{\partial p}{\partial T}\right)_0\frac{\partial\overline{T}}{\partial x}=0\,,$$
$$\rho_0\left(\frac{\partial u}{\partial\rho}\right)_0\frac{\partial\overline{\rho}}{\partial t}+\rho_0\left(\frac{\partial u}{\partial T}\right)_0\frac{\partial\overline{T}}{\partial t}=-p_0\frac{\partial\overline{w}}{\partial x}\,.$$

By $\frac{\partial \overline{w}}{\partial x} = \frac{1}{\rho_0}\frac{\partial \overline{\rho}}{\partial t}$ and the integrability condition (4.24), and $c_\upsilon = \frac{\partial u}{\partial T}$ we obtain

$$\frac{\partial \overline{\rho}}{\partial t} + \rho_0 \frac{\partial \overline{w}}{\partial x} = 0\,,$$

$$\rho_0 \frac{\partial \overline{w}}{\partial t} + \left(\frac{\partial p}{\partial \rho}\right)_0 \frac{\partial \overline{\rho}}{\partial x} + \left(\frac{\partial p}{\partial T}\right)_0 \frac{\partial \overline{T}}{\partial x} = 0\,, \tag{11.38}$$

$$\rho_0 c_\upsilon^0 \frac{\partial \overline{T}}{\partial t} - \frac{T_0}{\rho_0}\left(\frac{\partial p}{\partial T}\right)_0 \frac{\partial \overline{\rho}}{\partial t} = 0\,.$$

From this system of differential equations of first order we derive *one* differential equation of second order as follows. We differentiate the first equation with respect to t, the second one with respect to x, and eliminate $\rho_0 \frac{\partial^2 \overline{w}}{\partial t \partial x}$, thus obtaining

$$\frac{\partial^2 \overline{\rho}}{\partial t^2} - \left(\frac{\partial p}{\partial \rho}\right)_0 \frac{\partial^2 \overline{\rho}}{\partial x^2} - \left(\frac{\partial p}{\partial T}\right)_0 \frac{\partial^2 \overline{T}}{\partial x^2} = 0 \tag{11.39}$$

Integration of $(10.38)_3$ leads to

$$\overline{T} = \frac{T_0}{\rho_0^2}\frac{1}{c_\upsilon^0}\left(\frac{\partial p}{\partial T}\right)_0 \overline{\rho}\,, \tag{11.40}$$

if one sets the constant of integration equal to zero; this only assumes that $\overline{T}$ and $\overline{\rho}$ are zero at the position x at some time. Insertion of (11.40) into (11.39) gives a single differential equation for $\overline{\rho}$, *viz.*

$$\frac{\partial^2 \overline{\rho}}{\partial t^2} - \left[\left(\frac{\partial p}{\partial \rho}\right)_0 + \frac{T_0}{\rho_0^2}\frac{1}{c_\upsilon^0}\left(\frac{\partial p}{\partial T}\right)_0^2\right]\frac{\partial^2 \overline{\rho}}{\partial x^2} = 0\,. \tag{11.41}$$

The coefficient in brackets may be simplified as follows

$$\begin{aligned}\left(\frac{\partial p}{\partial \rho}\right)_0 + \frac{T_0}{\rho_0^2}\frac{1}{c_\upsilon^0}\left(\frac{\partial p}{\partial T}\right)_0^2 &= \left(\frac{\partial p}{\partial \rho}\right)_0 + \frac{T_0}{\rho_0^2}\frac{1}{c_\upsilon^0}\left(\frac{\partial \rho}{\partial T}\right)_0^2\left(\frac{\partial p}{\partial \rho}\right)_0^2 \\ &= \frac{1}{c_\upsilon^0}\left(\frac{\partial p}{\partial \rho}\right)_0\left[c_\upsilon^0 - \frac{T_0}{\rho_0^2}\left(\frac{\partial \rho}{\partial T}\right)_0\left(\frac{\partial p}{\partial T}\right)_0\right] \\ &= \frac{c_p^0}{c_\upsilon^0}\left(\frac{\partial p}{\partial \rho}\right)_0 \frac{\partial \overline{\rho}}{\partial t} = 0\,.\end{aligned} \tag{11.42}$$

With $\kappa = \frac{c_p}{c_\upsilon}$ we may thus write the differential equation in the form

$$\frac{\partial^2 \overline{\rho}}{\partial t^2} - \kappa_0\left(\frac{\partial p}{\partial \rho}\right)_0 \frac{\partial^2 \overline{\rho}}{\partial x^2} = 0\,. \tag{11.43}$$

This is a wave equation and we conclude that the solutions of the linearized equations of thermodynamics are waves. We call them sound waves.

11.4.2 *Solution of the wave equation, d'Alembert method*

In order to see that the equation (11.43) has waves as solutions and for the determination of the speed of propagation we solve the equations for an infinitely long rod. For this purpose we substitute variables $(x,t) \Leftrightarrow (\zeta,\eta)$ as follows

$$\begin{matrix} \zeta = x + at \\ \eta = x - at \end{matrix} \quad \text{or} \quad \begin{matrix} x = \frac{1}{2}(\zeta+\eta) \\ t = \frac{1}{2a}(\zeta-\eta) \end{matrix},$$

where a is an arbitrary constant to begin with. We obtain

$$\frac{\partial}{\partial x} = \frac{\partial}{\partial \zeta} + \frac{\partial}{\partial \eta}, \qquad \frac{\partial^2}{\partial x^2} = \frac{\partial}{\partial \zeta^2} + 2\frac{\partial^2}{\partial \zeta \partial \eta} + \frac{\partial^2}{\partial \eta^2},$$

$$\frac{\partial}{\partial t} = a\frac{\partial}{\partial \zeta} - a\frac{\partial}{\partial \eta}, \quad \frac{\partial^2}{\partial t^2} = a^2\frac{\partial}{\partial \zeta^2} - 2a^2\frac{\partial^2}{\partial \zeta \partial \eta} + a^2\frac{\partial^2}{\partial \eta^2}$$

and therefore the differential equation reads

$$\left[a^2 - \kappa_0\left(\frac{\partial p}{\partial \rho}\right)_0\right]\frac{\partial^2 \overline{\rho}}{\partial \zeta^2} - 2\left[a^2 + \kappa_0\left(\frac{\partial p}{\partial \rho}\right)_0\right]\frac{\partial^2 \overline{\rho}}{\partial \zeta \partial \eta} + \left[a^2 - \kappa_0\left(\frac{\partial p}{\partial \rho}\right)_0\right]\frac{\partial^2 \overline{\rho}}{\partial \eta^2} = 0.$$

Now we choose $a^2 = \kappa_0\left(\frac{\partial p}{\partial \rho}\right)_0$, because this leads us to the very simple differential equation $\frac{\partial^2 \overline{\rho}}{\partial \zeta \partial \eta} = 0$ with the easy-to-find solution

$$\overline{\rho}(x,t) = g(\zeta) + h(\eta), \text{ or, more explicitly}$$

$$\overline{\rho}(x,t) = g\left(x + \sqrt{\kappa_0\left(\frac{\partial p}{\partial \rho}\right)_0}\, t\right) + h\left(x - \sqrt{\kappa_0\left(\frac{\partial p}{\partial \rho}\right)_0}\, t\right). \tag{11.44}$$

This solution is the superposition of two waves, one moving to the left and the other one moving to the right. The arbitrary functions g and h have to be determined from initial conditions: If the density at time 0 is given as shown in Fig. 11.14 a and if $\frac{\partial \overline{\rho}}{\partial t}$=0 holds at time 0, we obtain

$$\begin{matrix} g(x) + h(x) = \rho_0(x) \\ g'(x) - h'(x) = 0 \end{matrix} \quad \text{hence} \quad \begin{matrix} g(x) = \frac{\rho_0(x)}{2} + \frac{c}{2} \\ h(x) = \frac{\rho_0(x)}{2} - \frac{c}{2} \end{matrix},$$

where c is a constant of integration. Hence follows the solution of the initial value problem

$$\bar{\rho}(x,t) = \frac{1}{2}\left[\rho_0\left(x + \sqrt{\kappa_0\left(\frac{\partial p}{\partial \rho}\right)_0} t\right) + \rho_0\left(x - \sqrt{\kappa_0\left(\frac{\partial p}{\partial \rho}\right)_0} t\right)\right].$$

This solution is represented in Fig. 11.14 b: Half of the initial triangle moves to the left, the other half moves to the right, and both move with the speed

$$V = \sqrt{\kappa_0\left(\frac{\partial p}{\partial \rho}\right)_0}, \tag{11.45}$$

which is called the speed of sound.

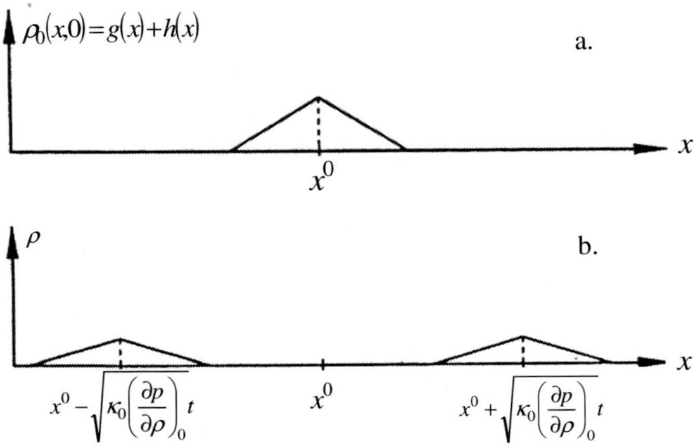

Fig. 11.14 Propagation of a sound pulse

For ideal gases we obtain from (11.45)

$$V = \sqrt{\kappa_0 \frac{R}{M} T_0}, \tag{11.46}$$

i.e. the speed of sound depends on temperature and it is independent of the density. For air of 20°C we obtain with $\kappa_0 = \frac{7}{5}$ and $M = 29\frac{\text{g}}{\text{mol}}$

$$V = 343\frac{\text{m}}{\text{s}}.$$

For less compressible fluids the speed of sound is larger. For water V has a value of about $1500\frac{\text{m}}{\text{s}}$. If an incompressible liquid existed, its speed of sound would be ∞.

11.4.3 *Plane harmonic waves*

A field of the form

$$S(x_i,t) = \bar{S}\exp(k_\text{I} n_i x_i)\cos(\omega t - k_\text{R} n_i x_i + \varphi) \tag{11.47}$$

is called a plane harmonic wave with

amplitude $\bar{S}$

angular frequency ω ($\omega = 2\pi\nu$, where ν is the frequency)

wave number k_R ($k_R = 2\pi/\lambda$, where λ is the wave length)

attenuation ($-k_I$)

phase shift φ

direction of propagation n_i ($n_i n_i = 1$).

The wave is called a plane wave, because its phase $\omega t - k_R n_i x_i + \varphi$ is constant on the planes $\omega t - k_R n_i x_i + \varphi = \text{const}$ perpendicular to n_i. A fixed value of the phase moves with the velocity υ_{Ph} in the direction n_i and we obviously have

$$\upsilon_{Ph} = n_i \frac{dx_i}{dt} = \frac{\omega}{k_R}. \tag{11.48}$$

υ_{Ph} is called the phase velocity of the wave. The wave is called harmonic, because at a fixed position $S(x_i,t)$ is given by a harmonic oscillation. For $k_I < 0$ the wave is damped in the direction of propagation.

The study of harmonic waves is important because – according to FOURIER, *cf.* Section 7.1 – the propagation of arbitrary disturbances may be represented as a superposition of propagating harmonic waves.

To be sure, the form (11.47) is not the simplest representation of a plane harmonic wave. In fact, if we introduce

the complex amplitude $\hat{\bar{S}} = \bar{S}\exp(\mathrm{i}\varphi)$ and

the complex wave number $k = k_R + \mathrm{i}k_I$,

we may write (11.47) as the real part of a complex function

$$S(x_i,t) = \mathrm{Re}\left\{\hat{\bar{S}}\exp[\mathrm{i}(\omega t - k_I n_i x_i)]\right\}. \tag{11.49}$$

This form is useful, because the exponential function reproduces itself upon differentiation or integration. Equation (11.49) is called the complex representation of a plane harmonic wave and, indeed, mostly one writes

$$S(x_i,t) = \hat{\bar{S}}\exp[\mathrm{i}(\omega t - k\, n_i x_i)], \tag{11.50}$$

recalling tacitly that the wave is represented by the real part of this complex function.

11.4.4 *Plane harmonic sound waves*

We recall the linearized system of field equations (11.38) and assume the solution to have the form of plane harmonic waves propagating in the x-direction

$$\begin{aligned}\bar{\rho}(x,t) &= \hat{\bar{\rho}}\exp[\mathrm{i}(\omega t - kx)],\\ \bar{w}(x,t) &= \hat{\bar{w}}\exp[\mathrm{i}(\omega t - kx)],\\ \bar{T}(x,t) &= \hat{\bar{T}}\exp[\mathrm{i}(\omega t - kx)].\end{aligned} \tag{11.51}$$

Insertion of this ansatz into the system (11.38) provides the algebraic system of equations

$$\begin{array}{llll} \mathrm{i}\omega\hat{\hat{\rho}} & -\mathrm{i}k\rho_0\hat{\hat{w}} & & =0 \\ -\mathrm{i}k\left(\dfrac{\partial p}{\partial \rho}\right)_0 \hat{\hat{\rho}} & +\mathrm{i}\omega\rho_0\hat{\hat{w}} & -\mathrm{i}k\left(\dfrac{\partial p}{\partial T}\right)_0 \hat{\hat{T}} & =0 \\ \mathrm{i}\omega\dfrac{T_0}{\rho_0}\left(\dfrac{\partial p}{\partial T}\right)_0 \hat{\hat{\rho}} & & +\mathrm{i}\omega\rho_0 c_v^0\hat{\hat{T}} & =0. \end{array} \tag{11.52}$$

This is a *homogeneous* linear algebraic system which has non-vanishing solutions for $\hat{\hat{\rho}}$, , $\hat{\hat{w}}$, and $\hat{\hat{T}}$ only, if the determinant vanishes. Therefore we must require

$$\begin{vmatrix} \mathrm{i}\omega & -\mathrm{i}k\rho_0 & 0 \\ -\mathrm{i}k\left(\dfrac{\partial p}{\partial \rho}\right)_0 & +\mathrm{i}\omega\rho_0 & -\mathrm{i}k\left(\dfrac{\partial p}{\partial T}\right)_0 \\ \mathrm{i}\omega\dfrac{T_0}{\rho_0}\left(\dfrac{\partial p}{\partial T}\right)_0 & 0 & +\mathrm{i}\omega\rho_0 c_v^0 \end{vmatrix} = 0 \Rightarrow \frac{\omega}{k} = \sqrt{\left(\frac{\partial p}{\partial \rho}\right)_0 + \frac{T_0}{\rho_0^2}\frac{1}{c_v^0}\left[\left(\frac{\partial p}{\partial T}\right)_0\right]^2}. \tag{11.53}$$

It follows, that the plane harmonic wave (11.51) is a solution of the equations *only*, if ω and k are related as in (11.53). We recall that ω/k is equal to the phase velocity v_{Ph} and we also recall that the root in (11.53) is equal to the speed of sound, *cf.* (11.42), (11.45). Thus we have

$$v_{\mathrm{Ph}} = \sqrt{\kappa_0\left(\frac{\partial p}{\partial \rho}\right)_0}. \tag{11.54}$$

and conclude that – not surprisingly – the phase speed of a harmonic wave is equal to the speed of sound. From (11.53) we further conclude that k is real, so that the harmonic waves of the system (11.36) are not damped. Friction and heat conduction, which were neglected here, introduce damping. That more general case will be considered in Paragraph 12.1.3.

For the vanishing determinant the homogeneous system (11.52) may by solved, *i.e.* for instance we may calculate $\hat{\hat{\rho}}$ and $\hat{\hat{T}}$, provided that $\hat{\hat{w}}$ is given. We obtain

$$\frac{1}{\rho_0}\hat{\hat{\rho}} = \frac{1}{v_{\mathrm{Ph}}}\hat{\hat{w}} \quad \text{and} \quad \frac{1}{T_0}\hat{\hat{T}} = \frac{1}{\rho_0 c_v^0}\left(\frac{\partial p}{\partial T}\right)_0 \frac{1}{v_{Ph}}\hat{\hat{w}}. \tag{11.55}$$

For a diatomic ideal gas this gives

$$\frac{1}{\rho_0}\hat{\hat{\rho}} = \frac{1}{\sqrt{\frac{7}{5}\frac{R}{M}T_0}}\hat{\hat{w}} \quad \text{and} \quad \frac{1}{T_0}\hat{\hat{T}} = \frac{2}{5}\frac{1}{\sqrt{\frac{7}{5}\frac{R}{M}T_0}}\hat{\hat{w}}$$

and the pressure amplitude follows from the ideal gas law

$$\hat{\hat{p}} = \rho_0\sqrt{\frac{7}{5}\frac{R}{M}T_0}\,\hat{\hat{w}}.$$

A powerful – nowadays common – disco-sound with 120 decibel has the pressure amplitude

$$\overline{p} = 20\,\text{Pa}\,.$$

Hence follows

$$\overline{w} = 0.064\frac{\text{m}}{\text{s}}\,, \quad \overline{\rho} = 2.4\cdot 10^{-4}\frac{\text{kg}}{\text{m}^3}\,, \quad \overline{T} = 0.022\,\text{K}\,.$$

It is only $\overline{p}$ that we register with our senses, since our ear drums are sensitive organs, able to register even much smaller pressure amplitudes. The temperature amplitude coupled to the pressure wave cannot be detected by our senses.

11.5 Landau theory of phase transitions

11.5.1 *Free energy and load as functions of temperature and strain: Phase transitions of first and second order*

LANDAU has invented a mathematical model which may be applied to the thermomechanical behavior of a one-dimensional rod, *cf.* Fig. 5.3. In this model the free energy F is given as a function of temperature T and of strain $D = L - L_0$ by a simple polynomial function

$$F(D,T) = F_0(T) + \tfrac{1}{2}a(T - T_0)D^2 - \tfrac{1}{4}bD^4 + \tfrac{1}{6}cD^6\,. \tag{11.56}$$

The load P is the derivative of F with respect to D, so that the load-deformation of the Landau model reads

$$P(D,T) = a(T_0 - T)D - bD^3 + cD^5\,. \tag{11.57}$$

a, b, and c are constants. There are two important cases

i.) Phase transitions of *first order* for which all constants are positive
ii.) Phase transitions of *second order* for which $b < 0$ holds.

We consider these cases separately.

Lev Davidovich LANDAU (1908-1968) was the foremost representative of Soviet physics. He did much to clarify the phase transition of liquid helium at low and lowest temperatures. And he is well known to physics students for his excellent series of text books on theoretical physics which he wrote together with his student Evgenii Michailovich LIFSHITZ (1915-1985).

LAUDAU's work was rewarded with the Nobel prize for physics in 1962. From January of that year, and for the whole year, he lay in a coma after a car accident. They say he passed away several times in that year and was brought back to life by drastic methods.

11.5.2 *Phase transitions of first order*

We draw isotherms $F(D,T)$ and $P(D,T)$ and obtain the graphs of Fig. 11.15 which correspond to the ranges of temperature given in the figure. The limits of these ranges are determined by the coefficients T_0 and a, b, c and they read

$$T_a = T_0 + \frac{3}{16}\frac{b^2}{ac}\,, \quad T_b = T_0 + \frac{1}{5}\frac{b^2}{ac}\,, \quad T_c = T_0 + \frac{1}{4}\frac{b^2}{ac}\,, \quad T_d = T_0 + \frac{9}{20}\frac{b^2}{ac}\,. \tag{11.58}$$

Below T_0 we have two lateral minima of the free energy, while between T_0 and T_c there are three minima – two lateral ones and one at $D = 0$. Obviously the minima correspond to equilibria of the load-free body. We call these the load-free phases and distinguish the phase (0) with $D_{(0)} = 0$ and the phases $(\pm)$ for $D_{(\pm)} \neq 0$. *Stable* are the phases with the lowest free energy. The phases $(\pm)$ may be called *twin phases*, since they are energetically identical, although they are not identical in strain.

We recall from Chaps. 4 and 5 that the (P, D)-branches with a negative slope are unstable and from the considerations of Paragraph 4.2.5 we know how a body avoids such states, namely by forming two phases, whose proportion changes linearly along horizontal lines in the (P, D)-diagram. The height of the horizontal lines is determined by the "equal area rule," *cf.* Fig. 4.5. In Fig. 11.15 the horizontal lines of phase equilibrium are indicated in some graphs by dashed lines. The equilibrium strains as functions of temperature result from (11.57) by setting $P = 0$, and solving for D. We obtain

$$D_{(0)} = 0 \quad \text{and} \quad D_{(\pm)} = \pm\sqrt{\frac{1}{2}\frac{b}{c} \overset{(+)}{\underset{-}{}} \sqrt{\frac{1}{4}\left(\frac{b}{c}\right)^2 - \frac{a}{c}\left(T - T_0\right)}} \; . \tag{11.59}$$

The corresponding graphs $D(T)$ are indicated in Fig. 11.16 a by the thick lines; the dashed lines correspond to unstable load-free equilibria, *i.e.* either maxima of the free energy function or local minima.

If we start at some large value of T, and lower the temperature, we have initially only one phase, namely phase (0); this phase is stable, as long as the central minimum is the deepest one. That situation changes when T_a is reached. At that temperature the phases $(\pm)$ become stable. Usually, however, the phase transition is not accompanied by a change of strain. It is true that locally the strain jumps to the value $D_{(+)}$ or $D_{(-)}$, but different parts of the body jump to $D_{(+)}$ and $D_{(-)}$ with equal probability, so that *on average* the strain remains zero.

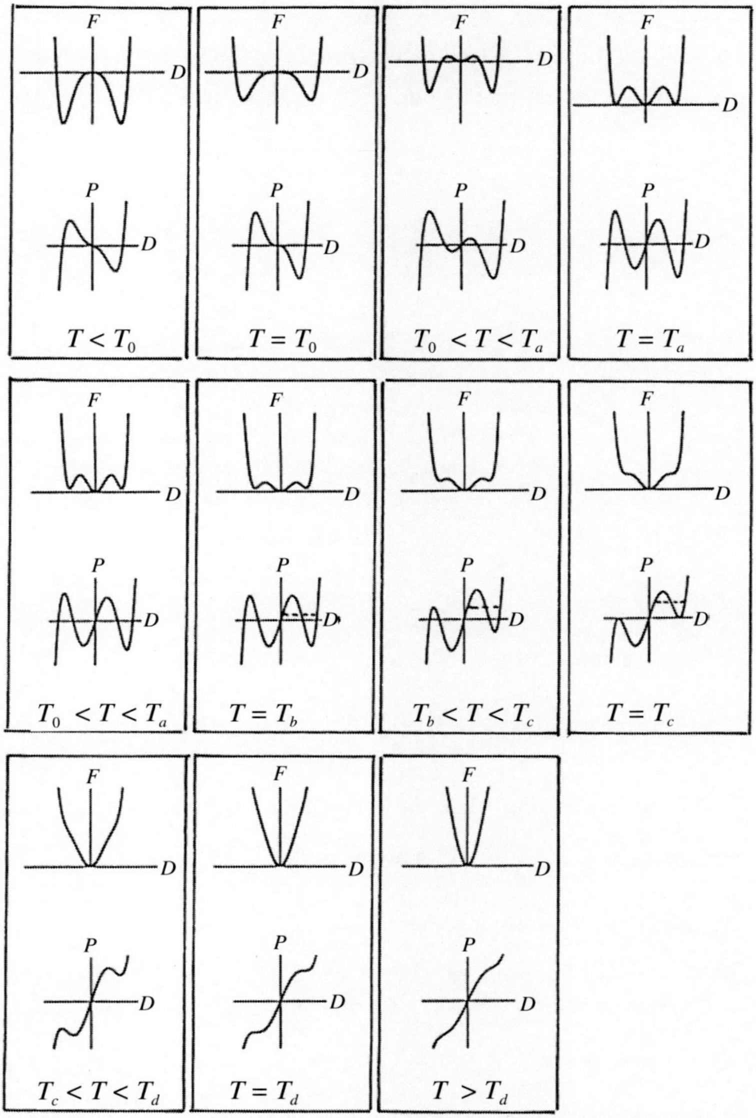

Fig. 11.15 Free energies and load-strain curves in a phase transition of first order

We define the elastic moduli of the phases by

$$E = \left.\frac{\partial P}{\partial D}\right|_{P=0} = a(T - T_0) - 3bD^2\big|_{P=0} + 5cD^4\big|_{P=0}. \tag{11.60}$$

For phase (0) we thus obtain an elastic modulus which is linear in T, since we have to set $D_{(0)} = 0$ in (11.60). The elastic moduli $E_{(\pm)}$ result when we insert

$D_{(\pm)}$ from (11.59) into (11.60). What results is a fairly complex function of T which has been plotted in Fig. 11.16 b along with $E_{(0)}$. Since the phase transition $(0) \to (\pm)$ occurs at T_a, the elastic modulus jumps at that temperature.

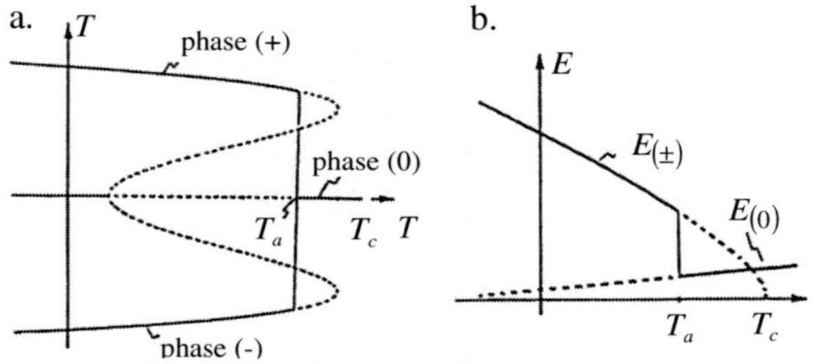

Fig. 11.16 Phase transition of first order
a. $D(T)$ for the phases $(\pm)$ and (0)
b. Elastic modulus as a function of temperature

The entropy S results from $S = -\left(\frac{\partial F}{\partial T}\right)$ and we obtain

$$S = S_0(T) - \frac{1}{2} a D^2 . \tag{11.61}$$

Thus, by (11.59), the phases (0) and $(\pm)$ have the entropies

$$S = S_0(T) \text{ and } S_{(\pm)} = S_0(T) - \frac{a}{2}\left[\frac{1}{2}\frac{b}{c} \overset{(+)}{\underset{-}{}} \sqrt{\frac{1}{4}\left(\frac{b}{c}\right)^2 - \frac{a}{c}(T - T_0)}\right]. \tag{11.62}$$

We conclude that S has a jump when the transition occurs; therefore we cannot define the heat capacity $C = T\frac{\partial S}{\partial T}$ at that point.

11.5.3 *Phase transitions of second order*

If the coefficient b in (11.56) or (11.57) is smaller than zero, we obtain only three qualitatively different isotherms $F = F(D,T)$ and $P = P(D,T)$, namely those for the temperature values $T \overset{<}{\underset{>}{=}} T_0$. The corresponding isotherms are plotted in Fig. 11.17. Again we recognize the phase (0) and the twin phases $(\pm)$ as minima of the free energy. The deepest minima are the stable ones as before. The deformations of the load-free phases are again given by (11.59).
Since $b < 0$ holds, we have at most three real solutions of (11.59): For $T \geq T_0$ there is only one real solution, namely $D = 0$, and for $T < T_0$ there are three, among them $D = 0$; the latter solution, however, is unstable, since it corresponds to a maximum. In the neighborhood of $T = T_0$ the three solutions read

$$D_{(0)} = 0 \quad \text{and} \quad D_{(\pm)} = \pm\sqrt{\frac{a}{-b}(T_0 - T)}\,. \tag{11.63}$$

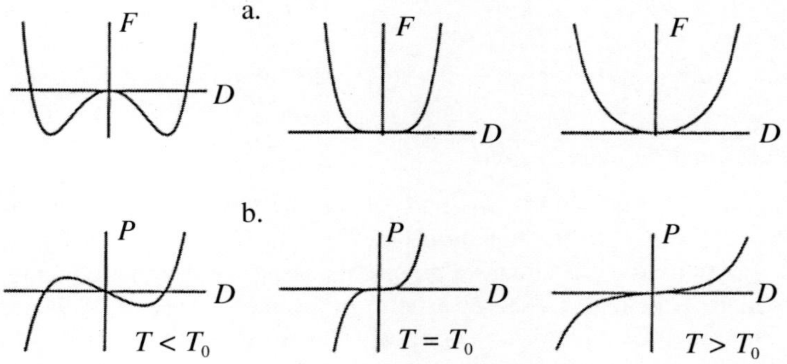

Fig. 11.17 Phase transition of second order
a. Free energies at different temperatures
b. Load-strain diagrams

The graphs $D(T)$ are shown in Fig. 11.18 a; those for $D_{(\pm)}(T)$ are parabolic at $T \le T_0$ at least for $T \approx T_0$.

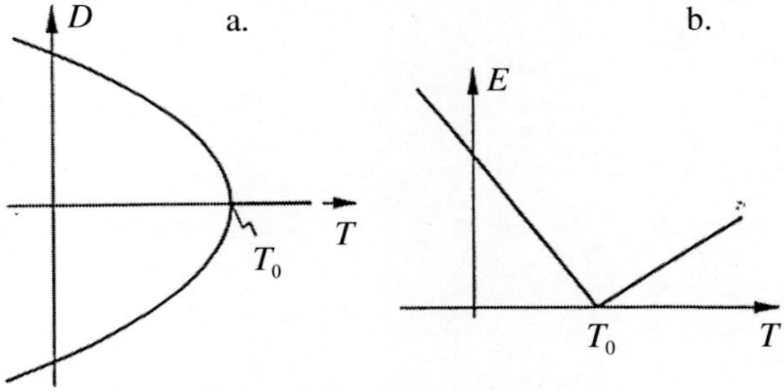

Fig. 11.18 Phase transition of second order
a. $D(T)$ for the phases $(\pm)$ and (0)
b. Elastic modulus as a function of temperature

The elastic moduli result from (11.60) and (11.63)

$$E_{(0)} = a(T - T_0) \quad \text{and} \quad E_{(\pm)} = -2a(T - T_0); \tag{11.64}$$

the latter one has an additional term which, however, is small near $T = T_0$. Fig. 11.18 b exhibits the T-dependence of the elastic moduli. For $T = T_0$ we have $E = 0$, which means that the body is "soft" at the transition $(0) \leftrightarrow (\pm)$. It is also noteworthy that the slopes of E in the two phases near $T = T_0$ differ by a factor 2.

The entropy follows from (11.62). Insertion of (11.63) gives for the neighborhood of $T = T_0$

$$S = S_0(T) \text{ and } S_{(\pm)} = S_0(T) - \frac{a^2}{2b}(T_0 - T). \qquad (11.65)$$

We conclude that $S(T)$ has a kink at $T = T_0$, and therefore the heat capacity $C = T\frac{\partial S}{\partial T}$ has a jump: It decreases at $T = T_0$ by the value $\frac{a^2}{2|b|}T_0$.

11.5.4 *Phase transitions under load*

In Paragraphs 11.5.2 and 11.5.3 we have considered phase transitions in the unloaded body. In that case the equation (11.57) with $P = 0$ was easily solved. However, if $P \neq 0$ holds, this equation represents a genuine algebraic equation of fifth order which is difficult to solve, except, of course, numerically. Fig. 11.19 shows such solutions for a phase transition of first order (a.) and of second order (b.).

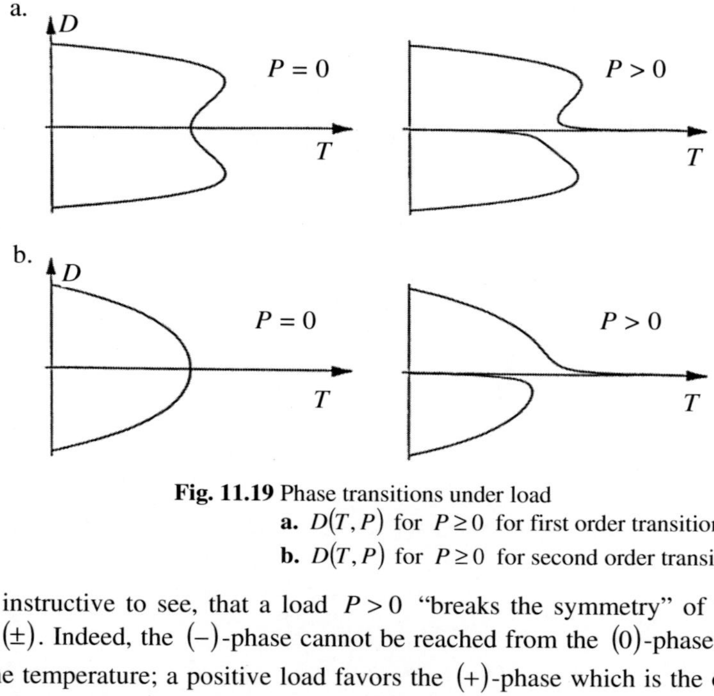

Fig. 11.19 Phase transitions under load
a. $D(T,P)$ for $P \geq 0$ for first order transitions
b. $D(T,P)$ for $P \geq 0$ for second order transitions

It is instructive to see, that a load $P > 0$ "breaks the symmetry" of the twin phases $(\pm)$. Indeed, the $(-)$-phase cannot be reached from the (0)-phase by lowering the temperature; a positive load favors the $(+)$-phase which is the only one to appear upon a decrease of temperature. It follows that the first order transition is now connected with a large jump of the strain.

11.5.5 *A remark on the classification of phase transitions*

The classification of phase transitions into those of first and second order is older than the simple analytic Landau model and more general. The original definition distinguished between transitions, in which the first derivatives of the Gibbs free energy – here S and D – have a jump (first order) or

a kink (second order)

as functions of temperature. Looking back on the arguments of the previous paragraphs we see that this lack of smoothness in $S(T)$ and $D(T)$ may be found in the Landau model at the appropriate places.

In this book we have only considered phase transitions of first order and only such transitions that concerned load, deformation, and temperature, or – in the case of fluids – pressure, volume, and temperature.

It must be said, however, that phase transitions are universal phenomena which occur in many fields of physics, in particular for electro-magnetic properties of bodies. In those cases the electro-magnetic fields and magnetization or polarization replace the mechanical quantities load and strain. Indeed, the best known example of a phase transition of second order is ferromagnetism, where the atomic magnetic dipoles of a body align themselves below a characteristic temperature.

11.6 Swelling and shrinking of gels

11.6.1 *Phenomenon*

Gels consist of long chain-molecules which are joined to form a network in a manner similar to the one described earlier in context of rubber, *cf.* Sect. 5.5. In a *polyelectrolytic* gel an occasional link of the chain molecule is ionized. The standard example is polybisacrylamide, in which some $CONH_2$-groups have been modified into $COOH$-groups by hydrolysis. Such groups ionize in a water solution and split off an H^+-ion which attaches itself to a water molecule to form an H_3O^+-ion; this H_3O^+-ion is called a counter ion in the jargon of gel thermodynamics.

Gels swell in water solution by absorbing water and they shrink by losing water. Swelling and shrinking are particularly pronounced in polyelectrolytic gels. The volume change may be triggered by

- a change in temperature,
- a change of the degree of ionization,
- the application of a load.

In a typical state the gel consists of three types of particles: chain links of which some are ionized, counter ions, and water molecules. It is convenient and customary to represent the observations in a (T,ν)-diagram, where T is the temperature and ν is the *shrink factor*, defined as the quotient of the number of chain links and the total number of particles in the gel. Thus ν has the value 1 when the gel is completely shrunk, otherwise we have $\nu < 1$.

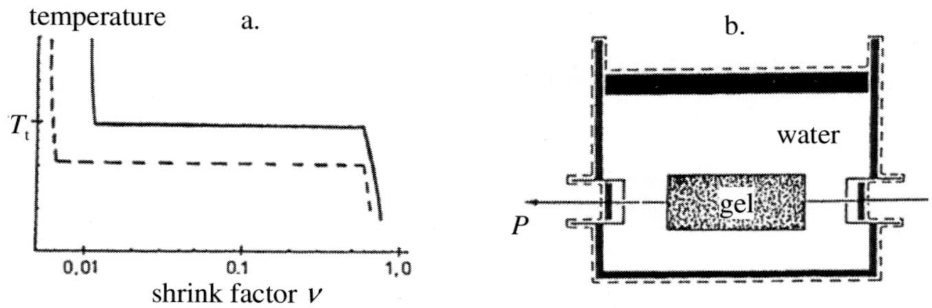

Fig. 11.20 Gel in water solution
a. (T,ν)-diagram for two polyelectrolytic gels. The dashed graph correspond to a higher degree of ionization of the gel
b. Gel in water reservoir, loaded by a tensile load P

The figure shows a schematic (T,ν)-diagram. At low temperatures the gel is shrunk and it swells only a little upon a small rise in temperature. However, at the transition temperature T_t, the swelling becomes dramatic and continues to reach values $\nu << 1$ such that the volume increases by a factor 300 or even more. Afterwards a further rise in temperature produces only minimal further swelling. A subsequent decrease of temperature lets the state of the gel run backwards along the same curve. In particular the essential shrinking occurs at the temperature T_t again.

Obviously the phenomenon is reminiscent of a phase transition, *e.g.* an exchange of stability between the vapor state and the liquid state (say). As such it should be connected with a replacement of energy by entropy as the dominant influence, or *vice versa*. And indeed, we shall describe and understand the swelling and shrinking of a gel as a competition between three thermodynamic effects, two of them entropic and one energetic. These forces are

- the osmotic pressure of the counter ions, an expansive entropic force,
- the network elasticity, a contractive entropic force,
- the molecular interaction which represents an energetic contractive force for the gel.

At the transition temperature the expansive tendency of the osmotic pressure overcomes the two contractive tendencies. This leads to the abrupt swelling. We proceed to consider these forces in detail and discuss their origins.

The electrostatic interaction of the network-ions and of the counter-ions enforces electro-neutrality in each point of the gel. Consequently, since the network-ions are fixed on the gel, the counter-ions cannot leave the gel either. On the other hand the counter-ions are virtually free particles and have the tendency to fill the whole volume of the gel *and the water reservoir* homogeneously. Therefore they push against the surface of the gel from inside and expand its network by partially disentangling the long-chain molecules of the network. The volume thus gained in the gel is filled by water molecules. These can enter and leave the gel freely. Thus the surface of the gel acts like a semi-permeable membrane – permeable for water – and the pushing by the counter-ions may be considered as an osmotic pressure.

The expansion of the gel forces the long chain-molecules into the improbable disentangled state. They have the tendency to entangle – *cf.* Section 5.3 – and thus cause the network elasticity, an entropic force which opposes the osmotic pressure. Both forces depend linearly on temperature.

The network is energetically in its most favorable state when only the links of the network are present so that there is no energetic *malus* for the formation of unequal next neighbors. There *is* such a malus, however, for the formation of pairs of unequal next neighbors. This effect is largely independent of temperature and therefore it determines the behavior at low temperature – where the entropic forces have a small effect – and leads to shrinking.

The whole system of the gel and the water reservoir may be considered as incompressible.

11.6.2 *Gibbs free energy*

We focus the attention on the system shown in Fig. 11.20 b and consider its stability in a manner as previously described in Paragraph 4.2.7. Because of the assumed incompressibility the available free energy has the form

$$A = E - TS - Pl \,. \tag{11.66}$$

The constant load P can change the length l of the gel.

In order to characterize the system we introduce notation as follows:

- n_1 – No. of water molecules
  - n_1^{I} – No. of water molecules in the gel
  - n_1^{II} – No. of water molecules in the water reservoir
- n_2 – No. of polymer chains
  - xn_2 – No. of chain links of "molecular size"
- n_3 – No. of counter ions
- $n^{\mathrm{I}} = n_1^{\mathrm{I}} + xn_2 + n_3$ – No. of particles of "molecular size" in gel.

In each chain we thus introduce x chain links of "molecular size." This is motivated by the desire to be able to think of particles in the gel which all have the same size, at least approximately: water molecules, counter-ions, and chain links. In this case the shrink factor

$$\nu = \frac{xn_2}{n^{\mathrm{I}}} \tag{11.67}$$

is equal to the *volume* ratio of the shrunk and swollen gel.

The deformation gradient has the form

$$F = \begin{bmatrix} \alpha & 0 & 0 \\ 0 & \beta & 0 \\ 0 & 0 & \beta \end{bmatrix} = \begin{bmatrix} l/L & 0 & 0 \\ 0 & b/B & 0 \\ 0 & 0 & b/B \end{bmatrix}, \tag{11.68}$$

where l, b and L, B are length and width of the swollen and of the shrunk gel in the load-free configuration, respectively. Therefore it follows that $\alpha\beta^2$ is equal to the volume ratio of the swollen and the shrunk gel and we have

$$\alpha\beta^2 = \frac{1}{\nu}. \tag{11.69}$$

The quantities x, n_2, and n_3 are fixed by the production process of the gel. Therefore there are only two variables, *viz.*

$$n_1^{\mathrm{I}} \text{ (or } \nu \text{) and } \alpha. \tag{11.70}$$

ν has the value 1, when the gel is free of water, *i.e.* when it is completely shrunk. Otherwise its value is smaller than 1.

Apart from ν and α we introduce the degree of ionization

$$R = \frac{n_3}{x n_2} \tag{11.71}$$

as a parameter. R obviously characterizes the fraction of the ionized chain links. Typically it has values 0.2 or 0.3.

We proceed to calculate the available free energy $\mathcal{A}$ as a function of the variables ν and α. First we consider the gel which contains the three constituents: water molecules, counter-ions, and chain links. The temperature is T and the pressure is p.

The free energy $E-TS$ of the gel is equal to the free energies of the unmixed constituents at T and p plus the free energy of mixing, *cf.* Paragraphs 8.2.1 through 8.2.4. We write

$$\begin{aligned}(E-TS)_{\mathrm{Gel}} &= n_1^{\mathrm{I}} f_1(T,p) + x n_2 f_2(T,p) + n_3 f_3(T,p) + \\ &+ e_{12} n_1^{\mathrm{I}} \frac{x n_2}{n^{\mathrm{I}}} + e_{13} n_1^{\mathrm{I}} \frac{n_3}{n^{\mathrm{I}}} + e_{23} n_3 \frac{x n_2}{n^{\mathrm{I}}} + \\ &+ kT\left(n_1^{\mathrm{I}} \ln \frac{n_1^{\mathrm{I}}}{n^{\mathrm{I}}} + n_2 \ln \frac{x n_2}{n^{\mathrm{I}}} + n_3 \ln \frac{n_3}{n^{\mathrm{I}}} \right).\end{aligned} \tag{11.72}$$

f_α $(\alpha = 1,2,3)$ are the specific free energies of the constituents (free energy per particle). The second line represents the heat of mixing and e_{12} is the *malus* for the formation of a pair water molecule–chain link; e_{12} and e_{23} are defined accordingly; we have assumed that the number of unequal pairs of next neighbors is equal to their expectation values, namely

$$n_1^{\mathrm{I}} \frac{x n_2}{n^{\mathrm{I}}},\ n_1^{\mathrm{I}} \frac{n_3}{n^{\mathrm{I}}},\ n_3 \frac{x n_2}{n^{\mathrm{I}}}. \tag{11.73}$$

The third line in (11.72) is the entropy of mixing.*

---

* In Chap. 8 we have discussed and used the entropy of mixing extensively. It may be calculated from the number of possibilities to combine the available particles to form a mixture. However, in one point the present entropy of mixing deviates from previous expressions: Note that the second logarithmic term is multiplied by n_2 rather than $x n_2$. The

But this is not all. The network with its entangled chains also contributes to the entropy. This contribution has been calculated in Chap. 5 and again in Sect. 11.3 for different cases of loading. Here we obtain by a slight modification of the previous arguments

$$S_{\text{Network}} = k\frac{n_2}{2}\left(\alpha^2 + 2\beta^2 - 3\right). \tag{11.74}$$

Finally the free energy of the water reservoir is given by

$$(E - TS)_{\text{Water reservoir}} = n_1^{\text{II}} f_1(T, p). \tag{11.75}$$

We combine the equations (11.72) through (11.75) and obtain an explicit form of the available free energy (11.66), namely

$$\begin{aligned}
\mathcal{A} &= n_1 f_1(T,p) + x n_2 f_2(T,p) + n_3 f_3(T,p) + \\
&+ e_{12} n_1^{\text{I}} \frac{x n_2}{n^{\text{I}}} + e_{13} n_1^{\text{I}} \frac{n_3}{n^{\text{I}}} + e_{23} n_3 \frac{x n_2}{n^{\text{I}}} + \\
&+ kT\left(n_1^{\text{I}} \ln \frac{n_1^{\text{I}}}{n^{\text{I}}} + n_2 \ln \frac{x n_2}{n^{\text{I}}} + n_3 \ln \frac{n_3}{n^{\text{I}}} \right) + \\
&+ k\frac{n_2}{2}\left(\alpha^2 + 2\beta^2 - 3\right) - \\
&- PL\alpha .
\end{aligned} \tag{11.76}$$

We recall that n_1, n_2, n_3, and x are all constants and thus confirm that $\mathcal{A}$ indeed depends only on two variables, namely ν and α.

In the following calculation we assume that all three *mali* e_{ij} have the same value *e*; we divide by $x n_2 kT$ and move all constant values in (11.76) to the left hand side. We also introduce the dimensionless force $f = \frac{PL}{n_1 kT}$ and obtain

$$\begin{aligned}
\overline{\mathcal{A}} &= \frac{1-(1+R)\nu}{\nu} \ln[1-(1+R)\nu] + R \ln \nu + \frac{e}{kT}\left[R\nu - (1+R)^2 \nu\right] + \\
&+ \frac{1}{2x}\left(\alpha^2 + \frac{2}{\alpha\nu} + 2\ln \nu\right) - \frac{1}{x} f\alpha ,
\end{aligned} \tag{11.77}$$

where $\overline{\mathcal{A}}$ is the variable part of $\mathcal{A}$, made dimensionless through division by $x n_2 kT$.

---

reason lies in the fact that the chain-links in the mixture cannot be distributed arbitrarily; rather they must occupy neighboring sites because, after all, they are to form a chain and the network. FLORY has calculated the number of possibilities for creating the mixture under these circumstances and he has obtained the expression given in (11.72) (*cf.*, FLORY, P.J. Principles of Polymer Science, Cornell University Press, Ithaca, N.Y., London.)

11.6.3 *Swelling and shrinking as function of temperature*

First of all we consider the load-free gel and set $f = 0$. Because of the isotropy of the gel the components of the deformation gradient (11.68) are then all equal and we conclude from (11.69) that we have

$$\alpha = \nu^{-\frac{1}{3}}. \tag{11.78}$$

Thus $\overline{A}$ in (11.77) becomes a function of the single variable ν, the shrink factor. We find the minimum of the function by differentiation with respect to ν and obtain

$$\frac{kT}{e} = \left(1 + R + R^2\right)\frac{\nu^2}{\ln[1-(1+R)\nu] + \left(1+\frac{1}{x}\right) + \frac{1}{x}\nu^{\frac{1}{3}}}. \tag{11.79}$$

Fig. 11.21 a shows two graphs of this function $T(x)$ for $x = 100$ and $R = 0.2$ and $R = 0.3$, respectively. The most conspicuous feature of the graphs is their non-monotonicity. From previous experience – with the van der Waals gas, the crystallization of rubber, and the Landau model – this feature lets us assume that, at some T, there may be a sudden transition of the gel to a swollen state.

The temperature at which such a transition will occur may be read off from Fig. 11.21 b, which shows how the available free energy depends on ν for different values of kT/e. In a certain range of this parameter there are two minima. The interpretation is clear: As long as the right minimum is lower than the left one, the gel will be shrunk. Swelling occurs when the two minima exchange their relative depth. For $R = 0.2$ this is the case for $kT/e = 0.6$ whereas for $R = 0.3$ (not shown in the figure) it happens at $kT/e = 0.45$. The corresponding horizontal transitions in the (T,ν)-diagram are shown in Fig. 11.21 a. We conclude from this observation that swelling occurs for a lower temperature in a more strongly ionized gel. The reason is obvious: Stronger ionization means more counter ions and therefore a higher expansive osmotic pressure.

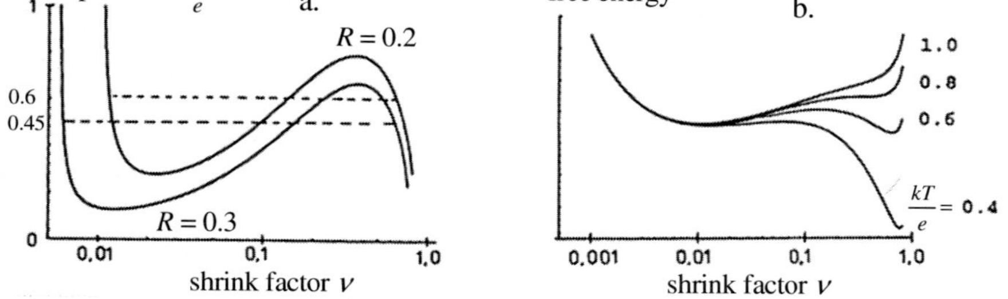

Fig. 11.21 The "phase transition" swollen • shrunk
a. Temperature as a function of the shrink factor for two degrees of ionization.
b. Available free energy for different temperatures.

Next we consider the uniaxially loaded gel, *i.e.* we set $f \neq 0$. In this case the available free energy (11.77) is a function of ν and α, and its minima are obtained where the derivatives with respect to both variables vanish. We obtain

$$\frac{kT}{e} = \left(1+R+R^2\right)\frac{\nu^2}{\ln[1-(1+R)\nu]+\left(1+\frac{1}{x}\right)+\frac{1}{x}\frac{1}{\alpha}} \quad \text{and} \quad f = \alpha - \frac{1}{\nu}\frac{1}{\alpha^2}. \tag{11.80}$$

For given values of f and T these are two equations for the unknowns ν and α. Equation $(11.80)_2$ is a cubic equation for α which, however, contains only one real solution, *viz.*

$$\alpha = \frac{f}{3} + \frac{2^{\frac{1}{3}} f^2 \nu^{\frac{1}{3}}}{3\left(27+2f^3\nu+3^{\frac{1}{3}}\sqrt{27+4f^3\nu}\right)^{\frac{1}{3}}} + \frac{\left(27+2f^3\nu+3^{\frac{3}{2}}\sqrt{27+4f^3\nu}\right)^{\frac{1}{3}}}{3\cdot 2^{\frac{1}{3}}\nu^{\frac{1}{3}}}. \tag{11.81}$$

If this value of α is inserted into $(11.80)_1$, one obtains kT/e as a function of the shrink factor with the load as a parameter. Fig. 11.22 shows the corresponding graphs – for $R = 0.2$ and $x = 100$ – for five different values of f.

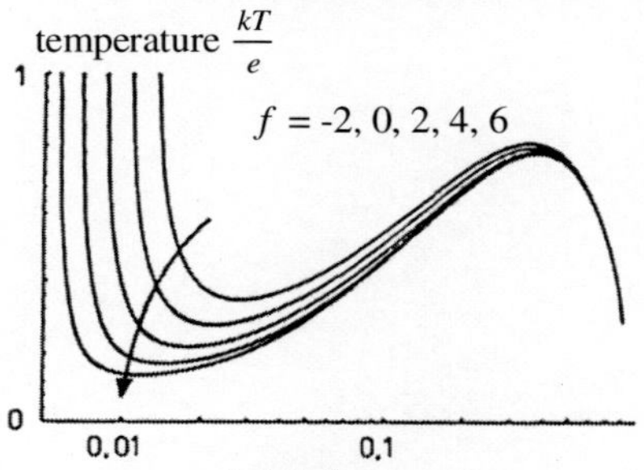

Fig. 11.22 Temperature as a function of the shrink factor for different loads

We see from the figure that the branches with positive slopes in the curves $T(\nu; f)$ are lower for tensile loads $f > 0$ than for zero load and for compressive loads $f < 0$. Hence we conclude that the swelling is facilitated by a tensile load and made more difficult by a compressive one. This phenomenon is easily understood intuitively.

A suggestive illustration for swelling and shrinking of the loaded gel is shown by the contour plots of Fig. 11.23. This figure shows the lines of constant $\mathcal{A}$, according to (11.77), on a plane spanned by ν and α. All pictures refer to the temperature $kT/e = 0.66$ and to different loads: Two compressive ones, the load-free case, and a tensile load.

We recognize two minima that correspond to the swollen and the shrunk configuration. For compressive loads $f < 0$, the shrunken states are stable, since their minima are deeper; for $f = 0$ the swelling has occurred, because now the deeper minimum lies at small values of ν. And for $f = 2$ the deeper minimum has progressed to larger values of α. In summary it follows that the swelling may be induced by a tensile load.

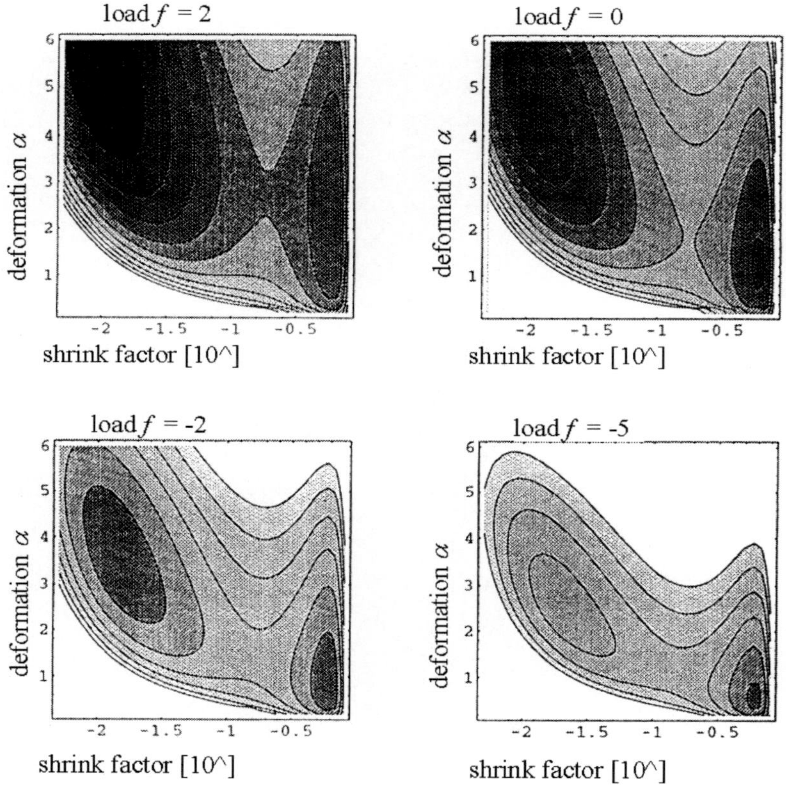

Fig. 11.23 Contour plots of $\overline{A}(\nu,\alpha)$ for a fixed temperature and different loads

12 Thermodynamics of irreversible processes

12.1 Single fluids

12.1.1 *The laws of* FOURIER *and* NAVIER-STOKES

We recall that – at the very beginning of this book – we have defined the determination of the five fields

$$\text{mass density } \rho(\underline{x},t)\ ,\ \text{velocity } w_i(\underline{x},t)\ ,\ \text{temperature } T(\underline{x},t) \qquad (12.1)$$

as the objective of thermodynamics of fluids. For this purpose we need field equations and these are derived from the equations of balance of fluid mechanics and thermodynamics, namely the conservation laws of mass, momentum, and energy, *cf.* (1.12), (1.19), (1.44), which read, without radiation and gravitation

$$\begin{aligned}
&\frac{\partial\rho}{\partial t}+\frac{\partial\rho w_i}{\partial x_i}=0 && \dot{\rho}+\rho\frac{\partial w_i}{\partial x_i}=0\\
&\frac{\partial\rho w_j}{\partial t}+\frac{\partial\left(\rho w_j w_i-t_{ji}\right)}{\partial x_i}=0 && \rho\dot{w}_j-\frac{\partial t_{ji}}{\partial x_i}=0 \qquad (12.2)\\
&\frac{\partial\rho\left(u+\frac{1}{2}w^2\right)}{\partial t}+\frac{\partial\left(\rho\left(u+\frac{1}{2}w^2\right)w_i-t_{ji}w_j+q_i\right)}{\partial x_i}=0 && \rho\dot{u}+\frac{\partial q_i}{\partial x_i}=t_{ji}\frac{\partial w_i}{\partial x_j}.
\end{aligned}$$

In $(12.2)_2$ we have introduced *material* derivatives $\dot{a}=\frac{\partial a}{\partial t}+w_i\frac{\partial a}{\partial x_i}$, *i.e.* rates of change of a quantity a as seen by an observer who moves – locally – with the flow of the fluid.

It is true that the equations (12.2) – in either form – provide five equations, but they are not field equations for the five fields (12.1). Indeed, the temperature T does not even occur in (12.2) and instead there are additional quantities, namely

the symmetric stress t_{ij}
the heat flux q_i, and (12.3)
the specific internal energy u.

In order to close the system (12.2) of five equations, we need relations between t_{ij}, q_i, and u, and the fields (12.1), the so-called *constitutive equations*, which represent the material properties of the fluid.

In thermodynamics of irreversible processes – a theory universally known as TIP – such equations are derived from an entropy balance which is based on the Gibbs equation (4.19) of reversible thermodynamics. Referred to the unit mass and with $\upsilon=1/\rho$ this equation may be written as

$$\dot{s}=\frac{1}{T}\left(\dot{u}-\frac{p}{\rho^2}\dot{\rho}\right). \qquad (12.4)$$

u and p are considered as functions of ρ and T, and these functions are supposed to be given by the caloric and thermal equations of state, as if the considered processes were reversible.*

If $\dot{u}$ and $\dot{\rho}$ are eliminated between the Gibbs equation and the equations of balance of mass and internal energy we obtain after an easy rearrangement**

$$\rho\dot{s}+\frac{\partial}{\partial x_i}\left(\frac{q_i}{T}\right)=q_i\frac{\partial 1/T}{\partial x_i}+\frac{1}{T}t_{\langle ij\rangle}\frac{\partial w_{\langle i}}{\partial x_{j\rangle}}+\frac{1}{T}\left(\frac{1}{3}t_{ii}+p\right)\frac{\partial w_k}{\partial x_k}. \tag{12.5}$$

This equation may obviously be interpreted as a balance of entropy; *cf.* (1.9) for the generic form of a balance equation. For this interpretation we must assume that

$\dfrac{q_i}{T}$ is the non-convective entropy flux, and (12.6)

$q_i\dfrac{\partial 1/T}{\partial x_i}+\dfrac{1}{T}t_{\langle ij\rangle}\dfrac{\partial w_{\langle i}}{\partial x_{j\rangle}}+\dfrac{1}{T}\left(\dfrac{1}{3}t_{ii}+p\right)\dfrac{\partial w_k}{\partial x_k}$ is the density of entropy production.

We observe that the entropy production density is a sum of products of

thermodynamic fluxes	and	*thermodynamic forces*	
heat flux q_i		temperature gradient $\dfrac{\partial T}{\partial x_i}$	(12.7)
deviatoric stress $t_{\langle ij\rangle}$		deviatoric velocity gradient $\dfrac{\partial w_{\langle i}}{\partial x_{j\rangle}}$	
dynamic pressure $\pi=-p-\dfrac{1}{3}t_{ii}$		divergence of velocity $\dfrac{\partial w_k}{\partial x_k}$.	

The Second Law requires that the entropy production be non-negative for irreversible processes. This requirement is satisfied in TIP by letting the fluxes be linear functions of the forces, namely

$$q_i=-\kappa\frac{\partial T}{\partial x_i},\qquad \kappa\geq 0$$

$$t_{\langle ij\rangle}=2\eta\frac{\partial w_{\langle i}}{\partial x_{j\rangle}},\qquad \eta\geq 0 \tag{12.8}$$

* This important reference to the thermodynamics of reversible processes is known as the *principle of local equilibrium.* It is sufficient for our purposes, also not strictly true as revealed by *extended thermodynamics*.

** Angular brackets represent deviatoric, *i.e.* symmetric and trace-free tensors, *e.g.* $t_{\langle ij\rangle}=t_{ij}-\frac{1}{3}t_{kk}\delta_{ij}$ or $\dfrac{\partial w_{\langle i}}{\partial x_{j\rangle}}=\dfrac{1}{2}\left(\dfrac{\partial w_i}{\partial x_j}+\dfrac{\partial w_j}{\partial x_i}-\dfrac{2}{3}\dfrac{\partial w_k}{\partial x_k}\delta_{ij}\right)$.

$$\pi = -\lambda \frac{\partial w_k}{\partial x_k}, \qquad \lambda \geq 0 .$$

In the jargon of TIP these relations are called phenomenological equations, because they were postulated – long before TIP – on the basis of observation of the phenomena of heat conduction and viscous friction. The pioneers of the phenomenological equations were NEWTON, FOURIER, NAVIER, and STOKES. We have presented these laws without any motivation in Chap. 2. κ is the thermal conductivity, and η and λ are shear- and bulk-viscosity, respectively. In general these coefficients are functions of ρ and T, which must be determined by measurement.

The assumption $(12.6)_1$ on the entropy flux is obviously the natural extension of the formula (4.18) of CLAUSIUS for the case when the heating $\dot{Q}$ is applied to a volume of whose surface the temperature is not homogeneous. This extension is due to Pierre Maurice Marie DUHEM (1861-1916) and the inequality (12.5) – with a non-negative production – is therefore known as the Clausius-Duhem inequality.*

If the thermal and caloric equations of state are known, and if the coefficients $\kappa(\rho,T)$, $\eta(\rho,T)$, $\lambda(\rho,T)$ have been measured, we may eliminate q_i, $t_{\langle ij \rangle}$, and π between the equations of balance (12.2) and the phenomenological equations (12.8), and obtain explicit field equations for the fields $\rho(x_i,t)$, $w_j(x_i,t)$, and $T(x_i,t)$. For their solution, of course, we need initial and boundary values.

12.1.2 *Shear flow and heat conduction between parallel plates*

We recall from Paragraph 2.2.1 the consideration of a stationary shear flow, *cf.* Fig. 2.1. At that time we had a priori assumed that the temperature is homogeneous and this assumption had led to a difficulty with the energy balance, see final remarks in Paragraph 2.2.1.

Now we consider the same situation, but we allow T as well as ρ and w_2 to depend on x_3, while w_1 and w_3 are still assumed to be zero. In other words, our semi-inverse assumption now reads

$$\rho = \rho(x_3), \; w_2 = w_2(x_3), \; T = T(x_3), \tag{12.9}$$

and these three functions are to be determined from the balance equations of mass, momentum, and energy so that the following boundary conditions are satisfied

$$w_2(0) = 0, \; w_2(D) = V, \; T(0) = T_0, \; T(D) = T_D, \; p(0) = p_0 . \tag{12.10}$$

We assume that the fluid is an ideal gas with $p = \rho \frac{R}{M} T$ and constant shear viscosity η and thermal conductivity κ. We set the bulk viscosity equal to zero and obtain from (12.8)

---

* TIP was first formulated in the shape presented here by Carl ECKART (1902-1973).

$$q_i=-\kappa\begin{bmatrix}0\\0\\\mathrm{d}T/\mathrm{d}x_3\end{bmatrix},\ t_{ij}=-p(\rho,T)\delta_{ij}+\eta\begin{bmatrix}0&0&0\\0&0&\mathrm{d}w_3/\mathrm{d}x_3\\0&\mathrm{d}w_3/\mathrm{d}x_3&0\end{bmatrix},\ \pi=0 \qquad (12.11)$$

The mass balance is identically satisfied and the balance equations of momentum and energy read

momentum 1st component: identically satisfied

$$\text{momentum } 2^{\text{nd}} \text{ component: } \frac{\mathrm{d}^2 w_2}{\mathrm{d}x_3^2}=0 \qquad (12.12)$$

$$\text{momentum } 3^{\text{rd}} \text{ component: } \frac{\mathrm{d}p}{\mathrm{d}x_3}=-\rho g$$

$$\text{energy: } -\kappa\frac{\mathrm{d}^2T}{\mathrm{d}x_3^2}=2\eta\left(\frac{\mathrm{d}w_3}{\mathrm{d}x_3}\right)^2. \qquad (12.13)$$

From $(12.12)_2$ and the boundary conditions $(12.10)_{1,2}$ we get

$$w_2(x_3)=V\frac{x_3}{D}. \qquad (12.14)$$

Using this result we obtain by integration of the energy equation

$$T(x_3)=\frac{\eta}{\kappa}V^2\frac{x_3}{D}\left(1-\frac{x_3}{D}\right)+T_0\left(1-\frac{x_3}{D}\right)+T_D\frac{x_3}{D}. \qquad (12.15)$$

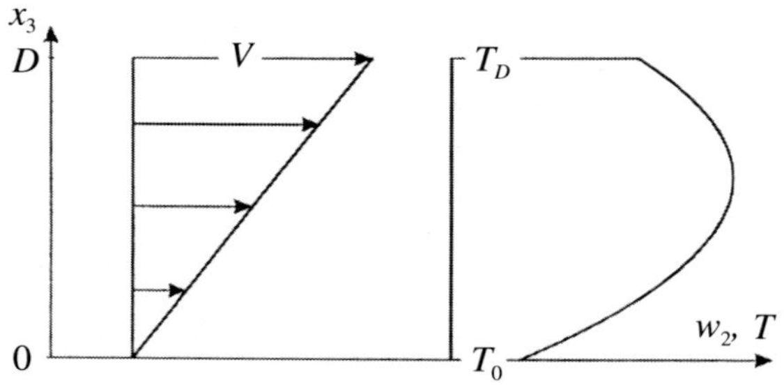

Fig. 12.1 Velocity and temperature profiles of a shear flow

Finally we obtain by integration of $(12.12)_3$, and using (12.15)

$$p(x_3)=p_0\left(\frac{\frac{x_3}{A_-}+1}{\frac{x_3}{A_+}+1}\right)^{\frac{g}{R/M}\frac{\kappa D^2}{\rho V^2}\frac{1}{A_+-A_-}}, \qquad (12.16)$$

where the abbreviations $A_\pm$ were introduced

$$A_\pm=\frac{D}{2}\left[-\left(1+\frac{\kappa(T_D-T_0)}{\eta V^2}\right)\pm\sqrt{1+2\frac{\kappa(T_D-T_0)}{\eta V^2}+\left(\frac{\kappa(T_D-T_0)}{\eta V^2}\right)^2}\right]. \qquad (12.17)$$

Thus the problem is solved. Fig. 12.1 shows the fields $w_2(x_3)$ and $T(x_3)$ qualitatively; the profiles of w_2 and T are linear and parabolic, respectively. It is clear that at the boundaries we must withdraw heat – or better: internal energy – in order to maintain the prescribed temperatures. This internal energy is produced between the plates by the shear flow as put in evidence by the production term in the balance of internal energy.

12.1.3 *Absorption and dispersion of sound*

We recall Sect. 11.4, where we have considered the propagation of sound in a non-viscous and non-conducting fluid. Now we shall investigate the effects of viscosity and heat conduction on sound propagation. In particular, we shall determine the attenuation of sound and the dependence of the phase speed on frequency.

As in Sect. 11.4 we consider the one-dimensional case $\rho(x,t)$, $w_1 = w(x,t)$, $w_2 = w_3 = 0$, and $T = T(x,t)$ but, unlike before, we now take the heat flux and the viscous stress into account, albeit without the bulk viscosity. Therefore we put

$$q_i = -\kappa \begin{bmatrix} \partial T/\partial x \\ 0 \\ 0 \end{bmatrix} \quad \text{and} \quad t_{ij} = -p\delta_{ij} + \eta \begin{bmatrix} \frac{4}{3}\frac{\partial w}{\partial x} & 0 & 0 \\ 0 & -\frac{2}{3}\frac{\partial w}{\partial x} & 0 \\ 0 & 0 & -\frac{2}{3}\frac{\partial w}{\partial x} \end{bmatrix}. \tag{12.18}$$

Instead of the equations (11.36) we thus obtain

$$\begin{aligned}
&\frac{\partial \rho}{\partial t} + \frac{\partial \rho w}{\partial x} = 0 \\
&\frac{\partial \rho w}{\partial t} + \frac{\partial \rho w^2}{\partial x} + \frac{\partial p}{\partial x} - \tfrac{4}{3}\eta \frac{\partial^2 w}{\partial x^2} = 0 \\
&\frac{\partial \rho u}{\partial t} + \frac{\partial \rho u w}{\partial x} - \kappa \frac{\partial^2 T}{\partial x^2} = -p\frac{\partial w}{\partial x} + \tfrac{2}{3}\eta \left(\frac{\partial w}{\partial x}\right)^2 .
\end{aligned} \tag{12.19}$$

As before we linearize about a homogeneous and time-independent state ρ_0, $w_0 = 0$, T_0, *i.e.* we put

$$\begin{aligned}
\rho(x,t) &= \rho_0 + \bar{\rho}(x,t) \\
w(x,t) &= \quad\;\; \bar{w}(x,t) \\
T(x,t) &= T_0 + \bar{T}(x,t)
\end{aligned} \tag{12.20}$$

and neglect all products of the barred quantities and their derivatives. Thus we obtain the linearized system of equations in the form

$$\begin{aligned}
&\frac{\partial \bar{\rho}}{\partial t} + \rho_0 \frac{\partial \bar{w}}{\partial x} = 0 \\
&\rho_0 \frac{\partial \bar{w}}{\partial t} + \left(\frac{\partial p}{\partial \rho}\right)_0 \frac{\partial \bar{\rho}}{\partial x} + \left(\frac{\partial p}{\partial T}\right)_0 \frac{\partial \bar{T}}{\partial x} - \tfrac{4}{3}\eta \frac{\partial^2 \bar{w}}{\partial x^2} = 0
\end{aligned} \tag{12.21}$$

$$\rho_0 c_v^0 \frac{\partial \overline{T}}{\partial t} - \frac{1}{\rho_0} T_0 \left(\frac{\partial p}{\partial T} \right)_0 \frac{\partial \overline{\rho}}{\partial t} - \kappa \frac{\partial^2 \overline{T}}{\partial x^2} = 0 .$$

We investigate plane harmonic waves of small amplitudes, which move in the x-direction, and employ the complex representation of such waves, *cf.* Paragraph 11.4.4

$$\overline{\rho}(x,t) = \hat{\overline{\rho}}\, \mathrm{e}^{\mathrm{i}(\omega t - kx)} , \quad \overline{w}(x,t) = \hat{\overline{w}}\, \mathrm{e}^{\mathrm{i}(\omega t - kx)} , \quad \overline{T}(x,t) = \hat{\overline{T}}\, \mathrm{e}^{\mathrm{i}(\omega t - kx)} . \tag{12.22}$$

Insertion into (12.21) with $p = \rho \frac{R}{M} T$ and $c_v = \frac{3}{2}\frac{R}{M}$ – for a monatomic ideal gas – provides a linear homogeneous algebraic system of the form

$$\begin{bmatrix} \omega & -k\rho_0 & 0 \\ k\frac{R}{M}T_0 & \rho_0\omega - \mathrm{i}\frac{4}{3}\eta k^2 & -k\frac{R}{M}\rho_0 \\ -\omega T_0 \frac{R}{M} & 0 & -\omega\rho_0 \frac{3}{2}\frac{R}{M} - \mathrm{i}k^2\kappa \end{bmatrix} \begin{bmatrix} \hat{\overline{\rho}} \\ \hat{\overline{w}} \\ \hat{\overline{T}} \end{bmatrix} = \begin{bmatrix} 0 \\ 0 \\ 0 \end{bmatrix} . \tag{12.23}$$

Non-vanishing solutions exist only, if the determinant of the matrix in (12.23) vanishes, *i.e.* if – for a given real ω – the complex wave number k is the solution of the bi-quadratic equation

$$\left[\mathrm{i}\frac{3}{2} \frac{\kappa\omega}{\rho \frac{3}{2}\frac{R}{M}} - \frac{\frac{4}{3}\frac{\eta\omega}{\rho} \frac{\kappa\omega}{\rho\frac{3}{2}\frac{R}{M}}}{V^2(0)} \right] \left(\frac{k}{\omega/V(0)} \right)^4 - \tag{12.24}$$

$$- \left[V^2(0) + \mathrm{i}\left(\frac{\kappa\omega}{\rho\frac{3}{2}\frac{R}{M}} + \frac{4}{3}\frac{\eta\omega}{\rho} \right) \right] \left(\frac{k}{\omega/V(0)} \right)^2 + V^2(0) = 0 .$$

Here $V^2(0) = \frac{5}{3}\frac{R}{M}T_0$ is the "normal" speed of sound, which we know already. The equation (12.24) is called *dispersion relation*; it determines the wave number as a function of ω. Because of the complex coefficients in (12.24) the wave number k is also a complex number and we know from Paragraph 11.4.3 that its real part determines the phase speed, while the damping or attenuation is determined by the imaginary part of k.

The full solution of the equation (12.24) with its double roots is little instructive; therefore we restrict the attention to the case of small frequencies. "Small" frequencies in this context are those of the usual audible sound frequencies with $10^2\ ^1/_\mathrm{s} < \omega < 10^5\ ^1/_\mathrm{s}$. We may consider these as small, because they must be compared with the collision frequencies of the atoms of the gas which amount to approximately $10^9\ ^1/_\mathrm{s}$ for normal pressures and temperatures. Note that in the coefficients of the dispersion relation ω occurs only as a factor of κ, or η, and both of these are inversely proportional to the collision frequency. A proof of this statement cannot be furnished in this book.

For $\omega \to 0$ the dispersion relation reduces to $\frac{k_R}{\omega} = \frac{1}{V(0)}$, and it follows that

$$-k_{\mathrm{I}}=0\,,\ \upsilon_{\mathrm{Ph}}=\frac{\omega}{k_{\mathrm{R}}}=\sqrt{\tfrac{5}{3}\tfrac{R}{M}T_0}\,. \tag{12.25}$$

In this case neither damping nor dispersion do occur, because $k_{\mathrm{I}}=0$, and because the phase speed is independent of ω.

The leading term in the damping is linear in ω, and the leading ω-dependent term in υ_{Ph} is of second order in ω. A calculation provides

$$-k_{\mathrm{I}}=\frac{2}{5p_0}\left(\frac{1}{5}\frac{\kappa}{\frac{R}{M}}+\eta\right)\omega \tag{12.26}$$

$$\upsilon_{\mathrm{Ph}}=\frac{\omega}{k_{\mathrm{R}}}=\sqrt{\tfrac{5}{3}\tfrac{R}{M}T_0}\left[1-\frac{1}{25p_0^2}\left(\frac{6}{25}\left(\frac{\kappa}{R/M}\right)^2-4\frac{\kappa}{R/M}\eta-6\eta^2\right)\omega^2\right].$$

Thus we conclude that thermal conductivity and viscosity lead to damping and dispersion; the phase speed increases with growing frequency. Therefore a group of waves with different frequencies is *dispersed* because the individual waves move with different speeds.

There is little sense in the evaluation of the dispersion relation for terms higher than second order in ω, because the laws of Fourier and Navier-Stokes become invalid for higher frequencies. We are then in the range of steep gradients and rapid rates of change where ordinary thermodynamics is no longer valid. It must be replaced by Extended Thermodynamics.*

Of course, the bi-quadratic dispersion relation (12.24) has *two* relevant solutions, of which we have discussed only the one which tends to the un-damped *ordinary sound* in the limit of small frequencies. The other solution is damped so strongly that it cannot be heard when it reaches our ears.

12.1.4 *Eshelby tensor*

The generic equation of balance (1.2) applied to a boundary between two phases – with unit normal e_i and velocity $u_i=ue_i$ – requires that the normal flux of Ψ into the boundary at one side comes out on the other side unchanged in amount, provided that no production occurs on the boundary. We write

$$[\rho\psi(w_i-ue_i)e_i+\phi_i e_i]=0\,,$$

where the square bracket $[a]\equiv a^+-a^-$ denotes the difference of a generic quantity a on the + and – sides of the boundary.

For Ψ as mass, momentum, and energy we thus obtain from the table in Sect. 1.7**

mass $\quad[\rho(w_i-ue_i)e_i]=0\,,$

momentum $\quad[\rho w_j(w_i-ue_i)e_i-t_{ji}e_i]=0\,,$

* An account of that theory is given in the book: I. Müller, T. Ruggeri, Rational Extended Thermodynamics. Springer Lecture Notes on Natural Philosophy **37**. 2nd edition (1998).

** The specific internal energy is denoted by ε in this paragraph, because the letter u is now reserved for the normal speed of the phase boundary.

energy $\left[\rho\left(\varepsilon+\tfrac{1}{2}w^2\right)(w_i-ue_i)e_i-w_jt_{ji}e_i+q_ie_i\right]=0$.

Mass, momentum, and energy are all conserved quantities so that the productions vanish. This is different for entropy which has a non-negative production density. Therefore the efflux of entropy from the boundary may be greater than the influx and, by (12.6), we write

entropy $\left[\rho s(w_i-ue_i)e_i+\frac{q_i}{T}e_i\right]=\Sigma\geq 0$,

where Σ is the entropy production in the surface.

We assume that T is continuous across the boundary. It is then possible to eliminate $[q_ie_i]$ between the energy equation and the entropy inequality and we obtain

$$\left[\rho\left(\varepsilon-Ts+\tfrac{1}{2}w^2\right)(w_i-ue_i)e_i-w_jt_{ji}e_i\right]=-T\Sigma\leq 0\,.$$

By use of the momentum equation and repeated use of the mass balance we may write the inequality as

$$\left[\rho\left(\varepsilon-Ts+\tfrac{1}{2}(\underline{w}-u\underline{e})^2\right)(w_i-ue_i)e_i-(w_j-ue_j)t_{ji}e_i\right]=-T\Sigma\leq 0\,.$$

And, if the tangential components of w_j is continuous we obtain after some juggling

$$\left[\varepsilon-Ts+\tfrac{1}{2}(\underline{w}-u\underline{e})^2-\frac{e_jt_{ji}e_i}{\rho}\right]\rho(w_ie_i-u)=-T\Sigma\leq 0\,.$$

We conclude that – once again – the entropy production is a product of

a *thermodynamic fluxes*	and	a *thermodynamic forces*
$\rho(w_ie_i-u)$		$\left[\varepsilon-Ts+\tfrac{1}{2}(\underline{w}-u\underline{e})^2-\frac{e_jt_{ji}e_i}{\rho}\right]$.

We satisfy the inequality by letting the flux be linearly dependent on the force and obtain the *phenomenological equation*

$$\rho(w_ie_i-u)=-L\left[\varepsilon-Ts-\frac{e_jt_{ji}e_i}{\rho}\right]\quad\text{with}\quad L\geq 0\,.$$

Thus the mass transfer rate $\rho(w_ie_i-u)$ across the phase boundary stops when the Eshelby tensor

$$(\varepsilon-Ts)\delta_{ij}-\frac{t_{ji}}{\rho}\quad\text{with}\quad L\geq 0\,.$$

has the same normal component on both sides of the boundary. In particular, if the stress reduces to $-p\delta_{ij}$ – as it might in fluids with negligible viscosities – the Eshelby tensor is reduced to the specific Gibbs free energy $g=\varepsilon-Ts+\frac{p}{\rho}$. We recall from Paragraph 4.2.4 that, indeed, the free energy is continuous in equilibrium at a phase boundary between two fluids, *e.g.* a liquid and its vapor.

It is clear from the phenomenological equation that the growth of one phase in another one in a solid generally depends on the direction of the boundary. Thus an inclusion of one phase in the bulk of another phase may become penny-shaped or needle-shaped depending on the stress components in the phases.

The above analysis is somewhat more complex, if we cannot take it for granted that the tangential components of w_j are continuous across the phase boundary. That complication occurs in moving phase boundaries between the austenitic phase and the austensitic one in shape memory alloys. We go not into details here.*

12.2 Mixtures of Fluids

12.2.1 *The laws of Fourier, Fick, and Navier-Stokes*

We consider homogeneously mixed fluids of ν constituents in which all constituents are present in all points of the mixture albeit in different concentrations. One may say that it is the objective of the thermodynamics of such mixtures to determine the $\nu+1$-fields

$$\begin{aligned} &\text{mass densities } \rho_\alpha(\underline{x},t) \quad (\alpha=1,2,\cdots,\nu) \\ &\text{velocity } w_i(\underline{x},t) \\ &\text{temperature } T(\underline{x},t). \end{aligned} \tag{12.27}$$

In thermodynamics of irreversible processes (TIP) we use the equations of balance of all masses and of momentum and energy as a basis for the field equations. The equations of balance have the forms

$$\begin{aligned} &\frac{\partial \rho_\alpha}{\partial t}+\frac{\partial \rho_\alpha w_i^\alpha}{\partial x_i}=\tau_\alpha \ , \ \alpha=1,2,\cdots,\nu \qquad && \begin{cases} \dot{\rho}+\rho\dfrac{\partial w_i}{\partial x_i}=0 \\ \rho\dot{c}_\alpha+\dfrac{\partial J_i^\alpha}{\partial x_i}=\tau_\alpha \ , \alpha=1,2,\cdots,\nu-1 \end{cases} \\ &\frac{\partial \rho w_j}{\partial t}+\frac{\partial\left(\rho w_j w_i - t_{ji}\right)}{\partial x_i}=0 && \rho\dot{w}_j-\frac{\partial t_{ji}}{\partial x_i}=0 \\ &\frac{\partial \rho\left(u+\frac{1}{2}w^2\right)}{\partial t}+\frac{\partial\left(\rho\left(u+\frac{1}{2}w^2\right)w_i-t_{ji}w_j+q_i\right)}{\partial x_i}=0 && \rho\dot{u}+\frac{\partial q_i}{\partial x_i}=t_{ji}\frac{\partial w_i}{\partial x_j}. \end{aligned} \tag{12.28}$$

As before the dot represents the rate of change as seen by an observer who moves with the velocity $w_j=\sum_{\alpha=1}^{\nu}\frac{\rho_\alpha}{\rho}w_j^\alpha$ of the mixture. ρ is the sum of the partial densities ρ_α, and $c_\alpha=\frac{\rho_\alpha}{\rho}$ is the concentration of constituent α. $J_i^\alpha=\rho_\alpha\left(w_i^\alpha-w_i\right)$

* The interested reader is referred to the paper: BISCARI, G., HUO, Y., MÜLLER, I. Eshelby tensor as a tensor of free enthalpy. J. of Elasticity **72**, (2003).

is the diffusion flux; it represents the mass flux of constituent α with respect to the mixture.*

τ_α is the mass production density of constituent α, just as in Sect. 9.1 and again, as in Chap. 9, we may employ stoichiometric arguments to derive

$$\tau_\alpha = \sum_{a=1}^{n} \gamma_\alpha^a m_\alpha \lambda^a . \tag{12.29}$$

In contrast to Chap. 9 we are now allowing for more than one independent reaction, – n of them – each one with its own stoichiometric coefficients γ_α^a and with its own reaction rate density λ^α $(a = 1, 2, \cdots, n)$.

In order to close the system (12.28) we need constitutive equations for

the diffusion fluxes J_i^α $(\alpha = 1, 2, \cdots, \nu - 1)$
the reaction rate densities λ^a $(a = 1, 2, \cdots, n)$
the symmetric stress t_{ji} (12.30)
the specific internal energy u
the flux of internal energy q_i.

In TIP the constitutive equations are motivated from the Gibbs equation for mixtures (8.9). With $G = U - TS + pV$ and, referred to the mass, this equation may be written in the form

$$\dot{s} = \frac{1}{T}\left(\dot{u} - \frac{p}{\rho^2}\dot{\rho} - \sum_{\alpha=1}^{\nu} \mu_\alpha \dot{c}_\alpha \right). \tag{12.31}$$

s is the specific entropy and p is the pressure of the mixture, μ_α are the chemical potentials. u, p, and μ_α are considered as known functions of ρ, c_α, and T of the same form that these functions have in equilibrium. Once again this assumption represents the principle of local equilibrium, *cf.* Paragraph 12.1.1 now applied to mixtures.

We eliminate $\dot{u}$, $\dot{\rho}$, and $\dot{c}_\alpha$ between (12.31) and the equations of balance (12.28) and obtain after some rearrangement

$$\rho \dot{s} + \frac{\partial}{\partial x_i} \frac{q_i - \sum_{\alpha=1}^{\nu} \mu_\alpha J_i^\alpha}{T} = -\frac{1}{T} \sum_{a=1}^{n} \left(\sum_{\alpha=1}^{\nu} \mu_\alpha \gamma_\alpha^a m_\alpha \right) \lambda^a + \tag{12.32}$$
$$+ q_i \frac{\partial \frac{1}{T}}{\partial x_i} - \sum_{\alpha=1}^{\nu} J_i^\alpha \frac{\partial \frac{\mu_\alpha}{T}}{\partial x_j} + \frac{1}{T} t_{\langle ij \rangle} \frac{\partial w_{\langle i}}{\partial x_{j \rangle}} + \frac{1}{T}\left(\frac{1}{3} t_{ii} + p \right) \frac{\partial w_k}{\partial x_k} .$$

---

* Note that there are only $\nu - 1$ independent concentrations c_α and diffusion fluxes J_i^α, because, by definition, we must have $\sum_{\alpha=1}^{\nu} c_\alpha = 1$ and $\sum_{\alpha=1}^{\nu} J_i^\alpha = 0$.

This relation may be interpreted as the equation of balance of entropy, if we consider

$$\frac{1}{T}\left(q_i - \sum_{\alpha=1}^{\nu} \mu_\alpha J_i^\alpha\right) \text{ as the non-convective entropy flux, and} \qquad (12.33)$$

$$-\frac{1}{T}\sum_{a=1}^{n}\left(\sum_{\alpha=1}^{\nu}\mu_\alpha \gamma_\alpha^a m_\alpha\right)\lambda^a + q_i\frac{\partial \frac{1}{T}}{\partial x_i} + \sum_{\alpha=1}^{\nu} J_i^\alpha \frac{\partial -\left(\frac{\mu_\alpha - \mu_\nu}{T}\right)}{\partial x_j} +$$

$$+\frac{1}{T}t_{\langle ij\rangle}\frac{\partial w_{\langle i}}{\partial x_{j\rangle}} + \frac{1}{T}\left(\frac{1}{3}t_{ii} + p\right)\frac{\partial w_k}{\partial x_k} \text{ as entropy production density.}$$

Inspection shows that the density of entropy production is a sum of products of

thermodynamic fluxes	and	*thermodynamic forces*
flux of internal energy q_i		temperature gradient $\frac{\partial \frac{1}{T}}{\partial x_i}$
diffusion fluxes J_i^α		chemical potential gradients $\frac{\partial -\left(\frac{\mu_\alpha - \mu_\nu}{T}\right)}{\partial x_i}$
deviatoric stress $t_{\langle ij\rangle}$		deviatoric velocity gradient $\frac{\partial w_{\langle i}}{\partial x_{j\rangle}}$
dynamic pressure $\pi = -p - \frac{1}{3}t_{ii}$		divergence of velocity $\frac{\partial w_k}{\partial x_k}$
reaction rate densities λ^a		chemical affinities $\sum_{\alpha=1}^{\nu}\mu_\alpha \gamma_\alpha^a m_\alpha$.

By the Second Law the entropy production must be non-negative. This requirement is satisfied in TIP by making the fluxes linear functions of the forces. Thus one obtains the phenomenological equations for mixtures

$$\lambda^a = -\sum_{b=1}^{n} L_{ab}\left(\sum_{\alpha=1}^{\nu}\mu_\alpha \gamma_\alpha^b m_\alpha\right) + L_a\frac{\partial w_k}{\partial x_k}$$

$$\frac{1}{3}t_{ii} + p = -\sum_{b=1}^{n}\tilde{L}_b\left(\sum_{\alpha=1}^{\nu}\mu_\alpha \gamma_\alpha^b m_\alpha\right) + \lambda\frac{\partial w_k}{\partial x_k} \qquad \begin{bmatrix} L_{ab} & L_a \\ \tilde{L}_b & \lambda \end{bmatrix} \text{ - non-neg. def.}$$

$$q_i = \kappa T^2 \frac{\partial \frac{1}{T}}{\partial x_i} + \sum_{\beta=1}^{\nu-1} B_\beta \frac{\partial -\left(\frac{\mu_\beta - \mu_\nu}{T}\right)}{\partial x_i}$$

$$J_i^\alpha = \widetilde{B}_\alpha \frac{\partial \frac{1}{T}}{\partial x_i} + \sum_{\beta=1}^{\nu-1} B_{\alpha\beta} \frac{\partial -\left(\frac{\mu_\beta - \mu_\nu}{T}\right)}{\partial x_i} \qquad \begin{bmatrix} \kappa T^2 & B_\beta \\ \widetilde{B}_\alpha & B_{\alpha\beta} \end{bmatrix} \text{- non-neg. def.}$$

$$t_{\langle ij\rangle} = 2\eta \frac{\partial w_{\langle i}}{\partial x_{j\rangle}} . \tag{12.34}$$

The terms with κ, η, and λ are already known from TIP of single fluids, *cf.* Paragraph 12.1.1. The coefficients represent the thermal conductivity and the viscosities, just like before. Here, however, the phenomenological equations are considerably more complex than previously: Indeed, according to $(12.34)_3$ the flux of internal energy may be driven by a gradient of chemical potentials, – not only by a temperature gradient. This is known as the diffusion-thermo effect. And the dynamic pressure $\pi = -\frac{1}{3}t_{ii} - p$ may not only depend on the divergence of velocity, but also on the chemical affinities $\sum_{\alpha=1}^{\nu} \mu_\alpha \gamma_\alpha^a m_\alpha$ of the reactions $(a = 1, 2, \cdots, n)$.

In addition we have new equations, namely those for λ^a and J_i^α which do not occur in a single fluid. The primary thermodynamic forces of the diffusion fluxes are the gradients of chemical potentials. This makes immediate sense when we recall from Chap. 8 that chemical potentials become homogeneous when equilibrium is approached and diffusion stops. However, $(12.34)_4$ implies that a diffusion flux may also be caused by a temperature gradient. This phenomenon is indeed observed and it is known as the thermo-diffusion effect.

The reaction rates λ^a depend primarily on the chemical affinities $\sum_{\alpha=1}^{\nu} \mu_\alpha \gamma_\alpha^a m_\alpha$. Indeed, without the divergence term in $(12.34)_1$ we conclude that the reaction rates vanish when the affinities are zero. In this case we have chemical equilibrium characterized by the law of mass action

$$\sum_{\alpha=1}^{\nu} \mu_\alpha\big|_{\mathrm{E}} \gamma_\alpha^a m_\alpha = 0 \ , \ (a = 1, 2, \cdots, n). \tag{12.35}$$

We know this law from Chap. 9, although in that earlier chapter we had restricted the attention to a single reaction.

12.2.2 *Diffusion coefficient and diffusion equation*

In practice diffusion phenomena are usually observed under isobaric and isothermal conditions and when the mixture is at rest, *i.e.* $w_i = 0$. In such a case the equations $(12.34)_4$ reduce to

$$J_i^\alpha = -\sum_{\beta=1}^{\nu-1} \frac{B_{\alpha\beta}}{T} \frac{\partial(\mu_\beta - \mu_\nu)}{\partial x_i}$$

and the chemical potential difference depends only on the concentrations c_δ of the constituents. Therefore we may write

$$J_i^\alpha = -\sum_{\delta=1}^{\nu-1}\left[\sum_{\beta=1}^{\nu-1} \frac{B_{\alpha\beta}}{T} \frac{\partial(\mu_\beta - \mu_\nu)}{\partial c_\delta}\right]\frac{\partial c_\delta}{\partial x_i} \equiv -\sum_{\delta=1}^{\nu-1} \rho D_{\alpha\delta} \frac{\partial c_\delta}{\partial x_i} \tag{12.36}$$

$D_{\alpha\delta}$ is called the matrix of diffusion coefficients.*

Of course, $\sum_{\alpha=1}^{\nu} D_{\alpha\delta} = 0$ must hold so as to satisfy the constraint $\sum_{\alpha=1}^{\nu} J_i^\alpha = 0$. For a binary mixture – with $\nu = 2$ – (12.36) represents a single independent equation for $J_i \equiv J_i^1$ and we may write, with $D_{11} = D$ and $c_1 = c$

$$J_i = -\rho D \frac{\partial c}{\partial x_i}. \tag{12.37}$$

This equation – a special case of (12.34) is often called FICK's equation after Adolf FICK (1829-1901) who wrote it down first.

The diffusion coefficient D has been measured for many binary mixtures and Table 12.1 shows some values for the diffusion of gases. It is true that D depends weakly on p and c, but this is ignored in the table which refers to $p = 1\,\text{atm}$.

Table 12.1 Diffusion coefficient D in $10^{-4}\,\text{m}^2/\text{s}$ for some mixtures of ideal gases

1st gas	2nd gas	300K	400K	500K
H_2	O_2	0.887	1.425	2.040
N_2	O_2	0.243	0.400	0.587
O_2	CO_2	0.245	0.401	0.585
H_2	CO_2	0.806	1.272	1.807

If J_i is eliminated between (12.37) and $(12.28)_2$ for $\alpha = 1$, with $w_i = 0$, and without chemical reaction, *i.e.*

$$\rho \frac{\partial c}{\partial t} + \frac{\partial J_i}{\partial x_i} = 0, \tag{12.38}$$

we obtain

* The factor ρ in (12.36) is made explicit so that ρ drops out from the diffusion equation below.

$$\frac{\partial c}{\partial t} = D\frac{\partial^2 c}{\partial x_i \partial x_i}. \tag{12.39}$$

Mathematically this is the same differential equation as in (7.3). The latter determines the field of temperature through "heat diffusion" and it has been called the heat conduction equation. In the mathematical literature these equations are called *diffusion equations* even when they apply to heat conduction. The equation is the prototype of *parabolic* differential equations.

12.2.3 *Stationary heat conduction coupled with diffusion and chemical reaction*

We consider a diatomic gas A_2 between two impermeable plates at $x = \pm D$, whose temperatures $T_\pm$ are different. The temperature difference causes a heat flux. At the same time there is diffusion on account of the differences of concentrations between the gas A_2 and the monatomic gas A into which A_2 decomposes, primarily in the neighborhood of the hotter plate. The decomposition – and recombination – of A_2 is kept going by the diffusion. We proceed to analyze this coupled process qualitatively.

We assume stationarity and one-dimensionality so that all fields depend only on the spatial coordinate x. Also we assume that the mixture as a whole is at rest between the plates and that the pressure p is homogeneous. Viscosity is neglected. In this case among the equations (12.28) the equations of balance of total mass and of momentum are identically satisfied and the "concentration balance" $(12.28)_2$ and the energy balance $(12.28)_4$ reduce to

$$\frac{\mathrm{d}J}{\mathrm{d}x} = \tau \text{ and } \frac{\mathrm{d}q}{\mathrm{d}x} = 0, \tag{12.40}$$

where J and q are the 1-components of the diffusion flux and of the energy flux, while τ is the mass production density of constituent 1. We describe the chemical reaction as $\frac{1}{2}A_2 \to A$ and call A_2 the constituent 1. Thus we have $\gamma_1 = -\frac{1}{2}$, $\gamma_2 = 1$ and $m_1 = 2m_2$. Consequently, the laws of mass action for equilibrium reads $\mu_1|_E = \mu_2|_E$. We write $\mu = \mu_1 - \mu_2$ and obtain for the phenomenological equations $(12.34)_{3,4}$ and $(12.34)_1$ with (12.29)

$$q = \kappa T^2 \frac{\mathrm{d}\frac{1}{T}}{\mathrm{d}x} - B_1\frac{\mathrm{d}\frac{\mu}{T}}{\mathrm{d}x}, \quad J = \tilde{B}_1\frac{\mathrm{d}\frac{1}{T}}{\mathrm{d}x} - B_{11}\frac{\mathrm{d}\frac{\mu}{T}}{\mathrm{d}x}, \quad \tau = -L_{11}m_2^2\mu. \tag{12.41}$$

We linearize about an equilibrium state with $J = \bar{J} = 0$, $T = \bar{T}$, and $\mu = \bar{\mu} = 0$ and obtain by insertion of (12.41) into (12.40)

$$\frac{\mathrm{d}^2\mu}{\mathrm{d}x^2} = P\mu \text{ and } \frac{\mathrm{d}^2T}{\mathrm{d}x^2} = -\frac{B_1}{\kappa\bar{T}}\frac{\mathrm{d}^2\mu}{\mathrm{d}x^2}, \tag{12.42}$$

where the positive constant $P=\frac{L_{11}\kappa\overline{T}^3 m_2^2}{\det B}$ has been introduced; $\det B$ is the determinant of the matrix in $(12.34)_{3,4}$ for the present case. Integration provides

$$\begin{aligned}
\mu(x) &= \alpha\cosh\left(\sqrt{P}x\right)+\beta\sinh\left(\sqrt{P}x\right)\\
T(x) &= -\frac{B_1}{\kappa\overline{T}}\left[\alpha\cosh\left(\sqrt{P}x\right)+\beta\sinh\left(\sqrt{P}x\right)\right]+\gamma x+\delta \\
J(x) &= -\frac{\det B}{\kappa\overline{T}^3}\left[\alpha\sqrt{P}\sinh\left(\sqrt{P}x\right)+\beta\sqrt{P}\cosh\left(\sqrt{P}x\right)\right]-\frac{\widetilde{B}_1}{\overline{T}^2}\gamma .
\end{aligned} \tag{12.43}$$

This solution must be fitted to the boundary conditions which require

$$J(\pm D)=0 \text{ and } T(\pm D)=T_\pm . \tag{12.44}$$

Hence follow the values α, β, γ, δ of the four constants of integration and we obtain for the final solution

$$\begin{aligned}
\frac{B_1}{\kappa\overline{T}^2}\mu(x) &= -\frac{\sinh\left(\sqrt{P}x\right)}{\sinh\left(\sqrt{P}D\right)+\frac{\det B}{B_1\widetilde{B}_1}\sqrt{P}D\cosh\left(\sqrt{P}D\right)}\frac{T_+-T_-}{T_++T_-}\\
\frac{T(x)}{\frac{T_++T_-}{2}}-1 &= \frac{\sinh\left(\sqrt{P}x\right)}{\sinh\left(\sqrt{P}D\right)+\frac{\det B}{B_1\widetilde{B}_1}\sqrt{P}D\cosh\left(\sqrt{P}D\right)}\frac{T_+-T_-}{T_++T_-}+\\
&\quad+\frac{\sqrt{P}\cosh\left(\sqrt{P}D\right)}{\frac{B_1\widetilde{B}_1}{\det B}\sinh\left(\sqrt{P}D\right)+\sqrt{P}D\cosh\left(\sqrt{P}D\right)}\frac{T_+-T_-}{T_++T_-}x\\
\frac{\overline{T}D}{\widetilde{B}_1}J(x) &= -\frac{\sqrt{P}D\left[\cosh\left(\sqrt{P}x\right)-\cosh\left(\sqrt{P}D\right)\right]}{\frac{B_1\widetilde{B}_1}{\det B}\sinh\left(\sqrt{P}D\right)+\sqrt{P}D\cosh\left(\sqrt{P}D\right)}\frac{T_+-T_-}{T_++T_-} .
\end{aligned} \tag{12.45}$$

In order to be able to draw quantitative graphs of the solution we represent the dimensionless quantities $\frac{B_1}{\kappa\overline{T}^2}\mu(x)$, $\frac{T}{\overline{T}}$, and $\frac{\overline{T}}{\widetilde{B}_1}J(x)$, and we have chosen $\overline{T}=\frac{T_++T_-}{2}$.

We conclude that, of course, it is the temperature difference T_+-T_- that drives the whole process. Inspection of the solutions shows that there are only two dimensionless parameters, namely

$$\sqrt{P}D \quad \text{and} \quad \frac{\det B}{B_1\widetilde{B}_1} .$$

Being interested in a quantitative result we arbitrarily put $\det B=0.1B_1\widetilde{B}_1$ and plot the solution for two values of $\sqrt{P}D$, namely 2 and 5.

Since P is proportional to the coefficient L_{11} which determines the size of the reaction rate for a given μ, we may therefore conclude that P is larger in a dense gas than in a rarefied one. Accordingly the graphs in Fig. 12.2 are tagged "dense" and "rarefied" for the two values of P.

We conclude from the figure that for the dense gas chemical equilibrium prevails on much of the range between the plates, since the chemical potential difference vanishes. Also the temperature field is fairly linear except near the boundaries and the diffusion flux is nearly homogeneous. Chemical non-equilibrium occurs only in a narrow boundary layer near the plates. For rarefied conditions, however, the boundary layer becomes wider, or is not even recognizable as a boundary layer.

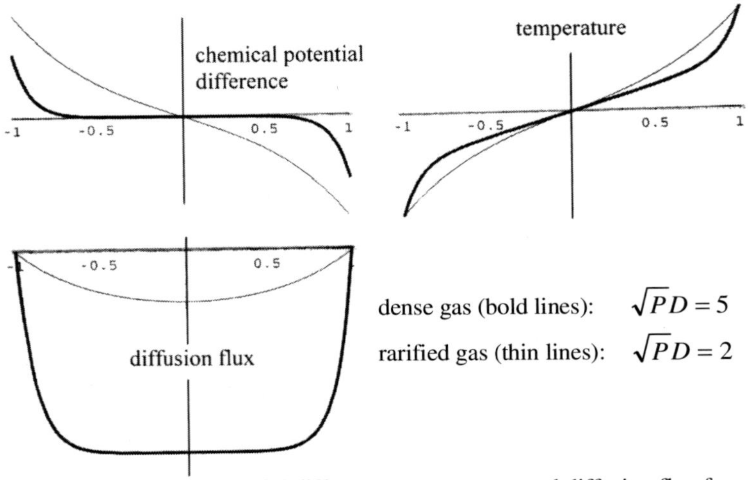

Fig. 12.2 Chemical potential difference, temperature and diffusion flux for $\sqrt{PD}=2$ (rarefied gas) and $\sqrt{PD}=5$ (dense gas). The ordinate scale is arbitrary

12.3 Flames

12.3.1 *Chapman-Jouguet equations*

We consider a flame which propagates with constant speed in the x-direction. The flame is a very narrow layer – thin in the x-direction – which moves into the fuel, burns it, and leaves the burned-out fuel – the reaction product – behind. If, for simplicity, we move with the flame, we may consider the process as stationary so that the equations of balance of mass, momentum, and energy read, *cf.* (12.28)

$$\begin{aligned} &\frac{\mathrm{d}\rho w}{\mathrm{d}x}=0 \\ &\frac{\mathrm{d}\left(\rho w^2+p-\tau_{11}\right)}{\mathrm{d}x}=0 \\ &\frac{\mathrm{d}\left(\rho\left(h+\frac{1}{2}w^2\right)w+q-\tau_{11}w\right)}{\mathrm{d}x}=0\,, \end{aligned} \tag{12.46}$$

where τ_{11} is the (1,1)-component of the viscous stress.

We denote the initial state far in the front of the flame by i and the final state far behind by f. In these positions the flux of internal energy q and the viscous stress τ_{11} are zero and thus we obtain by integration of (12.46) between i and f

$$\begin{aligned} \rho_f w_f &= \rho_i w_i \\ \rho_f w_f^2 + p_f &= \rho_i w_i^2 + p_i \\ h_f + \tfrac{1}{2} w_f^2 &= h_i + \tfrac{1}{2} w_i^2 . \end{aligned} \tag{12.47}$$

We assume that the fuel in front of the flame and the reaction product behind the flame are both ideal gases, so that the enthalpies h_f and h_i are both given by $(9.31)_1$. For simplicity we also assume that the specific heats are equal.* Then we use $(12.47)_1$ to eliminate w_f from the other two equations (12.47). After some algebraic rearrangement we thus obtain for the equations of balance of momentum and energy

$$\begin{aligned} \frac{p_f}{p_i} - 1 &= \kappa M_i^2 \left(1 - \frac{\rho_i}{\rho_f}\right) , \\ \frac{p_f}{p_i}\frac{\rho_i}{\rho_f} - 1 - Q &= \frac{\kappa - 1}{2\kappa}\left(\frac{\rho_i}{\rho_f} + 1\right)\left(\frac{p_f}{p_i} - 1\right) , \end{aligned} \tag{12.48}$$

where M_i is the Mach number of the gases moving toward the flame far in front of it. The coefficient κ is the ratio of the specific heats and Q is defined by

$$Q = -\frac{h_f^R - h_i^R}{c_p T_i} . \tag{12.49}$$

Obviously Q would be zero, if the flame did not change the nature of the gas by a chemical reaction. Q may be considered as a dimensionless heat of reaction, since it is proportional to the enthalpy difference of the in- and out-flowing gas, *cf.* (9.32).

The equations (12.48) are the basis for the Chapman-Jouguet theory. The equations determine the pair (p_f, ρ_f) which solves the problem for given values of p_i, ρ_i, M_i and Q. We proceed to explain the result using the Hugoniot diagram of Fig. 12.3. The plot shows

- two straight lines in a $\left(\frac{p}{p_i}, \frac{\rho_i}{\rho}\right)$-diagram which represent the linear equation $(12.48)_1$ – the momentum balance – for two Mach numbers. These lines are called Rayleigh lines. The steep one corresponds to $M_i > 1$ while the flat one corresponds to $M_i < 1$;

* These assumptions may not be realistic, but they simplify the mathematics. It is entirely possible, however, to perform the evaluation of (12.47) with more realistic assumptions, except that this becomes more cumbersome.

- several hyperbolic, so-called Hugoniot curves which represent the energy balance $(12.48)_2$ for different non-negative values of Q.

Obviously the point $(1,1)$ of the diagram reflects the starting point far in front of the flame. The final point $\left(\frac{p_f}{p_i}, \frac{\rho_i}{\rho_f}\right)$ must lie on the appropriate Rayleigh line *and on* the appropriate Hugoniot curve. We discuss the possibilities permitted by the balance laws; they embrace phenomena other than flames, namely detonations and shock waves.

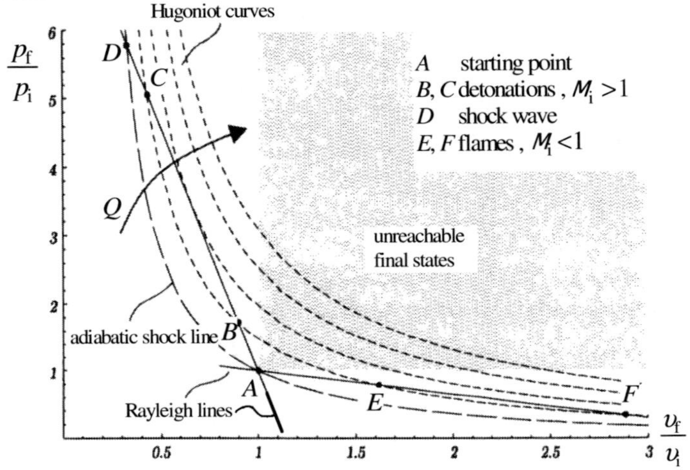

Fig. 12.3 Hugoniot diagram

12.3.2 *Detonations and flames*

For a given Q, *i.e.* a given Hugoniot-curve, the value of the Mach number of the incoming flow generally determines a Rayleigh line which intersects the Hugoniot curve in two points. There are two possibilities:

- If the incoming flow is supersonic, *i.e.* $M_i > 1$, we speak of a *detonation*. The density grows and the velocity decreases. In Fig. 12.3 possible end-states are denoted by B and C. Note that as Q tends to zero, the point B eventually coincides with the initial state A, and the point C approaches D on the hyperbola with Q=0. This limit state characterizes the Rankine-Hugoniot solution behind a shock wave in a single gas. A shock wave is thus seen as the limiting case of a detonation when no chemistry in involved. We do not pursue this case further, nor will we mention detonations again.

- If the incoming flow is subsonic, we speak of a *flame*. In this case the density of the gas decreases and accordingly, by $(12.47)_1$, the velocity increases. In Fig. 12.3 the points E and F characterize possible end-states behind the flame. The corresponding density ratios may easily be computed from (12.48) as solutions of a quadratic equation and we obtain

$$\frac{\rho_{\mathrm{i}}}{\rho_{\mathrm{f}}^{\mathrm{E,F}}}=\frac{1}{M_{\mathrm{i}}^2(\kappa+1)}\left[1+\kappa M_{\mathrm{i}}^2\pm\sqrt{\left(M_{\mathrm{i}}^2-1\right)^2-2(\kappa+1)M_{\mathrm{i}}^2Q}\right],\quad\text{hence}$$

$$\frac{p_{\mathrm{f}}^{\mathrm{E,F}}}{p_{\mathrm{i}}}=\frac{1}{\kappa+1}\left[1+\kappa M_{\mathrm{i}}^2\mp\kappa\sqrt{\left(M_{\mathrm{i}}^2-1\right)^2-2(\kappa+1)M_{\mathrm{i}}^2Q}\right] \tag{12.50}$$

$$\frac{T_{\mathrm{f}}^{\mathrm{E,F}}}{T_{\mathrm{i}}}=\frac{p_{\mathrm{f}}^{\mathrm{E,F}}}{p_{\mathrm{i}}}\frac{\rho_{\mathrm{i}}}{\rho_{\mathrm{f}}^{\mathrm{E,F}}}.$$

We pursue, only the second case, *i.e.* the flames. It is obvious from $(12.48)_1$ that the Rayleigh line becomes steeper for growing – always subsonic – Mach numbers, and the figure shows that the Rayleigh line eventually becomes tangential to the given Hugoniot curve. The point where the graphs touch is called the Chapman-Jouguet point; the square roots in $(12.50)_{1,2}$ vanish in that case, and there are no flames for larger Mach numbers.

The natural question as to whether the gas behind the flame adopts the state E or F does not seem to have a definite answer within the Chapman-Jouguet theory. Experiments indicate that the pressure change is usually small across the flame, so that point E – on the left-hand side of the Chapman-Jouguet point – is preferred over point F.

12.3.3 *Equations of balance inside the flame*

Considering only initial and final states – far in front of the flame and far behind – the Chapman-Jouguet theory ignores the flame structure entirely. This theory cannot provide any information about the structure of the flame, *i.e.* for instance about the temperature profile as the temperature grows from its small initial value to a large final value. If we wish to obtain such information, we must discuss the equations of balance inside the flame and this is what we proceed to do now. For the presentation we simplify the problem as much as possible without making it entirely trivial.

The mass balance reads as before, *cf.* $(12.46)_1$

$$\frac{\mathrm{d}\rho w}{\mathrm{d}x}=0. \tag{12.51}$$

We neglect the viscous stress so that the momentum balance $(12.46)_2$ reduces to the form

$$\frac{\mathrm{d}\left(\rho w^2+p\right)}{\mathrm{d}x}=0. \tag{12.52}$$

The other balance equations – those for the partial masses and for energy – require that we give some thought to the chemical reaction that occurs in the flame. We simplify the argument by recognizing only two constituents, namely the fuel F and the reaction product P. Thus the "stoichiometric equation" reads

$$\mathrm{F}\rightarrow\mathrm{P},$$

and it is only inside the flame that *both* constituents are present.

- ***Balance of fuel mass***

Apart from the total mass balance (12.51) we need only the balance of mass of one constituent, the fuel F (say), in the binary mixture; by $(12.28)_1$ it reads

$$\frac{\mathrm{d}(\rho_F w_F)}{\mathrm{d}x} = \tau_F \ , \ \text{or} \ \frac{\mathrm{d}(\rho c w + J)}{\mathrm{d}x} = \tau_F \ , \tag{12.53}$$

where c is the fuel concentration and J is the *diffusion flux of the fuel*. Concerning the mass production density τ_F of the fuel we must realize that the fuel is not in chemical equilibrium – not even far in front of the flame. It is chemically *metastable*, lying fairly safely behind an energetic barrier E with a negligible reaction rate until a rise in temperature ignites it, *i.e.* enables it to overcome the barrier.

Concerning τ_F we assume that it is negative, of course, and proportional to the fuel concentration so that $\tau_F = -K c$ holds. The metastable character of the fuel before the flame is represented by the assumption that the rate constant K is given by

$$K = a \exp\left(-\frac{E}{kT}\right), \text{ hence } \tau_F = -a\, c \exp\left(-\frac{E}{kT}\right); \tag{12.54}$$

a is some positive rate factor, and E is called the activation energy needed for making the fuel burn. kT is a typical value for the kinetic energy of atoms at temperature T, *cf.* (P.10). Thus τ_F is negligible before the flame, because the temperature is small, and it is zero behind the flame because c vanishes there. Accordingly the mass balance of the fuel reads

$$\frac{\mathrm{d}(\rho c w + J)}{\mathrm{d}x} = -a\, c \exp\left(-\frac{E}{kT}\right). \tag{12.55}$$

For the diffusion flux J we assume that it is given by Fick's law (12.37) so that it is proportional to the concentration gradient. Therefore we set

$$J = -D \frac{\mathrm{d}c}{\mathrm{d}x} . \tag{12.56}$$

- ***Energy conservation***

By (12.46) the balance of energy reads in the non-viscous case

$$\frac{\mathrm{d}\left[\rho\left(h + \frac{1}{2} w^2\right) w + q\right]}{\mathrm{d}x} = 0 \tag{12.57}$$

so that the energy flux is constant throughout the flame. Of course, q is given by $(12.34)_3$. However, in a mixture of non-viscous ideal gases we can be more specific as follows: The energy flux is equal to the sum of the partial energy fluxes of the ideal gas constituents so that we have

$$\rho\left(h + \tfrac{1}{2} w^2\right) w + q = \sum_{\alpha=F}^{P} \left[\rho_\alpha \left(h_\alpha + \tfrac{1}{2} w_\alpha^2\right) w_\alpha + q_\alpha\right],$$

where the summation extends over the fuel F and the reaction product P. We introduce $\rho_\alpha w_\alpha = \rho_\alpha w + J_\alpha$ and use $h_\alpha = h_\alpha^{\mathrm{R}} + c_p(T - T_R)$. Also we neglect terms with J_α^2. In this manner we obtain

$$\rho\left(h+\tfrac{1}{2}w^2\right)w + q = \sum_{\alpha=\mathrm{F}}^{\mathrm{P}} h_\alpha^{\mathrm{R}}\left(\rho c_\alpha w + J_\alpha\right) + \rho w c_p (T - T_R) + \tfrac{\rho}{2} w^3 + \sum_{\alpha=\mathrm{F}}^{\mathrm{P}} q_\alpha \,. \quad (12.58)$$

The last term $\sum_{\alpha=\mathrm{F}}^{\mathrm{P}} q_\alpha$ is the heat flux and we assume Fourier's law for it*

$$\sum_{\alpha=\mathrm{F}}^{\mathrm{P}} q_\alpha = -\lambda \frac{\mathrm{d}T}{\mathrm{d}x}\,. \quad (12.59)$$

In summary we have the conservation laws of mass, momentum, and energy *and* the equation of balance of the fuel mass with Fick's law. These equations read

$$\begin{aligned}
&\rho w = \rho_{\mathrm{i}} w_{\mathrm{i}} \\
&\rho w^2 + \rho \tfrac{R}{M} T = \rho_{\mathrm{i}} w_{\mathrm{i}}^2 + \rho_{\mathrm{i}} \tfrac{R}{M} T_{\mathrm{i}} \qquad (12.60)\\
&\left(h_{\mathrm{R}}^{\mathrm{F}} - h_{\mathrm{R}}^{\mathrm{P}}\right)\left(\rho c w + J\right) + \rho w c_p T + \tfrac{\rho}{2} w^3 - \lambda \frac{\mathrm{d}T}{\mathrm{d}x} = \\
&\qquad = \left(h_{\mathrm{R}}^{\mathrm{F}} - h_{\mathrm{R}}^{\mathrm{P}}\right)\left(\rho_{\mathrm{i}} c_{\mathrm{i}} w_{\mathrm{i}} + J_{\mathrm{i}}\right) + \rho_{\mathrm{i}} w_{\mathrm{i}} c_p T_{\mathrm{i}} + \tfrac{\rho_{\mathrm{i}}}{2} w_{\mathrm{i}}^3 - \lambda \left.\frac{\mathrm{d}T}{\mathrm{d}x}\right|_{\mathrm{i}} \\
&\frac{\mathrm{d}\left(\rho c w + J\right)}{\mathrm{d}x} = -a\, c \exp\left(-\frac{E}{kT}\right) \\
&J = -D \frac{\mathrm{d}c}{\mathrm{d}x}\,.
\end{aligned}$$

The system (12.60) is an algebro-differential system for the fields

$$\rho(x), w(x), T(x), c(x), J(x)\,.$$

Given the constitutive parameters $h_R^F - h_R^P$, λ, c_p, a, E, and D we may solve the system for the initial value problem

$$\rho_{\mathrm{i}},\ w_{\mathrm{i}},\ T_{\mathrm{i}},\ c_{\mathrm{i}} \approx 1,\ J_{\mathrm{i}} = 0,\ \left.\frac{\mathrm{d}T}{\mathrm{d}x}\right|_{\mathrm{i}} \quad \text{so that } c_{\mathrm{f}} = 0 \text{ holds.}$$

12.3.4 *Dimensionless equations*

Clearly, because of the ubiquitous non-linearities the solution of the system (12.60) must be found numerically. It is then convenient or, in fact, imperative that we introduce dimensionless quantities. We put

$$\hat{\rho} = \frac{\rho}{\rho_{\mathrm{i}}},\quad \hat{w} = \frac{w}{w_{\mathrm{i}}},\quad \hat{T} = \frac{T}{T_{\mathrm{i}}} \quad \text{and} \quad \hat{J} = \frac{J}{\rho_{\mathrm{i}} w_{\mathrm{i}}} \quad (12.61)$$

---

* Here we denote the thermal conductivity by λ, since the customary letter κ has already been used for the ratio of specific heats.

and choose the dimensionless position variable

$$\hat{x} = \frac{c_p \rho_\mathrm{i} w_\mathrm{i}}{\lambda} x\,. \tag{12.62}$$

In addition we define dimensionless parameters as follows

$$\begin{aligned}
&\text{heat of reaction} && Q = \frac{h_\mathrm{R}^\mathrm{F} - h_\mathrm{R}^\mathrm{P}}{c_p T_\mathrm{i}}\,,\\
&\text{activation temperature} && \hat{T}_\mathrm{act} = \frac{E}{kT_\mathrm{i}}\\
&\text{Mach number} && M_\mathrm{i} = \frac{w_\mathrm{i}}{\sqrt{\kappa \frac{R}{M} T_\mathrm{i}}}\\
&\text{Lewis number} && \mathrm{Le} = \frac{\lambda}{D c_p}\\
&\text{flame eigen-value} && \mu = \frac{a}{\rho_\mathrm{i} w_\mathrm{i}} \frac{\lambda}{c_p \rho_\mathrm{i} w_\mathrm{i}}\,.
\end{aligned} \tag{12.63}$$

The algebraic equations $(12.60)_{1,2}$ may be used to calculate ρ and T, or $\hat{\rho}$ and $\hat{T}$ in terms of $\hat{w}$. We obtain

$$\hat{\rho} = \frac{1}{\hat{w}} \quad \text{and} \quad \hat{T} = \left(\kappa M_\mathrm{i}^2 + 1\right)\hat{w} - \kappa M_\mathrm{i}^2 \hat{w}^2 \tag{12.64}$$

and, if we use these relations to eliminate $\hat{\rho}$ and $\hat{T}$ from the differential equations $(12.60)_{3,4,5}$ we obtain

$$\begin{aligned}
&\left[\left(\kappa M_\mathrm{i}^2 + 1\right)\hat{w} - 2\kappa M_\mathrm{i}^2 \hat{w}^2\right]\frac{\mathrm{d}\hat{w}}{\mathrm{d}\hat{x}} = -\tfrac{\kappa+1}{2} M_\mathrm{i}^2 + \left(\kappa M_\mathrm{i}^2 + 1\right)\hat{w} + Q(c + J - 1) - \left(1 + \tfrac{\kappa-1}{2} M_\mathrm{i}^2\right)\\
&\frac{\mathrm{d}c}{\mathrm{d}\hat{x}} + \frac{\mathrm{d}\hat{J}}{\mathrm{d}\hat{x}} = \mu c \exp\left[-\frac{\hat{T}_\mathrm{act}}{\left(\kappa M_\mathrm{i}^2 + 1\right)\hat{w} - \kappa M_\mathrm{i}^2 \hat{w}^2}\right]\\
&\frac{\mathrm{d}c}{\mathrm{d}\hat{x}} = -\mathrm{Le}\,\hat{J}\,.
\end{aligned} \tag{12.65}$$

Given the parameters κ, M_i, Q, μ, $\hat{T}_\mathrm{act}$, and Le these are three first order differential equations for the determination of $\hat{w}$, c, and $\hat{J}$. After the solution has been obtained, $\hat{\rho}$ and $\hat{T}$ follow from the algebraic equations (12.64).

12.3.5 *Solutions*

The most natural procedure for the solution of the set of equations (12.65) starts in the initial state in front of the flame – with $c_\mathrm{i} \approx 1$ – and proceeds by stepwise integration into and through the flame. The flame eigen-value may be chosen as a shooting parameter so that behind the flame, in the final state, the fuel concentration assumes the value $c_\mathrm{f} = 0$.

This procedure, however, is quite impractical. Indeed, behind the flame our system of equations has a saddle point and it is never possible – by the shooting method – to reach a saddle point.* If this procedure is attempted, we may see the solution *c*(*x*) (say) "try" to trace out a sharp drop of fuel concentration followed by a new plateau close to zero, – the expected behavior –, but then the graph veers off sharply upward or downward so that the state $c_\mathrm{f} = 0$ cannot be reached.

The resolution of dilemma is quite easy: We need to shoot backwards from state f to state i ! The "initial" conditions for the integration in this case are $c_\mathrm{f} \approx 0$, $\hat{J}_\mathrm{f} \approx 0$ and $\hat{w}_\mathrm{f}$ – the final values in the physical problem. A major difficulty is that we do not *a priori* know the value of $\hat{w}_\mathrm{f}$. However, for given κ, M_i, and Q, this value may be determined from the Chapman-Jouguet equations (12.50) from which we conclude, *cf.* $(12.64)_1$

$$\hat{w}_\mathrm{f} = \frac{1}{M_\mathrm{i}^2(\kappa+1)}\left(1+\kappa M_\mathrm{i}^2 - \sqrt{\left(M_\mathrm{i}^2-1\right)^2 - 2(\kappa+1)M_\mathrm{i}^2 Q}\right).$$

For $\kappa = \frac{4}{3}$, $M_\mathrm{i} = 0.1$, and $Q = 10$ we thus obtain $\hat{w}_\mathrm{f} = 12.7196$.

Accordingly we now start the backward integration with the initial value

$$c_\mathrm{f} = 0.00001\ ,\quad \hat{J}_\mathrm{f} = 0.00001\ ,\quad \hat{w}_\mathrm{f} = 12.7196$$

and choose the remaining parameters – all except μ – as $L = 1$, $\hat{T}_\mathrm{act} = 22$. The flame eigen-value μ is again a shooting parameter; it must be chosen so as to provide $c_\mathrm{i} \approx 1$.

For the actual integration we have used Mathematica® and we have obtained the graphs of Fig. 12.4 for c, $\hat{w}$, and $\hat{J}$ by integration of (12.65) and the graphs for $\hat{T}$ and $\hat{p} = \hat{\rho}\hat{T}$ from the algebraic equations (12.64). The eigen-value μ turns out to be equal to 28.72 by trial and error. For this value c_i turns out to be 1.

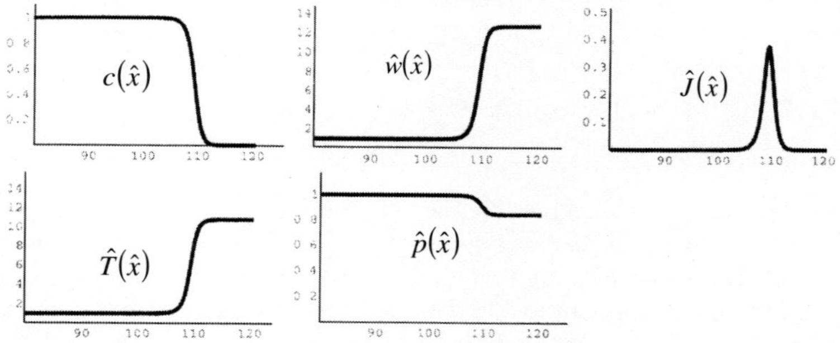

Fig. 12.4 Fields in a flame

---

* We cannot go here into the mathematical theory of systems of first order differential equations, where saddles, knots, and whirlpools are defined and discussed. We refer the interested reader to the mathematical literature.

Inspection of the graphs of the figure shows the decrease of the fuel concentration c and the corresponding increase of temperature. The pressure drops slightly and the velocity increases. The diffusion flux is non-zero only inside the flame, because it is only there that the two constituents coexist and diffusion can occur.

The usual interpretation of the existence of stationary flames is that the reaction raises the temperature. And thus the heat flux, which points into the fuel, carries the increased temperature into the incoming fuel so that it can be ignited.

12.3.6 *On the precarious nature of a flame*

The delicate role of the shooting parameter μ is confirmed in the backward integration. It must be chosen so as to provide $c_{\mathrm{i}} \approx 1$. And Fig. 12.5 indicates how the solution $c(x)$ depends on the choice of μ.

By its definition $(12.63)_5$ the flame eigen-value represents the ratio of the number of reactions per length to the incoming fuel mass. If μ is too large, the flame will be starved of fuel and, if μ is too small, the flame will be blown out. Thus a flame is a precarious phenomenon that exists only, when the consumed fuel just has the correct proportion to the fuel supplied.

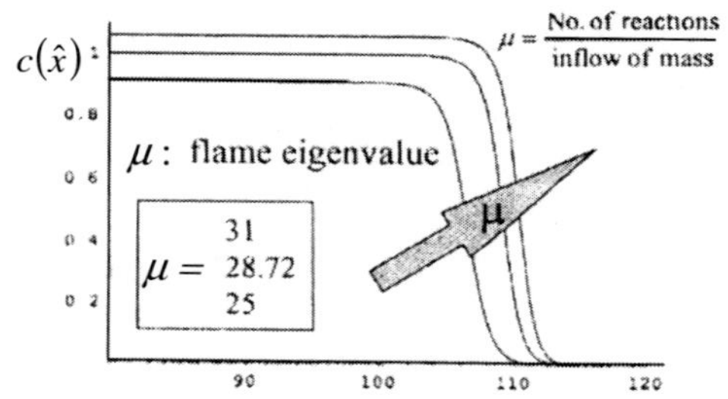

Fig. 12.5 On the choice of the flame eigenvalue μ

$(\kappa = \frac{4}{3},\ M_{\mathrm{i}} = 0.1,\ Q = 10,\ L = 1, \hat{T}_{\mathrm{act}} = 22)$

12.4 A model for linear visco-elasticity

12.4.1 *Internal variable*

In elastic bodies the stress depends on the present value of the strain. And in visco-elasticity the stress depends also on the past history of the strain. When this occurs, *i.e.* when past values of strain affect the present stress, the cause is usually some delayed internal adjustment inside the body. We proceed to investigate this phenomenon in a specific case which may serve as a prototype for viscous behavior of elastic bodies. The case is specific enough that it is easily accessible to a thermodynamic treatment.

First of all we consider a crystalline, incompressible *elastic* solid under a uniaxial load P which changes the original length L_0 to L, and the original cross-section A_0 to A by lateral contraction. By incompressibility the volume V is equal to $AL = A_0 L_0$. Thus we may define

$$\text{strain } \varepsilon = \frac{L}{L_0} - 1 \text{ and stress } \sigma = \frac{P}{A_0}.^{*}$$

The lattice cells of the crystalline body are supposed to be subject to the same stress and the same strain, so that the original cubes are deformed into rectangular boxes, *cf.* Fig. 12.6.

The working of the load is $P\frac{\mathrm{d}L}{\mathrm{d}t} = V\sigma\frac{\mathrm{d}\varepsilon}{\mathrm{d}t}$, and this is converted into elastic energy which is due to tension and lateral compression. The stress is a function of ε so that we have $\sigma = \sigma(\varepsilon)$. For linear elasticity the function is linear and we obtain Hooke's law

$$\sigma = E\varepsilon, \tag{12.66}$$

where E is the elastic modulus, or Young's modulus.

The work expended in the deformation may be regained upon lowering the load to zero. No heating is involved in the loading-unloading process, nor is there a change of temperature: The elastic process is therefore reversible.

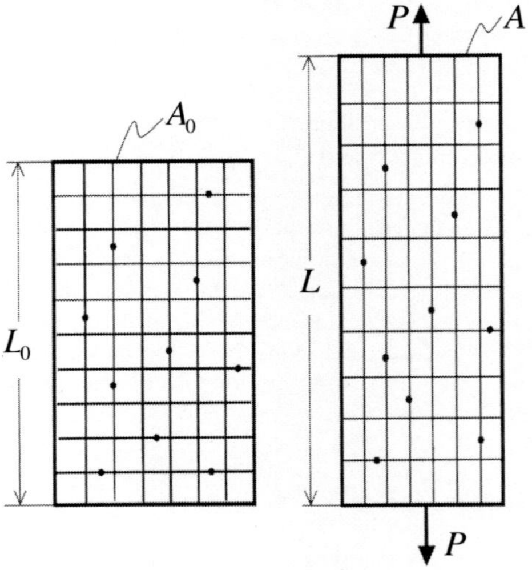

Fig. 12.6 Model of a visco-elastic body

Now, however, we suppose that there are a few interstitial atoms in the lattice, *i.e.* atoms sitting between the lattice atoms in a manner shown in Fig. 12.6. In the un-deformed state we may assume that there is an equal number N_V and N_H on vertical and horizontal lattice sites. But when the body is deformed, the H-atoms,

---

* σ is the load referred to the area A_0 of the unloaded body. It is sometimes called the engineering stress.

sitting on horizontal sites, are "squeezed" out of their positions by the lateral contraction of the lattice and thus become V-atoms.

For brevity we introduce the internal variable $\zeta = N_V\text{-}N_H$. We call ζ an *internal variable*, because it cannot be changed directly by the experimenter. However, one may conjecture that the stress σ is no longer a function of ε alone, as in elasticity. Indeed, it seems reasonable to assume that – for a fixed strain – the stress is affected, while the adjustment of ζ to a new value proceeds. Therefore σ should depend on ζ as well as on ε but, of course, in such a manner that σ vanishes when both ε and ζ are zero. In a linear approximation we may thus write without loss of generality

$$\sigma = E\varepsilon + \beta\zeta . \tag{12.67}$$

The elastic modulus E and the coefficient ζ may depend on the temperature T. Comparison with (12.66) shows that Hooke's law has acquired an additional term. This term distinguishes linear visco-elasticity from linear elasticity; it represents the effect of the internal variable on the stress, and we might call $\beta\,\zeta$ an internal stress contribution.

Concerning ζ we assume that the rate of change of ζ is given by a mutilated balance equation – a balance equation without flux – of the form

$$\frac{\mathrm{d}\zeta}{\mathrm{d}t} = A(T, \varepsilon, \zeta). \tag{12.68}$$

A is the production of ζ and, as indicated, it may depend on T, ε, and ζ; in a linear theory A is a linear function of ε, and ζ with T-dependent coefficients.

Internal variables may be of very different types. The above value ζ, the difference of N_H and N_V, is the prototype of an internal variable, which is responsible for the small visco-elastic effects in α-iron, where a few carbon atoms occupy interstitial positions in a cubic iron lattice. Another, physically very different example is an elastomer, like rubber, which consists of a network of long chain molecules in the manner of Fig. 5.3: An originally isotropic distribution of molecules is forced into anisotropy by the load. But when the load is kept constant over a long time, the joints between the molecules move slowly so as to make the network isotropic again. That slow process is accompanied by creep, a slow increase in length.*

12.4.2 *Rheological equation of state*

We employ the strategy of TIP, explained in Sect. 12.1 and 12.2, and start with the Gibbs equation – written as the rate of change of S in a process passing through equilibria characterized by U, ε, and ζ, *cf.* Sect. 12.2. Thus the Gibbs equation reads

$$\frac{\mathrm{d}S}{\mathrm{d}t} = \frac{1}{T}\left(\frac{\mathrm{d}U}{\mathrm{d}t} - V\sigma\frac{\mathrm{d}\varepsilon}{\mathrm{d}t} - \mu\frac{\mathrm{d}\zeta}{\mathrm{d}t}\right) \quad \text{or, for the free energy } F=U\text{-}TS$$

$$\frac{\mathrm{d}F}{\mathrm{d}t} = -S\frac{\mathrm{d}T}{\mathrm{d}t} + V\sigma\frac{\mathrm{d}\varepsilon}{\mathrm{d}t} + \mu\frac{\mathrm{d}\zeta}{\mathrm{d}t} . \tag{12.69}$$

* That visco-elastic effect in rubber is **not** described in Chap. 5.

This Gibbs equation is adapted from the theory of mixtures – *cf.* (8.9) and (12.31) for the present case of a solid with two "constituents" namely the V-atoms and the H-atoms. From its position in the Gibbs equation it is clear that μ is the difference of the chemical potentials of the constituents.

In the spirit of TIP we eliminate $\frac{\mathrm{d}U}{\mathrm{d}t}$ and $\frac{\mathrm{d}\zeta}{\mathrm{d}t}$ from $(12.69)_1$ by use of the energy balance $\frac{\mathrm{d}U}{\mathrm{d}t} = \dot{Q} + V\sigma\frac{\mathrm{d}\varepsilon}{\mathrm{d}t}$ and the ζ-balance (12.68) to obtain an equation in the form of an entropy balance, namely

$$\frac{\mathrm{d}S}{\mathrm{d}t} - \frac{\dot{Q}}{T} = -\mu A \,. \tag{12.70}$$

The right-hand side must be interpreted as the entropy production, and that production is a product of the *thermodynamic force A* and the *thermodynamic flux* μ. Since the entropy production ought to be non-negative, we set

$\mu = \alpha A$ with $\alpha \le 0$ or, by (12.68)

$$\mu = \alpha\frac{\mathrm{d}\zeta}{\mathrm{d}t} \quad \alpha \le 0 \,. \tag{12.71}$$

Thus μ vanishes in equilibrium and therefore in a linear theory it has the representation – with possibly T-dependent coefficients –

$$\mu = \gamma\varepsilon + \delta\zeta \,. \tag{12.72}$$

Integrability of $(12.69)_2$ requires $\frac{\partial\mu}{\partial\varepsilon} = V\frac{\partial\sigma}{\partial\zeta}$ so that the coefficient γ in (12.71) is equal to $V\beta$. We use this relation to replace γ in the sequel, wherever it occurs.

We shall now eliminate μ and ζ between the equations (12.67), (12.71), and (12.72). The elimination results in a differential equation for σ and ε, *viz.*

$$\underbrace{\frac{\alpha}{\delta}}_{-\tau_\sigma}\frac{\mathrm{d}\sigma}{\mathrm{d}t} - \sigma = \underbrace{E\frac{\alpha}{\delta}}_{-E_0\tau_\varepsilon}\frac{\mathrm{d}\varepsilon}{\mathrm{d}t} - \underbrace{E\left(1 - \frac{V\beta^2}{\delta E}\right)}_{E_0}\varepsilon \,. \tag{12.73}$$

With the indicated new coefficients E_0 and τ_σ, τ_ε we obtain

$$\tau_\sigma\frac{\mathrm{d}\sigma}{\mathrm{d}t} + \sigma = E_0\left(\tau_\varepsilon\frac{\mathrm{d}\varepsilon}{\mathrm{d}t} + \varepsilon\right). \tag{12.74}$$

This is the (σ,ε)-relation of linear visco-elasticity. Because of the rate terms, it is called *rheological equation of state* (Greek: "rheo" flow and "logos" doctrine). τ_σ and τ_ε are called relaxation times, and E_0 is the *static* elastic modulus, appropriate to very small – *i.e.* negligible – rates of change of σ and ε.

12.4.3 *Creep and stress relaxation*

We consider the initial value problem

$$\sigma(t) = 0 \text{ and } \varepsilon(t) = 0 \text{ for } t \le 0 \,, \tag{12.75}$$

and prescribe $\sigma(t)$ for $t>0$. The process is supposed to be isothermal. Thus follows $\varepsilon(t)$ as the solution of the differential equation (12.74). Mathematically the equation is one of the simplest differential equations solved in a course on differential equations: First we solve the homogeneous equation to obtain

$$\varepsilon(t)=C\exp\left(-\frac{t}{\tau_\varepsilon}\right)$$

and then we allow the constant of integration to depend on t. The method is therefore known as the *variation of the constant.* We get

$$\varepsilon(t)=\frac{1}{E_0}\frac{\tau_\sigma}{\tau_\varepsilon}\sigma(t)+\frac{1}{E_0\tau_\varepsilon}\left(1-\frac{\tau_\sigma}{\tau_\varepsilon}\right)\int_0^t\sigma(t-u)\exp\left(-\frac{u}{\tau_\varepsilon}\right)du\,. \tag{12.76}$$

Inspection shows that $\varepsilon(t)$ has two terms: One that follows $\sigma(t)$ immediately and another one that depends on the past values of the stress, *i.e.* the stress history. The influence of the past values is weakened exponentially by their "distance" in the past. Laws of this type are sometimes said to describe *hereditary behavior*, or also *fading memory*.

An instructive example is the response of the strain to a sudden jump in stress so that

$$\sigma(t)=\begin{cases}0 & t\le 0\\ \sigma_0 & t>0.\end{cases}$$

In this case (12.76) reduces to

$$\varepsilon(t)=\frac{1}{E_0}\frac{\tau_\sigma}{\tau_\varepsilon}\sigma_0+\frac{1}{E_0}\left(1-\frac{\tau_\sigma}{\tau_\varepsilon}\right)\sigma_0\left[1-\exp\left(-\frac{t}{\tau_\varepsilon}\right)\right]. \tag{12.77}$$

The "cause" $\sigma(t)$ and the "effect" $\varepsilon(t)$ for this case are plotted in Fig. 12.7 a. The exponential approach of $\varepsilon(t)$ to the final value $\frac{1}{E_0}\sigma_0$ is called creep.

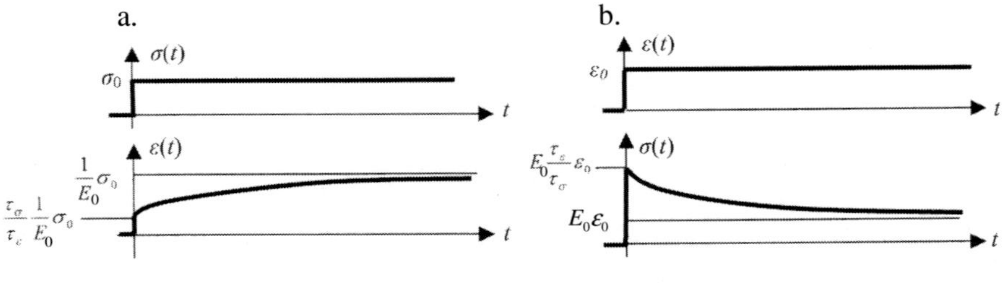

Fig. 12.7 a. Creep
b. Stress relaxation

An analogous calculation gives the solution of the initial value problem when $\varepsilon(t)$ is prescribed for $t>0$. In this case $\sigma(t)$ comes out as

$$\sigma(t)=E_0\frac{\tau_\varepsilon}{\tau_\sigma}\varepsilon(t)+E\frac{1}{\tau_\sigma}\left(1-\frac{\tau_\varepsilon}{\tau_\sigma}\right)\int_0^t\varepsilon(t-u)\exp\left(-\frac{u}{\tau_\sigma}\right)\mathrm{d}u\,.$$

which represents the stress having fading memory for past values of $\varepsilon(t)$.

The special case of a jump of strain from 0 to ε_0 at time 0 has the solution

$$\sigma(t)=E_0\frac{\tau_\varepsilon}{\tau_\sigma}\varepsilon_0+E\left(1-\frac{\tau_\varepsilon}{\tau_\sigma}\right)\varepsilon_0\left[1-\exp\left(-\frac{t}{\tau_\sigma}\right)\right].$$

The graphs of $\varepsilon(t)$ and $\sigma(t)$ in this case are shown in Fig. 12.7 b. The phenomenon is known as *stress relaxation*: The stress-response to a sudden strain is a sudden large stress which subsequently relaxes, obviously in response to the readjustment of the internal variable ζ.

12.4.4 *Stability conditions*

None of the coefficients τ_σ, τ_ε, E, and E_0 defined in (12.74) has a definite sign so far. Of course, by (12.71), α is less than 0, but what about E and E_0? And in particular, can we be sure that $\tau_\varepsilon > \tau_\sigma$ holds, as implied by the graphs of Fig. 12.7. If τ_ε were smaller than τ_σ, the responses $\varepsilon(t)$ and $\sigma(t)$ should be reversed and we should observe "strain-relaxation" and "stress creep." This, however, is impossible: The Second Law forbids it by its stability conditions. We proceed to consider this. The issue is a subject of thermodynamic stability, by which – for fixed L, or ε, and fixed T – the free energy has a minimum in equilibrium, *cf.* Paragraph 4.2.7. Integration of $(12.69)_2$, with σ and μ from (12.67) and (12.72), provides an explicit form for the free energy, *viz.*

$$\frac{F(T,\varepsilon,\zeta)}{V}=\frac{1}{2}E\varepsilon^2+\varepsilon(\beta\zeta)+\frac{1}{2}\frac{1}{E-E_0}(\beta\zeta)^2+f_0(T)\,, \tag{12.78}$$

where f_0 is an unimportant additive constant, – constant with respect to ε and ζ. Thus F is minimal for ε=0 and ζ=0, if we have

$$E>0\,,\ E_0>0,\ \text{and}\ E>E_0\,,\ \text{hence}\ \tau_\varepsilon>\tau_\sigma\,.$$

Those are the results that we have anticipated in drawing the graphs of Fig. 12.7. No strain-relaxation is possible, nor any stress-creep.

Nothing can be learned about β from the stability criterion. Indeed, quite naturally, the theory does not permit the calculation of $\zeta = N_V - N_H$. Rather it determines what we have called the internal stress contribution $\beta\,\zeta$.

12.4.5 *Irreversibility of creep*

During the creep process under a fixed value σ_0 the internal variable ζ obeys the relation $\beta\zeta(t)=\sigma_0-E\varepsilon(t)$ and, by (12.76) this means that the internal stress contribution $\beta\,\zeta$ decreases exponentially, starting with zero. We have

$$\beta\zeta(t)=\sigma_0-E\varepsilon(t)=\left(1-\frac{E}{E_0}\right)\sigma_0\left[1-\exp\left(-\frac{t}{\tau_\varepsilon}\right)\right]\quad\text{so that} \tag{12.79}$$

$$\beta\zeta(\infty)=\left(1-\frac{E}{E_0}\right)\sigma_0 \qquad \text{(in creep).}$$

Now, let us assume that in the past we have reached the asymptotic state σ_0, $\varepsilon(\infty)=\frac{1}{E_0}\sigma_0$, and $\beta\zeta(\infty)=\left(1-\frac{E}{E_0}\right)\sigma_0$ of the creep process. Then at time 0 we suddenly remove the stress σ_0. Easy calculations – similar to those that led to (12.76) and (12.77) – show that the strain will immediately jump to a smaller, albeit still positive value and will then creep back, or recover, to zero. The internal variable will recover exponentially as follows

$$\beta\zeta(t)=-E\varepsilon(t)=\left(1-\frac{E}{E_0}\right)\sigma_0\exp\left(-\frac{t}{\tau_\varepsilon}\right), \text{ so that} \tag{12.80}$$

$$\beta\zeta(\infty)=0 \qquad \text{(in recovery).}$$

So, in a manner of speaking the creep is reversible. But that is not the manner of speaking of thermodynamics, where reversibility is connected with a reversal of heating, *cf.* Paragraph 1.5.6. Let us therefore compare the heating during creep and recovery.

According to the entropy balance (12.70) the heating is given by

$$\dot{Q}=T\frac{\mathrm{d}S}{\mathrm{d}t}+T\mu A\,.$$

By $(12.69)_2$ the entropy S is given by

$$S=-\left(\frac{\partial F}{\partial T}\right)_{\varepsilon,\zeta}=-\frac{1}{2}V\frac{\mathrm{d}E}{\mathrm{d}T}\varepsilon^2-V\frac{\mathrm{d}\beta}{\mathrm{d}T}\varepsilon\zeta-\frac{1}{2}\frac{\mathrm{d}\delta}{\mathrm{d}T}\zeta^2-\frac{\mathrm{d}f_0}{\mathrm{d}t}\,.$$

If the loading process of Fig. 12.7 a is followed by sudden unloading and recovery, obviously there is no residual change of change of entropy, since the initial and final states are both characterized by $\varepsilon=0$ and $\zeta=0$. So, the $\frac{\mathrm{d}S}{\mathrm{d}t}$-part of the heating is reversible. However, of course, it is the non-negative entropy *production* which creates irreversibility. We proceed to calculate the irreversible heating from (12.70) and (12.68).

$$\dot{Q}_{\text{irr}}=T\mu\frac{\mathrm{d}\zeta}{\mathrm{d}t} \qquad \text{and, by (12.72)}$$

$$\dot{Q}_{\text{irr}}=T(V\beta\varepsilon+\delta\zeta)\frac{\mathrm{d}\zeta}{\mathrm{d}t} \qquad \text{or by (12.73)}$$

$$\dot{Q}_{\text{irr}}=TV\left(\varepsilon\frac{\mathrm{d}\beta\zeta}{\mathrm{d}t}+\frac{1}{2}\frac{1}{E-E_0}\frac{\mathrm{d}(\beta\zeta)^2}{\mathrm{d}t}\right).$$

During creep we have $\varepsilon=\frac{1}{E}(\sigma_0-\beta\zeta)$ and during reverse creep, or recovery, we have $\varepsilon=-\frac{1}{E}\beta\zeta$, where $\beta\,\zeta(t)$ is given by (12.79) and (12.80), respectively.

Therefore the irreversible part of the heating, integrated over the duration of creep and recovery, is equal to

$$Q_{\text{irr}} = \frac{TV}{E}\sigma_0^2 \frac{1}{2}(1-\frac{E}{E_0}) < 0$$

in both cases. Thus creep and recovery are truly irreversible in the thermodynamic sense of the word.

12.4.6 *Frequency-dependent elastic modulus and the complex elastic modulus*

From Fig. 12.7 we may read off that the response to the jump-type loading suffices to determine the coefficients E_0, τ_ε, and τ_σ in the rheological equations of state. However, in practice this is difficult: The sudden jump of the load and the initial jump of the response are difficult or impossible to realize in an experiment, because inevitably inertial effects do occur which we have ignored throughout the argument.

It is much more practical to apply a harmonically oscillating stress (say) with frequency ω of the type

$$\sigma(t) = \begin{cases} 0 & t \le 0 \\ \sigma_0 \sin(\omega t) & t > 0. \end{cases} \quad (12.81)$$

If this is done, and when the effect of the initial kink has died out, the strain oscillates with the same frequency, albeit with a phase shift. It is given by

$$\varepsilon(t) = \varepsilon_0 \sin(\omega t + \varphi). \quad (12.82)$$

It is fairly easy to measure the ratio σ_0/ε_0 and the phase shift φ for different frequencies and both contain information about the desired quantities E_0, τ_ε, and τ_σ. We proceed to show this.

Insertion of (12.81) and (12.82) into the rheological equation of state (12.74) provides an equation which must be valid at all times after the initial adjustment. Therefore the factors of $\sin\omega t$ and $\cos\omega t$ in the equations must both vanish and we obtain

$$\begin{aligned} \frac{\sigma_0}{E_0\varepsilon_0} &= -\tau_\varepsilon\omega\sin\varphi + \cos\varphi \\ \tau_\sigma\omega\frac{\sigma_0}{E_0\varepsilon_0} &= \sin\varphi + \tau_\varepsilon\omega\cos\varphi. \end{aligned}$$

Hence follows

$$\tan\varphi = \frac{(\tau_\sigma - \tau_\varepsilon)\omega}{1+\tau_\varepsilon\tau_\sigma\omega^2} \quad \text{and} \quad \sigma_0 = \underbrace{E_0\sqrt{\frac{1+\tau_\varepsilon^2\omega^2}{1+\tau_\sigma^2\omega^2}}}_{E(\omega)}\varepsilon_0. \quad (12.83)$$

The ratio of the amplitudes σ_0 and ε_0 is sometimes called the *ω-dependent elastic modulus* $E(\omega)$ as indicated in (12.83). For $\omega \to 0$ it tends to the static modulus E_0 and for $\omega \to \infty$ it tends to $E = \frac{\tau_\varepsilon}{\tau_\sigma} E_0$. In both limits the phase shift vanishes.

σ_0 may be considered as given. Therefore measurements of the amplitude ε_0 and of the phase shift angle φ may be used to determine E_0, τ_ε, and τ_σ, the parameters of the rheological equation.

An elegant alternative to the expressions (12.83) for the phase shift and the ratio σ_0/ε_0 of amplitudes results when we write $(12.81)_2$ and (12.83) as

$$\sigma(t) = \mathrm{Im}\,\sigma_0 \mathrm{e}^{\mathrm{i}\omega t} \quad \text{and} \quad \varepsilon(t) = \mathrm{Im}\,\hat{\varepsilon}_0 \mathrm{e}^{\mathrm{i}\omega t},$$

where $\hat{\varepsilon}_0 = \varepsilon_0 \mathrm{e}^{\mathrm{i}\varphi}$ defines the complex strain amplitude. We drop the "Im" and consider a complex $\sigma(t) = \sigma_0 \mathrm{e}^{\mathrm{i}\omega t}$ and a complex $\varepsilon(t) = \hat{\varepsilon}_0 \mathrm{e}^{\mathrm{i}\omega t}$. Insertion into the rheological equation provides

$$\sigma_0 = \underbrace{E_0 \frac{1+\mathrm{i}\omega\tau_\varepsilon}{1+\mathrm{i}\omega\tau_\sigma}}_{E_\mathrm{c}(\omega)} \hat{\varepsilon}_0 . \tag{12.84}$$

$E_\mathrm{c}(\omega)$, defined in (12.84), is called the complex elastic modulus. An easy check confirms that its absolute value is the $E(\omega)$ of $(12.83)_2$ and its argument – *i.e.* $\arctan\left(\frac{\mathrm{Im}\,E_\mathrm{c}(\omega)}{\mathrm{Re}\,E_\mathrm{c}(\omega)}\right)$ – is the phase shift angle φ of $(12.83)_1$.

12.5 Shape memory alloys

12.5.1 Phenomena and applications

In some metallic alloys a plastic deformation can be reversed by heating: Upon heating they return to their un-deformed shape. In a manner of speaking the metal "remembers" that shape and the alloys, which have this property, are called shape-memory alloys. Fig. 12.8 illustrates that memory effect schematically.

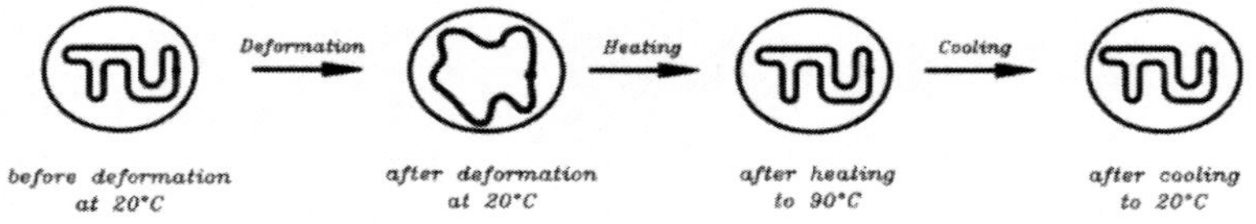

Fig. 12.8 Shape memory effect in a wire representing the logogram of our institution, the Technical University Berlin

The shape memory effect is a consequence of a rich thermo-mechanical behavior of such alloys. This behavior is shown – again schematically – in Fig. 12.9 which exhibits load-deformation diagrams at different fixed temperatures.

For a small temperature T_1 the load-deformation diagram is similar to the one of a plastic specimen: A loading-unloading cycle with a small load leads to a small deformation along the elastic line through the origin. However, as soon as a load reaches the yield limit, the strain increases without an increase in load and we say that the specimen yields. This phenomenon occurs also in a plastic specimen. However, in contrast to true plasticity, the yield process comes to an end on a second elastic line along which the specimen may be loaded elastically far beyond the original yield load. Afterwards unloading leaves a residual deformation D_1, *cf.* Fig. 12.9.

For a somewhat higher temperature T_2 this kind of behavior is qualitatively unchanged, except that the yield limit is lower.

Under compression, *i.e.* for negative loads, one observes a similar behavior and thus upon alternating between tension and compression one may run through a hysteresis loop around the origin. Because of the similarity of such load-deformation curves to those of a plastic body one calls this type of deformation quasiplastic as indicated in Fig. 12.9.

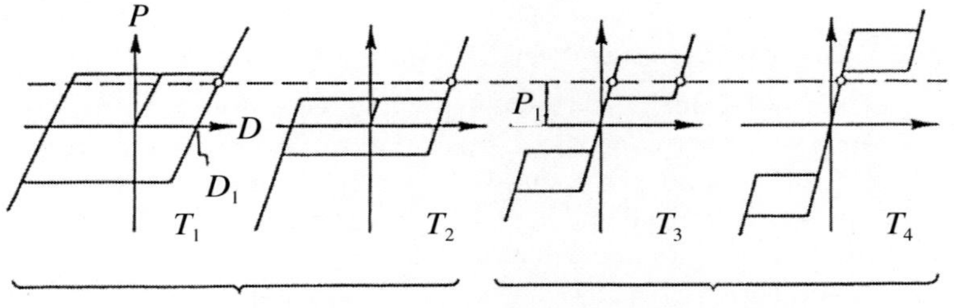

Fig. 12.9 Load-deformation diagrams for increasing temperature
Quasiplasticity and pseudoelasticity

At the yet higher temperature T_3 one observes a quite different behavior. To be sure, there is still an elastic line through the origin, and a yield load, and a lateral elastic line on the left and right hand sides. However, unloading along that elastic line will lead – at a positive load – to a recovery of the deformation that was yielded before; and this happens as soon as the load falls below the *recovery limit*. Complete unloading will thus lead back to the origin. This kind of behavior is called *pseudoelastic*; it is elastic in the sense that after loading and unloading the body is back at the origin. But it is only pseudoelastic because a loading-unloading process runs through a hysteresis loop. At still higher temperature this pseudoelastic behavior is qualitatively unchanged. It is true that the yield limit and the recovery limit move to higher loads, but they grow equally fast, so that the height of the hysteresis remains essentially unchanged.

It is clear that the diagrams of Fig. 12.9 imply the shape memory effect. Indeed, if – at the low temperature T_1 – we lead the body into the load-free deformation D_1, *cf.* figure, it must return to the origin upon heating to T_3 or T_4, because at these temperatures $D = 0$ is the only possible load-free configuration.

Typically the temperature interval for the graphs of Fig. 12.9 has a width of 40 K about room temperature and the maximal recoverable deformation amounts to 6 through 8% of the original length.

If one records sufficiently many load-deformation diagrams for different temperatures, one can construct a deformation-temperature diagram for some fixed load. Fig. 12.10 shows such a diagram for a positive load in schematic form. The dots at T_1 through T_4 indicate how the relevant points can be obtained from the (*P*,*D*)-diagrams of Fig. 12.9, where the fixed load $P_1 > 0$ is marked by a dashed line. It is thus clear that the hysteresis loop in the (*P*,*D*)-diagram reflects the hysteresis loops in the (P, D)-diagrams.

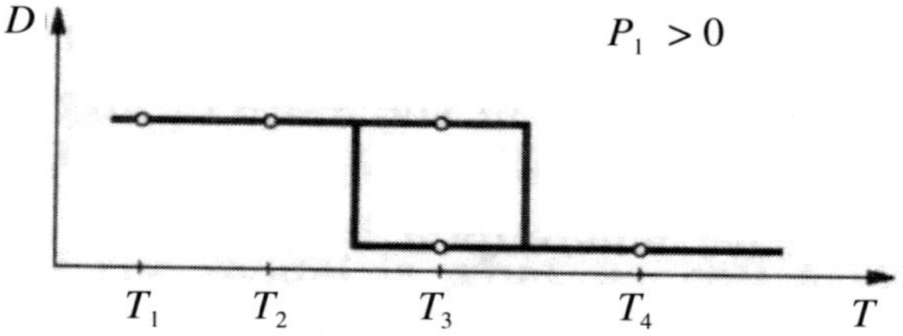

Fig. 12.10 Deformation-temperature diagram for a fixed tensile load

Inspection of the (*D*,*T*)-diagram shows that a memory alloy may be said to store two deformations in its memory: a small one for high temperature and a large one for small temperature. If the specimen is subjected to an alternating temperature it will alternately expand and contract. Therefore memory alloys may be used to build temperature-controlled actuators. Or else, if we install a suitable mechanical power transmission, one may construct a heat engine. Fig. 12.11 shows two interesting – and functioning – constructions.

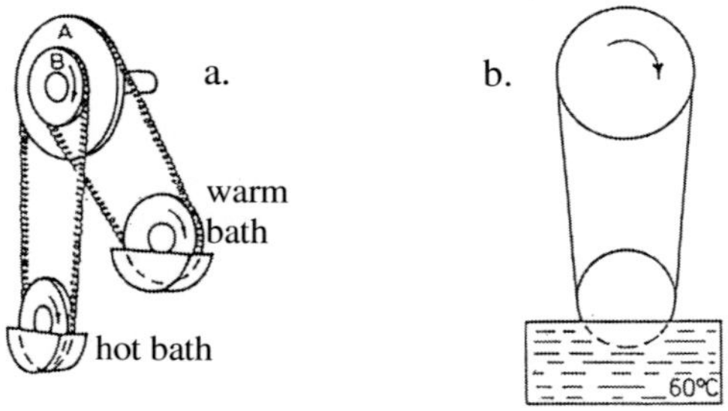

Fig. 12.11 Two memory engines
a. Engine of A.D. Johnson
b. Engine of F.E Wang

Often repeated loading and unloading may induce internal stress fields which forces the *externally unloaded* specimen to alternate between two deformations for alternating temperatures. This phenomenon is called the *two-way shape memory effect*.

Interesting applications occur in the medical field; they are illustrated in Fig. 12.12. Figure 12.12 a. shows a broken jaw bone which must be fixed by a splint. During the installation of the splint it is important to ensure that the bone-ends are tightly pressed against each other. This requires much skill of the surgeon when he uses a steel splint, and even then the compressive force is smaller than is desirable.

This is where a shape memory alloy can help. The memory splint is screwed into place in the state of large deformation on the upper border of the hysteresis loop and – naturally – at the body temperature of 36°C. The implant is then heated to 45°C (say), whereupon it contracts and pushes the bone ends firmly together. The contraction persists even after the splint returns to body temperature, because now the state of the material lies on the lower border of the hysteresis loop. The relevant states are marked by circles in Fig. 12.12 a.

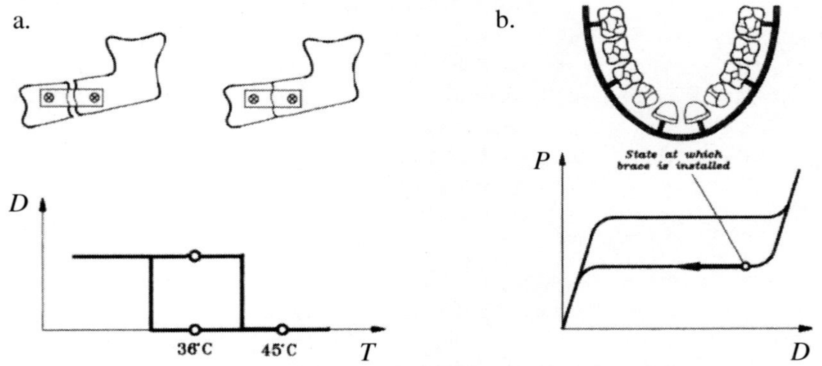

Fig. 12.12 a. Applying a shape memory splint to a broken jaw
b. Shape memory for braces in dentistry

Another medical application occurs in dentistry when braces are used for adjusting the position of teeth. The brace is pseudoelastic at body temperature and pre-deformed and pre-stressed so as to lie on the recovery line upon installation, *cf.* Fig 12.12 b. The teeth move under the load and the deformation decreases; and as it does so, the load is unchanged for some considerable period of contraction. Therefore a readjustment of the brace is not necessary for a long while.

The key to the understanding of the observed phenomenon lies in the observation that the metallic lattice of the alloy undergoes a phase transition. At high temperatures the highly symmetric austenitic phase prevails, whereas at low temperatures the lattice assumes the less symmetric martensitic phase. And martensite may form twins, *cf.* Figs. 12.13 and 12.14.

In the original state at low temperature different twins are ideally present in equal proportions. By uniaxial loading *one* of the twins is favored by the direction of the load. That twin will then form at the expense of the others in a process called twinning, which is the process underlying quasiplastic yielding. At high temperatures no martensite occurs in an unloaded specimen. But the application of a load may force the prevailing austenite into a martensitic twin variant. This happens on the pseudoelastic yield line. On the recovery line the reverse transition occurs.

On the basis of these observations we may – in imagination – construct a simple model that is capable of simulating the observed phenomena, and which is amenable to a thermodynamic treatment.

12.5.2 *A model for shape memory alloys*

The basic element of the model is what we call a lattice particle, a small piece of the metallic lattice. Fig. 12.13 a shows this lattice particle in three equilibrium configurations. They are denoted by A and $M_{\pm}$ which stands for austenite and two martensitic twins, respectively. In the present simple model there are only *two* twins.

One may imagine that the martensitic twins are sheared versions of the austenitic particle. The postulated form of the potential energy as a function of the shear length $\varDelta$ is characterized by stable lateral minima which correspond to the martensitic twins and by a metastable central minimum for the austenitic particle. In-between those minima there are energetic barriers, *cf.* Fig. 12.13 a. For simplicity we represent this three-well-potential by a train of three parabolae, *viz.*

$$\Phi(\varDelta)=\begin{cases}\Phi_A &= \Phi_0 + K_A\varDelta^2 \\ \Phi_{M_\pm} &= K_M(\varDelta \mp J)^2\end{cases} \text{ for } \begin{matrix}|\varDelta|\le\varDelta_s \\ |\varDelta|>\varDelta_s .\end{matrix} \tag{12.85}$$

It is important for the proper simulation of the observed phenomena to choose $\Phi_0 > 0$, and $K_M > K_A$, so that the lateral potential wells are narrower and deeper than the central one.

If a load P is exerted on the lattice particles, *cf.* Fig. 12.13 b, its potential energy must be added to $\Phi(\varDelta)$. The potential energy of the load is proportional to the shear length $\varDelta$ with P as the factor of proportionality. Thus the effective potential energy is deformed as shown in Fig. 12.13 b.

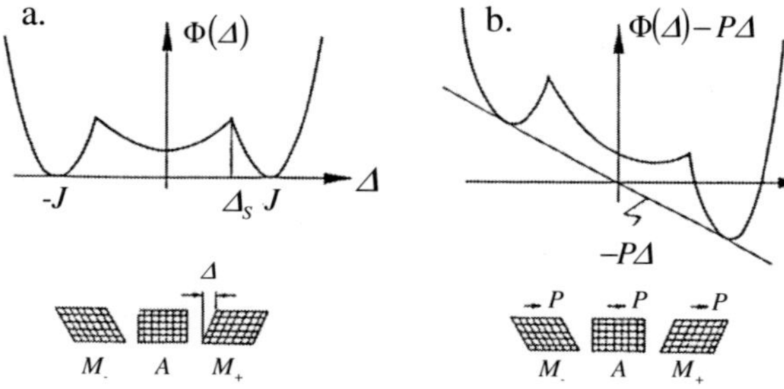

Fig. 12.13 Lattice particle and its potential energy
a. unloaded particle
b. particle under a load P

A model for a tensile specimen results by assembling lattice particles into layers and forming stacks of layers – as indicated in Fig. 12.14 – oriented at 45° to the direction of the tensile force. On the left-hand side of the figure the specimen with alternating layers of $M_{\pm}$ particles corresponds to the unloaded specimen at low temperature. We proceed to discuss the comportment of the model under vertical loading.

If the tensile force is small, the layers are subjected to a small shear force which makes the M_+-layers flatter and the M_--layers steeper. The vertical components of the shear lengths Δ_i of all N layers add up to the deformation D of the specimen

$$D = L - L_0 = \frac{1}{\sqrt{2}} \sum_{i=1}^{N} \Delta_i \,. \tag{12.86}$$

Upon unloading the layers fall back to their original positions, which means that the deformation was elastic. If, however, the load is increased, there comes the point when the M_--layers flip over into the M_+-position and thus increase their shear length drastically. Consequently the deformation grows drastically at the load where flipping occurs. Subsequent unloading does not induce the layers to flip back. Rather they will all move to the unloaded M_+-position and therefore a large residual deformation results.

In this plausible and suggestive manner we understand the initial elastic deformation, the yield as a consequence of flipping, and the residual deformation – all at low temperature.

Upon heating the austenitic phase is formed. Consequently the layers straighten up and the specimen will shorten as shown on the right-hand side of Fig. 12.14. For the naked eye the specimen has then already recovered its original shape. To be sure, it still has a different internal structure and a smooth surface, but that can only be detected under the microscope. Subsequent cooling reproduces the martensitic twins and, in all probability, equal numbers of M_+ and M_-. Therefore the body may return to its old state, both with respect to shape and internal structure, and surface structure.

Thus we have led the model specimen through the full cycle of elastic-plastic deformation and restitution of the initial shape. All steps are easy except one: We must ask why the austenite can be stable at high temperatures although its potential minimum is only meta-stable? The answer is *entropic stabilization* which we proceed to discuss.

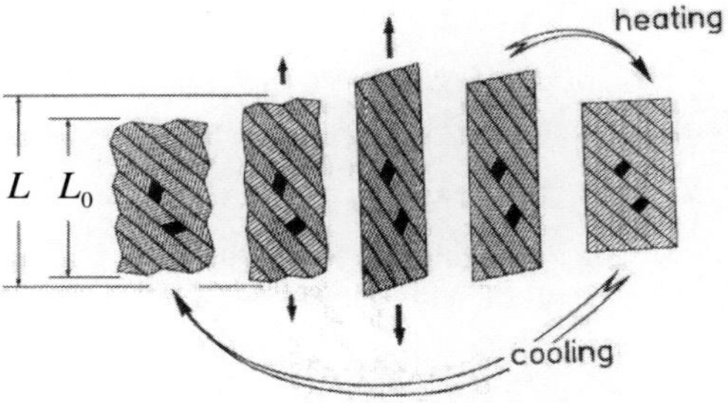

Fig. 12.14 A model for a tensile specimen

12.5.3 *Entropic stabilization*

Why is the austenite stable at high temperature when martensite is energetically more favorable? It is true that the martensitic minima are lower, but they are also narrower – recall $K_M > K_A$ – and that feature stabilizes the austenite for high temperatures. We discuss this phenomenon.

The lattice layers fluctuate about their minima and that motion is livelier for a higher temperature. Thus occasionally a layer will be able to surmount the barriers. It is easier to pass the small $A \to M$ barrier than the large $M \to A$ barrier. Therefore we should expect martensite to be dominant. However: Since the M-wells are narrower than the A-well, the M-layers hit the barrier more often than the A-layers hit theirs; and each time they have the chance to overcome the barrier. This advantage may compensate – for the M-layers – the disadvantage of the deeper well so that the layers tend to assemble in the A-well.

This is a case of entropic stabilization which we proceed to discuss in terms of the competition of energy and entropy.

According to Paragraph 4.2.7 energy "tries" to become minimal, here by assembling the layers in the depth of their potential wells. And entropy "tries" to become maximal, here by distributing the layers evenly over the available shear lengths. In this situation it is the free energy

$$F = U - TS \,. \tag{12.87}$$

which in fact becomes minimal. For low temperatures the second term may be neglected and the free energy assumes a minimum, because the energy becomes minimal. For high temperatures the first term of F may be neglected and the free energy becomes minimal because the entropy becomes maximal.

In the present case the entropic tendency toward a maximum by homogeneous distribution in the biggest available space means that the layers tend to jump out of their narrow lateral potential wells at high temperatures and assemble in the flat and wide central well. In this manner their distribution is "more homogeneous" than before and a greater space is filled in terms of shear lengths Δ.

We discuss this situation more formally and introduce the distribution function N_Δ which characterizes the number of layers with the shear length Δ. Thus for the deformation, potential energy and entropy of the model body we have*

$$D = \sum_{\Delta} \Delta N_\Delta \,, \quad U = \sum_{\Delta} \Phi_\Delta N_\Delta \,, \quad S = k \ln \frac{N!}{\prod_{\Delta} N_\Delta !} \,, \tag{12.88}$$

where the summation extends over all possible shear lengths. Deformation, potential energy and entropy of the phases A and $M_\pm$ are similarly defined except that the Δ's run over Δ_A, *i.e.* $|\Delta| < \Delta_s$, and $\Delta_{M_\pm}$, *i.e.* $\Delta > +\Delta_s$ and $\Delta < -\Delta_s$, respectively. We use the Stirling formula and write the free energies, numbers of layers and deformations of each phase as functions of N_Δ

$$F_A = \sum_{\Delta_A} \left[\Phi_A(\Delta) + kT \ln \frac{N_\Delta}{N_A} \right] N_\Delta \,, \qquad N_A = \sum_{\Delta_A} N_\Delta \,, \quad D_A = \sum_{\Delta_A} \Delta N_\Delta \,,$$

$$F_{M_\pm} = \sum_{\Delta_{M_\pm}} \left[\Phi_{M_\pm}(\Delta) + kT \ln \frac{N_\Delta}{N_{M_\pm}} \right] N_\Delta \,, \quad N_{M_\pm} = \sum_{\Delta_{M_\pm}} N_\Delta \,, \quad D_{M_\pm} = \sum_{\Delta_{M_\pm}} \Delta N_\Delta \,. \tag{12.89}$$

---

* In $(12.88)_2$ and subsequently we incorporate the factor $\sqrt{2}$ from (12.86) into the shear length. This simplifies the relation.

In order to find the equilibrium values of the numbers of layers in the phases A and $M_\pm$ we minimize F_A and $F_{M_\pm}$ under constraints which fix N_A, D_A and $N_{M_\pm}$, $D_{M_\pm}$, respectively. Thus we obtain

$$N_\Delta = N_A \frac{\mathrm{e}^{-\frac{\Phi(\Delta)-\beta_A\Delta}{kT}}}{\sum\limits_{\Delta_A}\mathrm{e}^{-\frac{\Phi(\Delta)-\beta_A\Delta}{kT}}} \text{ and } N_\Delta = N_{M_\pm} \frac{\mathrm{e}^{-\frac{\Phi(\Delta)-\beta_{M_\pm}\Delta}{kT}}}{\sum\limits_{\Delta_{M\pm}}\mathrm{e}^{-\frac{\Phi(\Delta)-\beta_{M_\pm}\Delta}{kT}}}. \tag{12.90}$$

for $\Delta \le \Delta_s$ and $\Delta > +\Delta_s$, $\Delta < -\Delta_s$, respectively. The β's are Lagrange multipliers that take care of the constraints of fixed deformations. Their values follow from the constraints. In order to exploit these we need to convert the sums into integrals. This is done by assuming that the number of shear lengths between Δ and $\Delta + \mathrm{d}\Delta$ is equal to $Y\,\mathrm{d}\Delta$, where Y is some factor of proportionality. We simplify the integrations by carrying them out between $-\infty$ and $+\infty$ for all wells instead of only about the proper ranges of the parabolic wells; this procedure may be justified, if the layers are all situated near the minima of the wells. Thus we obtain from (12.89)

$$\beta_A = 2K_A \frac{D_A}{N_A} \text{ and } \beta_{M_\pm} = 2K_M\left(\frac{D_{M_\pm}}{N_{M_\pm}} \mp J\right). \tag{12.91}$$

Insertion into (12.90) provides the distribution functions in explicit form and, when these are used in $(12.89)_1$, we obtain

$$\frac{F_A}{N_A} = \underline{\Phi_0 + K_A\left(\frac{D_A}{N_A}\right)^2} + \tfrac{1}{2}kT\ln K_A - kT\ln\left(Y\sqrt{\pi kT}\right) \tag{12.92}$$

$$\frac{F_{M_\pm}}{N_{M_\pm}} = \underline{K_M\left(\frac{D_{M_\pm}}{N_{M_\pm}} \mp J\right)^2} + \tfrac{1}{2}k\ln K_M - kT\ln\left(Y\sqrt{\pi kT}\right).$$

From thermodynamics we recall that $P = \frac{\partial F}{\partial D}$ is the load. Therefore, by (12.91) and (12.92), we conclude that β_A and $\beta_{M_\pm}$ are the loads on the phases. These loads are needed to maintain the deformations D_A and $D_{M_\pm}$.

We recognize that for $T = 0$ in (12.92) we retain only the underlined parts. As functions of D/N – the contribution of one layer to the deformation – these parts are equal to the potential energy functions (12.85). With increasing temperature the functions F_A and $F_{M_\pm}$ are shifted vertically at different rates: $F_{M_\pm}$ more so than F_A since $K_M > K_A$ holds. Thus for increasing temperature we obtain the graphs of Fig. 12.15, where $F(0)$ has arbitrarily – and without loss of generality – been set equal to zero in all cases.

Inspection shows that for small temperatures the martensitic twins are stable, *i.e.* that they have the lowest minima of the free energy. For some higher temperature all minima are equally deep; but then for high temperatures the austenitic minimum is deepest. The phase change $M_{\pm} \Leftrightarrow A$ occurs when all minima have the same depth. Here we have the analytic – and graphical – version of the previously discussed entropic stabilization. Indeed, the graphs of Fig. 12.15 differ only by the entropic, T -dependent contributions, *cf.* (12.92).

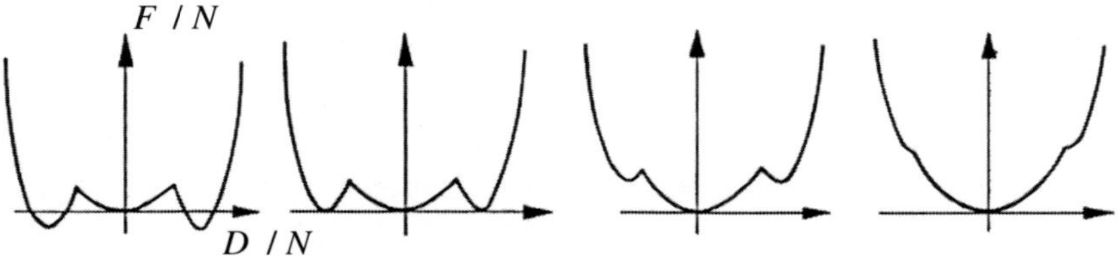

Fig. 12.15 Free energies of the phases A and $M_{\pm}$ as functions of D/N for increasing temperature, *cf.* (12.92)

For the benefit of those readers who recall the Landau-Devonshire theory of phase transitions we mention that the sequence of graphs of Fig.12.15 qualitatively resembles free energies of that theory, *cf.* Sect. 11.5. To be sure, in Landau's case each graph is represented by a single temperature-dependent sixth-order polynomial, whereas here they are given by a train of parabolae; but the relative positions of the minima and the non-convex character of the free energy are essentially the same.

12.5.4 *Pseudoelasticity*

Neither here nor in the Landau-Devonshire model do hystereses appear. Therefore the model is not yet capable of describing the load-deformation curves of Fig. 12.9. We suggest that the proper representation of the hysteresis loops requires that a penalty for the formation of phase interfaces be taken into account. This is the idea which we pursue in this section, albeit, for simplicity, not for the full spectrum of possibilities represented in Fig. 12.9 but only for pseudoelasticity.

The pseudoelastic hysteresis is richer than we have discussed so far. Indeed, when one interrupts the yield process by unloading or the recovery process by reloading, one observes the behavior shown in Fig. 12.16. The steep lines inside the hysteresis loop are elastic, *i.e.* reversible. However, upon unloading, and when the load falls below the diagonal line – from the upper left to the lower right corner of the loop –, recovery occurs inside the hysteresis loop at constant load. Analogously one can observe internal yield, if the diagonal line is approached from below. By conducting a loading-unloading process properly, one may thus trace out inner loops, *cf.* Fig. 12.16 c.

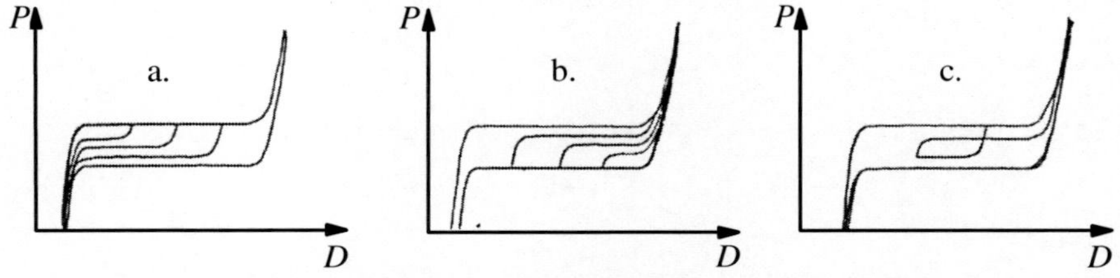

Fig. 12.16 a. Internal recovery, **b.** internal yield, and **c.** internal loop

In the pseudoelastic range martensite is not stable in the unloaded state. This is to say that the free energy is represented by one of the two graphs on the right hand side of Fig. 12.15. In Fig. 12.17 a this function is redrawn for positive values of D.

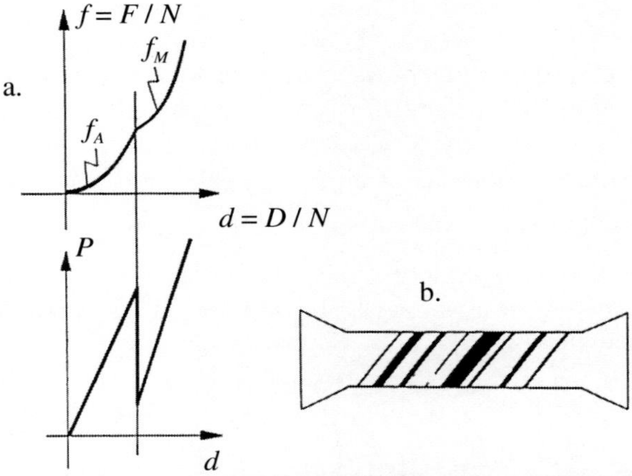

Fig. 12.17 a. Specific free energy and load as functions of specific deformation
b. Schematic view of a tensile specimen during pseudoelastic yielding

The lower parabola corresponds to austenite, the upper one to martensite. If both phases are present, as they are in the pseudoelastic range, the tensile specimen alternates between austenitic and martensitic stripes, *cf.* Fig. 12.17 b. Hundreds or thousands of such stripes can occur and we assume that their number may be set proportional to $N_A \frac{N_M}{N}$, where N_A and N_M are the number of layers in the austenitic and martensitic phase, respectively, and $N = N_A + N_M$. This expression is plausible, since it represents the expectation value of the number of interfaces, if their arrangement is random.

Each interface is endowed with a certain energetic penalty A because the lattice must be deformed for the formation of an interface in a coherent lattice. We call A the coherency factor and write the total energy penalty in the form

$$AN_A \frac{N_M}{N} \quad \text{with } A > 0 \text{ as the } \textit{coherency factor}. \tag{12.93}$$

Thus the free energy and the deformation of the specimen are given by

$$F = F_A + F_M + AN_A \frac{N_M}{N} \quad \text{and} \quad D = D_A + D_M . \tag{12.94}$$

and for the specific values $f = F/N$ and $d = D/N$, referred to a layer, we have

$$f = x_A f_A(d_A) + (1 - x_A) f_M(d_M) + A x_A (1 - x_A) \quad \text{and} \tag{12.95}$$
$$d = x_A d_A + (1 - x_A) d_M .$$

x_A is the fraction of austenitic layers. The variable T in f_A and f_M has been suppressed for brevity.

In order to determine the phase equilibrium we must minimize f in $(12.95)_1$ under the constraint that d in $(12.95)_2$ is fixed. We use a Lagrange multiplier λ to take care of the constraint and obtain as equilibrium conditions

$$\lambda = \frac{\partial f_A}{\partial d_A} = \frac{\partial f_M}{\partial d_M} = \frac{f_M - f_A + A(1 - 2x_A)}{d_M - d_A} . \tag{12.96}$$

Together with the constraint $(12.95)_2$ we thus have four equations to determine equilibrium values of the four quantities d_A, d_M, x_A, and λ. Since f_A and f_M are analytically given by (12.92), it is possible to solve these equations analytically. It is more instructive, however, to apply a graphical procedure as follows.

Since $\frac{\partial f}{\partial d}$ in equilibrium is equal to the load P, $(12.96)_{1,2}$ show that the Lagrange multiplier λ is the *equal* load on the two phases. The load as the derivative of the free energy is shown in Fig.12.17. In the model it is represented by a straight graph in both phases.

For $\lambda = P$ and $f = \int P(d)\mathrm{d}d$ $(12.96)_3$ implies

$$P(d_M - d_A) - \int_{d_A}^{d_M} P(d)\mathrm{d}d = -A(1 - 2x_A) . \tag{12.97}$$

This is the condition for phase equilibrium and we shall use it to determine the load and the corresponding equilibrium deformations d_A and d_M. Equation (12.97) may be used for a graphical construction of these values. They will depend on x_A.

We consider $x_A \approx 1$, *i.e.* just a little martensite in equilibrium with the predominant austenite. In that case the right-hand side of (12.97) is equal to A and the rectangle $P(d_M - d_A)$ must be chosen so as to be larger by the amount A than the integral $\int_{d_A}^{d_M} P(d)\mathrm{d}d$ under the graph $P(d)$. Figure 12.18 a shows how the corresponding values P^0, d_M^0, d_A^0 may be found: Obviously the triangular areas I and II must differ by A. Next we consider the case $x_A \approx 0$, *i.e.* just a little austenite coexists with the prevailing martensite. The right-hand side of (12.97) equals $-A$ in this case and the second graph of Fig. 12.18 shows how the corresponding values P^1, d_M^1, d_A^1 may be found: The triangular areas III and IV must differ by A.

It stands to reason that the loads P^0 and P^1, which characterize the beginning phase transitions, represent the observed yield load and the recovery load, respectively. If this is the case, the hysteresis loop has the area 2*A* as indicated in Fig.12.18 d.

The equilibrium load P^x that characterizes a phase equilibrium with x_A somewhere in-between 0 and 1 follows from (12.97) in much the same manner as P^0 and P^1 and, once found, the corresponding values d_M^x, d_A^x may be calculated from $(12.96)_{1,2}$. Thus follows the (P,d)-curve of phase equilibrium which in the present case is a descending straight line. The negative slope of that line indicates that the phase equilibria are unstable. And indeed, it can be shown that

- in fact the free energy on this line possesses a minimum in comparison with other states of the same deformation,
- but that the free enthalpy $f - Pd$ has a maximum on this line in comparison with other states of the same load.

Thus the points on the equilibrium line are similar to saddle points. A state on this line can "slide off" laterally as indicated by the horizontal arrows in Fig. 12.18 c. It seems plausible to relate the phenomena of internal yield and internal recovery, *cf.* Fig. 12.16, to this instability.

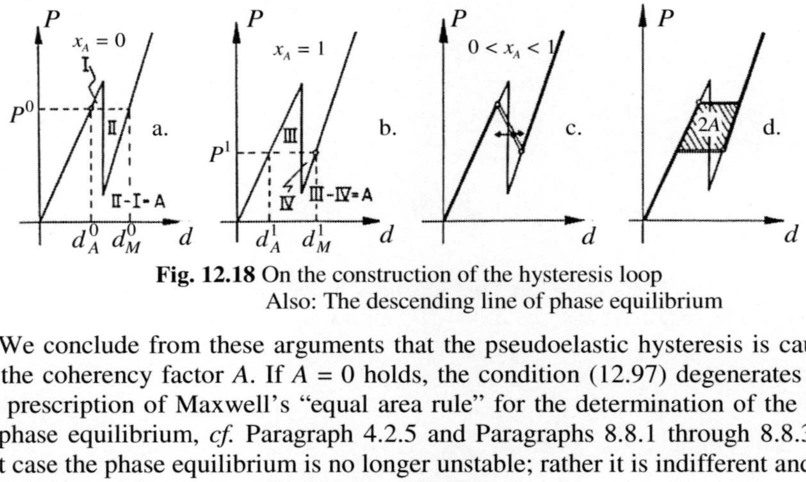

Fig. 12.18 On the construction of the hysteresis loop
Also: The descending line of phase equilibrium

We conclude from these arguments that the pseudoelastic hysteresis is caused by the coherency factor *A*. If $A = 0$ holds, the condition (12.97) degenerates into the prescription of Maxwell's "equal area rule" for the determination of the load of phase equilibrium, *cf.* Paragraph 4.2.5 and Paragraphs 8.8.1 through 8.8.3. In that case the phase equilibrium is no longer unstable; rather it is indifferent and the hysteresis loop shrinks to a single horizontal line.

12.5.5 *Latent heat*

The austenitic ⇔ martensitic phase transition during yield and recovery in the pseudoelastic range does not proceed along equilibria. Indeed, yield and recovery occur along horizontal lines in the (P,d)-diagram rather than along the descending diagonal equilibrium line of Fig. 12.18. We proceed to calculate the specific heating $\dot{q}$ associated with yield and recovery.

The specific heating occurs in the First and Second Laws, *viz.*

$$\frac{\mathrm{d}u}{\mathrm{d}t}-\dot{q}-P\frac{\mathrm{d}d}{\mathrm{d}t}=0 \quad \text{and} \quad \frac{\mathrm{d}s}{\mathrm{d}t}-\frac{\dot{q}}{T}=\sigma\geq 0, \tag{12.98}$$

where σ is the non-negative specific entropy source. With $s=-\frac{\partial f}{\partial T}$ and f from (12.95), and f_A, f_M from (12.92) we obtain

$$\frac{\mathrm{d}s}{\mathrm{d}t}=-\tfrac{1}{2}k\ln\left(\frac{K_A}{K_M}\right)\frac{\mathrm{d}x_A}{\mathrm{d}t}\text{ , hence by } (12.98)_2$$

$$\dot{q}=-\tfrac{1}{2}kT\ln\left(\frac{K_A}{K_M}\right)\frac{\mathrm{d}x_A}{\mathrm{d}t}-T\sigma\,. \tag{12.99}$$

Therefore it remains to calculate the entropy production σ. Elimination of $\dot{q}$ between $(12.98)_{1,2}$ provides with $f=u-Ts$ – always for an isothermal process –

$$-T\sigma=\frac{\mathrm{d}f}{\mathrm{d}t}-P\frac{\mathrm{d}d}{\mathrm{d}t} \quad \text{or by (12.95)}$$

$$-T\sigma=(1-x_M)\underline{\left(\frac{\partial f_A}{\partial d_A}-P\right)}\frac{\mathrm{d}d_A}{\mathrm{d}t}+x_M\underline{\left(\frac{\partial f_M}{\partial d_M}-P\right)}\frac{\mathrm{d}d_M}{\mathrm{d}t}- \tag{12.100}$$
$$-\left[f_M-f_A-P(d_M-d_A)+A(1-2x_A)\right]\frac{\mathrm{d}x_A}{\mathrm{d}t}\,.$$

We assume that force equilibrium prevails, even along the yield and recovery lines, where phase equilibrium does not prevail. By $(12.96)_{1,2}$ – and with $\lambda=P$ – this means that the underlined terms in (12.100) vanish so that (12.100) may be written in the form

$$-T\sigma=\left[P(d_M-d_A)-\int_{d_A}^{d_M}P(d)\mathrm{d}d-A(1-2x_A)\right]\frac{\mathrm{d}x_A}{\mathrm{d}t}\,. \tag{12.101}$$

On the yield- and recovery lines we have $P=P^0$ and $P=P^1$, respectively, *cf.* Paragraph 12.5.4 and these load values are given by

$$P(d_M-d_A)-\int_{d_A}^{d_M}P(d)\mathrm{d}d=\pm A\,,$$

respectively. It follows from (12.101) that we have

$$-T\sigma=\begin{cases}+2A(1-x_A)\dfrac{\mathrm{d}x_A}{\mathrm{d}t} & \text{yield line} \\ & \text{on the} \\ -2A\quad x_A\quad\dfrac{\mathrm{d}x_A}{\mathrm{d}t} & \text{recovery line}.\end{cases} \tag{12.102}$$

Insertion into $(12.99)_2$ determines the heating during yield and recovery

$$\dot{q} = \begin{cases} \left(-\frac{1}{2}kT\ln\frac{K_A}{K_M} + 2Ax_M\right)\frac{\mathrm{d}x_A}{\mathrm{d}t} & \text{yield} \\ \left(-\frac{1}{2}kT\ln\frac{K_A}{K_M} + 2A(1-x_M)\right)\frac{\mathrm{d}x_A}{\mathrm{d}t} & \text{recovery}. \end{cases} \tag{12.103}$$

If we run only partly through the yield line, beginning at $x_A = 1$ and proceeding to a generic value $x_A = 1 - x_M$, the integrated heating is given by

$$q^Y(x_M) = \frac{1}{2}kT\ln\left(\frac{K_A}{K_M}\right)x_M - Ax_M^2 . \tag{12.104}$$

Similarly for partial recovery starting at $x_M = 1$ we obtain

$$q^R(x_M) = \frac{1}{2}kT\ln\left(\frac{K_A}{K_M}\right)(x_M - 1) - A(x_M - 1)^2 . \tag{12.105}$$

The first terms in (12.104), (12.105) are different in sign so that the heat removed during yield is restored during recovery. This is the well-known reversible heat exchange during an isothermal phase transition, the so-called latent heat. For $A = 0$ this is the only contribution to heating. If, however, $A \neq 0$ holds, the phase transition is accompanied by an irreversible heating which depends quadratically upon the phase fraction.

The total heat production for a complete transition of the hysteresis loop – yield and recovery and all – comes out as:

$$q^Y(x_M = 1) + q^R(x_M = 0) = -2A . \tag{12.106}$$

Not surprisingly it corresponds to the area inside the loop, *cf.* Fig. 12.18 d.

12.5.6 *Kinetic theory of shape memory*

The basis of the kinetic theory of shape memory is the formulation of rate laws for the phase fractions of the austenitic lattice layers and of the martensitic twins. We continue to consider the potential energy (12.85) with its three minima, *cf.* Fig. 12.13 and Fig. 12.19. On the left-hand side of the left barrier and on the right-hand side of the right barrier we have the M_--layers and M_+-layers, respectively. Between the barriers we have A-layers. We ask for rates of change of the phase fractions $x_{M_\pm}$ and x_A. We assume that the phases are in equilibrium within themselves so that (12.90) holds. Also we suppose that there is equilibrium of forces so that $\beta_\pm$ and β_A are all equal to the load P.

We let the rates be determined by *rate laws* as follows

$$\begin{aligned} \dot{x}_{M_-} &= -p^{-0}x_{M_-} + p^{0-}x_A \\ \dot{x}_A &= p^{-0}x_{M_-} - p^{0-}x_A - p^{0+}x_A + p^{+0}x_{M_+} \\ \dot{x}_{M_+} &= p^{0+}x_A - p^{+0}x_{M_+} . \end{aligned} \tag{12.107}$$

For motivation of this ansatz we discuss the rate $\dot{x}_{M_-}$: It has two parts, a loss and a gain. The loss is due to the layers that jump out of the left potential well;

their number is proportional to the number of layers in that well and the factor of proportionality is p^{-0}, the *transition probability* from left to center. The gain is due to the layers that jump from the central potential well into the left one and it is proportional to the fraction of central layers with the factor of proportionality p^{0-}. The other rates in (12.107) are similarly constructed except that $\dot{x}_A$ contains four terms, because the central well can exchange layers with both lateral ones.

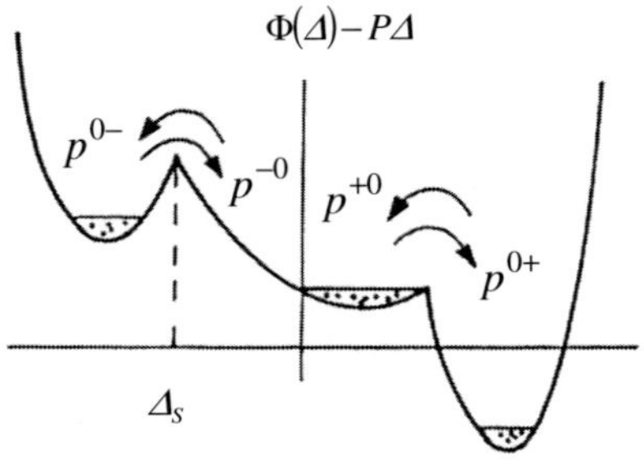

Fig. 12.19 Exchange of lattice layers between potential wells

The transition probabilities are constructed by the principles of theories *of activated processes* which were first developed in chemistry. In the present case the transition probability p^{-0} thus results as

$$p^{-0}=\sqrt{\frac{kT}{2\pi m}}\frac{\mathrm{e}^{-\frac{\Phi(\Delta_s)+P\Delta_s}{kT}}}{\sum\limits_{\Delta_{M_-}}\mathrm{e}^{-\frac{\Phi(\Delta)-P\Delta}{kT}}}\mathrm{e}^{-\frac{A}{kT}(1-2x_A)}\,. \tag{12.108}$$

By (12.90) the fraction in this expression determines the probability that a layer of phase M_- has an energy as large as the potential energy at $\Delta=-\Delta_s$. The idea is that a layer cannot surpass the barrier unless it has reached that height at least. Actually the layer needs a little more energy than that: Its energy must reach the value $\Phi(-\Delta_s)+P\Delta_s+A(1-2x_A)$, because the coherency energy $NAx_A(1-x_A)$ changes by the amount $Ax_A(1-2x_A)$ when a layer passes a barrier, *cf.* Paragraph 12.5.4. This interfacial contribution is taken care of in (12.108) by the exponential term connected with A. The square root in the equation represents the mean velocity with which M_--layers move toward the barrier, where m is an effective mass of the layer.

The sum in the denominator of the fraction in (12.108) may be converted into an integral in the manner described previously in Paragraph 12.5.3 and we obtain

$$p^{-0}=\frac{1}{2\pi Y}\sqrt{\frac{2K_M}{m}}\mathrm{e}^{-\frac{P^2/4K_M-PJ}{kT}}\mathrm{e}^{-\frac{\Phi(-\Delta_s)+P\Delta_s}{kT}}\mathrm{e}^{-\frac{A(1-2x_A)}{kT}}\,. \tag{12.109}$$

We see in this formula an explicit confirmation of our previous argument about entropic stabilization: When the curvature K_M of the martensitic potential wells becomes larger, *i.e.* when the well becomes narrower, the exit probability grows.

The other transition probabilities may be calculated in the same manner as (12.108) or (12.109). We do not list them. Suffice it to say that all transition probabilities are explicit functions of *P*, *T* and $x_{M_\pm}$, x_A. Therefore the rate laws (12.107) permit us to solve the following problem:

Given *P(t)*, *T(t)* and the initial values $x_{M_\pm}(0)$, $x_A(0)$, we may calculate $x_{M_\pm}(t)$, $x_A(t)$.

The solution results by stepwise integration in a simple numerical scheme. An analytic solution is impossible because of the ubiquitous non-linearities.

As soon as $x_{M_\pm}(t)$, $x_A(t)$ have been determined we may calculate the deformation $D(t)$. For this purpose we write $D = D_{M_-} + D_A + D_{M+}$ and calculate D_{M+} and D_A from (12.91). We recall that β_{M_+} and β_A are all equal to P, *cf.* $(12.96)_{1,2}$, and obtain

$$D = N\left[x_{M_-}\left(\frac{P}{2K_M}+J\right)+x_A\frac{P}{2K_A}+x_{M_-}\left(\frac{P}{2K_M}-J\right)\right]. \qquad (12.110)$$

Fig. 12.20 shows some solutions $x_{M_\pm}(t)$, $x_A(t)$ and $D(t)$ thus obtained. Obviously for numerical solutions we need to introduce dimensionless variables. As such we have chosen

$$\text{temperature } \theta = \frac{kT}{\Phi(\Delta_s)}, \text{ load } P = \frac{J}{\Phi(\Delta_s)}P, \text{ deformation } l = \frac{D}{NJ}. \qquad (12.111)$$

Furthermore we need to fix the parameters of the model. We set

$$\frac{A}{\Phi(\varDelta_s)} = 0.35, \quad \frac{\Phi_0 + K_A\varDelta_s^2}{\Phi_0} = 10.8, \quad \frac{K_M}{K_A} = 100, \quad \frac{1}{\sqrt{2\pi}Y}\sqrt{\frac{K_A}{m}} = 100. \qquad (12.112)$$

It must be admitted that these parameters have been chosen by a mix of reasonable expectation and hindsight. They provide us with the "right" type of curves in Fig. 12.20. We proceed to discuss those curves.

In all situations represented by Fig. 12.20 the load is prescribed as a triangular function alternating between tension and compression. The temperature is kept constant: low on top and high at the bottom.

In the upper two tableaux we start with $x_{M_\pm} = \pm\frac{1}{2}$ and observe that x_{M_+} gains at the expense of x_{M_-} as long as the load is a tensile load. When the load becomes compressive, x_{M_-} is formed. The deformation alternates between positive and negative values along with the load. The lower tableaux refer to higher temperatures. Accordingly we start with $x_A(0) = 1$. Under a load of sufficient size the deformation is still alternating between positive and negative values for tensile and compressive loads, but it remains small as long as the austenite prevails.

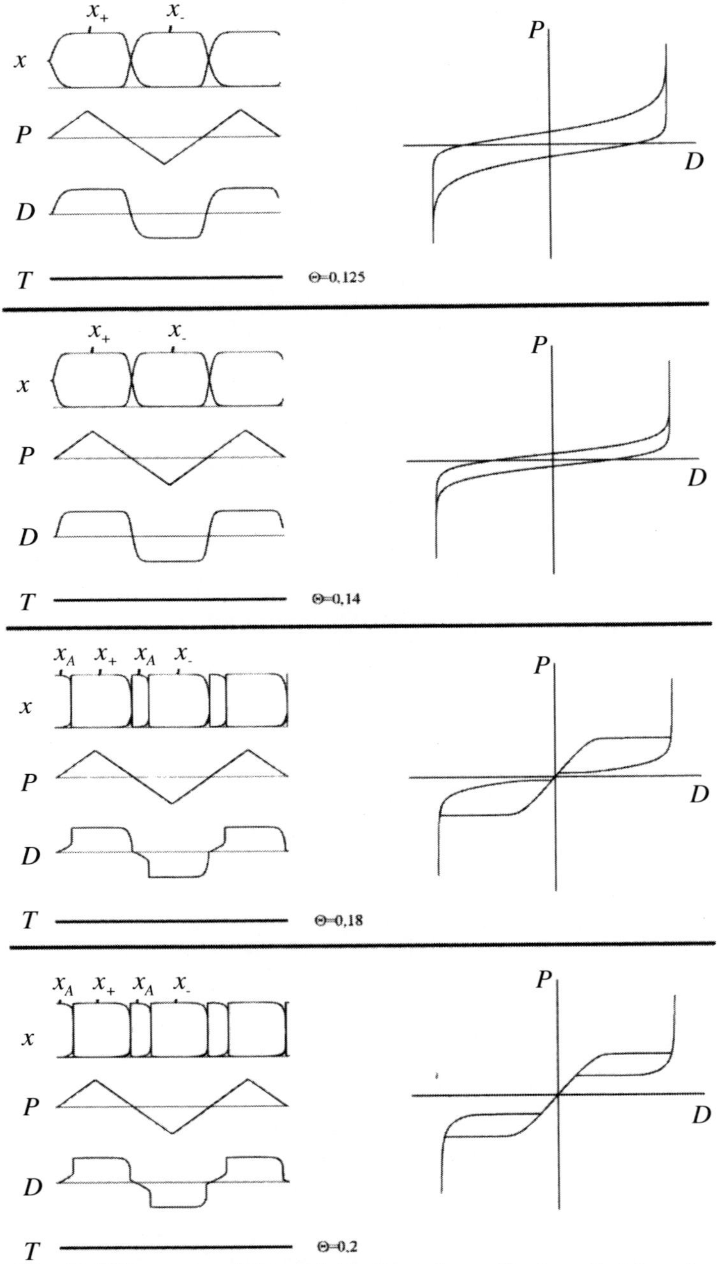

Fig. 12.20 Response of the model body to the application of a triangular alternating tension and compression at different constant temperatures

Particularly instructive are the (P,D)-diagrams of Fig. 12.20 that follow from $P(t)$ and $D(t)$ by the elimination of time. Comparison with the schematic graphs of Fig. 12.9 shows that the model body traces out the quasiplastic hystereses at low temperatures and the pseudoelastic hystereses at high temperature.

12.5.7 *Molecular dynamics*

Shape memory may also be understood and described through the behavior of the atoms in the crystalline lattice of the alloy. Fig. 12.21 shows a two dimensional model of a lattice particle, – a small piece of the metallic lattice –, where two constituents form a square and a hexagonal lattice respectively. The two constituents of the alloy are shown in black and gray. An atom, denoted by a Greek index α (α = 1,2,…N) interacts with all others. The interaction potential is assumed to be a Lennard-Jones potential which depends on the distance $x_{\beta\alpha} = |x_\alpha - x_\beta|$ of atoms α and β; it has the form

$$V_{\alpha\beta}(x_{\alpha\beta}) = 4\varepsilon_{\alpha\beta}\left(\left(\frac{\sigma_{\alpha\beta}}{x_{\beta\alpha}}\right)^{12} - \left(\frac{\sigma_{\alpha\beta}}{x_{\beta\alpha}}\right)^{6}\right). \quad (\alpha,\beta = 1,2,\ldots N) \qquad (12.113)$$

The Lennard-Jones potential is a popular analytic form for the schematic van der Waals potential in Fig. 2.12 a. Here we have three such potentials, two for like atoms – black and gray – and one between a black and a gray atom. Accordingly there are six parameters $\varepsilon_{\alpha\beta}$ and $\sigma_{\alpha\beta}$.

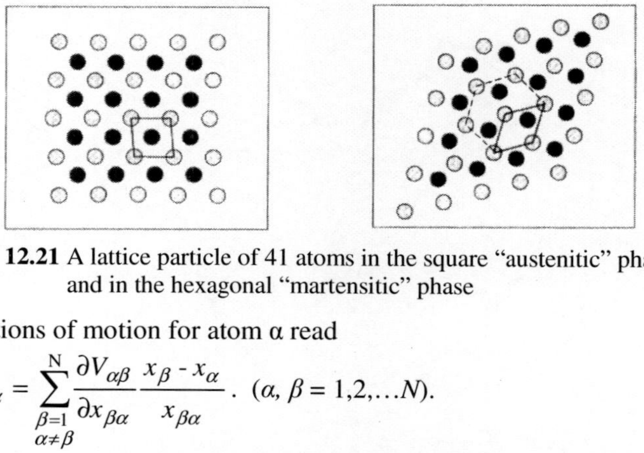

Fig. 12.21 A lattice particle of 41 atoms in the square "austenitic" phase and in the hexagonal "martensitic" phase

The equations of motion for atom α read

$$m_\alpha \ddot{x}_\alpha = \sum_{\substack{\beta=1 \\ \alpha\neq\beta}}^{N} \frac{\partial V_{\alpha\beta}}{\partial x_{\beta\alpha}} \frac{x_\beta - x_\alpha}{x_{\beta\alpha}}. \quad (\alpha,\ \beta = 1,2,\ldots N). \qquad (12.114)$$

Given initial positions and initial velocities for all atoms these equations may be integrated to give the positions and velocities of the atoms as functions of time. Hence follow the kinetic energy, *i.e.* the temperature, and the potential energy of the atoms.

$$U_{\text{kin}} = \sum_{\alpha=1}^{N} \frac{m_\alpha}{2} \dot{x}_\alpha^2 = NkT \quad \text{and} \quad U_{\text{pot}} = \frac{1}{2} \sum_{\substack{\beta=1 \\ \beta\neq\alpha}}^{N} V_{\alpha\beta}(x_{\beta\alpha}) \qquad (12.115)$$

For the method of integration and the choice of parameters we refer the interested reader to a paper by KASTNER.* Suffice it to say here that the parameters were chosen in a realistic manner for atomic interaction.

The calculations show that the square phase, the austenite, is stable at high temperature and that the hexagonal one, the martensite, is stable at low temperature. Fig. 12.22 shows that the model is capable of simulating the austenite ↔ martensite transitions and that there is a hysteresis: The a → m transition occurs at approximately 400 K, while the reverse transition m → a occurs at about 700 K. This is qualitatively as observed and, of course, we must not expect quantitative agreement with experiment in this two-dimensional model.

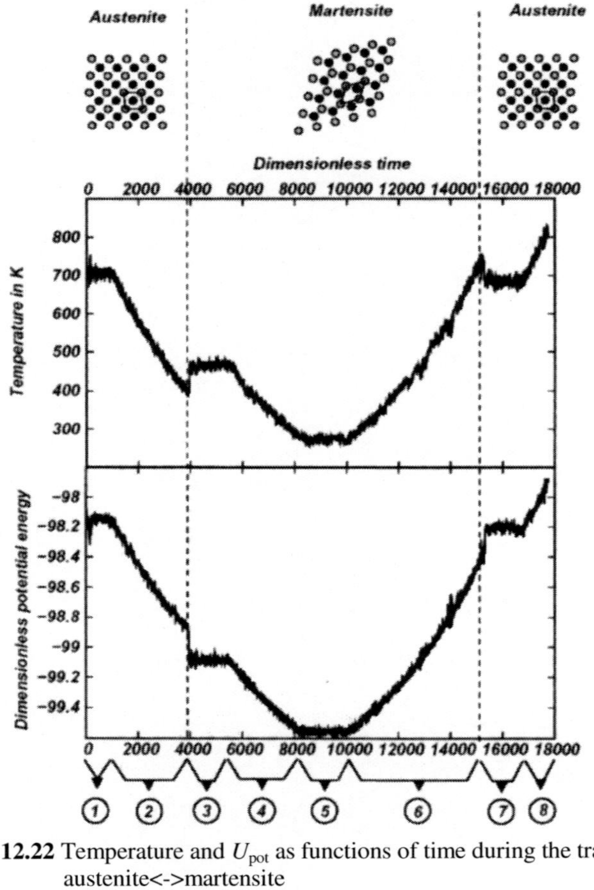

Fig. 12.22 Temperature and U_{pot} as functions of time during the transitions austenite<->martensite

* O. KASTNER. Zweidimensionale molekular-dynamische Untersuchung des Austenit ↔ Martensit Phasenübergangs in Formgedächtnislegierungen. Dissertation TU Berlin, Shaker Verlag (2003)

When the lattice particles are arranged in layers and the layers are stacked in the manner shown in Fig. 12.23, the a → m transition leads to different martensitic twins. The twins occur when – during the transition – the central particle in a square cell is pushed into a corner of the emerging rhombic cell; twins differ by the different choice of corner.

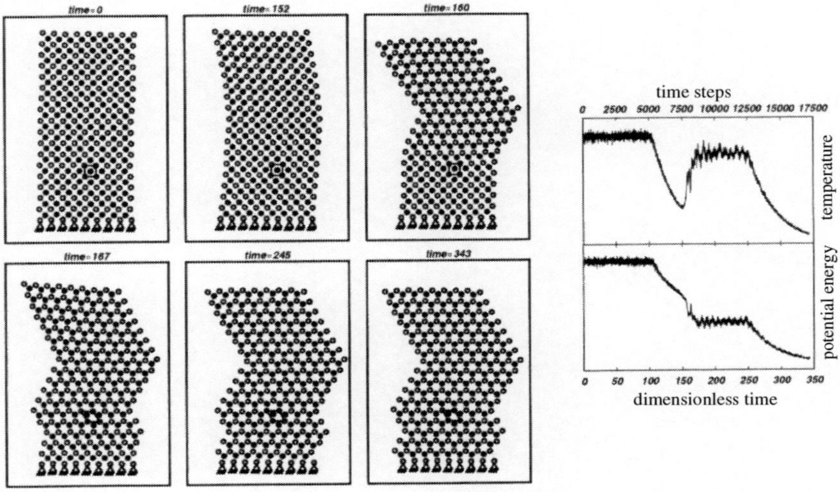

Fig. 12.23 Formation of martensitic twins during an a → m transition in a model of 298 atoms. The atoms at the bottom are fixed in position

Thus we see that molecular dynamics is capable of representing the gross features which characterize a shape memory alloy. Also, in a rough and ready manner, which is made explicit in Kastner's calculations, we see that the m → a transition leads from a phase of low entropy – martensite – to a phase of high entropy – austenite. Indeed, entropy grows with volume so that the black atoms in a gray square have a higher entropy than those in the corner of the rhombus. Thus, once again, the square phase is entropically stabilized.

Name and subject index

Absolute zero 3
absolute temperature 1, 5, 141
– as integrating factor 108
– interpretation 4-6, 141
absorption of sound 347-349
attenuation of sound 327, 349
absorption number
– spectral 186
– averaged 187, 189
activation energy 362, 388
activated process 388
activity 231
– coefficient 231, 234
– determination 232
ADAMS, Henry 133
adiabatic 30
adiabatic equation of state 36
d'Alembert solution of wave equation 325
alloys
– phase diagrams 243-247
– shape memory 374-393
ammonia synthesis 260, 270-272
Andrews, Thomas 82
anomaly of water 29, 70, 73, 90
Apollo 19
ARCHIMEDES 17
– buoyancy law of 16
ASIMOV, Isaac 179
atmosphere
– physical atm 16
– technical at 16
– of the earth 2, 15, 17, 298
– CO_2 in atmosphere 237
austenite 377 pp.
available free energy 123
– generic 123
– for droplets / bubbles 301 pp.
– for fog / clouds 307
– for balloons 315
– for gels 337
– relation to Helmholtz and Gibbs free energy 124
AVOGADRO, Amadeo 4
– law of 2, 206
– number 2
azeotropy 235

Balance equations
see equations of balance
balloons 312-322
– pressure-mass relation 315, 318
– pressure-radius relation 314
– stability of 315-31
– interconnected balloons 320-322
bar 16
barometric altitude equation 18
barometric step 18
barrier
– energetic 362, 378, 380
BÉNARD, Henri 128
BERNOULLI, Daniel 6, 179
Bernoulli equation 23, 24
BISCARI, G. 351
black body 185
– radiation 185, 186
body force 12
boiling line, evaporation line
– in single fluid 74, 75, 77, 159
– in binary solution 220-225, 233, 235, 236, 238
boiling temperature
– increase of 222
BOLTZMANN, Ludwig E. 152
– law of Stefan-Boltzmann 188, 193
Boltzmann constant 2, 185
BOYLE, Robert 3
bubbles 301-307
buoyancy
– force of Archimedes 16
burning glass 198

CAILLETET, Louis, B. 123
CARNOT, Sadi N.L. 131, 132
carnotization 162
cascade
– separating cascade 227-229
cavity radiation 193
CELSIUS, Aulus C. 8
Celsius scale 7
LE CHALELIER, Henri L. 264
Chapman-Jouguet theory 358-360
chemical constant 255, 257-260
chemical reactions 251-284
chemical potential 199-204
– general 201-203
– ideal mixtures 204 -208
– measurability 200, 203
chimney 48-52
circulation 25
CLAPEYRON, Benoît P.E. 118
CLAUSIUS, Rudolf J.E. 132
Clausius
– formulation of 2nd law 105
– inequality 112
Clausisus-Clapeyron equation 117
– approximate analytic form 118
Clausius-Rankine process 157-160
cloud base 295-297, 299
coherency 383
– energy 383-389
– factor 383
collectors of radiation 52, 198
common tangent rule 119, 240, 242
communicating balloons 320-322
compression ratio 98, 99
compressor 84-86
– two stage 86
concentration 199
condensation of vapor 68-70
condenser 154
conode 248
conservation laws 9, 10
– mass 11,12
– momentum 12-25
– energy 26-52
consolute temperature 248
constitutive equations 57-82
continuity equation 9, 12
convection 127-130
convective momentum flux 20
convexification of free energy 119
cooling limit 292-295
correspondence principle 81
counter ions 335-337, 340
creep 369-373
critical point 70-77, 238
critical radius
– droplet 304
– bubble 306
crystallizing rubber 146-151
cumulus cloud 295
cycles 89-104
– efficiency of 89-104
– Brayton 102
– Carnot 91
  – modified with heat conduction 92-94
– Diesel 99
– Ericson 94
– gas-vapor 164
– Joule 90
– mercury-water 163
– Otto 96
– Stirling 96

DAIMLER, Gottlieb 101
Dalton's law 206
Damping of sound 327, 349
deformation-temperature diagram
– shape memory alloys 377
– Landau theory 332, 333
degree
– of moisture 285
– of saturation 289
– of freedom 217, 247, 249
desalination by osmosis 215
detonations 360

dew line
– in binary solution 220, 225, 238, 242
– in moist air 288
– in single fluid 74, 77, 159, 163
dew point 298
diffusion
– coefficients 354
– equation 356
– in reacting mixture 356-358
DIESEL, Rudolf 101
Diesel engine 99
– efficiency 101
diffusion coefficients 354
digester 79
dispersion of sound 347-349
dissociation 211, 335
distillation 223- 230
– batch 224
– cascade 227
– continuous 227
double glazed window 61
droplets 301-312
– vapor pressure of 307
dry ice 72

ECKART, Carl 345
efficiency
– of cycles 90-104
– of fuel cell 277
– of heat pump 168
– of Newcomen engine 153
– of power station 161
– propulsive 104
– of refrigerator 165
– of thermal power station 52
– of Watt steam engine 154, 155
elastic modulus 331-333, 367
– complex 374
– frequency dependence 373
– static 369
electro-neutrality 336
emission coefficient
– spectral 186
– averaged 187, 189, 190
energy
– absolute value 267
– barrier 362, 378, 380
– chemical 27
– elastic 27
– kinetic 26
– internal 26-29
– nuclear 29
– potential 31
– solar 190, 194
energy conservation 26
enthalpy 34
entropic
– elasticity 144, 145
– stabilization 379, 380, 393
entropy 105-133, 135-152
– absolute value 267
– flux and production
  – in mixtures of fluids
  – in single fluids 344
– of a gas and a polymer molecule 135-139
– as $k \ln W$ 135-152, 207
– as a measure of disorder 139-140
– growth 108-109, 112, 123, 125, 137
– of mixing 204-207, 214, 268, 284
– molecular interpretation 137
– of a rubber rod 141-143
– statistical interpretation 135
equal area rule 119
equations of balance
– generic 9
  – for closed system 10
  – for open system at rest 10
  – local in regular points 10
– of mass 11, 343
– of momentum 12, 343
– of energy 26, 280, 343
– short forms for closed systems 33
– of internal energy 32
  – of pdV-thermodynamics 34

– of entropy 282, 344, 351, 352
– of internal variable in visco-elasticity 368
– of phase fractions in shape memory alloys 387
equation of state
– adiabatic for ideal gases 36
– caloric
  – generic for fluids 64-68
  – ideal gas 1
  – incompressible fluid 2
  – van der Waals 114
– thermal
  – generic for fluids 63
  – ideal gas 1
  – incompressible fluid 1
  – rubber 145
  – van der Waals 79-81, 119-122
– rheological 368,369
Ericson process 94, 95
– efficiency 95
Eshelby tensor 349-351
Euler theorem on homogeneous functions 202
eutectic
– point 236, 246
– pressure 243
EVANS, Oliver 154
evaporation
– in single fluid 74, 75, 77
– in binary solution 220-225, 233, 235, 236, 238
– limit 291, 292, 294
– under load 78,79
extent of reaction 252

FAHRENHEIT, Daniel 8
Fahrenheit scale 8
fan 35
fanjet engine 104
Faraday, Michael 81
fat 28
feed pump 155- 157, 163
– work of 157
FICK, Adolf 355
Fick's law 355
fields of thermodynamics 7, 57
First Law of thermodynamics 9, 26-55
– history of 53
fixed points 8
flames 358-366
flame eigenvalue 364, 366
fluid mechanics
– objective of 7
fluorized hydrocarbons 167
flux
– convective 9
– non-convective 9
– heat flux 30
focusing collectors 198
föhn 298
fog 306-312
– region 288
– droplets in air 306-312
FORD, Henry 101
FOURIER, Baron de 179
Fourier
– law 58
– series 179
free energy
  see Helmholtz free energy
free enthalpy
  see Gibbs free energy
fugacity 231
fugacity coefficient 231
– definition 231
– determination 234

GALILEI, Galileo 7
gas
– ideal 1
– real 79
– permanent 82
gas turbine 102-104
Gauß's theorem 10, 16
GAY LUSSAC, Joseph 3
Gay-Lussac 27

– experiment of 38
– and carbohydrates 28
gels 343-350
GIBBS, Willard J. 155
Gibbs
– equation 111,148
– paradox 212
– phase rule 223, 224, 225, 238, 244, 253, 256
Gibbs-Duhem relation 208, 223
glass
– thermal conductivity 61, 62
– transparency 203
glucose 28
– synthesis 266, 277-279, 285
gray body 185
green house 203
GULDBERG, Cato M. 261

HABER, Fritz 272
Haber-Bosch synthesis 270-272
Hagen-Poiseuille
–law of 196
harmonic waves 326
harmonic sound waves
– amplitudes 326, 328
– with viscosity and heat conduction 347-349
– without viscosity and heat conduction 327
heat
– of evaporation 69, 76
– as integrated heating 84
– latent, in pseudoelasticity 385
– of melting of water 76
– of mixing 205
– of reaction 262, 359
– of sublimation of water 76
heat conduction
– instationary 171-179
  – in finite rod 172-175
  – in infinite rod 175, 176
  – in the soil 177,178
– in heat exchangers 179-184
– through window 61, 62
heat conduction equation 171
heat death 132
heat exchangers 179-184
– parallel and antiparallel 180-183
– reversible 184
heating 30
heating a room 40-42
heat pole 175, 176
heat pump 168
heat of radiation 184
heat
– transport coefficient 179
– transfer coefficient 179
HELMHOLTZ, Hermann L.F. von 54
Henry coefficient 236
history
– of equations of state 3
– of first law of thermodynamics 53
– of Haber-Bosch synthesis 271
– of heat conduction 179
– of heat radiation 193
– of internal combustion engine 101
– liquefying gases and solidifying solids 81
– of molecular interpretation of entropy 151
– of pressure and pressure units 15
– of Second Law of thermodynamics 131
– of steam engine 153
– of temperature 7
hot air engine 87
Hugoniot diagram 360
humidity
– relative 297
HUO, Y. 351
hysteresis
– in rubber 146-150
– in balloons 321, 322
– in shape memory alloys 375, 376, 382-390, 392

Ideal gases 1
ideal mixtures 207

injection ratio 100, 101
integrability condition 113, 115, 204, 202
integrating factor 111
internal combustion engines 96-102
– Otto engine 96
– Diesel engine 99
– history of 101
inversion curve 122

Jet engine 104
thrust of 19, 20
jet propulsion process 103
JOUKOVSKI, Nikolai E. 25
JOULE, James P. 4
Joule
– process 90
– heating 274

KASTNER, Oliver 392
KELVIN, Lord 4
Kelvin
– temperature 4
– formulation of 2nd law 105
KIRCHHOFF, Gustav R. 187
– law of radiation 186
kinetic theory
– of gases 4., 6
– of shape memory alloys 387
KUTTA, Martin W. 25
– law of Kutta-Joukovski 24

LANDAU, Lev D. 329
Landau theory of phase transition 329-335
latent heat 265, 385
LAVAL, Patrik C.G. de 46
Laval nozzle 46
Laws of thermodynamics
– Zeroth 7
– First 26-36
– Second 105-113
– Third 264-267

load-deformation diagram
– rubber 146, 149
– shape memory alloy 375, 383, 390
LAVAL, Gustaf P. 46
Laval nozzle 46
LAVOISIER, Antoine L. 53
LENOIR, Jean J.E. 101
LEIBNIZ, Gottfried, W. 26
Leibniz rule 10, 18
LIFSHITZ, Evgenii M. 329
lift force of an airfoil 24, 25
LILIENTHAL, Otto von 25
LINDE, Carl, Ritter von 101, 123
liquefaction of gases 81,82
LOSCHMIDT, Johann J. 132

MARIOTTE, Edmé 1, 3
martensite 377 pp.
martensitic twins 378, 382
mass action, law of 253-260
– ideal mixtures 254
– mixtures of ideal gases 254
– history of 255
– applications of 256-260
mass transfer rate 12
MAXWELL, James C. 152
Maxwell
– line 120
– distribution 140
– demon 152
MAYER, Robert J. 54
mechanical equivalent of heat 53, 54
melt 243, 246
melting line 74, 77
memory engines 376
metastability 265, 362, 378
MIDDLETON, Knowles W.E. 8
microwave 184
– oven 184, 185
miscibility gap
– in binary solutions and alloys 243, 244
– in ternary solutions 248-250

mixtures 199-250
– ideal 207
– of ideal gases 206
– chemically reacting 251-284
moisture content 285
mol 2
– fraction 199
– as a unit 223
– mass 1, 223
– number 199, 223
MÜLLER, Ingo 177, 319, 349

Navier-Stokes law 57, 58
NERNST, Walter, H. 264, 266
NEWCOMEN, Thomas 153
Newton, Isaac
– law of motion 12
– law of friction 60
Newtonian fluid 58
nozzle flow
– mass balance 11,12
– momentum balance 21-23
– energy balance 42-46
nucleation barrier
– of droplets 304, 311
– of bubbles 305

Osmosis 208-216
– desalination 215
– energetic interpretation 213-215
– in Pfeffer tube 210-213
– physiological salt solution 213
– van't Hoff law 208-210
osmotic pressure 209
– of sea water 212
OTTO, Nikolaus A. 101
Otto engine 96
– efficiency 98
overheating 306
ozone
– hole 168
– good and bad 194, 167

PAPIN, Denis 79, 153
Pascal Pa 16
PFEFFER, Wilhelm 211
Pfeffer tube 210, 215
phase diagrams
– graphical construction 238-246
– (g,X)-diagram 239, 241, 242, 244, 245
– (h,T)-diagram 76, 77
– (p,T)-diagram 70-73, 75, 223, 262
– (p, υ)-diagram 74, 75
– (p,X)-diagram for binary solution 220, 222, 233, 235, 236, 238, 242, 244
– (r,T)-diagram 76
– (T,X)-diagram for binary solution 220,225, 228, 229, 245, 246
– for ternary solutions 247, 250
phase equilibrium
– for van der Waals gas 119, 120
– for Landau theory 330, 331
phase shift 327, 373
phase transitions
– first and second order 329
  – under load 334
– in single fluid 68-82
  – solid-liquid 70, 74, 77
  – solid-vapor 70, 74, 77
  – liquid-vapor 68-82
– in shape memory alloys 377-391
– in solutions and alloys
  – liquid vapor 219-243
  – solid-liquid 243-247
– in ternary solutions 247-250
– Landau theory 329-335
– reversible and hysteretic 150, 382-385
phase speed 327, 328
phenomenological equations 345, 350, 353, 354
photosynthesis 278-284
physiological salt solution 213
plait point 249
PLANCK, Max K. 186

– Third law 266
– formula for black body radiation 185
– constant 185, 267
pneumatic engine 86
Poincaré, Jules H. 151
polybisacrylamide 335
polyelectrolytic gels 335-342
potential
– chemical 199-204
– gravitational 31
– thermodynamic 115
preheating 163
preservation
– jar 77
– alcohol 78
pressure cooker 79
principle
– of least constraint 264
– of local equilibrium 344, 352
process
– Brayton 102
– Carnot 91
– Clausius-Rankine 155-160
– Diesel 99
– Ericson 94
– Joule 90
– Otto 96
– Stirling 96
propjet 104
proteins 28
pseudo-elasticity 375, 382-387

Quasi-plasticity 375

Radiation 184-198
– between two plates 187, 188-190
– black and gray 185-190
– and conduction 192
– between sun and planets 190, 191
– infrared 185
– ultraviolet 168
radiative energy flux density 185-190
– from cavity 193
– of the sun 191
RAOULT, Francois M. 219
Raoult´s law and applications
– ideal 218-227
– real 232
rate laws
– for viscoelasticity 369
– for shape memory alloys 387
Rayleigh line 359, 360
reaction rate 274, 275, 279-281
reaction rate density 252
real gas 79
recovery 375
–load 375, 377
rectification column 229
recurrence objection 151
refrigerants 167
refrigerator 164-167
– compression 164
– absorption 166
regeneration 95, 96, 103
relative humidity 297
RENOIR, Etienne J.J. 101
residual deformation 375
respiratory quotient RQ 253
reversibility objection 151
reversible process 34, 83-104
– heat and work 83, 89
– cycle 89-96
Reynold's transport theorem 10
rocket equation 19
RUGGERI, Tommaso 177, 349
RUMFORD, Graf von 53

SAGREDO, Gianfrancesco 7
Saturn 1B 19
saturated steam 68-72
saturation pressure 70-72
– decrease of 222
sauna 294

Second Law of thermodynamics 90, 105-133
– history of 131
separating cascade 227-229
separation of variables 171,172
semipermeable membrane 200, 210-216
shape memory 374 -393
– applications 376, 377
– model 378-379, 391-393
– two-way shape memory 376
shear flow 59-61, 345
shrink factor 335 pp.
solar constant 29, 53, 191, 280
solar energy 29, 53, 191
– utilization of 196-198
solid solution 246
sound 322-329
– speed 45, 46, 47, 326
  – in ideal gases 326
specific heat (capacity) 64 pp.
– ideal gases 66
– liquids and vapors 65, 126
– water 67
– singularity at critical point 126
specific values 9
SPENGLER, Oswald 133
stability
– of balloons 315 pp.
– conditions 125, 126, 371
– criteria 123
standard atmosphere 17
steam
– saturated 68, 69
– wet 68,69
STEFAN, Josef
– law of Stefan-Boltzmann 188, 193
stoichiometry 251
stoichiometric
– equations 251
– coefficients 251
– mixture 258
streamline 23
STREHLOW, Peter 319
stress
– tensor 12,13
  – isotropic 14
– force 12
stress relaxation 369-371
STRÖMER, Märten 8
sublimation
– line 74, 77
– region 74, 77
suction pump 15
superheating 156-159, 162
surface energy 301
surface tension 301
swelling and shrinking of gels 335-342
system
– closed 10
– open and at rest 10

Temperature
– absolute 1
– amplitude in sound wave 328, 329
– astronomical 191
– consolute 248
– molecular interpretation 5, 141
– as integrating factor 108
– of sun and planets 190, 191
– waves in the soil 177
temperature gradient
– in standard atmosphere 17
– dry adiabatic 298
– moist adiabatic 298
thermodynamics
– objective of 7, 57, 343
thermodynamics of irreversible processes (TIP) 343-393
– single fluids 343-351
– mixtures 351-358
thermodynamic forces and fluxes 344, 350, 353, 369
thermal conductivity 58
thermal diffusivity 171
thermals 130, 295
thermal power station 50

thermosiphon 195
THILORIER, Charles S.A. 82
Thomson formulae 304, 306
THOMPSON, Benjamin
see Graf RUMFORD
THOMSON, William
see Lord KELVIN
throttling 40, 120-123, 163, 165-167
thrust
– of a rocket 18
– of a jet engine 19
tie line 248
tin
– pest 265
– white and gray 265
TORRICELLI, Evangelista 15
Torricelli vacuum 15,
traction 12
transition probability 388
TREVITHICK, Richard 154
trioleine 28
triple point 9, 70-76
triple line 74, 75, 77
turbine 47, 50, 102
turbofan engine 104

Undercooling 306
unrestricted miscibility 241

VAN DER WAALS, Johannes D. 80
– equation 79-81
– potential 79
– internal energy and entropy 114
– phase equilibrium 119
– throttling 120
VAN'T HOFF, Jakobus H. 209
– law of 209, 212, 216
visco-elasticity 366- 374
viscosity 58, 60, 344
– shear 58,60, 344
– bulk 58, 345
volume of mixing 205, 207
volume fraction 2, 199

WAAGE, Peter 255
WATT, James 155
wave equation 322-325
– d'Alembert solution 325, 326
waves
– plane harmonic 327
– of temperature 177
– of sound 325-329
work
– as integrated working 83
working 30
– internal 33
– of stress 33

Young's modulus 367

ZERMELO, Ernst 152